International Conference on Nanoscience and Nanotechnology (ICONN 2021)

IOP Conference Series: Materials Science and Engineering
Volume 1219

Chennai, India
1 - 3 February 2021

Editors:

T. Vijaya Kumar
Debabrata Sarkar
Jaivardhan Sinha
Sandeep Kumar Lakhera
Junaid Masud Laskar
Renil Kumar
V. Eswariah
Suresh Perumal
Saurabh Ghosh
Tushar Rana

ISBN: 978-1-7138-4678-9
ISSN: 1757-8981

Printed from e-media with permission by:

Curran Associates, Inc.
57 Morehouse Lane
Red Hook, NY 12571

proceedings.com

Some format issues inherent in the e-media version may also appear in this print version.

Institute of Physics
Dirac House, Temple Back
Bristol BS1 6BE UK

Phone: 44 1 17 929 7481
Fax: 44 1 17 920 0979

techtracking@iop.org

Additional copies of this publication are available from:

Curran Associates, Inc.
57 Morehouse Lane
Red Hook, NY 12571 USA
Phone: 845-758-0400
Fax: 845-758-2633
Email: curran@proceedings.com
Web: www.proceedings.com

TABLE OF CONTENTS

IOP Conf. Series: Materials Science and Engineering 1219 (2022) 011001 doi:10.1088/1757-899X/1219/1/011001

Preface

6th International Conference on Nanoscience and Nanotechnology (ICONN) – 2021 has been organized virtually during 01 to 03 Feb 2021 jointly by the Department of Physics and Nanotechnology, SRM Institute of Science and Technology (SRM IST), India along with Shizuoka University- Japan and NCTU-Taiwan, GNS-New Zealand, University of Rome Tor Vergata, Italy, ACCMS, Indian Physics Association, ICS India, Tata Institute of Fundamental Research India, Materials Research Society of India, RMIT University Australia and Springer Nature. ICONN 2021 emerged as a grand event by bringing many researchers to a common platform across the world to discuss and deliberate the front line areas of nanoscience and nanotechnology. Organizing committee of ICONN 2021 received around 2000 abstracts for oral and poster presentations along with the keynote, plenary and invited talks.

The full length papers of conference are brought out as a special volume with the IOP Conference Series : Materials Science and Engineering. Pool of experts acted as technical committee, editorial board members along with the reviewers helped in reviewing the full length manuscripts. The members of the committees are given below.,

Main editor : Dr. T.V. Lakshmi Kumar, SRM Institute of Science and Technology

Technical Committee:

Dr. Rudra Banerjee, SRM Institute of Science and Technology

Dr. Jitendra Kumar Tripathi, SRM Institute of Science and Technology

R. Ajay Rakkesh, SRM Institute of Science and Technology

Dr. P. Justin Jesuraj, SRM Institute of Science and Technology

Dr. Meenal S. Deo, SRM Institute of Science and Technology

Dr. Venkata Ravindra A, SRM Institute of Science and Technology

IOP Conf. Series: Materials Science and Engineering 1219 (2022) 011001 doi:10.1088/1757-899X/1219/1/011001

Themewsie Editorial Teams (All are from SRM Institute of Science and Technology) :

Dr. Debabrata Sarkar & Dr. T. Vijaya Kumar

- ☐ Nanocomposites and catalysts
- ☐ Environmental nanotechnology
- ☐ Nanobiotechnology, nanotoxicology and nanomedicine

Dr. Jaivardhan Sinha & Dr. Sandeep Kumar Lakhera

- ☐ Nanotechnology for energy harnessing, transport and storage
- ☐ Manipulation and characterization of materials at nanoscale/atomic scale
- ☐ Nanomagnetism and spintronics

Dr. Junaid Masud Laskar & Dr. Renil Kumar

- ☐ Surfaces, interfaces and thin films
- ☐ Complex and meta-nanophotonics: classical to quantum ,
- ☐ Industrial nanotechnology: products and commercialization

Dr.V. Eswariah & Dr. Suresh Perumal

- ☐ Low-dimensional and 2D materials
- ☐ Nanostructured solar cells and thermoelectrics
- ☐ Nanotribology

Dr. Saurabh Ghosh & Dr. Tushar Rana

- ☐ Multifunctional nanomaterials
- ☐ Molecular and nanoelectronics
- ☐ Nanoscale modelling and applications
- ☐ Computational materials, methods and numerical techniques

IOP Conf. Series: Materials Science and Engineering 1219 (2022) 011001 doi:10.1088/1757-899X/1219/1/011001

The disclaimer should have the following information:

Reasons why the meeting was changed to a virtual format (covid-19/travel restrictions, etc).
Due to COVID

Why the meeting was virtual instead of being postponed. (What would it mean to the community if the meeting could not take place in a virtual format?)

The ICONN is a biennial conference and the present one is 6th edition. The postponement of conference may lead to missing of on-time deliberations

Virtual Conference Dates.

01 -03 Feb 2021

Location of organisers.

Department of Physics and Nanotechnology, SRM Institute of Science and Technology, Kattankulathur – 603 203, Tamilnadu, India

Conference model: Information on plenary sessions. Amount of time that authors/presenters were given to talk.

More than 70 scientific sessions were held with eminent speakers from renowned international institutions such as University of California, Berkeley-CA, Tokyo Institute of Technology-Japan, The Australian National University, Canberra-Australia, University of Cambridge-UK, University of Maryland-USA, Institut für Experimental physik, Berlin-Germany and also from premier Institutes of India such as IISc Bangalore, Bhabha Atomic Research Centre (BARC) Mumbai, Anna University-Chennai, Indira Gandhi Centre for Atomic Research (IGCAR)-Kalpakkam. The total number of participants in the conference was about 2400. For more information, please visit https://www.srmist.edu.in/iconn-2021/

Information on discussions and Q&A sessions.

Yes, it has been carried out for every session along with the poster sessions.

Location of participants and overall participant numbers.

Participation from 14 countries and the number of total participants are 2400

Technology that was used to deliver the conference.

Zoom Video Conferencing
Any reports on overall technical quality and success of delivery of the conference.

Report is enclosed.

IOP Conf. Series: Materials Science and Engineering 1219 (2022) 011001 doi:10.1088/1757-899X/1219/1/011001

SRM Institute of Science and Technology
6th International Conference on Nanoscience and Nanotechnology (ICONN 2021)
February 1-3, 2021

REPORT ON THE CONFERENCE

The Sixth Biennial International Conference on Nanoscience and Nanotechnology (ICONN-2021) is organized by Department of Physics and Nanotechnology, SRM Institute of Science and Technology (SRMIST), Kattankulathur, in association with Shizuoka University-Japan, National Chiao Tung University (NCTU)-Taiwan, GNS Geological and Nuclear Sciences (GNS) Science-New Zealand, University of Rome Tor Vergata-Italy, RMIT University-Australia, Tata Institute of Fundamental Research (TIFR)-India, Asian Consortium on Computational Materials Science (ACCMS), Indian Physics Association (IPA), Materials Research Society of India (MRSI), Indian Carbon Society (ICS)-India and Springer Nature from 1st Feb 2021 to 3rd Feb 2021.

The conference is aimed at providing a platform for researchers and scientists to assess recent trends and developments in Nanoscience and Nanotechnology for promising applications. Scientific Sessions of the conference includes Low-dimensional and 2D materials, Surfaces, interfaces and thin films, Nanocomposites and catalysts, Multifunctional nanomaterials, to mention a few. More than 70 scientific sessions were held with eminent speakers from renowned international institutions such as University of California, Berkeley-CA, Tokyo Institute of Technology-Japan, The Australian National University, Canberra-Australia, University of Cambridge-UK, University of Maryland-USA, Institut für Experimental physik, Berlin-Germany and also from premier Institutes of India such as IISc Bangalore, Bhabha Atomic Research Centre (BARC) Mumbai, Anna University-Chennai, Indira Gandhi Centre for Atomic Research (IGCAR)-Kalpakkam. The total number of participants in the conference was about 2400.

The conference welcome address was delivered by Dr. D. John Thiruvadigal, Convener, ICONN 2021, SRMIST, India, followed with the presidential address delivered by Dr. Sandeep Sancheti, Vice Chancellor, SRMIST. His address was focused on emerging research and development (R&D) in the field of Nanoscience and Nanotechnology. Dr. C Muthamizhchelvan, Pro Vice Chancellor (Engineering and Technology) has released the proceedings of the ICONN 2021 conference and greeted the guests and participants. The inaugural address was delivered by

IOP Conf. Series: Materials Science and Engineering 1219 (2022) 011001 doi:10.1088/1757-899X/1219/1/011001

Prof. Ashutosh Sharma, Secretary, Department of Science and Technology (DST) Govt. of India, CV Seshadri Chair Professor, Department of Chemical Engineering, IIT Kanpur. In his address, he has mentioned about the initiatives and schemes framed by DST, India, to develop the R&D infrastructure in various institutions across the country through funding research projects.

Prof. Ramamoorthy Ramesh, University of California, Berkeley, CA, had announced SRM-ICONN Award of excellence in Nanoscience and Nanotechnology to Prof. Ashutosh Sharma, Secretary, Department of Science and Technology (DST) Govt. of India, CV Seshadri Chair Professor, Department of Chemical Engineering, IIT Kanpur. The inaugural function was concluded with a vote of thanks by Dr. S. Ponnusamy, Controller of Examinations, SRMIST.

SRMIST is also honouring eminent scientists with SRM-ICONN Innovation Award, SRM-ICONN Outstanding Woman Researcher Award, with SRM-ICONN Young Researcher Award.

The second day of the conference will feature an exclusive program on "Seeding nanoscience in Young Minds – Live interaction with Nanoscientists", co-organized by the Indian Science Congress Association (ISCA) – Chennai Chapter in celebration of DST, Govt. of India Golden Jubilee Year (2021. Around 300 hundred students across India from various colleges/schools have registered for this program. Prof. B. S. Murty, Director IIT Hyderabad will deliver the inaugural address. Prof. Vijay Laxmi Saxena, General President of the Indian Science Congress Association (ISCA), will deliver the Presidential address.

IOP Conf. Series: Materials Science and Engineering 1219 (2022) 011002 doi:10.1088/1757-899X/1219/1/011002

Peer review declaration

Type of peer review: Single-blind

Single-anonymous: authors' identities are known to the reviewers, reviewers' identities are hidden from authors

Double-anonymous: author and reviewer identities are hidden to each other

Triple-blind: author and reviewer identities are hidden to each other, and from the Editor(s)

Open: author and reviewer identities are known to each other

- **Describe criteria used by Reviewers when accepting/declining papers. Was there the opportunity to resubmit articles after revisions?**

  Reviewers accepted the manuscripts when the authors successfully carried out the revisions based on the comments. If it is not satisfactory, the manuscript is declined. Yes, there was an opportunity to submit the revised versions of the manuscripts. Before sending to the reviewers, the editorial team has considered the following points to proceed for the review.

  - Novelty of the work (we did not consider any routine work)
  - Plagiarism (less than 25% has been considered)
  - Quality of English
  - Proper research data presentation

- **Conference submission management system:**

  The submission of papers has been done with an email (manuscripticonn2021@srmist.edu.in). The main editor placed the manuscript to the technical committee for the initial plagiarism check. If the plagiarism is found less than 25%, then the manuscript has been forwarded to the respective editorial team who had expertize in the relevant field. The editorial team identified the suitable reviewers and the manuscript had been sent to two reviewers. The comments had been sent back to the authors for the revision. The editorial team receives the revised version and sent them again to reviewers. Provisional acceptance given by the editorial teams once the reviewers recommended the manuscript for publication. Finally, the main editor sent the acceptance of the manuscript to the corresponding author.

- **Number of submissions received: 69**

- **Number of submissions sent for review: 69**

- **Number of submissions accepted: 53**

- **Acceptance Rate (Number of Submissions Accepted / Number of Submissions Received X 100): 77%**

IOP Conf. Series: Materials Science and Engineering 1219 (2022) 011002 doi:10.1088/1757-899X/1219/1/011002

- **Average number of reviews per paper: 2**

- **Total number of reviewers involved: 90**

- **Any additional info on review process (ie plagiarism check system):**

  Turnitin available with SRM Institute of Science and Technology. Similarity index less than 25% was considered.

- **Contact person for queries:**
Prof. D. John Thiruvadigal
Chairperson
School of Applied Sciences
SRM Institute of Science and Technology
Kattankulathur – 603 203
Tamilnadu, India

dean.sciences.ktr@srmist.edu.in

IOP Conf. Series: Materials Science and Engineering 1219 (2022) 012001 doi:10.1088/1757-899X/1219/1/012001

Hypothetically Inspired Electro-Optic Chromophores for Laser Space Communication Applications

Tanmay Dashaputre[1], Y.Sheeba Sherlin[2] and T.Vijayakumar[3,*]

[1]Department of Electronics and Communication Engineering, College of Engineering and Technology, Faculty of Engineering and Technology, SRM Institute of Science and Technology, SRM Nagar, Kattankulathur-603203, Kanchipuram, Chennai, Tamil Nadu, India.
[2]Department of Physics, Scott Christian College (Autonomous), Nagercoil-629003, Tamil Nadu, India.
[3]Futuristic Materials Research Centre for Planetary Exploration, Department of Physics and Nanotechnology, College of Engineering and Technology, Faculty of Engineering and Technology, SRM Institute of Science and Technology, SRM Nagar, Kattankulathur-603203, Kanchipuram, Chennai, Tamil Nadu, India.

[*]E-mail: vijayakt1@srmist.edu.in

Abstract: This paper describes the theoretically engineered electro optic chromophore (bromine based) and various characteristics of the chromophore. The simulation of halogen based (chlorine and bromine) electro optic chromophore is done. This chromophore can be especially used in the prospective applications due to its exceptional properties and conclusion drawn from the theoretically engineered electro optic chromophore. The halogens because of their electronic configuration tend to attain the stability with the other compound due to inductive effect. The halogen based chromophore enables the stable charge transfer due to change in the internal energy level The factors responsible for the enhancement of hyperpolarizability of the simulated chromophores which can lead to further advancement in the laser space communications applications are discussed.

Keywords: Electro-Optics; NLO Chromophore; Electro-Optic Modulator; Laser Communications

1. Introduction

Nonlinear optical (NLO) materials in which the change in the optical response due to interaction in the delocalized electrons with applied electric fields that ultimately causes change in the chemical and physical properties making those materials as promising candidate for the applications in optical signal processing, ultrafast switches, ultra-short pulsed lasers, laser amplifiers sensors and various other applications. At room temperature bromine exists as fuming red brown liquid and it is obtained from salt brines which exist in the United States, China and Dead Sea found in Israel and Japan [4-7]. The halogen-based chromophore (Tribrdibam) is almost a planar structure having substituted benzene ring on the either side having an arrangement of (2, 3, 5 - Tribromo benzene). The characteristic spectra and HOMO-LUMO energy band gap calculation are performed for the chromophore (2,3,5-Tribrdibam). The present work describes the molecular structural features, vibrational spectral studies and hyperpolarizability of the NLO chromophore (2,3,5-Tribromo) benzene (TBR) emphasizing on the mechanism of intramolecular charge transfer (ICT).

2. Theoretical background

The optical response in organic compound is due to non-linearity that arises from the molecular origin and hyperpolarization of π electrons. The force ($F = qE$) is caused by interaction of the electric field with charge distribution of system results in displacement of electron density. Hence, the displacement of electron density from the nuclear framework causes the separation of positive and negative charges and consequently in an induced dipole (Equation 1). The formulation of nonlinear polarization starts with initial expansion of total dipole as Taylor series.

$$\mu_i = \mu_i^0 + \alpha_{ij}E_j + \frac{1}{2!}\beta_{ijk}E_jE_k + \frac{1}{3!}\gamma_{ijkl}E_jE_kE_l \text{.........} \ (1)$$

Here μ_i^0 represents the permanent ground state dipole moment, $\alpha_{ij}$is linear polarizability tensor, β_{ijk} is second order polarizability and γ_{ijkl} is third order polarizability.

When the electric field E is applied at the material the response is given by volume polarization (Equation 2)

$$P_i = P_i^0 + \chi_{ij}^{(1)}E_j + \frac{1}{2!}\chi_{ijk}^{(2)}E_jE_k + \frac{1}{3!}\chi_{ijkl}^{(3)}E_jE_kE_l + \text{..................}(2)$$

Here P_i^0represents permanent polarization, $\chi_{ij}^{(1)}$represents linear susceptibility of molecule and$\chi_{ijk}^{(2)}$, $\chi_{ijkl}^{(3)}$ represents second order and third order susceptibility.

To get more idea on the frequency doubling, the required electric field strength of plane light wave is expressed as in equation 3

$$P = \left(P_0 + \frac{1}{4}\chi^{(2)}E_0^2\right) + \chi^{(1)}E_0 cos\omega t + \frac{1}{4}\chi^{(2)}E_0^2 cos2\omega t \text{..................}(3)$$

The above equation indicates that during polarization, the second order DC field is contributed to state polarization, frequency component ω and a new frequency double component 2ω. Thus an inverse light beam passing through the NLO material of order 2 produces the light which is twice of input frequency. The second order susceptibility (Equation 4) can be written more precisely as tensor of rank 3, including the frequency of incident light.

$$\chi^{(2)} = \chi_{ijk}^{(2)}(-\omega_3; \omega_1; \omega_2)\text{................}(4)$$

Thus the process associated with pockel effect $\omega+0\rightarrow\omega$ can be viewed as particular case of sum frequency generation one light beam is substituted by static DC field of magnitude E_2. The second order polarization (Equation 5) becomes $\chi^{(2)}E_1E_2cos\omega_1t$ with voltage applied to non-linear material.

$$P = [\chi^{(1)}+\chi^{(2)}E_2]E_1cos\omega_1t\text{................}(5)$$

The above expression indicates that the electric field changes the effective linear susceptibility and the application of voltage to material causes the optical beam to look out for the different refractive index. The Pockel's effect is used for modulation of polarization by changing the applied voltage.

Computational methodology

Geometry optimization of all non-covalent complexes is performed with Gaussian 09W software utilizing the Density Functional Theory (DFT) at B3LYP/6-311++G(D,P) basis set. The optimized structure of molecule ensures that the structure corresponds to the true energy minima.

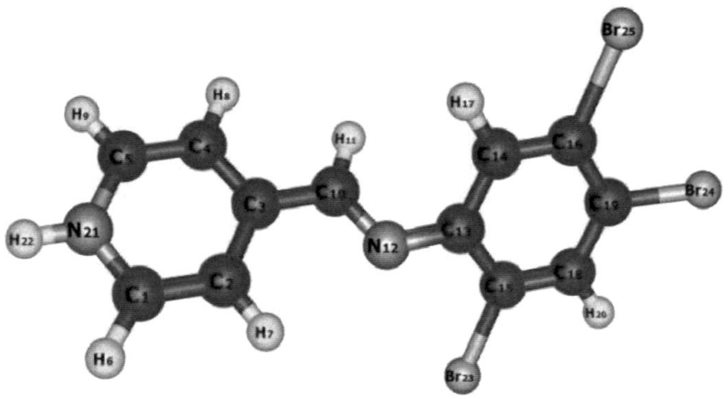

Fig. 1 Optimized Molecular Structure of TBR

3. Results and discussion

First order hyperpolarizability

The optimized structural parameters of TBR at B3LYP/6-311++G(d,p) are given in Table 1. This preliminary vibrational analysis of TBR is performed based on the characteristic vibrations of the benzene group, pyridine group and azo bridge and bromine. The chromophores with large first order hyperpolarizability β values promise for compact high-energy bandwidth and highly efficient electro-optic devices. The TBR is engineered with the aid of B3LYP,6-311G (d,p) and the first order hyperpolarizability β is calculated as 3.00714×10^{-29} e.s.u. The higher β value for TBR is due to the reason that the arrangement of Br in the ring which is responsible for the conjugation in the structure. It is observed that the structure is mostly planar except the pyridine ring side of the structure which is mostly responsible due to resonance and hyperconjugation effect till the azo bridge.

Table 1 Optimized Bond Lengths (Å), Bond Angles (°) and Dihedral Angles (°) of TBR

Bond Angle	Calc.	Bond Length	Calc.	Dihedral Angle	Calc.
C16-C19-C18	120.0	C19-Br24	1.91	Br25-C16-C19-Br24	0.00
C19-C16-Br25	119.99	C19-C18	1.40	Br14-C16-C14-H17	0.00
Br24-C19-C18	119.99	C15-Br23	1.91	H9-C5N21-H22	-58.51
H20-C18-C15	119.99	C18-H20	1.07	C2-C1-N21-H22	-121.48
Br23-C15-C13	120.02	C16-Br25	1.91	Br24-C19-C16-C14	180.00
C13-N12-C10	119.95	C14-H17	1.07	C10-N12-C13-C15	180.00
C10-C3-C2	120.001	C13-N12	1.47	H11-C10-C3-C2	180.00
H8-C4-C5	120.002	C10-N12	1.29	H8-C4-C3-C10	0.00
C5-N21-H22	106.78	C3-C10	1.54	Br23-C15-C13-C12	0.00
C19-C16-C14	120.00	C4-C5	1.40	C16-C19-C18-H20	180.00
C13-C15-Br23	120.00	C5-N21	1.40	H17-C14-C13-C15	-180.00
H7-C2-C1	120.00	N21-H22	1.00	C2-C3-C1-H6	0.00
C2-H7-C1	34.427	C1-H6	1.07	C13-N12-C10-H11	0.00

HOMO and LUMO molecular orbital Analysis

The HOMO and LUMO molecular orbital and their energy are one of the important parameters which help to obtain highly reactive side in π-electron system that illustrates the different types of reactions occurring in the conjugated system. HOMO is the highest occupied molecular orbital and LUMO is the lowest molecular unoccupied molecular orbital and the separation between these two orbitals (HOMO-LUMO Gap) is due to the transfer of charges from donor group to acceptor group of electrons with the help of conjugated π bonds following the linkage of (N-H) between the two rings. The HOMO (-4.2128 eV) (Fig. 2) is localized over the benzene, azo moiety and bromine while LUMO (-6.479 eV) (Fig. 3) is located over pyridine and azo moiety. Hence, the energy difference in chemical reactivity is given by 2.2662 eV.

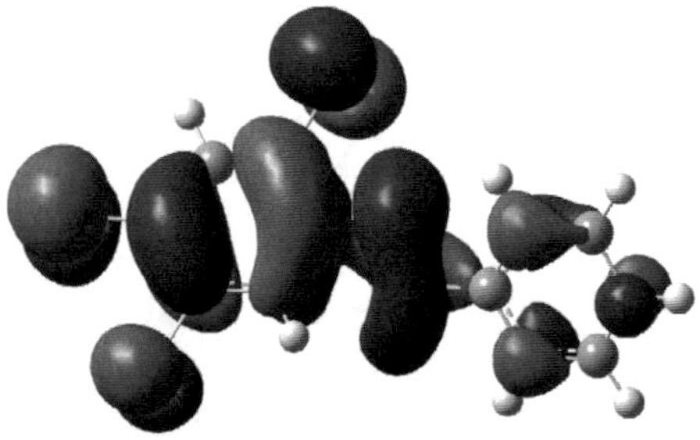

Fig. 2 HOMO plot of TBR

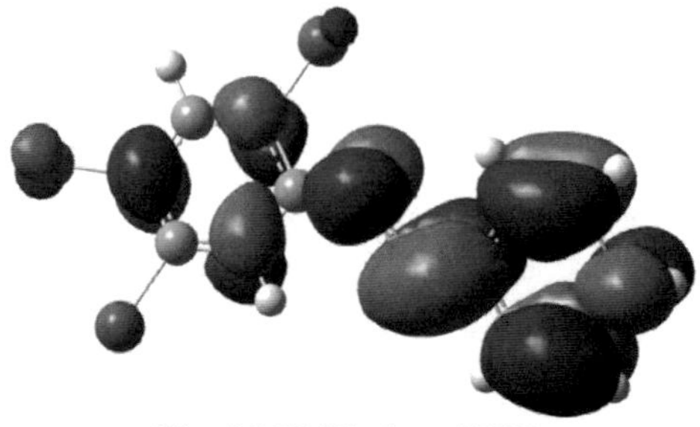

Fig. 3 LUMO plot of TBR

Electrostatic Potential Analysis

ESP (Electro Static Potential Analysis) is performed to obtain the distribution of negative charge within the compound which can lead to conclusion of overall stability of molecule. The ESP analysis is done to predict the reactive sites. The visual representation of chemically active sites using the Gauss View software is done and the generated ESP surfaces (Fig. 4) are represented by different colors.

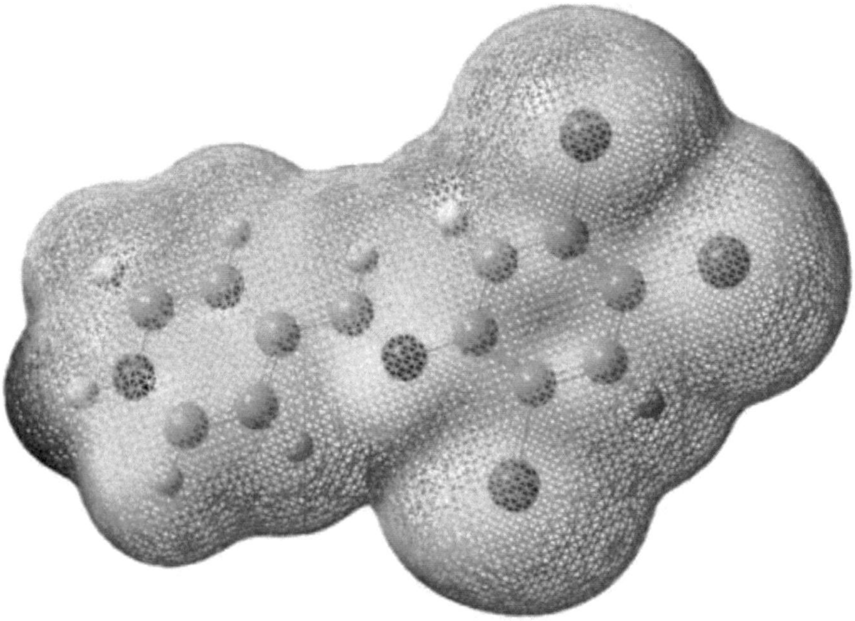

Fig. 4 Molecular electrostatic potential surface for TBR

In the following chromophore the lone pair of electrons is responsible for providing the stability to the structure and red color represents the negative charge which represents the electron negativity of structure. The region around the benzene ring and (N-H) linkage is the electron rich which is because of the arrangement of the Br atoms in the ring which are responsible for resonance in the ring from acceptor side hence it acts as a perfect binding site for nucleophiles.

Natural Bond Orbital and Vibrational Spectral Analysis

The natural bond orbital analysis (NBO) provides the method to study intramolecular in addition with intermolecular hydrogen bonding and provides information about orbital details regarding highest percentage of electron density. It also gives information regarding the interaction between virtual and filled orbital spaces that provides information in the form of vibrational spectrum analysis. Second order perturbation analysis of FOCK matrix in NBO is given in Table 2.

The vibrational spectral analysis is done to provide information regarding the molecular vibration and crystal structure. The theoretical FT-IR and FT-Raman spectra of TBR is taken using the B3LYP/6-311++G(D,P) basis set which are shown in Fig. 5 and 6. The values of every peak point indicate the frequency which is scaled against the Raman as well as IR. The simultaneous activation of few modes of benzene and azo group *ca* 1600 cm^{-1} indicates the intramolecular charge transfer of the TBR chromophore from donor to acceptor part through azo bridge.

Table 2 Second-order perturbation theory of Fock matrix in NBO basis

Acceptor NBO (j)	Donor NBO (i)	E(2) Kcal/Mol	F(i,j) A.U	E(j)-E(i) A.U
σ*(2) C10-N12	σ (2) C3-C4	13.85	0.077	0.27
σ*(2) C3- C4	N (1) C5	54.85	0.127	0.17
σ*(2) C1-C2	σ (2) C3 - C4	14.92	0.084	0.27
σ*(2) C13-C15	σ(2) C14-C16	10.36	0.072	0.29
σ*(2) C18-C19	σ (2) C13-C15	12.65	0.075	0.27
σ*(2) C18-C19	N (3)Br24	4.65	0.051	0.29
σ*(2)C18-C19	σ(2)C14-C16	7.90	0.061	0.28
σ*(2) C14-C16	N (3)Br25	5.38	0.055	0.29

Fig. 5 Theoretical Raman spectrum of TBR

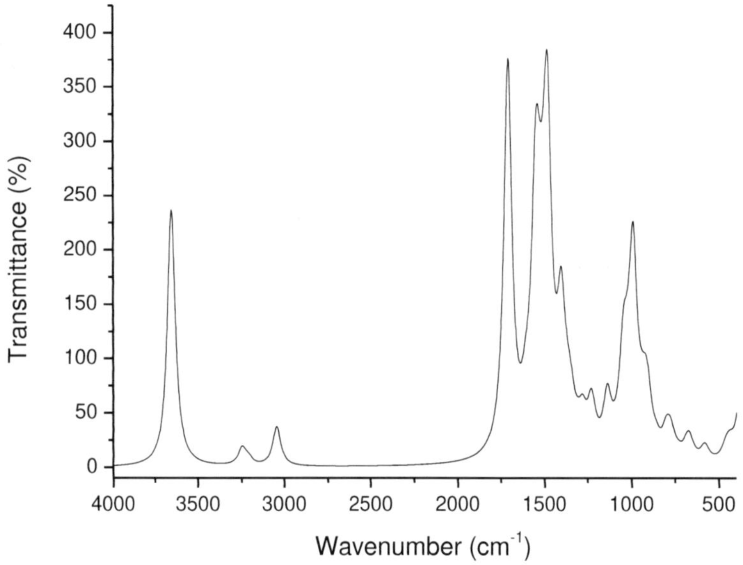

Fig. 6 Theoretical IR spectrum of TBR

Table 3 Global descriptors for TBR by B3LYP/6-311++G(d,p) method

Factors	B3LYP/6-311++G(d,p)
LUMO energy E_{LUMO} (eV)	-6.4790
HOMO energy E_{HOMO} (eV)	-4.21286
Ionization potential IP (eV)	4.21286
HOMO- LUMO energy gap ΔE_{GAP} (eV)	2.2662
Electronegativity χ (eV)	5.3459
Electron affinity EA (eV)	6.4790
Global softness S $(eV)^{-1}$	0.44128
Chemical potential μ (eV)	12.611
Electrophilicity index ω (eV)	-4.21286
Chemical hardness η (eV)	1.1331

Chemical Reactivity Descriptors

By Koopman's theorem, the calculation of ionization potential (I=-EHOMO) and electron affinity (A=-ELUMO) was performed. The electron affinity and ionization potential of the molecule are important for electron transfer from donor to acceptor thus serving as a key factor of determining the consequences. The calculation of electronegativity (χ) and chemical potential (μ) is done using the relation, χ = (IP+EA)/2 and μ = -(IP+EA)/2 respectively that is universally applicable. The chemical hardness (η) of TBR molecule has been simulated from the following relations, η = (I-A)/2 that act as the statistical information. The chemical reactivity of the molecule can be measured by its softness, and its calculation is performed with the help of the relation S= 1/η which necessitate support to the concluded points. The electrophilicity index ($\omega=\mu^2/2\eta$) is used to measure the power due to electrophilicity which tends to

IOP Conf. Series: Materials Science and Engineering 1219 (2022) 012001 doi:10.1088/1757-899X/1219/1/012001

explores the decrease in energy associated with the flow of electrons from donor to acceptor orbitals when they come in interaction with each other. The global reactivity descriptors for TBR molecule, the simulated values of chemical hardness, electronegativity, electrophilicity index, global softness, chemical potential are given in Table 3.

4. Conclusion

The characteristics of TBR is computed using GAUSSIAN 09W program package, Gauss View and Chemdraw and the change in the electronic distribution is studied by different analysis which confirms that TBR electro-optic chromophore is a prospective candidate for advancement in the field of future space laser communication systems.

Acknowledgements

Authors thank High Performance Computing Center, SRM Institute of Science and Technology for providing the computational facility. The author T.Vijayakumar acknowledges the financial support obtained from the Department of Space, Government of India [Grant no.B.19012,57,2016-II] through RESPOND project and the selective excellence initiative award received from SRM Institute of Science and Technology.

References

[1] Larry R. Dalton 2009 Thin Solid Films 518 428-431
[2] Vijayakumar T, Hubert Joe I, Reghunadhan Nair CP, Jayakumar VS 2008 Chem. Phys. 343 83-99
[3] Arivazhagan M 2012 Indian J. Pure Appl. Phys. 50 363-373
[4] Larry R. Dalton, Phillip A. Sullivan, Denis H. Bale 2010 Chem. Rev. 110 (1) 25-55
[5] Larry R Dalton 2003 J. Phys.: Condens. Matter 15(20) R897
[6] Hong Xi Zhanga, Dong Lu, Tao Liu, Masud Mansuripur, and Mahmoud Fallahi 2004 Appl. Phys. Lett. 85 4275
[7] Janaki A, Balachandran V, Lakshmi A 2013 Indian J. Pure Appl. Phys. 51 601-614

Investigating microwave absorption properties of cnfferric oxide nanocomposite

Namratha M[1], D Mouli Bhaskar[2], M. S. Nisha[3]
[1]Skolkovo Institute of Science and Technology, Moscow, Russia
[2]Indian Institute of Technology Kharagpur, Kharagpur, India
[3]Hindustan Institute of Technology and Science, Chennai, India

E-mail: [1]namrathamnk15@gmail.com

Abstract: Microwave absorbing materials plays a crucial role in reducing pollution created due to the electromagnetic radiation and several defence applications. This paper reports about the microwave absorption property of combined form of Carbon Nano fibres (CNF) and Ferric Oxide Nano powder (Fe_2O_3). Ferric Oxide Nano powder is deposited on the fibres of CNF using hydrothermal autoclave method. The samples of CNF- Ferric Oxide Nano powder with different weight percentages is fabricated and the microstructure and microwave absorbing properties are studied. By varying the amount of CNF and Ferric Oxide Nano powder the microwave absorbing and electromagnetic properties are modulated. For investigating the synergistic effects between CNF and ferric powders, different mass ratios of CNF to ferric oxide Nano powder have been synthesized and helped in enhancement of the effect between the magnetic loss and dielectric loss to achieve high microwave absorption. These results might be inspiration for the future to create microwave absorbing materials which are light weight.

Key Words: Hydrothermal Autoclave method,FE-SEM, X ray diffraction, Raman spectroscopy, Microwave absorption

1. Introduction:

In this modern information age, the swift development of wireless transmission leads to stray radiation in the atmosphere which in turn affects the standard of signal transmission and environment. The advancements in stealth technology and increase of EMI pollution in civil and defence areas resulted in fabrication of microwave absorbers. These materials have wide range applications in reduction of radar detection, electromagnetic pollution, interference, military, aerospace and automobile sectors [1-5].

The main factors that to be considered for a microwave absorbing material are reduction of weight and to achieve maximum efficiency in the operating bandwidth [6-8]. These can be achieved by composite microwave absorbers. Composite microwave absorbing materials are promising materials for microwave absorption. These materials are generally made up of two microwave absorbing materials one with high magnetic permeability and other with high electrical permit i.e., dielectric losses and magnetic losses [9].

In this search of composite microwave absorbing materials, carbon-based materials have been considerably reached due to their less density, length-diameter ratio, immense strength, good electrical properties, and resistance to corrosion [10]. Because of its poor impedance, immaculate carbon materials are unbefitting as an eligible absorber.

It is familiar that the unaccompanied absorber is hard to attain reflective losses and impedance matching, the convincing absorption band for a single component is comparatively tapered and only be used in a definite frequency. Multiple-component absorbers have exhibited considerably great impedance in comparison with single-component absorber because of variable polarization [11-12]. Under this circumstance, researchers are trying to blend carbon materials with different materials for attaining better microwave absorptive properties [13].

IOP Conf. Series: Materials Science and Engineering 1219 (2022) 012002 doi:10.1088/1757-899X/1219/1/012002

In this study, we chose Carbon Nanofibres (CNF) and ferric Nano powder (Fe_2O_3). CNFs have excellent thermal and electrical conductivities with high compatible nature which help them to absorb and store the energy [14]. Recently, adapted CNFs composites such as Co/CNFs [15], Fe/CNFs [16], Ni/ CNFs [17], FeCo/CNFs [18] and Fe_3O_4/CNFs [19], which are synthesized by the electrostatic spinning technique, have validated their significance. Generally ferric compounds and metal oxides are traditional microwave absorbing materials. Ferric Nano powder has relatively high magnetic permeability and produces good magnetic loss.

Ferric Nano powder is imposed over the CNFs using hydrothermal process and their reflection loss is estimated. The hybrid of ferric Nano powder and CNF displayed a most favourable reflection loss of -74.75 dB in the X(8.0-12.0GHz) band at a thickness of 2.8mm at 17wt%. This composite microwave absorption material has excellent structural flexibility with no fracture at effective cost.

2 Experimental methods

2.1 MATERIALS & REAGENTS

Ferric Nano powder (Average Particle Size (APS)-30nm, Assay-min 99%) and Carbon Nanofibres were purchased from Sisco Research Laboratory and Sigma Aldrich, respectively. Acetone and ethanol solvents were purchased from cero chemicals Pvt. Ltd. All the chemicals were used as received without any further distillation. Agilent N9917A network analyser within the frequency of 2-12 GHz is used to evaluate the reflection loss utilizing technique of waveguide.

2.2 SYNTHESIS OF CNF-Fe2O3 HYBRID

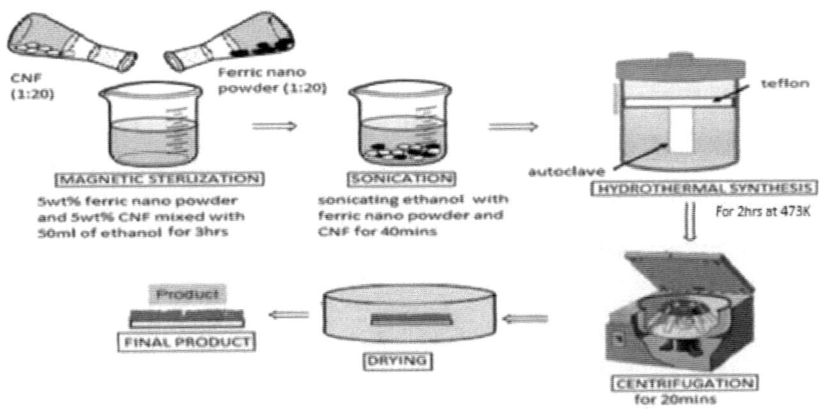

Fig.1 Process of synthesizing CNF- Fe_2O_3

Firstly, Fe_2O_3 and CNF were measured according to the weight percentage and completely mixed in 50 ml of ethanol solution. For ensuring the proper mixing of the compounds the solution was stirred using magnetic stirrer and then placed in ultra sonicator. The solution is stirred using a pellet at 800rpm for 30minutes in the magnetic stirrer and then shifted to ultra sonicator for an hour. Both the processes are done under room temperatures. The costefficient hydrothermal autoclave method is chosen for the synthesis. The solution from the ultra sonicator is transferred to the Teflon-lined stainless-steel autoclave with capacity of 100ml. The autoclave is left in the muffle furnace for 12 hours which is heated up to 200°C where the solution is converted into powder. The powder is again mixed with ethanol solution for centrifugation process. This is done for 30minutes at 3000rpm at room temperature. The retrieved solution from the centrifugation process is washed with ethanol and distilled water for several times. Finally, it is dried in the vacuum for 6 hours at 60°C. Thus, the hybrid material of

IOP Conf. Series: Materials Science and Engineering 1219 (2022) 012002 doi:10.1088/1757-899X/1219/1/012002

CNF- Fe_2O_3Nano composite is synthesized. For comparison, CNF- Fe_2O_3Nano composites were synthesized using the same method for all the various weight percentage configurations. The weight percentages are mentioned in the below table.

Samples	**Weight percentage(%)**	**Fe_2O_3 (in gms)**	**CNF (in gms)**	**Ethanol (in gms)**
S1	5wt% of Fe_2O_3 - CNF	1.25	1.25	47.5
S2	10wt% of Fe_2O_3 - CNF	2.5	2.5	45
S3	12wt% of Fe_2O_3 - CNF	3	3	44
S4	17wt% of Fe_2O_3 - CNF	4.25	4.25	41.5

Table 1: Different weight percentages considered

2.3 MATERIALS CHARACTERIZATION

The interrelation of the CNF- Fe_2O_3 hybrid were categorized on the Energy-dispersive X-ray spectroscopy and a field emission scanning electron microscopy (FE-SEM, MIRA3 TECSCAN) were used for the analysing of the microstructure and for the determination of the chemical composition of the samples. X-ray Diffractometer is used for the analysis of the power diffraction data and crystallographic structure of prepared samples (XRD, XPERTPRO) at 2θ angle. The scanning range of 20.00° to 80.00° (2θ) is used and with the scanning speed of 2.00°/min. The characteristics peak of the microwave absorbers was found using Raman Spectroscopy measurements.

3 RESULTS AND DISCUSSION

MICROSTRUCTURE PHASE ANALYSIS

The crystallographic structure of prepared samples was analysed with FE-SEM. Fig 2c shows resolution of 10 μm of 5 wt% of CNF- Fe_2O_3. The microstructure of the hybrid combination of Fe_2O_3 - CNF, can be clearly observed till 2 μm magnification as shown in fig 2a with an average diameter in the range of 18-33 nm, for 17 wt%.

The stage composition of the samples is analysed by XRD, as shown in Fig 3, which reveals the presence of the major compounds. All of the sharp peaks can be indexed to CNF and Fe_2O_3in the curves, with no presence of the impurities. The Fe_2O_3 - CNF Nano compositeswith diffraction peaks centred at 2θ = 18.8, 26.6, 29.09, 33, 46.4, 52.7 and 59.44 and can be assigned to the planes respectively.

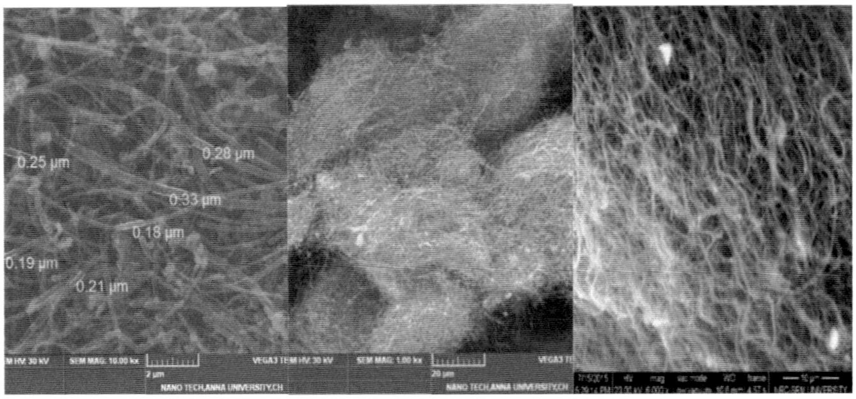

fig 2a fig2b fig2c

Fig.2(a,b,c) Microstructure of Fe_2O_3- CNF from SEM analysis fig2a) 2μm magnification fig2b) 20um magnification fig2c) 10um magnification

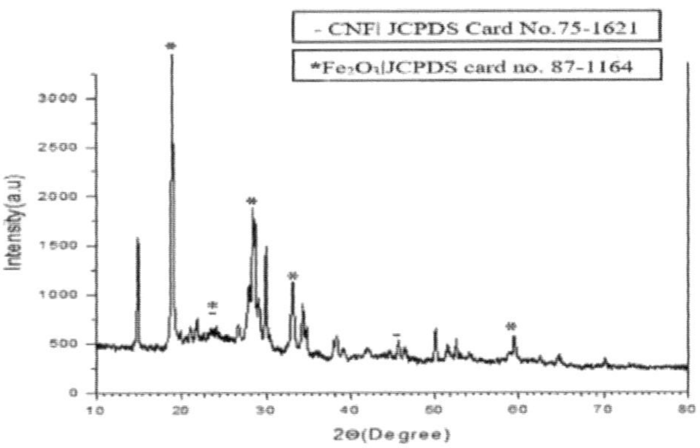

Fig.3 EDAX spectrum of hybrid Nano composites.

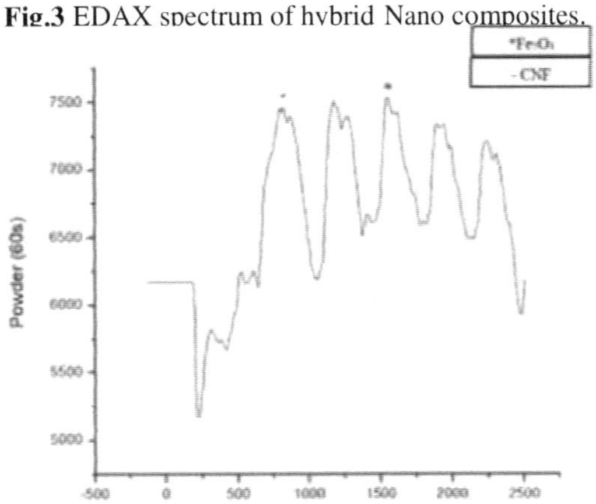

Fig.4 The XRD pattern of 5 wt% of hybrid CNF - Fe_2O_3 Nano composites

For explaining the chemical bonding of samples, Raman spectrum of thin film of Fe_2O_3 - CNF hybrid Nano composites using hydrothermal method was performed and distinguished by appearance of various ranges in chemical bonding of Nano powders. In Figure the peaks are presented in the wavelength of 1620 cm^{-1} gives CNF and the wavelength of 819 cm^{-1} represent Fe_2O_3. The peaks of the Raman spectra are sharp and strong. The peak of CNF results in the SP_2 hybridized carbon structure which is situated in the 2D band of Raman spectrum.

MICROWAVE ABSORPTION PERFORMANCE:

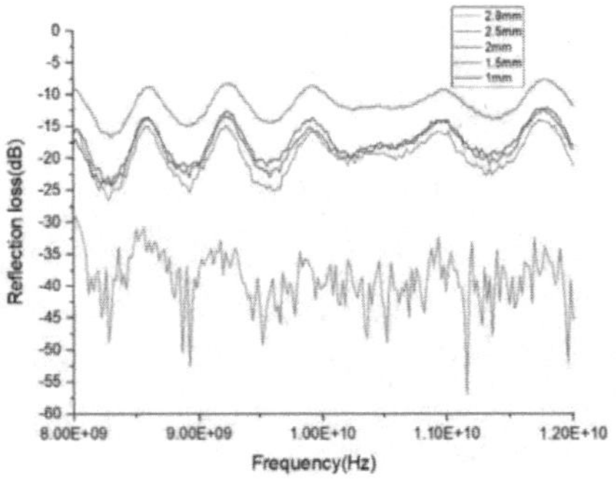

fig 5a

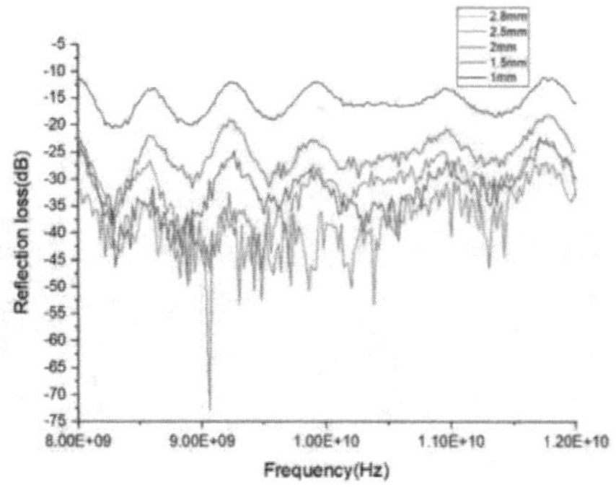

fig 5b

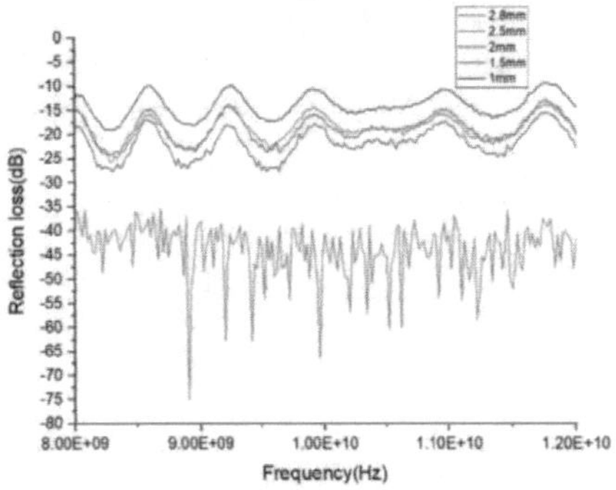

fig5c

Fig.5 RL vs frequency graphs (a). 10 wt% CNF - Fe_2O_3, (b). 12 wt% CNF - Fe_2O_3 (c). 17 wt% CNF - Fe_2O_3

The transmission line theory is used to find the reflection loss. The figures above show the reflection loss of absorbers with 10wt%, 12wt%, 17wt% of CNF-Fe2O3 hybrid Nano composites with thickness 1 mm, 1.5 mm, 2 mm, 2.5 mm, 2.8 mm at 8-12 GHz. From the above results it is evident that the reflection loss (RL) absolute value of CNF-Fe2O3 is significant than that of CNF which is −12.2 dB at 10.7 GHz [20]. By increasing the weight percentage from 5% to 17% for composite, the reflection loss value is found to be substantially increasing, results in the improved absorption performance. For the CNF-Fe2O3 composite the peak absorption value is noted at 17 wt%of 2.8 mm thickness and ata frequency of 8.9GHz, reflection loss value is -74.75 dB

Conclusions:

In summary, the Fe_2O_3-CNF hybrid Nano composites has shown fascinating magnetic properties, leading to better microwave absorption application. The crystal characterization of Fe_2O_3-CNF Nano composite was obtained and portrayed in SEM analysis with a diameter range of 18-33 nm at 17 wt%. The XRD peaks confirm the presence of ferric oxide Nano powder and CNF i.e., at $2\theta=33$ diffraction angle both the materials are discovered. The absolute bonding of Fe_2O_3-CNF was demonstrated using Raman spectra. The hybridization of the Nano materials helps in enhancing the reflection loss in the bandwidth of wide range. The maximum RL value of Fe_2O_3-CNF Nano composite can achieve -74.75 dB at 8.9GHz for a thickness of 2.8 mm. The results found that Fe_2O_3-CNF has a minimum RL value which is higher when compared to virgin forms of CNF and Fe_2O_3 Nano powder.

References:

1. Yuanliang Zhou,Ning Wangd,Javid Muhammada,Dongxing Wanga,Yuping Duana,Xuefeng Zhangc,Xinglong Donga,Zhidong Zhang"Graphene nano flakes with optimized nitrogen doping fabricated by arc discharge as highly efficient absorbers toward microwave absorption", (2019),issue 14 ,148,Pages 204-213
2. Yan Wang, Xiang Gao, Hongwei Zhou, Xinming Wu, Wenzhi Zhang, Qiguan Wang, Chunyan Luo,"Fabrication of biomass-derived carbon decorated with NiFe2O4 particles for broadband and strong microwave absorption", (2019), issues 3, Volume 345,Pages 370-378
3. Qin, Christian Brosseau, "A review and analysis of microwave absorption in polymer composite filled with carbonaceous particles", (2012)
4. Murowchick,Xiaobo Chen, "Microwave absorpt Michael Green,Anh Thi Van Tran, Russell Smedley,Adam Roach,James ion of magnesium/hydrogen-treated titanium dioxide nanoparticles", (2019),Issue 1,Volume 1, Pages 48-59
5. C. Hu, Z. Mou, G. Lu, N. Chen, Z. Dong, M. Hu, L. Qu, 3D graphene-Fe3O4 nano-composites with high-performance microwave absorption, Phys. Chem. Chem. Phys. 15 (2013) 13038–13043.
6. Ben Huang,Jianling Yue,Yongshan Wei,Xiaozhong Huang,Xiuzhi Tang,Zuojuan Du,Enhanced microwave absorption properties of carbon nanofibers functionalized by FeCo coatings,2019;483
7. Xu W, Wang G, Yin P. Designed fabrication of reduced graphene oxides/Ni hybrids for effective electromagnetic absorption and shielding. Carbon 2018;139:759
8. Liu W, Shao Q, Ji G, Liang X, Cheng Yan, Quan Bin, et al. Metal-organicframeworks derived porous carbon-wrapped Ni composites with optimized impedance matching as excellent lightweight electromagnetic wave absorber. Chem Eng J 2017;313:734–44.
9. Liu P, Yan J, Gao X. Construction of layer-by-layer sandwiched graphene/poly- aniline nanorods/carbon nanotubes heterostructures for high performance super- capacitors. Electrochim Acta 2018;272:77–87.
10. M. Zhu, G. Diao, Review on the progress in synthesis and application of magnetic carbon nanocomposites, Nanoscale 3 (2011) 2748–2767.
11. T. Wang, H. Wang, X. Chi, R. Li, J. Wang, Synthesis and microwave absorption properties of Fe–C nanofibers by electrospinning with disperse Fe nanoparticles parceled by carbon, Carbon 74 (2014) 312–318.
12. Jun Xiang, J. Li, X. Zhang, Q. Ye, J. Xua, X. Shen, Magnetic carbon nanofibers containing uniformly dispersed Fe/Co/Ni nanoparticles as stable and highperformance electromagnetic wave absorbers, J. Mater. Chem. A 2 (2014) 16905– 16914.
13. J. Xiang, X. Zhang, Q. Ye, J. Li, X. Shen, Synthesis and characterization of FeCo/C hybrid nanofibers with high performance of microwave absorption, Mater. Res. Bull. 60 (2014) 589–595.
14. T. Zhang, D. Huang, Y. Yang, F. Kang, J. Gu, Fe3O4/carbon composite nanofiber absorber with enhanced microwave absorption performance, Mater. Sci. Eng. B 178 (2013) 1–9.
15. R. Qiang, Y.C. Du, Y. Wang, N. Wang, C.H. Tian, J. Ma, P. Xu, X.J. Han, Rational design of yolk-shell C@C microspheres for the effective enhancement in microwave absorption, Carbon 98 (2016) 599-606.
16. J.W. Liu, R.C. Che, H.J. Chen, F. Zhang, F. Xia, Q.S. Wu, M. Wang, Microwave absorption enhancement of multifunctional composite microspheres with spinel Fe3O4 cores and anatase TiO2 shells, Small 8 (2012) 1214-1221.
17. X.H. Li, J. Feng, Y.P. Du, J.T. Bai, H.M. Fan, H.L. Zhang, Y. Peng, F.S. Li, One-pot synthesis of CoFe2O4/graphene oxide hybrids and their conversion into FeCo/graphene hybrids for lightweight and highly efficient microwave absorber, J. Mater. Chem. A 3 (2015) 5535-5546.

IOP Conf. Series: Materials Science and Engineering 1219 (2022) 012002 doi:10.1088/1757-899X/1219/1/012002

18. J. Xiang, X. Zhang, Q. Ye, J. Li, X. Shen, Synthesis and characterization of FeCo/C hybrid nanofibers with high performance of microwave absorption, Mater. Res. Bull. 60 (2014) 589–595.
19. T. Zhang, D. Huang, Y. Yang, F. Kang, J. Gu, Fe3O4/carbon composite nanofiber absorber with enhanced microwave absorption performance, Mater. Sci. Eng. B 178 (2013) 1–9.
20. Li, Guang, et al. "Microwave absorption enhancement of porous carbon fibers compared with carbon nanofibers." The Journal of Physical Chemistry C 116.16 (2012): 9196-9201.

Investigation of ab-inito, Photoluminescence and electrical studies of $Li_2Co(WO_4)_2$

D L Shruthi, G N Anil Kumar*and A Jagannatha Reddy

Department of Physics, Ramaiah Institute of Technology, (Autonomous Institute, Affiliated to Visveswaraya Technological University, Belgaum) Bengaluru , Karnataka 560054, India.

E-mail. anilgn@msrit.edu

Abstract. The optical properties of the material is depends on the site occupancy of the doping in the host lattice. In the present study, wereport the synthesis, characterisation, optical properties and ab-initio studies of $Li_2Co(WO_4)_2$microcrystals synthesized by traditional and easy solid state method. The x-ray diffraction study shows the compound which arecrystallized in the triclinic crystal system space group $P\bar{1}$.The average crystallite size was estimated from diffraction study found to be 109 nm. From UV-Visible spectroscopic study the calculated band gap is 2.69 eV for the phosphor. The $Li_2Co(WO_4)_2$ phosphors shows the Photoluminescence under an excitation wavelength of 331 nm. A strong blue green peak emission at 447 nm was observed due to the transition from ^3T_1u levels to the ^1A_1g level. The electrical impedance studies were carried out to enunciate the impedance behaviour of $Li_2Co(WO_4)_2$and to depict different conduction and relaxation processes.Further, the*ab-initio*studies were carried out in detail to understand electronic properties of the phosphor.

Keywords: Photoluminescence, ab-initio,Retiveld refinement

1. Introduction

Among many inorganic hosts, the metal tungstates shows application in different areas like acoustic-optics, solid state lasers, scintillators, and optoelectronic devices[1]. Particularly the Hosts such as doubletungstates are interest of many researchers due to their interesting structural, physical, and chemicalproperties [2].Especially the bimetallic tungstate with the general formula $XY(WO_4)_2$where X= Li, Na , K , Y=Co, Ni , Cu , Fe are commercial interest for many applications particularly in fluorescent devices and lasers. These types of materials were reported to exhibit excellent photoluminescence, magnetic, electrical conductivity properties [3-4].

The $Li_2Co(WO_4)_2$is consider as the excellent host which is used as electrochemical hydrogen storage and also used as magnetic material. The above particular host can also used in many other application such as the scintillator material, microwave application, photoanodes and as optical fiber. By using this type of compound the organic pollutants are removed from the water sample and act as a photo catalysts. Literature studiesshows the$A_2B(WO_4)_2$ compounds synthesized by hydrothermal,solgel method, co- perception and the direct chemical method. Compared tomany other synthesis methodsthe traditional solid state method have several advantages such as its shows intense luminescence, free form solvent and inexpensive [5-6].

In the present study,the investigationof$Li_2Co(WO_4)_2$ compound is prepared by the solid state method and explore the detail analysis of structural and optical properties of the synthesized samples.

2. Experimental Procedure

Synthesis of $Li_2Co(WO_4)_2$

The materials used for synthesis are Lithium carbonate (Li_2CO_3, 99% by Merck Ltd), tungsten trioxide (WO_3, 99% by Sigma Aldrich Ltd) and Cobalt carbonate (Co_2CO_3, 99% by Merck Ltd). In the present work, $Li_2Co(WO_4)_2$ compound are synthesised by traditional solid state method. The stoichiometric amounts of starting materials are homogenously mixed in a agate mortar which is kept in the alumina crucible. Initially temperature of reaction time was 500 °C for 4 hour in muffle furnace to decarbonisation the compound. After preheated the mixture is again calcinated at 650 °C for 130 hoursto obtain thecompound.

Characterisation

The X-ray diffraction patterns of the samples were recorded by using a RigakuUltimadiffractometerusing Cu Kα radiation. The diffraction data were recorded in the range of 10-80 °C with the step width of 0.02°. The Rietveldrefinementstudies carried out to determine the structure by using the TOPAS software.Morphology of the synthesised sample are recorded by the Scanning electron microscopy (SEM).The Fourier transform infrared analysis is monitoring in the range of 400-4000cm^{-1} and the samples are used in terms of pellet. The UV –visible measurements are recorded by using the Specord-210 plus analytic jean spectrometer in the wavelength range of 200 to 1000 nm. The excitation and emission spectra are recorded by using the Edinburgh FLS-980 fluorescencespectrofluorometer. The electrical characterization studies are carried out by using 4294A Impedance Analyzer.

3. Results and discussion

PXRD analysis

XRD analysis were performed in order to investigate the phase purity of the $Li_2Co(WO_4)_2$ sample. The PXRD pattern of synthesised samples here the experimentaldiffraction pattern are matched withthe standard JCPDS card number 96-400-2304 as shown in Fig 1.[7]

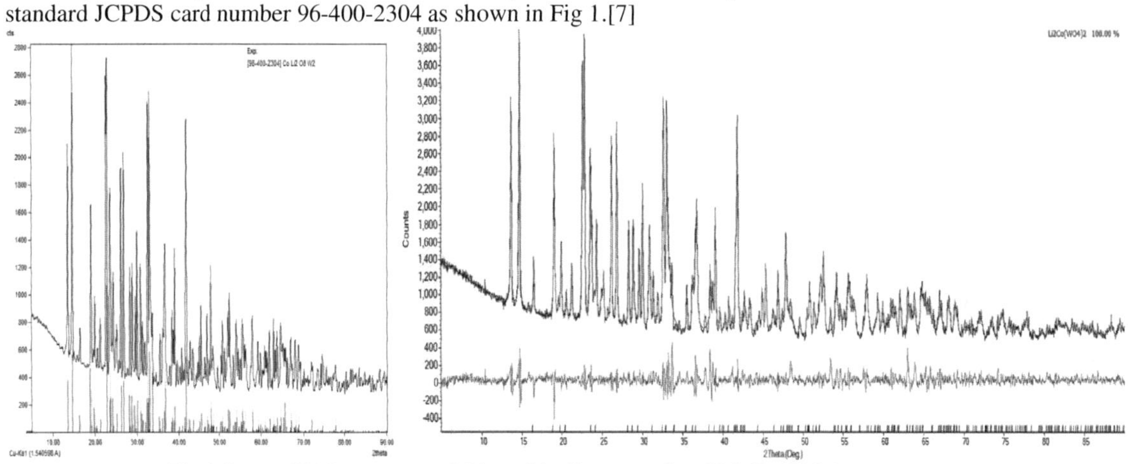

Fig 1 Power XRD pattern and Rietveldrefinementplotof $Li_2Co(WO_4)_2$

Therefined XRD pattern by Rietveld method of$Li_2Co(WO_4)_2$ compound using TOPAS software is shown in Fig 1. The results confirms the $Li_2Co(WO_4)_2$ compound in the space group $P\bar{1}$ and crystallize in the triclinic crystal system. Lattice parameters areobtained after refinement shows as a =4.91 Å, b=5.66 Å and c=5.87Å, α= 69.4°, β= 91.4° and γ= 116.1°. The reliability factors results are expected R-factor (R_{exp})= 3.26, weighted profile R –factor (R_{wp}) =6.18, Bragg R-factor (R_p)= 4.46, fit goodness

(GOF)=1.89. All the results indicate the $Li_2Co(WO_4)_2$ compound are formed without any impurity and show phase purity.
Fig 2 shows crystal structure of $Li_2Co(WO_4)_2$ compound. The structure of is $Li_2Co(WO_4)_2$ compound is composed of three dimensional framework consist of pyramidal WO_5 zigzag chain along c axis linked with the octahedral CoO_6. Each CoO_6 shares fourcorners with WO_5(three) pyramids in tungsten layer andtungsten layer share with LiO_6 Octahedron.This arrangement built upalternating layers of WO_5 pyramids and Co/LiO_6octahedra along [110] direction.

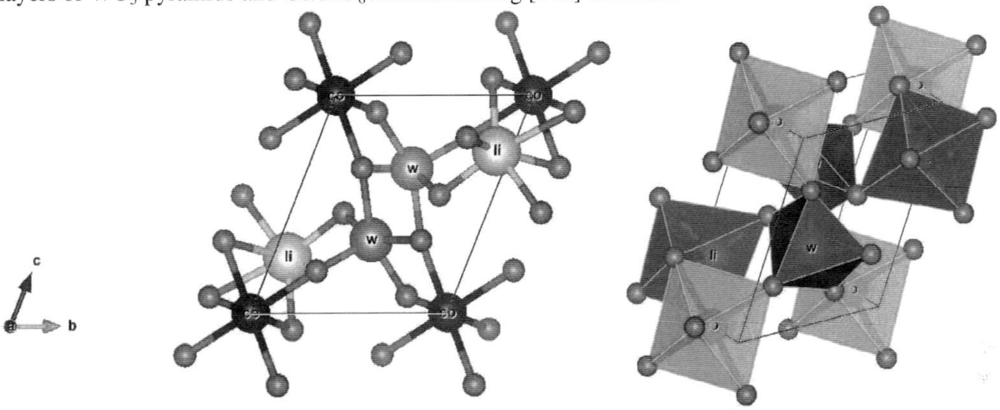

Fig. 2 Crystal structure and packing diagram of $Li_2Co(WO_4)_2$ compound.

Scherrer formula is used for the $Li_2Co(WO_4)_2$ compound to find the crystallite size by following equation[8]

$$D = \frac{k\lambda}{\beta cos\theta} \ldots\ldots\ldots\ldots\ldots\ldots..(1)$$

Wherepeak position is represent by θ, wavelength of x-ray source is given by λ, the full width half maxima is represent by β and Scherrer constant is represent by k.The crystallite size calculated which is found as 109nm. Scanning electron microscope of $Li_2Co(WO_4)_2$compound with different magnification are shown in Fig3.The morphology is clearly spherical shape with the average size of the particle (0.3-6 μm)

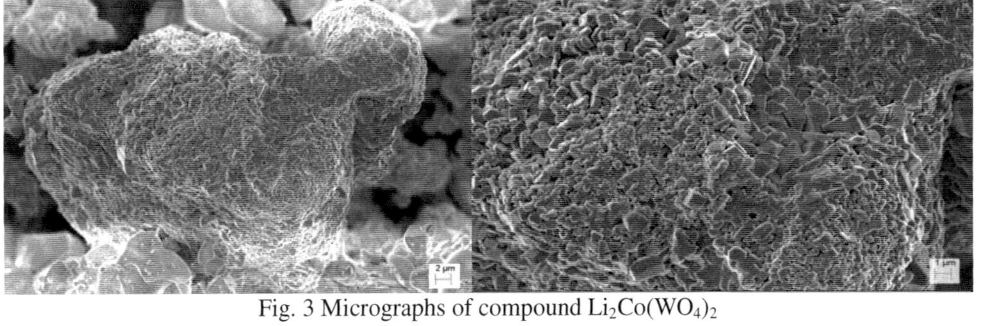

Fig. 3 Micrographs of compound $Li_2Co(WO_4)_2$

Fourier transform infrared spectra analysis (FTIR)

IOP Conf. Series: Materials Science and Engineering 1219 (2022) 012003 doi:10.1088/1757-899X/1219/1/012003

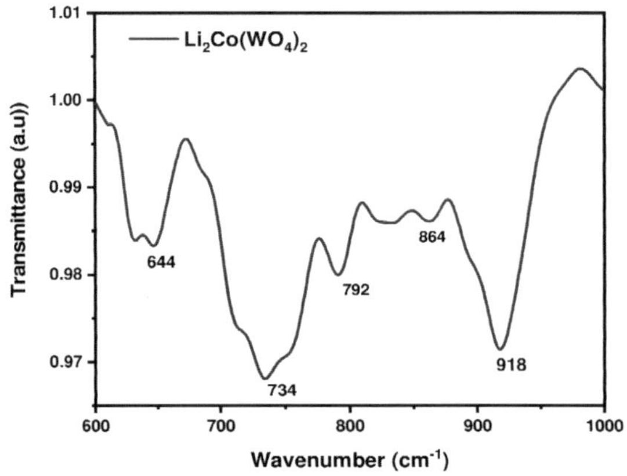

Fig. 4 FTIR spectra of $Li_2Co(WO_4)_2$

FTIR spectra of $Li_2Co(WO_4)_2$ compounds are shown in Fig4 which represents a characteristic functional groups of the compound. The absorption band shows the presence of 5 absorption bands (918, 864, 792, 734 and 644 cm^{-1}). Thestretching vibrations present in the WO_2in the W_2O_8group are indicates by the peak in the range of 710-870 cm^{-1} which is due to O-W-O stretching.The antisymmetric stretching of two oxygen bridge $W_2O_2$is related to the absorption band at the peak 644 cm^{-1}. The 918 cm^{-1} indicates the symmetric stretching vibrationof WO_4absorption band[9].

IRFrequency (cm^{-1})	Symmetry	Assignment
918	A_u	ν_s(W–O)
864	B_u	ν_{as}(W–O)
792	A_u	ν_{as}(WOOW)
734	B_u	ν_{as}(WOOW)
644	A_u	ν_{as}(WOOW)

Table -1 The assigned IR frequency vibrations of $Li_2Co(WO_4)_2$

UV-Visible spectroscopy

The $Li_2Co(WO_4)_2$ compound shows the diffuse reflectance spectra in the wavelength range between 200 – 800 nm.The charge transfer band in the range of 200-500nm are due to the absorption band of WO_6 transition in Fig 5(a).

Kubelka-Munk [10] relation is used to the calculate the $E_g$optical energy band gap

$F(R)h\upsilon = c(h\upsilon - E_g)^n$ ....(2)

Where energy of the photon is indicate by hυ, c is a constant, $E_g$indicate the optical energy band gap, n represent the pure number which is associated with the types of electronic transitions (n indicate the ½ and 2 which is corresponds to direct and indirect transition). Kubelka-Munk function is used to calculate the absorption coefficient which is given by

$F(R) = \frac{(1-R)^2}{2R} = \frac{k}{s}$ ....(3)

Here k indicates the molar absorption co-efficient and s represent the scattering co-efficient. Fig 5(b) shows the extrapolation of the line for $[F(R)h\upsilon]^{1/2}$ to [hυ] = 0represent the 2.69 eV optical band gap value for the $Li_2Co(WO_4)_2$.

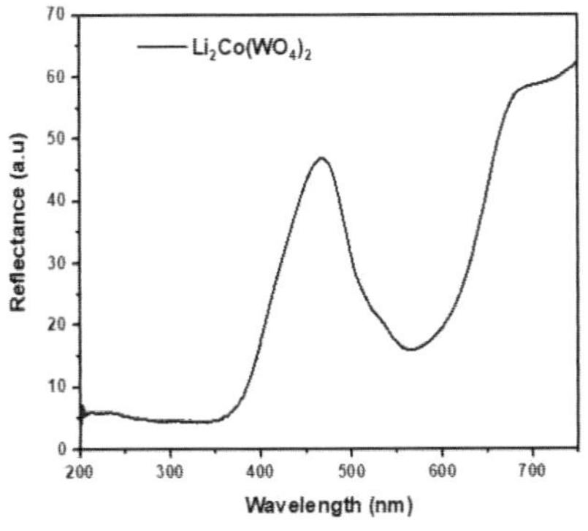

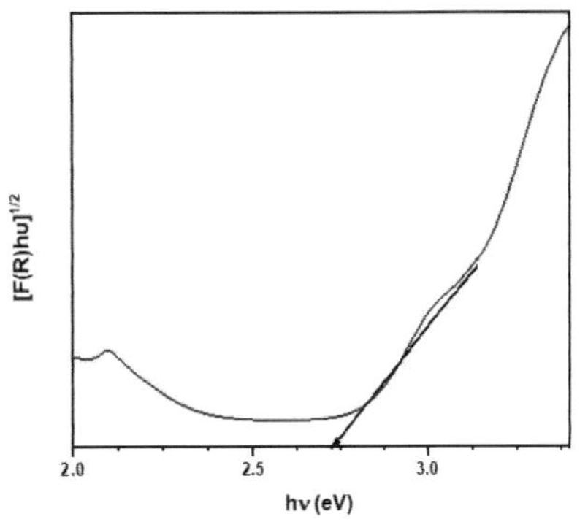

Fig5(a).Diffuse reflectance spectra Fig 5(b).Band gap energy of $Li_2Co(WO_4)_2$

Photoluminescence studies

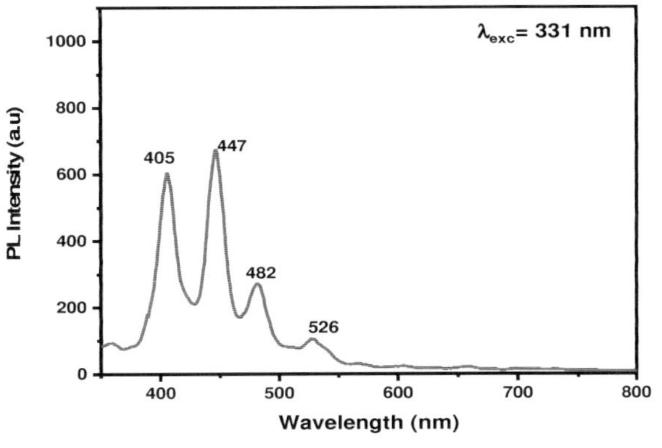

Fig. 6 PL emission spectra of $Li_2Co(WO_4)_2$compound.

The PL emission spectrum of $Li_2Co(WO_4)_2$ compound which is monitored atthe excitation wavelength of 331 nm is as shown in Fig 6.The photo-exited electrons and holes are generated in the Photoluminescence emission spectrum. The emission spectrum of$Li_2Co(WO_4)_2$ compound are represent in the wavelength between 300-800 nm.The emission spectra are due to four sharp emission peaks which are located at 405, 447, 482 and 526 nm.The multiphotonprocess are generated in the emission spectra, which are going to relaxed by various energy paths. The transitions peak between 400-530 nm is because oftransition from ^3T_1u levelsin the excited state to the ^1A_1g level in the ground state ofCobalt which is present in distorted octahedral co-ordination. The local low symmetry crystal field are also influenced by the degenerated excited state, and also the luminescence intensity is due to the trapping of exciton which is formed by the complex of $[WO_6]^{6-}$. Finally the holes and electronsare created and is caused by the charge difference betweenthe complex clusters of host lattice and dopant.

Commission InternationalDe'eclairage

The practical application of $Li_2Co(WO_4)_2$ compound is determine by calculating the CIE chromaticity coordinates.The chromaticity diagram is useful in lighting industry forestimating the color emitted by phosphor. In LED industry the CIE 1931 is the standard tool which significances the color mixing and

IOP Conf. Series: Materials Science and Engineering 1219 (2022) 012003 doi:10.1088/1757-899X/1219/1/012003

rendering.Fig 7 shows the CIE 1931 coordinates arecalculated by monitoring the emission spectra under the excitation of 331 nmwhich fall in the blue region as a value of(x= 0.173 , y= 0.1223) .

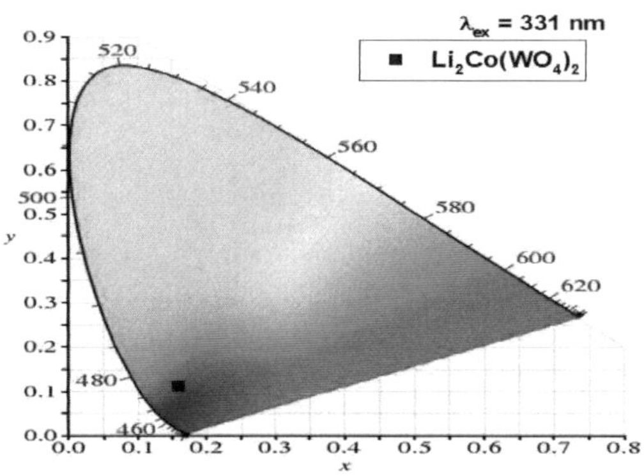

Fig.7 The CIE chromaticity diagram

Electrical properties Analysis

The electrical properties of the prepared sample and the various process like bulk, grain boundary effect with respect to the frequency domain is fund by impedance spectroscopy [11]. The impedance spectroscopy results gives the imaginary part (-Z″) and the real part (Z′) with respect to the frequency. The Fig.8(a) represent the variation of the real part of the impedance with the frequency. Here high value of impedance at low frequency is observed, this is due to different types of polarisation in material like atomic, ionic electronic , dipolar . As frequency increased for higher value the dipolar orientation and interfacial polarization decreases result in constant impedance.Fig.8(b) indicate the variation of frequency with respect to the imaginary part of impedance. The broad peak is observed in the material is due to the relaxation mechanism present in the compound.The *Z″* values achieve a maximum peak and that the value shows higher frequencies. These resultsindicate the mobile charge carriers take a long time to move from one site to another.

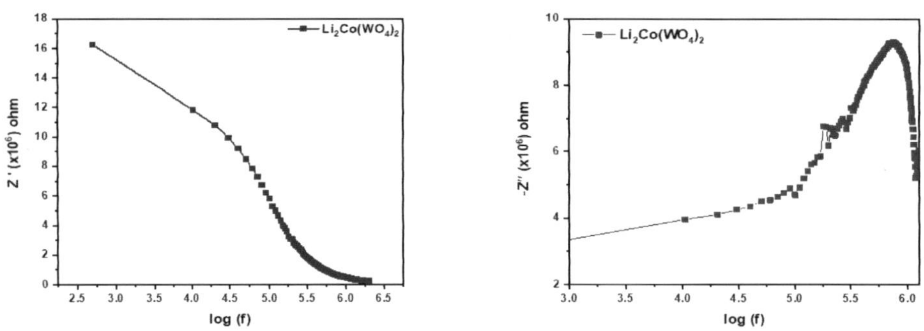

Fig 8(a)& 8(b) The Variation of the Z′ (real part) & Z″ (imaginary part) of impedance

The total impedance can be given by Z = Z′ + Z″. Fig 9 shows the Nyquist plot that is the variation ofreal part versus the imaginary part of impedance. A result indicates semicircle in the impedance plots shows the electrical conductivity in the synthesised sample. For polycrystalline materials during crystal formation some atoms gather around and form small groups in such a way that in each group atoms are arranged in a particular orientation. These groups of ordered atoms form a region called a

IOP Conf. Series: Materials Science and Engineering 1219 (2022) 012003 doi:10.1088/1757-899X/1219/1/012003

grain and the area between this type of regions is termed as the grain boundary. The interface which is occurred between the crystallites, 2 grains, polycrystalline material are the major reason for the grain boundary. The resistance of interior grain corresponds with high frequency, and resistance of grain boundary corresponding with lower frequency are occur in impedance spectroscopy. So, the high frequencies indicate the conduction of the grain and the low frequencies correspond to conduction of grain boundary. All the result indicatethe existence of the mechanism of conduction in the $Li_2Co(WO_4)_2$ compound.

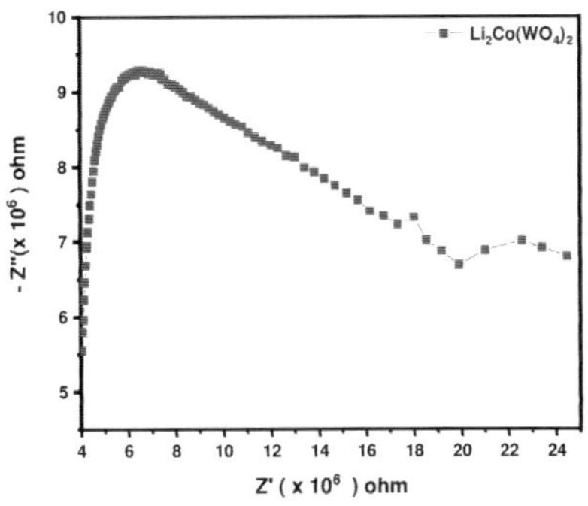

Fig 9.Nyquist plot $Li_2Co(WO_4)_2$ compound

Theoretical studies

The CRYSTAL17 code with the PBE hybrid functions is implemented to find the perioic DFT calculations [12]. The POB-TZVP basis set with the pseudo potential large core is used for describe the tungsten, Cobalt, lithium and oxygen atom.The 36 independent K-points brillouin zone is produced with respect the shrinking factor of 6 for the primitive cell of $Li_2Co(WO_4)_2$compound. The exchange integrals and the accuracy of coloumb are set to were set to be 10^{-14} (ITOL5) and to10^{-7} (ITOL1 to ITOL4), and self–consistent with the convergence on energy to 10^{-8}hartree for all geometry optimizations.

Band and Density of States:

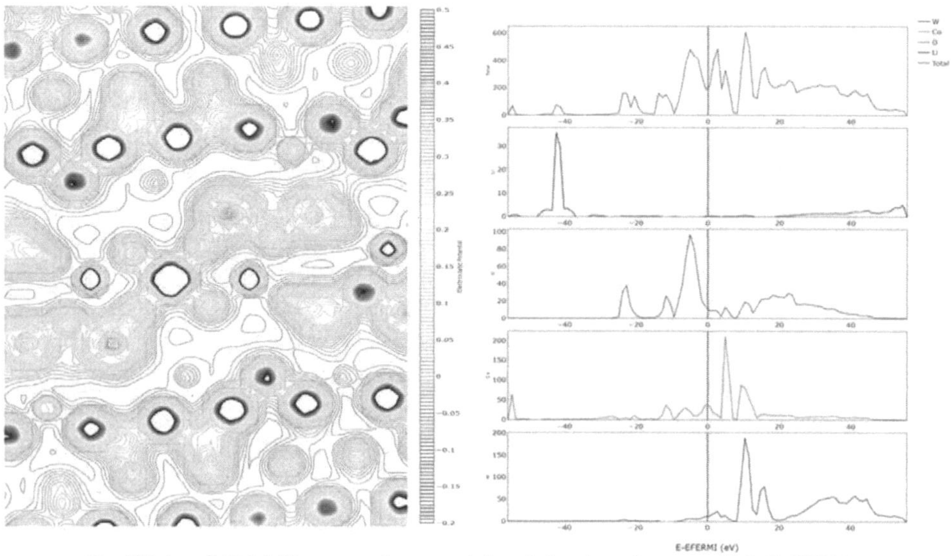

Fig 10(a) and 10(b) Electrostatic potential and density of states of $Li_2Co(WO_4)_2$

Fig 10(a). Shows the Electrostatic potential and electron density (ρ) distribution derived from crystalline density of $Li_2Co(WO_4)_2$. The chemical bonding of crystalline sample is represented by density distribution. In the electron density,dotted lines showsnegative values, solid lines positive and bold null contour. The negative values highlights fields of electron flow and atomic interactions and positive values represent charge accumulation. Larger charge accumulation on WO^{4-} group observed than Li-O, Co-O bonds. The density of states of $Li_2Co(WO_4)_2$ is shown in Fig. 10 (b). The $Li_2Co(WO_4)_2$is semi-metallic due to conduction band minimum (CBM) and a valence band maximum (VBM) close to each other. The valence band (VB) is mainly by oxygen p orbitals states, with little contribution from d orbitals of tungsten. The conduction band (CB) is formed from W and Co d states with same contributions of lithium and oxygenp states.

Sample	**Synthesis method**	**Calcination temperature**	**Lattice parameter for $Li_2Co(WO_4)_2$**	**Optical energy band gap**	**Reference number**
$Li_2Co(WO_4)_2$	Solid state method	650 °C for 130 h	a = 4.91 Å, b=5.66 Å c=5.87Å	2.96	**[Present work]**
$Li_2M(WO_4)_2$ (M = Co, Cu or Ni) materials	Solid state method	650 °C for 146 h	*a =4.916* Å *b =5.660* Å *c = 5.875* Å	-	**[13]**
$CoWO_4$	Hydrothermal method	180°C/8h	a= 4.90 Å, b=5.96 Å, c=5.84 Å	2.87	**[14]**
$Li_2Co(WO_4)_2$	Solidstate method	700 °C/160 h	a = 4.90 Å, b = 5.59 Å, c = 5.83 Å	-	**[15]**

$Li_2M(WO_4)_2$ M=Cu, Ni, Co	One step solid state method	700 °C for 160 h	a = 4.90 Å, b=5.60 Å, c=5.480 Å	1.65	**[16]**

Table-2 Comparative study $MX(WO_4)_2$ compounds

Table 2 shows the comparison of the structural properties, optical properties of the$MX(WO_4)_2$compounds reported by different synthetic methods. In present work, we use the solid state method for synthesis of $Li_2Co(WO_4)_2$,calcinated at 650 °C for 130 h which significantly less when compared to other methods. The calculated lattice parameters were good agreement with the literature. The detail studies of Photoluminescence and electrical properties are explored for the practical applications.

Conclusions

$Li_2Co(WO_4)_2$ compound is prepared by using thesimple solid state method with the calcination temperature at 650 °C for 130h. The phase purity of the sample is confirms by the X-ray diffractometerwhich is crystalline in the triclinic structure and spacegroup of $P\bar{1}$. The particle size of 0.3-6 μm with the morphology of spherical is identify by the analysis in the SEM. The FTIR spectroscopy confirms the functional groups which is present in the $Li_2Co(WO_4)_2$ compound. The Optical energy band gap value of 2.69 eV is found by the UV-visible spectroscopy. The broad blue emission band at 447 nm is confirmed by the photoluminescence studies. The emission spectra in PL show the transition from $^3T_1$ excited state to the ground state $^3A_2$. The Impedance spectroscopy shows the Conductivity mechanism which is formed by the grain boundaries in $Li_2Co(WO_4)_2$ compound. These observed results showsthat$Li_2Co(WO_4)_2$compound exhibit the excellent optical property.

Acknolwedgements

The authors thank CAMT MSRIT for characterization facilities. GNA thank VGST RGF/S GRD-689 scheme for grant. Shruthi D L thank RIT for PhD fellowship.

Reference

[1] Nikl, M. &Boháček, Pavel&Mihokova, Eva & Kobayashi, Masaaki & Ishii, M. &Usuki, Y. &Babin, V &Stolovich, A. &Zazubovich, S. &Bacci, Mauro 2000 Journal of Luminescence . 87. 1136-1139.

[2] Elisa Sani, Alfredo Brugioni, Luca Mercatelli, Daniela Parisi, Evgeny V. Zharikov, Denis A. Lis, Kirill A. Subbotin, 2019 Optical Materials.94. 415-422.

[3] Kuz'micheva G. &Kaurova I. 2020 Molecules. 25(10). 2.

[4] Brian L. Ellis, Kyu Tae Lee andLinda F. Nazar . 2010 Chem. Mater. 22. 691–714.

[5] Wang, Liping&Xu, Xudong& Yuan, Xin. 2010. Journal of Luminescence. 130. 137-140.

[6] Shruthi D L, Jagannatha Reddy A, Anilkumar G N, Jayasankar C K, 2020 J. Luminescence. 222 .11716.

[7] PanneerMuthuselvam, I. &Sankar, Raman &Ushakov, Alexey & G., NarsingaRao&Streltsov, S. & Chou, F.2014 PHYSICAL REVIEW B. 90.174430 .

[8] Shruthi D L, Jagannatha Reddy A, Anilkumar G N, Jayasankar C K, Harikrishna R, 2019 Material Research Express. 6 125053.

[9] D LShruthi, OjasShanker, LikithGowda, PranavKudige, A Jagannatha Reddyand G N Anil Kumar, 2020 J. Phys.: Conf. Ser. 1644 012049.

[10] D.L.Shruthi, A.Jagannatha Reddy, G.N.Anil Kumar, 2021 Ceramics International ,47 16342-16357.

[11] Schmidt, Rainer. (2007) Impedance Spectroscopy of Electroceramics. (Ceramic Materials Research Trends Nova Science Publishers) p 325-355.

[12] D.L. Shruthi, A. Jagannatha Reddy, G.N. Anil Kumar 2021 Optical Materials . 118 111243.
[13] KarimKaroui, Abdelfattah Mahmoud,Abdallah Ben Rhaiem and FrèdéricBoschini , 2019 RSC Adv.9. 6785-6792.
[14] Hui Zhang, Rui-Juan Bai, Chun Lu, Jun Li, Ying-GeXu, Ling-Bin Kong and Mao-Cheng Liu 2019Ionics. 25. 533–540.
[15] I. PanneerMuthuselvam, R. Sankar, A. V. Ushakov, G. NarsingaRao, Sergey V. Streltsovand F. C. Chou 2014 Phys. Rev. B. 90. 174430.
[16] Mahieddine, L. Adnane Amara, N. Gabouze, I. Belkhettab , International journal of hydrogen energy 2020. 08. 272.

Fabrication of cellular structures in thermoplastic polyurethane matrix using carbonaceous nanofillers

Adwaita SR Nair[a,b], Subhash Mandal[a*], Debmalya Roy[a], N. EswaraPrasad[c]

[a] Directorate of Nanomaterials and Technologies, DMSRDE, GT Road, Kanpur, India-208013
[b]Nanoscience and Technology, Central University of Jharkhand, Jharkhand, India- 835205

DMSRDE, GT Road, Kanpur, India-208013

E-mail. subhash@dmsrde.drdo.in, subh_ml@rediffmail.com

Abstract. In the present study, we have synthesized, graphene oxide (GO) by using modified Hummer's method and reduced graphene oxide(rGO) by using hydrazine hydrate as reducing agent. Since GO and rGO have high surface area and modification of surface is easier, they produce drastic changes in the matrix properties at a very low loading volume. Oxygen functionalities further allow increased interaction with polar polymer composites. Modified hummers method is the most commonly and widely used method of chemical reduction to synthesis graphene oxide as it is rapid and safe. Unlike other method, it is less hazardous and requires less reaction time. Sulfuric acid was used to disperse graphite and $NaNO_3$ and $KMNO_4$ as oxidizing agent. The use of $KMNO_4$ instead of $KClO_3$ reduced the chances of ClO_2 explosion and also accelerated the reaction. Characterization of graphene oxide and reduced graphene oxide was done using XRD, SEM, FTIR, Raman spectroscopy and TGA. The synthesized GO and rGO were used as nanofillers for the synthesis of polyurethane nanocomposite. Thermoplastic polyurethane is biodegradable and thus polyurethane nanocomposites have wide application. PU nanocomposites were prepared using thermo-chemical solvent mixing method and their microstructures were investigated using various characterization techniques.

Keywords – nanofillers, nanocomposites, hummers method, solvent mixing

1. Introduction

Intense interest from researchers have been generated towards the study of polyurethane(PU) due to its versatile application in various industries and unique properties. PU are semicrystalline thermoplastic copolymer alternately consisting of hard and soft segment in its domain. Diisocyanate constitutes the hard segment contributing to hardness of the polymer. Soft segment is attributed to polyols, either polyester or polyether, part which influence the elasticity of polymer. The unique property of PU arises due to incompatibilityand heterogeneity of both phases which leads to microphase separation. Crystallization of either or both of these segmentscontribute to microphase separation[1].Nature and size of segments, synthesis condition and manufacturing process affect microphase separation of PU.The chemical structure of polyol, diisocyanate and chain extenders affect the physical and thermal properties of PU[2]. Extensive study on the structural[3], mechanical[4], thermal[5] and rheological[6]properties of PU has been done by researchers. It has been investigated that crosslinking density also broadly effect properties[7].PU is a prominent candidate for fibers, coating, adhesives, biomedical items, manufacturing elastomers, binder resin etc.Versatile application of PU can be owed to its excellent flexibility, elasticity, damping ability, good stress-strain recovery, thermo-plasticity[8], biocompatibility and biodegradation.The use of PU for

industrial application has been limited due to its low stiffness, low tensile strength, easy combustibility[9],weak barrier properties, low water and gas permeability and deformability.
Incorporation of nanofillers can significantly impact the properties of nanocomposites due to high interfacial interaction between nanofillers and polymer, fiber – matrix adhesion and high aspect ratio. It has been reported that nanocomposites have higher thermal stability[10][11], mechanical properties[10] and electrical property[12]as compared to pure matrix material. As compared to micro-composites, interest in nanocomposites has escalated due to low percolation threshold, reduced distance between particles and particle – particle correlation at low volume. Moreover, significant enhancement in property is achieved at very low dispersion loading in nanocomposites.
Polyurethane nanocomposites with nano-clay as fillers have attracted a lot of researchers due to improved thermal stability, biodegradability (5-10-fold), tensile strength (2-5 times) and scratch hardness. Due to this PU/nano-clay nanocomposite can be used in biomaterials.[13][14] But the addition of inorganic fillers decreased elongation at break[11].Moreover, researchers tend to use fillers which does not deteriorate with time easily.
Apart from inorganic fillers carbonaceous fillers such as graphene, CNT, fullerenes due to its reinforcing effect and good compatibility with polymer[10]are prominently used as fillers in nanocomposites. Graphene nanoplatelets can cause increase in crystallization temperature which can improve the thermal properties.While at the same time functionalized graphene sheets usually hinder crystallization of soft segment[12].Among carbon nanofillers a widespread research is being done using graphene oxide and rGO as nanofillers in composites[15]. Since they have high surface area and modification of surface is easier, they produce drastic changes in the matrix properties at a very low loading volume.
Graphene oxide is highly stable and thus has wide range of applicability. Among the various synthesis method developed such as Brodie's method and Staudenmaier method, Hummer's method is the safest method considering the reduced risk of explosion as compared to other method.GO is an insulator due to the disruption of sp^2 carbon and its conductivity depends on the oxidation level. It is hydrophilic in nature and can be easily dispersed in water as well as other polar solvent like Dimethyl formaldehyde (DMF), ethylene glycol, Tetrahydrofuran (THF) etc. They are used as a membrane to measure osmotic pressure and membrane potential. Due to the attached functional groups, graphene oxide shows strong oxidizing properties. When functionalized with aromatic aliphatic amino acids by condensation reaction and nucleophilic substitution it shows stable dispersion in water and other organic solvents. Due to its high stability and the ability to control its properties according to its functionalities, they are used in nanocomposites. The zeta potential/surface charge measurement of graphene oxide shows negative charges on the surface when dispersed with water which suggests its high stability. The addition of graphene in composite can change transition temperature, chemical and physical properties of matrix. High dispersion ability of functionalized graphene helps in the synthesis of economically viable nanocomposites. Oxygen functionalities further allow increased interaction with polar polymer composites. Graphene polymer nanocomposites are widely being used to produce low weight, high strength material.
Intense interest has escalated among researches on PU nanocomposites due to improved Payne effect, shape memory effect[16],[17], self-healing effect, light absorption, improved gas barriers, biodegradability.These properties attributes to its versatile application in biomedical devices, thermo responsive shape memory, actuators.Particle size of nanofillers[18], method of synthesis and dispersion of nanofillers affects its property significantly. PU/graphene nanocomposites formed by film blowing has been reported by Russo et al [19]. Park et al[20]investigated that allyl isocyanate modified GO in PU matrix act as light actuators, nucleating agent and multifunctional cross-linkers. The increased near infrared absorption of PU/GO nanocomposites enabled light induced shape recoverywhich makes them suitable candidate for actuators. 10-fold increase in tensile stiffness and 90% decrease in nitrogen

permeability by the addition of 3 wt.% graphene reported by Kim shows high impact of nanofiller addition[21].Influence of rGO content on moisture permeability was also reported[22].Reduced nitrogen permeability was observed with increased graphene oxide content. By addition of modified GO in PU increased light absorption has been observed. This increased light absorption contributes to self-healing of polymer by accelerated thermal diffusion[23].PU/graphene nanocomposites formed by film blowing havebeen reported by Russo et al.[17]. Thus, PU nanocomposites can be tailored to meet he highly diversified demands of modern technology.
Nanocomposite synthesis can be done by various synthesis methods such as in situpolymerization, solvent method or extrusion.Whatever may be the chosen method for synthesis, good dispersion of nanofillers in polymer is critical to achieve high performance of nanocomposites. Solution mixing method is the most feasible method wherein good dispersion is attained by sonicating in a low viscous solution. The solvent used for the synthesis can easily penetrate in between nanofillers and hence agglomerationtendency is reduced and exfoliation is enhanced. The removal of solvent does not affect the exfoliated structure. Thus, this method is most suitable to attain polymer nanocomposites with high dispersion and low agglomeration. In this study authors have used solvent mixing method for synthesis. XRD, Raman spectroscopy, FTIR and SEM were used to chemically and structurally characterize nanocomposites. Authors believe that the improved property of nanocomposites can widen its application in industries[24].

2. Experimental

Graphene oxide (GO) synthesis[25]**:**For the synthesis of graphene oxide by modified hummer's method 3gm graphite was added to 75ml H_2SO_4. Further $NaNO_3$ was added and the temperature was maintained at 0°C by using an ice bath. This mixture was stirred for 2hrs while the temperature was constantly maintained at 0°C. Further 9gm of potassium permanganate ($KMNO_4$) was added very slowly to ensure homogenous mixture. While addition of KMNO4 temperature was maintained below 15°C, this mixture was continuously stirred for 1hr at 35°C. After 1hr, cooling bath was removed and mixture was left at room temperature. Once the temperature came down to room temperature 100mL deionized water (DI) was added. The beaker was put in an oil bath and heated up to 90C.After the reaction reaches 90°C, 300mL DI water was further added and this was stirred continuously for 1 ½ hour. This was treated with 30mL of 30% H_2O_2 and 3L of hot water addition for diluting. The mixture was left overnight to settle down and then acid wash was done. Next day sample was decanted and to the residue again 2L water was added. Further decant the sample next day and place in vacuum oven at 85°C for 2-3 hours. After drying, sample was collected and weighed.

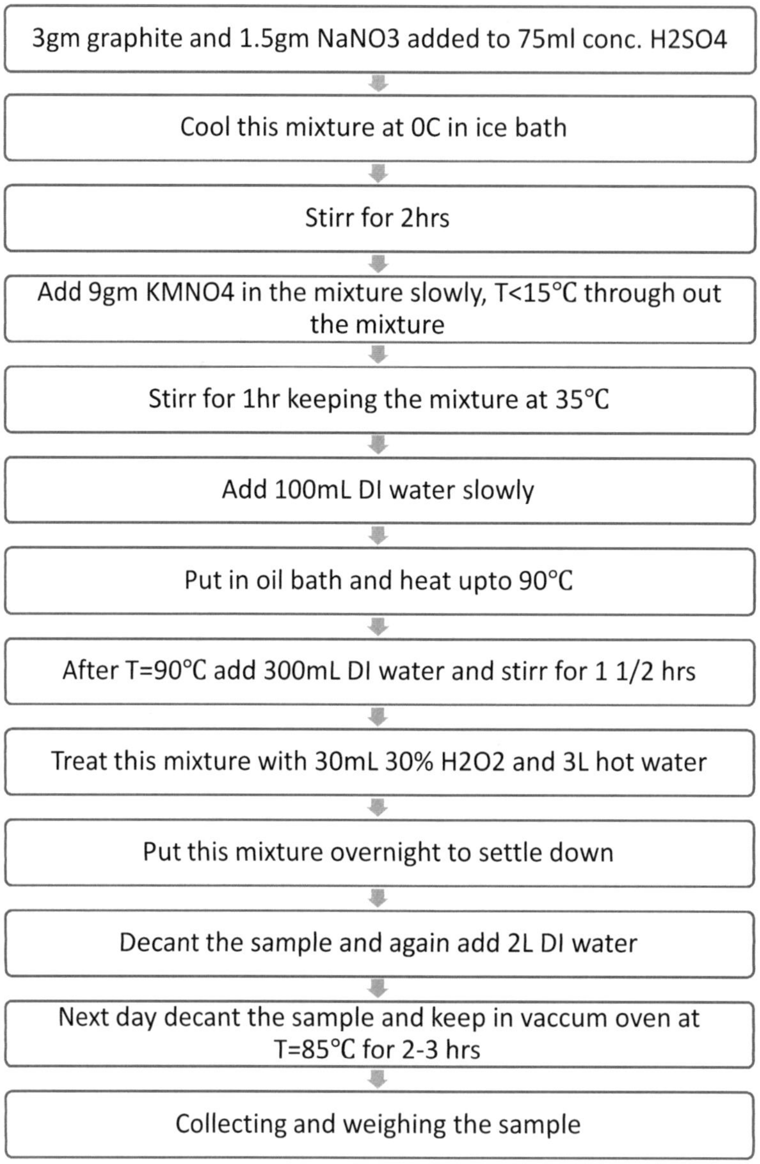

Reduced Graphene Oxide synthesis (rGO)[26]**:**250mg synthesized GO was added to 200mL DI water and sonicated for 3hr. To this dispersed GO solution 20mL hydrazine hydrate solution (10mL hydrazine hydrate added to 10mL DI) was added. Homogenization was done for 2hr at 80°C. Further it was cooled down and NaOH (1M) wash was done for several times to neutralize. The treated solution was the centrifuged at 5000rpm and heated in oven at 60°C.

IOP Conf. Series: Materials Science and Engineering 1219 (2022) 012004 doi:10.1088/1757-899X/1219/1/012004

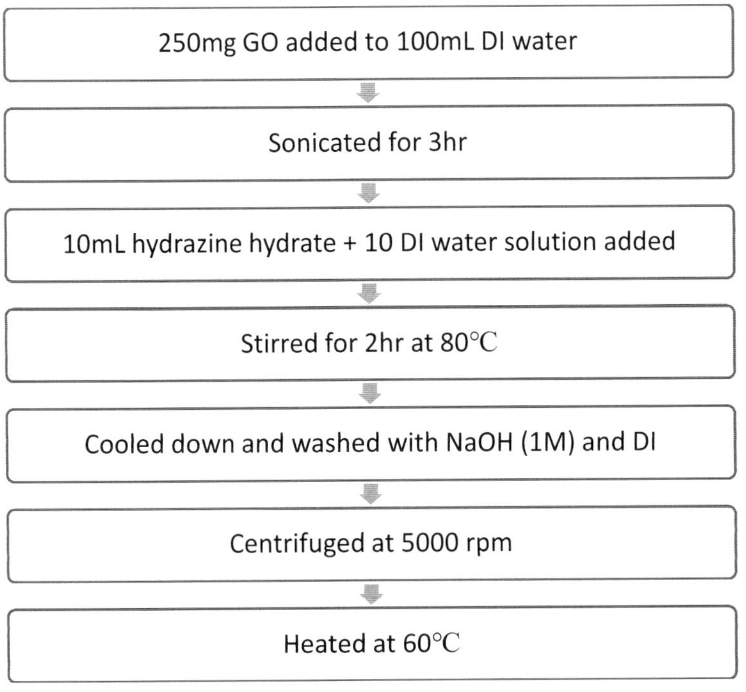

Nanocomposite synthesis[27]**:**GO and rGO fillers in varying concentration (0.5wt%, 1wt%, 5wt%) was added to 10mL dimethylformamide (DMF) and sonicated. This solution was centrifuged and redispersed in DMF. After complete dispersion the solution was heated to 90°C and added to hot polyurethane which was homogenized in 100mL DMF at 20mg/mL concentration. Further homogenization after solvent mixing was done for 1hr and then it was quenched to 30°C when precipitation starts to form. DMF solvent was removed by decantation. The gel formed where then soaked in ethanol and further heated in vacuum oven at 30°C to evaporate the solvent.

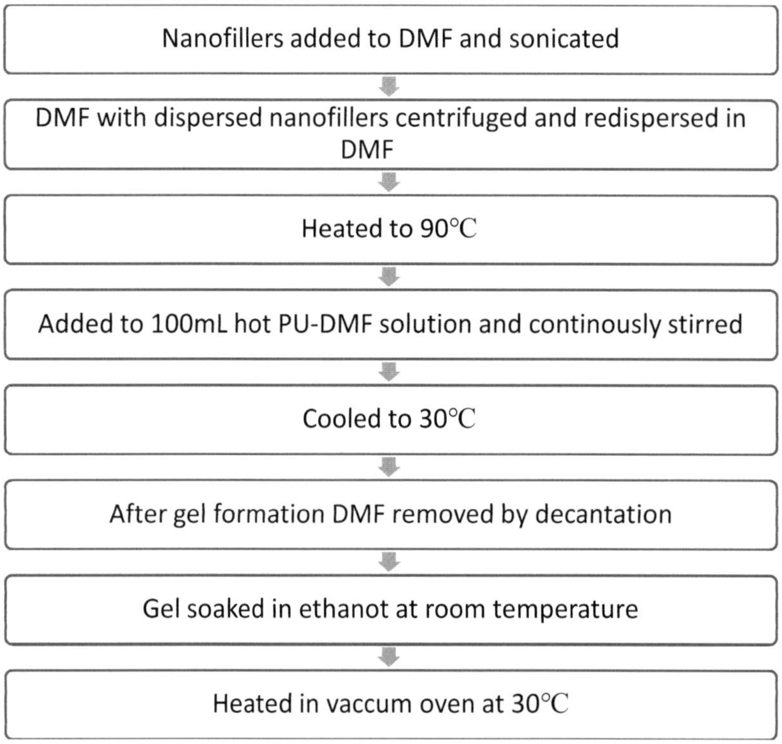

Results and discussions

XRD:

XRD was used to confirm the synthesis of GO and rGO. The graphite showed an intense peak at 26.34° and a broad peak at 54.47°. The disappearance of peak at 26.34° and appearance of peak in GO [Fig. 1 (b)] confirms the complete oxidation of graphite[28]. The reduction of GO to rGO was confirmed by the presence of two prominent peak, one at 24.32° and other less intense peak at 42.77°[29].

IOP Conf. Series: Materials Science and Engineering 1219 (2022) 012004 doi:10.1088/1757-899X/1219/1/012004

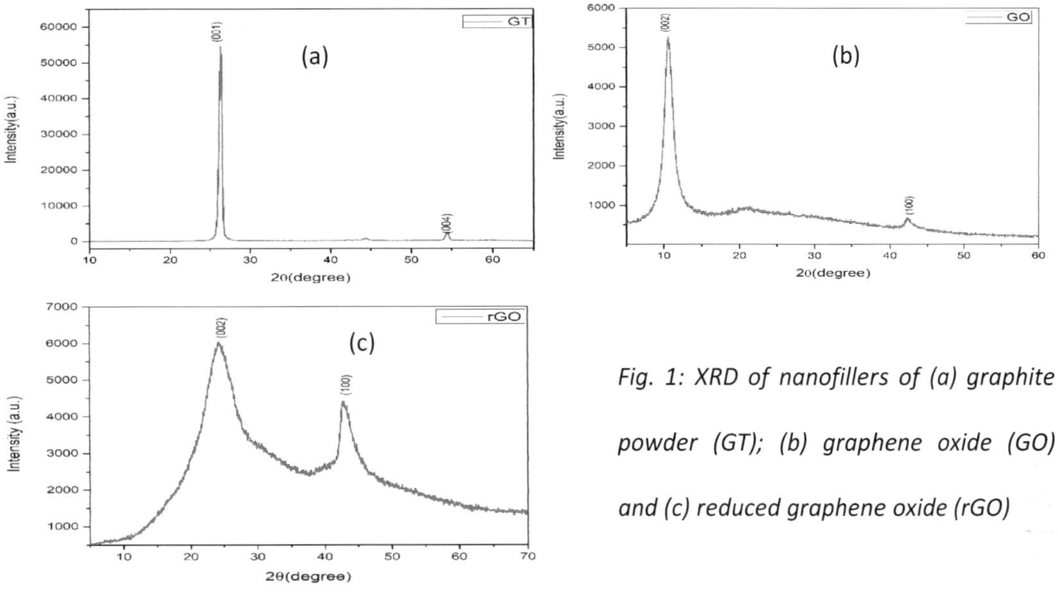

Fig. 1: XRD of nanofillers of (a) graphite powder (GT); (b) graphene oxide (GO) and (c) reduced graphene oxide (rGO)

Scanning electron microscopy:

Morphological analysis of nanofillers was done by using scanning electron microscopy (Fig 2). Both GO and rGO shows inhomogeneous and shows the presence of stacked layers. The presence of folds on the surface indicate that fillers are in the form of flakes. The flakes provide high surface area which makes modification easier hence these fillers can be modified according to applications. The nanofillers appear to be in the range of few to 10nm.

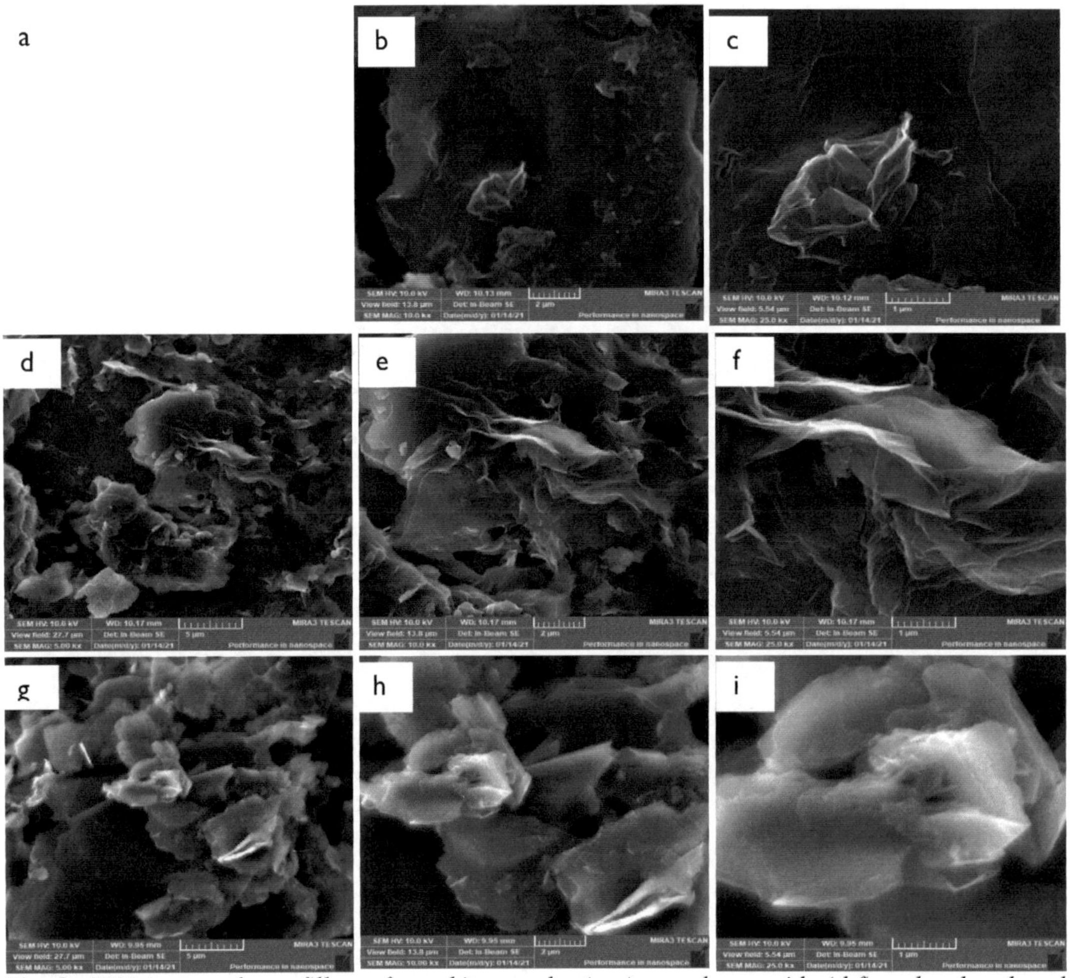

***Fig. 2:** SEM imagesof nanofillers of graphite powder (a-c); graphene oxide (d-f) and reduced grahene oxide (g-i)at varying magnification*

Microstructural analysis of the synthesized nanocomposites was also done at different magnification. It can be observed that nanocomposites with low loading showed higher dispersion and the nanofillers are embedded in the matrix indicating compatibility of polymer with nanofillers. The clear appearance of interface morphology at higher weight percentage suggests lower matrix-polymer interaction while at 0.5wt.% and 1wt.% higher interface interaction observed. The formation of agglomerate at higher loading results in collapse of pores. At low loading [Fig.3(c) and (f)] porous cellular structure formation is observed. The porosity of PU nanocomposite makes it a suitable candidate for various application such as water filtration membrane.

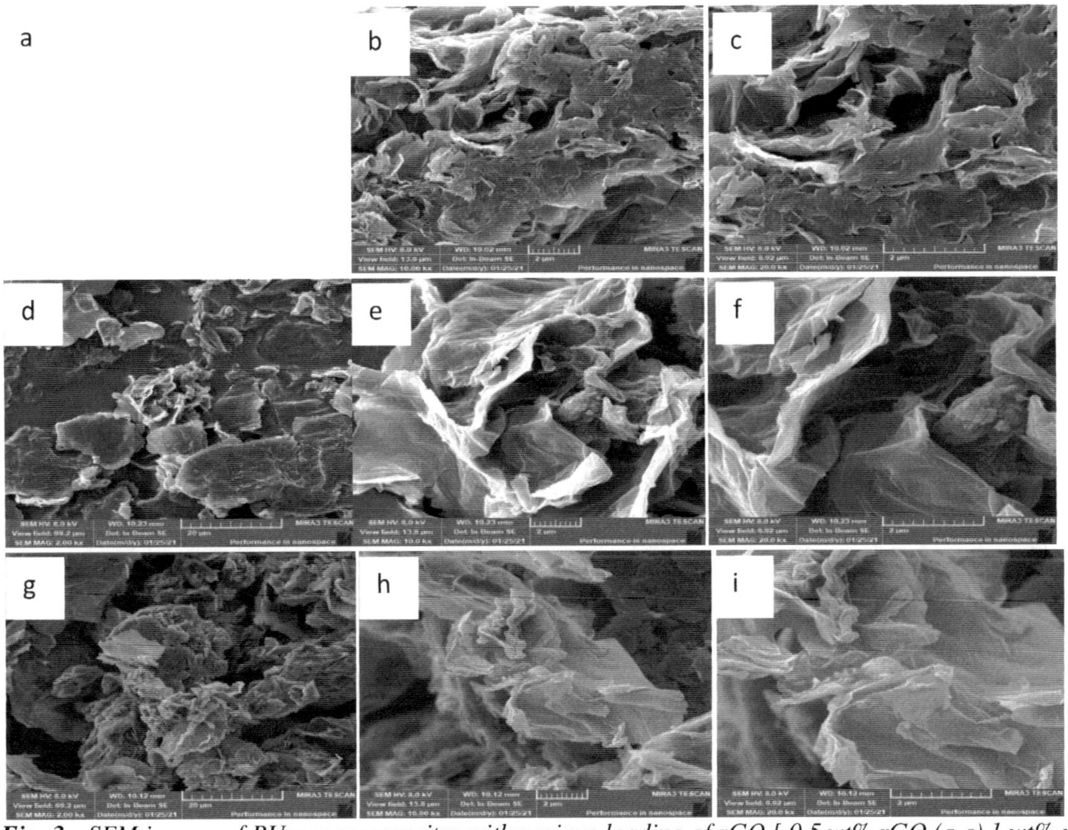

Fig. 3: *SEM images of PU nanocomposites with various loading of rGO [0.5 wt% rGO (a-c),1 wt% rGO (d-f) and 5wt% rGO(g-i)] at different magnification*

Raman spectroscopy:

The nanofillerswere characterized by using Raman spectroscopy. In graphite (GT)two prominent peaks are visible, one at 1571cm^{-1} and 2708 cm^{-1}. While in GO peaks at 1358.79cm^{-1} and 1599.2cm^{-1} each corresponding to D band and G band. The disordered D band is due to in plane vibrations of sp3 carbon while G band is due to out of plane vibration of sp2 carbon. Broadening of G band and narrowing of peak at D is indicative of reduction of GO into rGO.

IOP Conf. Series: Materials Science and Engineering 1219 (2022) 012004 doi:10.1088/1757-899X/1219/1/012004

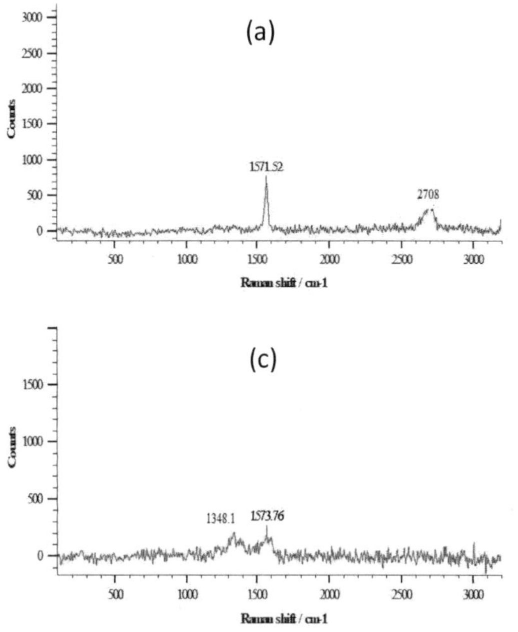

Fig. 4: *Raman spectroscopy of nanofillers of (a) graphite powder; (b) graphene oxide (GO) and (c) reduced graphene oxide (rGO)*

The Raman spectroscopy shown in Fig. 5 depicting the Raman shift in neat polyurethane (NPU)[Fig. 5a]; the nanocomposites of PU incorporated with graphite powder (GT)[Fig. 5b], (c) graphene oxide (GO)[Fig. 5c] and reduced graphene oxide (rGO)[Fig. 5d].
The D-band (disorder-inducedphonon mode), which corresponds to mode of theboundaries in the Brillouin zone and also having the high sensitivity todisordered of the structures in the carbon materials usually comes in the in the region of 1250–1450 cm^{-1} in the Raman spectroscopy; the G band is at thevicinity of 1580 cm^{-1} [30, 31].

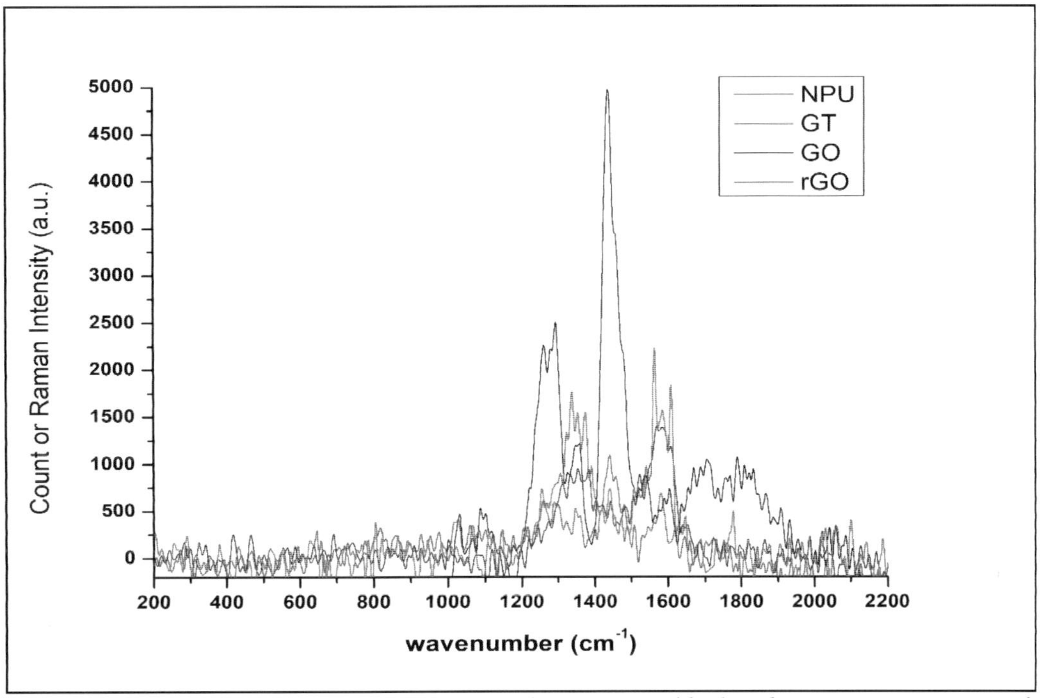

Fig. 5: Raman spectroscopy of (a) Neat polyurethane (NPU, black color); nanocomposites of PU reinforced by 1wt% fillers of (b) graphite powder(GT, red color), (c) graphene oxide (GO, black color) and (d) reduced graphene oxide (rGO, pink color).

TheRaman spectra of the fabricated nanocomposites showed significant changes due to the incorporation of graphite powder, GO and rGO to the PU matrix. The Raman bands related to phenol group (1616 cm^{-1}) and to urethane group (1537 cm^{-1}) of the PU are not visible due to the low intensities compared with the peaks of graphite powder, GO and rGO[32]. The above observation in Raman spectra of graphite powder, GO and rGO incorporated into the PU matrix were probably due to the π-π interaction between the functional groups (hydroxyl and urethane group) of PU chainsand the added fillers [33].

FTIR:

FTIR spectroscopy [Fig.6] was done to study the functional groups present in nanocomposite and to analyze changes that occurred after nanofiller addition. FTIR of neat PU shows broad peak at 3362 cm^{-1} and 1715 cm^{-1} corresponding to urethane N–H stretching and C=O stretching. The two peaks at 1553 and 1238 cm^{-1} are attributed to the N–H in-plane and C–N bond stretching. The characteristic peaks of neat PU are present in PU nanocomposites as well which indicate that no change in the functional group has taken place. A slight decrease in the absorption peak intensity was observed at the 3313-3362 cm^{-1}as compared to neat PU. This decrease indicated the H-bond formation between nanofillers and N-H group of polyurethane. The H-bond formation provides stability to the PU nanocomposite structure. As seen in PU with rGO[Fig.6(b)] with increasing weight percentage transmittance decreases due to the hydrogen bonding between the functional group of rGO and urethane group present in the hard segments of the PU.The distribution and dispersion of rGO is also better than GO due to its lower size which in turns increase the interfacial strength because of the hydrogen bonding.

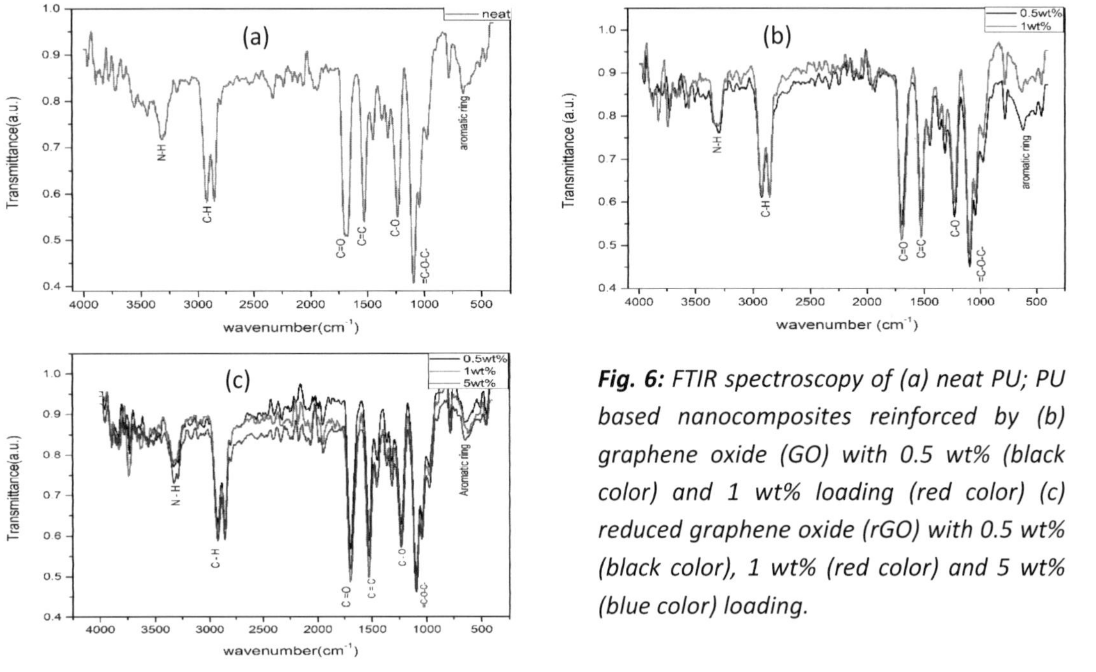

Fig. 6: FTIR spectroscopy of (a) neat PU; PU based nanocomposites reinforced by (b) graphene oxide (GO) with 0.5 wt% (black color) and 1 wt% loading (red color) (c) reduced graphene oxide (rGO) with 0.5 wt% (black color), 1 wt% (red color) and 5 wt% (blue color) loading.

Conclusions:

The synthesis of graphene oxide by modified Hummer's method and rGO by using hydrazine hydrate as reducing agent was carried out. The structural and microstructural changes that occur after synthesis of GO and rGO was showed in SEM images. The formation of GO and rGO was confirmed from the intensity of D-band and G-band in Raman analysis. Thus, modified Hummer's method and reduction by hydrazine hydrate was concluded to be feasible method for GO and rGO synthesis. Porous structure of polyurethane (PU) was obtained at low weight percentagei.e at 0.5wt% and 1wt% while agglomerate formation of nanofillers was observed at higher loadings i.e. 5wt%. FTIR confirmed that the nanofillers are bonded to PUby H-bonding and no other chemical change was found after the addition. Thethermo-chemical solvent mixing method yields stable nanocomposites with good nanofillers dispersion throughout at lower weight percentage of nanofillers. Thus, it can be adapted as a successful and reliable method for the preparation of PU nanocomposites with better dispersion than other reported method like melt mixing.

Acknowledgement:

The authors are thankful to the Director, DMSRDE, Defence Research and Development Organization for his encouragement and support. Authors are also thankful to Ms. SudeepaDevi for her help and support.

References:

[1] Velankar S and Cooper S L 2000 Microphase separation and rheological properties of polyurethane melts. 3. Effect of block incompatibility on the viscoelastic properties *Macromolecules***33** 395–403

[2] Liaw D J 1997 The relative physical and thermal properties of polyurethane elastomers: Effect of chain extenders of bisphenols, diisocyanate, and polyol structures *J. Appl. Polym. Sci.* **66** 1251–65
[3] Zlatanić A, Petrović Z S and Dušek K 2002 Structure and properties of triolein-based polyurethane networks *Biomacromolecules* **3** 1048–56
[4] Goods S H, Neuschwanger C L, Whinnery L L and Nix W D 1999 Mechanical properties of a particle-strengthened polyurethane foam *J. Appl. Polym. Sci.* **74** 2724–36
[5] Lefebvre J, Bastin B, Le Bras M, Duquesne S, Paleja R and Delobel R 2005 Thermal stability and fire properties of conventional flexible polyurethane foam formulations *Polym. Degrad. Stab.* **88** 28–34
[6] Velankar S and Cooper S L 1998 Microphase separation and rheological properties of polyurethane melts. 1. Effect of block length *Macromolecules* **31** 9181–92
[7] Blank W J and Tramontano V J 1996 Properties of crosslinked polyurethane dispersions *Prog. Org. Coatings* **27** 1–15
[8] Mothé C G and De Araújo C R 2000 Properties of polyurethane elastomers and composites by thermal analysis *Thermochim. Acta* **357–358** 321–5
[9] Wang J, Li G, Yang S and Jiang J 2004 New Intumescent Flame-Retardant Agent: Application to Polyurethane Coatings *J. Appl. Polym. Sci.* **91** 1193–206
[10] Xiong J, Zheng Z, Qin X, Li M, Li H and Wang X 2006 The thermal and mechanical properties of a polyurethane/multi-walled carbon nanotube composite *Carbon N. Y.* **44** 2701–7
[11] Xia H and Song M 2005 Preparation and characterization of polyurethane-carbon nanotube composites *Soft Matter* **1** 386–94
[12] Nguyen D A, Lee Y R, Raghu A V., Jeong H M, Shin C M and Kim B K 2009 Morphological and physical properties of a thermoplastic polyurethane reinforced with functionalized graphene sheet *Polym. Int.* **58** 412–7
[13] Dutta S, Karak N, Saikia J P and Konwar B K 2009 Biocompatible epoxy modified bio-based polyurethane nanocomposites: Mechanical property, cytotoxicity and biodegradation *Bioresour. Technol.* **100** 6391–7
[14] Dias R C M, Góes A M, Serakides R, Ayres E and Oréfce R L 2010 Porous biodegradable polyurethane nanocomposites: Preparation, characterization, and biocompatibility tests *Mater. Res.* **13** 211–8
[15] Ramanathan T, Abdala A A, Stankovich S, Dikin D A, Herrera-Alonso M, Piner R D, Adamson D H, Schniepp H C, Chen X, Ruoff R S, Nguyen S T, Aksay I A, Prud'Homme R K and Brinson L C 2008 Functionalized graphene sheets for polymer nanocomposites *Nat. Nanotechnol.* **3** 327–31
[16] Kausar A 2019 Review of fundamentals and applications of polyester nanocomposites filled with carbonaceous nanofillers *J. Plast. Film Sheeting* **35** 22–44
[17] Teymouri M, Kokabi M and Alamdarnejad G 2020 Conductive shape-memory polyurethane/multiwall carbon nanotube nanocomposite aerogels *J. Appl. Polym. Sci.* **137** 1–11
[18] Choi J T, Kim D H, Ryu K S, Lee H Il, Jeong H M, Shin C M, Kim J H and Kim B K 2011 Functionalized graphene sheet/polyurethane nanocomposites: Effect of particle size on physical properties *Macromol. Res.* **19** 809–14
[19] Russo P, Acierno D, Capezzuto F, Buonocore G G, Di Maio L and Lavorgna M 2015 Thermoplastic polyurethane/graphene nanocomposites: The effect of graphene oxide on physical properties *AIP Conf. Proc.* **1695** 1–6
[20] Park J H and Kim B K 2014 Infrared light actuated shape memory effects in crystalline polyurethane/graphene chemical hybrids *Smart Mater. Struct.* **23**
[21] Kim H, Miura Y and MacOsko C W 2010 Graphene/polyurethane nanocomposites for improved gas barrier and electrical conductivity *Chem. Mater.* **22** 3441–50

IOP Conf. Series: Materials Science and Engineering 1219 (2022) 012004 doi:10.1088/1757-899X/1219/1/012004

[22] Yousefi N, Gudarzi M M, Zheng Q, Lin X, Shen X, Jia J, Sharif F and Kim J K 2013 Highly aligned, ultralarge-size reduced graphene oxide/polyurethane nanocomposites: Mechanical properties and moisture permeability *Compos. Part A Appl. Sci. Manuf.* **49** 42–50

[23] Kim J T, Kim B K, Kim E Y, Kwon S H and Jeong H M 2013 Synthesis and properties of near IR induced self-healable polyurethane/graphene nanocomposites *Eur. Polym. J.* **49** 3889–96

[24] Lerf A, He H, Forster M and Klinowski J 1998 Structure of graphite oxide revisited *J. Phys. Chem. B* **102** 4477–82

[25] Hydrogel S G Self-Assembled Graphene Hydrogel **4** 4324–30

[26] Sharma N, Vyas R, Sharma V, Rahman H, Sharma S K and Sachdev K 2020 A comparative study on gas-sensing behavior of reduced graphene oxide (rGO) synthesized by chemical and environment-friendly green method *Appl. Nanosci.* **10** 517–28

[27] Mandal S, Roy D, Prasad N E and Joshi M 2021 Interfacial interactions and properties of cellular structured polyurethane nanocomposite based on carbonaceous nano-fillers *J. Appl. Polym. Sci.* **138** 1–16

[28] Kaushal A, Dhawan S K and Singh V 2019 Determination of crystallite size, number of graphene layers and defect density of graphene oxide (GO) and reduced graphene oxide (RGO) *AIP Conf. Proc.* **2115** 1–5

[29] Sharma R, Chadha N and Saini P 2017 Determination of defect density, crystallite size and number of graphene layers in graphene analogues using X-ray diffraction and Raman spectroscopy *Indian J. Pure Appl. Phys.* **55** 625–9

[30] DresselhausMS, Dresselhaus G, Saito Rand Jorio A2005 Raman spectroscopy of carbon nanotubes*Phys. Rep.* **409(2)** 47-99

[31] Dresselhaus MS, Dresselhaus G and Hofmann M 2007 The big picture of Raman scattering in carbon nanotubes*Vib.Spectrosc.* **45(2)**71-81

[32] Coleman JN, Curran S, Dalton AB, Davey AP, McCarthy B, Blau W andBarklie RC 1998 Percolation-dominated conductivity in a conjugated-polymer-carbon-nanotube composite *Phys. Rev. B.* **58(12)**R7492

[33] Naumenko A, Yashchuk V, Bliznyuk V andSingamaneni S 2012 Peculiarities of Raman spectra of polyurethane/carbon nanotube composite *Eur. Phys. J. B.* **85(4)**1-6.

Synthesis of Graphene Quantum dots from the fresh leaves extract of *Cynodon Dactylon* and its Photoluminescence studies

Asha R Pai[a,1], Silpa Sasi B[b], Arya J[c], Arjun K.S.[d]

[1]Department of Physics, Amrita School of Arts and Sciences, Amrita Vishwa Vidyapeetham, Amritapuri, Kollam, 690525, India
Corresponding author: vidugane2021@gmail.com; asharp@am.amrita.edu

Abstract. The current research is mainly focused on the synthesis of graphene quantum dots from the fresh extracts of Dhruva Grass (*Cynodon Dactylon*) by a One-pot, facile, solvothermal method at 60°C for 1 hr using a solvent mixture of Methanol and acetone in a ratio of 1:1. The high resolution transmission electron microscope image showed the formation of Graphene quantum dots with average size of 3-5 nm with the fluorescent quantum yield of 49.1%.
Keywords: Green synthesis, Graphene quantum dots (GQDs), *Cynodon Dactylon*, Transmission electron microscopy, Photoluminescence.

1. Introduction

The unique properties of Graphene have attracted its utility in wide range of applications [1, 2]. Graphene quantum dots (GQDs) like other traditional semiconductor quantum dots have potential applications in the biomedical field such as bioimaging, electrochemical biosensors, as well as in photovoltaic devices [3, 4, 5, 6]. As the various large-scale synthesis routes of GQDs such as electrochemical oxidation, lithographic patterning, hydrothermal, microwave, acidic oxidation and supercritical fluid treatment [7, 8] used, are more expensive and undesirable, so a novel route to GQD synthesis using a more facile, milder and more environmentally-friendly method needs to be used [9, 10]. Therefore, here we propose the green synthesis of GQDs using the fresh extract of the Dhruva grass (*Cynodon Dactylon*) which is novel and a more facile route.

2. Materials and Methodology

Dhruva Grass was initially cut into fine pieces and then subjected to constant stirring at 60°C for an 1hr in a solvent mixture of Methanol and acetone in a constant ratio. Further the mixture was centrifuged to separate the solid residue and the supernatant was dried at 100°C to remove the solvent to give dense brown colored powderwhich was mixed with acetone to be used for further studies. The HRTEM imaging was done to study the structure (average size) and morphology of the GQDs synthesized. The Fluorescence spectra was studied and quantum yield was calculated.

3. Results

The average size of the graphene quantum dots formed were confirmed as 3.6 nm from TEM micrographs as shown in Figure 1(a). The TEM images clearly reveal that the graphene dots are spherical in shape with a narrow size distribution ranging between 3 to 4.5 nm. The particle analyzed preserves the hexagonal structure, further confirming the graphene structure with an interatomic spacing, d as 0.34nm.

The synthesized graphene quantum dots in acetone showed blue - green luminescence on exposure of UV light. The Photoluminescence (PL) spectra of graphene dots with variation of excitation wavelength (350–400 nm) are shown in Figure 1(b). A strong PL emission peak located at 472 nm with an excitation wavelength of 370 nm. The emission peak was also shifted to a higher wavelength with the increase of the excitation wavelength, which is clearly shown in Figure 1(b).

The quantum yields (QY) of GQDs was calculated as:

$$\Phi_i = \frac{\Phi_s I_i f_s \mu_i^2}{I_s f_i \mu_s^2}$$

where Φi and Φs -----the photoluminescence QY of the sample and that of the standard, respectively.

I_i and I_s-------- the integrated intensities (areas) of sample and standard spectra, respectively (in units of photons);

f_x ---- the absorption factor, the fraction of the light impinging on the sample that is absorbed ($f_x = 1-10^{-A}$ x, where A = absorbance);

μ_i and μ_s--------the refractive indices of the sample and reference solution respectively.

The fluorescence quantum yield was calculated as 49.1% using quinine sulphate in 0.1 M H_2SO_4as standard and studying the absorbance at 340 nm.

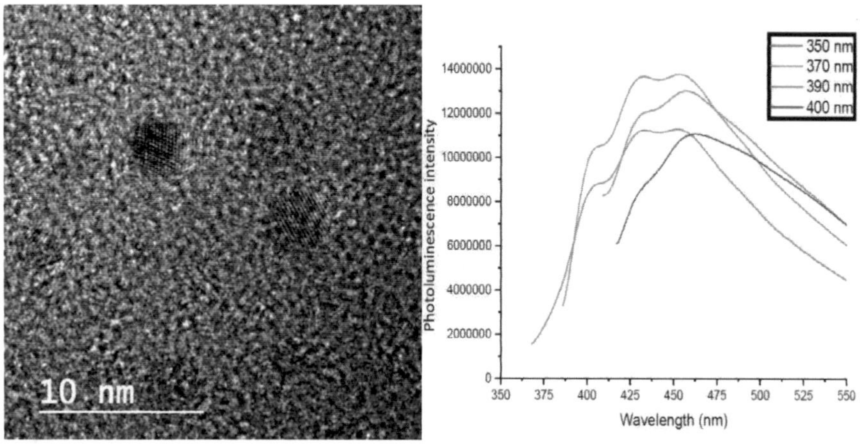

Fig 1. (a) The HRTEM image of the biosynthesized GQDs from fresh extracts of Dhruva Grass (b) PL spectra studied for the excitation wavelength ranging from 350-400 nm range.

4.Conclusion

The study shows the development of a novel, facile, green and efficient method for the fabrication of fluorescent GQDs using the green extract of Dhruva grass (*Cynodon dactylon*) in solvent mixture of methanol and acetone at $60^{O}C$ for 1 hr. The synthesis of GQDs as explained does not use any harsh reagents or elaborate synthesis conditions such as long reaction times or purification procedures. This simple, green process can result in mass production of fluorescent GQDs giving blue-green fluorescence. The graphene quantum dots are more stable and don't get photobleached but have a depth limit so good for in vitro and in vivo (small animal) studies only and also for sensing applications. Further studies of these green synthesized GQDs will be in the field of biosensing and bioimaging.

Acknowledgements

The Authors are very thankful to the Prof. Shanti Kumar V Nair and his staff of ACNSMM, Kochi for the instrumentation facilities provided. Also, we are grateful to Prof. Asoka Banerjee and Dr. Chinch Bose of Phytochemistry lab and also Prof. Bipin Nair, Dean, Amrita School of Biotechnology, Amritapuri for their kind support provided during this work.

References

[1.] Gem A K, 2009 *Science* **324**, 1530
[2.] Lee C, Wei X, Kysar J W, Hone J, 2009 *Science* **321**, 385
[3.] Bitounis D, Ali-Boucetta H, Hong B H, Min D, Kostarelos K, 2013 *Adv. Mater.* **25**, 2258
[4] Mehrdad-Vahdati B, Pourhashem S, Sedghi M, Vaezi Z, Shojaedin-Givi B, Rashidi A, Naderi-Manesh H, 2019 *Toxicol In Vitro.* **61**, 104649
[5] Chu X, Dai P, Liang S, Bhattacharya A, Dong Y, Epifani M, 2019 *Phys E.* **106**, 326
[6] Wang S, Li Z, Xu X, Zhang G, Li Y, Peng Q, 2019 *Adv. Mater. Interfaces.* **6**, 1801480
[7] Du Y, Guo S, 2016 Nanoscale **8**, 2532.
[8] Naik J P, Sutradhar P, Saha M, 2017 *J. Nanostruct. Chem.* **7**, 85
[9] Chen W, Lv G, Hu W, Li D, Chen S, Dai Z, 2018 *Nanotechnology. Rev.* **7**, 157
[10] Murugesan B, Sonamuthu J, Pandiyan N, Pandi B, Samayanan S, Mahalingam S, 2018 *J. Photochem. Photobiol. B.*, **178**, 371

Hydration of amino acids: An investigation using density functional theory

C. B. Sarath Kumar[1], Mayank[1], Rence P. Reji[1], M. Gobinath[1], Y.Sivalingam[2], and V. J. Surya[1,3*]

[1]Novel, Advanced, and Applied Materials (NAAM) Laboratory, Department of Physics and Nanotechnology, SRM Institute of Science and Technology, Kattankulathur 603203, Tamil Nadu, India.

[2]Laboratory of Sensors, Energy and Electronic devices (Lab SEED), Department of Physics and Nanotechnology, SRM Institute of Science and Technology, Kattankulathur 603203, Tamil Nadu, India.
[3]New Industry Creation Hatchery Center, Tohoku University, Aoba-Ku, Miyagi, Sendai 980-8579, Japan

E-mail.suryaj@srmist.edu.in

Abstract. Herein, we have investigated the effect of hydration of seven amino acids (AAs) such as alanine, aspartic acid,cysteine, glutamine, lysine, methionine, and phenylalanine with different number of water molecules using density functional theory. Among the hydrophobic AAs, cysteine has the highest interaction with water whereas in hydrophilic AAs, aspartic acid has the highest interaction with water. With the increase in number of water molecules, it is noted that alanine forms the most stable complex with water and methionine+water complex has least stability.In addition, we have calculated the chemical descriptors like chemical potential, electronegativity, hardness, softness, electrophilicity index of AAs to validate their interaction with water molecules. Interestingly, the gas phase interaction energy results match with the trend of electrophilicity index valuesof AAs.

1. Introduction

Amino acids (AAs) are the building blocks of proteins and they play an important role in the regulation of metabolism. AAs are widely used in many industrial and medical applications **[1].** They are made up of an amino group (NH_2^+), carboxylic acid (COOH) group, and a side chain attached to the alpha carbon. Based on the side chains, they are classified as polar, nonpolar, acidic, basic, hydrophobic and hydrophilic AAs. Side chains of polar AAs have functional groups such as acids, amides, alcohols, amines attached to α-carbon. Polar AAs are glutamine, tyrosine, cysteine and so on. Side chains of non-polar AAs have pure hydrocarbon alkyl or aromatic groups. Non-polar AAs are valine, methionine, glycine, proline, alanine etc. Acidic amino acids have side chain with carboxylic acid group such as aspartic acid and glutamic acid. Basic AAs such as arginine, histamine, lysine etc can accept an electron. Investigations on interaction between the AAs and water will be useful for understanding the physicochemical properties of AAs.There are many reports on behaviour of amino acids and peptides on solid surfaces like graphene and in solution **[1-7].** Experimentally, the solubility for many amino acids in water at different temperatures has been studied **[3-5].** Xu*et al.*have identified the number of water molecules required for the formation of Zwitter ion in AAs **[6].** Choi *et al.*have studied the effect of topological water network analysis in twenty AAs towards drug discovery and development through MD simulation **[7].**

In this work, we have examined the hydration effect of hydrophobic AAs such asalanine, phenylalanine, methionine, cysteine and hydrophilic AAs such as aspartic acid, glutamine, lysine with finite number of water molecules using density functional theory (DFT). Also, we have calculated the chemical descriptors such as chemical potential (μ), electronegativity (χ),

IOP Conf. Series: Materials Science and Engineering 1219 (2022) 012006 doi:10.1088/1757-899X/1219/1/012006

hardness (η), softness (S), electrophilicity index (ω), ionization potential (I), and electron affinity (A) to understand the chemical properties of AAs for validating the DFT results.

2. Computational details

The DFT calculations were carried out using Gaussian16 software with standard basis set and functional, 6-311G**/B3LYP[8]. The molecular structures of AAs and water (H_2O) were modeled and optimized to their ground state. Thereafter, the combined systems of AAs+nH_2O systems were built by addition of n=1-4 molecules, and they are optimized. The interaction energy, E_I of water with AAs is calculated using the following formula [9]:

1

where corresponds to the total energy of the constituents and is the number of molecules. In addition, highest occupied molecular orbital (HOMO) - lowest unoccupied molecular orbital (LUMO), electrostatic potential (ESP) analyses have been performed for all the systems. Chemicaldescriptors such as chemical potential (μ), electronegativity (χ), hardness (η), softness (S), and electrophilicity index (ω), ionization potential(I), electron affinity (A) were calculated using the following formulae:

$$\chi = \frac{(I + A)}{2}$$

where, $I = -E_{HOMO}$ and $A = -E_{LUMO}$

$$\eta = \frac{E_{LUMO} - E_{HOMO}}{2}$$

$$S = \frac{1}{2\eta}$$

$$\omega = \frac{\mu^2}{2\eta}$$

where E_{HOMO} and E_{LUMO} are the energy values of the HOMO and LUMO[10].

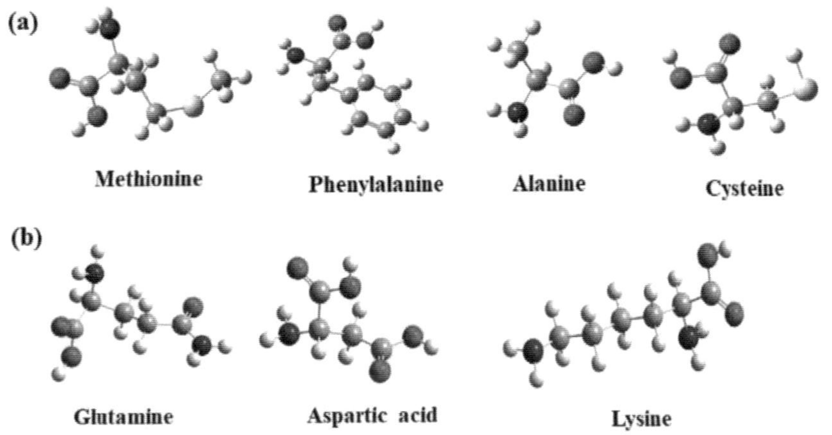

Figure 1. (a) Hydrophobic AAs, and (b) Hydrophilic AAs.

3. Results and Discussion

The optimized structures of all AAs are shown in Fig. 1 and the corresponding ESP plots are shown in Fig. 2. The combined AAs+H_2O systems are obtained by addition of water molecules. For instance, the optimized structures of AA+2H_2O are shown in the Fig. 3. The

interaction energies of AA+nH_2O (n=1-4) systems are reported in Table 1. There is no gradual rise or fall

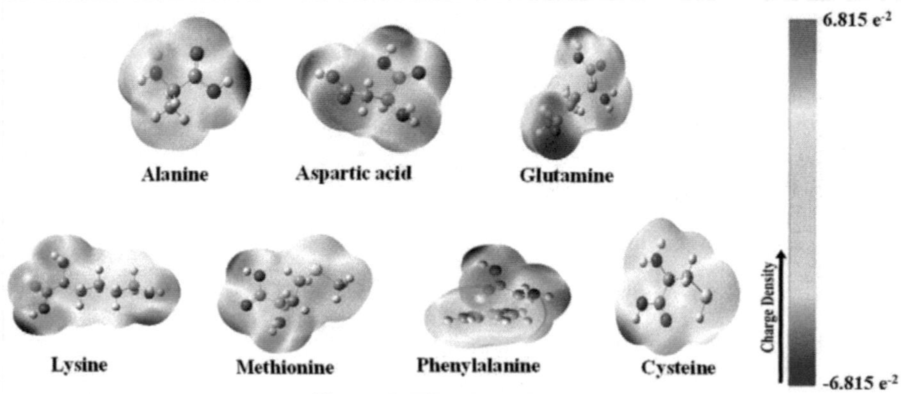

Figure 2. ESP plots of AAs.

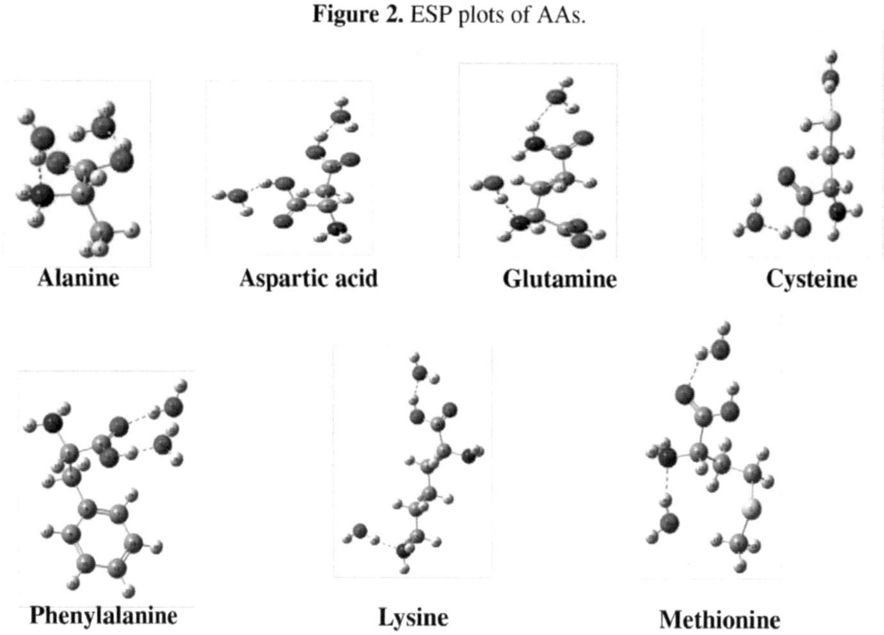

Figure 3. Interaction of AAs with 2H_2O molecules.

in the energies with the addition of water molecules. Hence, we have calculated the average interaction energies (Table 1). The trend is as follows, cysteine > aspartic acid>

IOP Conf. Series: Materials Science and Engineering 1219 (2022) 012006 doi:10.1088/1757-899X/1219/1/012006

methionine>alanine>glutamine>lysine>phenylalanine. It can be noted that cysteine and aspartic acid show higher interaction among the hydrophobic and hydrophilic amino acids, respectively.

Table 1. Interaction energies of water molecules with AAs.

AAs	Interaction energy (eV)				
	AA+H_2O	AA+2H_2O	AA+3H_2O	AA+4H_2O	Average value
Aspartic acid	-0.68	-0.66	-0.83	-0.49	-0.67
Alanine	-0.61	-0.46	-0.83	-0.67	-0.64
Cysteine	-0.68	-0.83	-0.44	-0.82	-0.69
Glutamine	-0.71	-0.69	-0.59	-0.49	-0.62
Methionine	-0.66	-0.56	-0.83	-0.57	-0.66
Phenylalanine	-0.66	-0.84	-0.44	-0.16	-0.53
Lysine	-0.38	-0.47	-0.43	-1.05	-0.59

The HOMO-LUMO values (Fig. 4a) are noted down and the energy gap values of pure AAs and AAs+nH_2O are calculated (Fig. 4b and Table 2). Among the hydrophobic and hydrophilic AAs, phenylalanineand aspartic acid have the highest energy gaps.Hence,they are chemically stable and less reactive. In the presence of water molecules,the change in HOMO-LUMO values is observed. The variations in energy gaps are plotted in Fig. 5a.

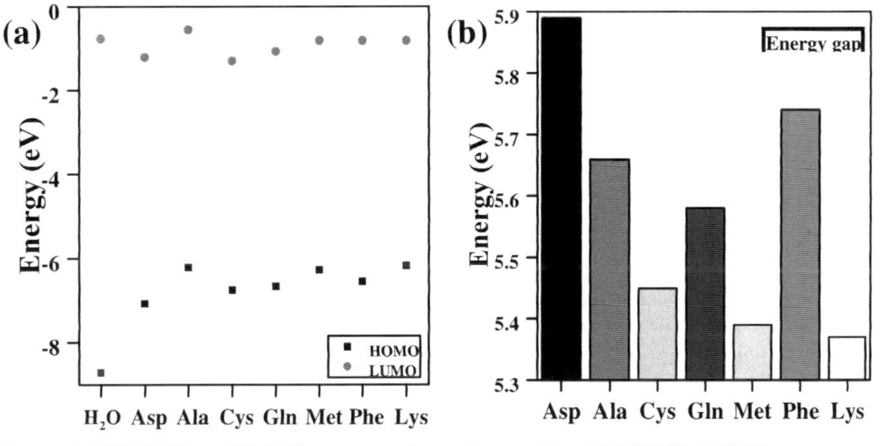

Figure 4. (a)HOMO and LUMO energy values of pure AAs, (b) HOMO-LUMO energy gap of pure AAs.

Table 2. HOMO-LUMO energy gap of pure and hydrated AAs.

AAs	Energy gap (eV)				
	Pure AAs	AA+H_2O	AA+2H_2O	AA+3H_2O	AA+4H_2O
Aspartic acid	5.89	5.92	5.92	5.92	5.59
Alanine	5.66	5.58	6.29	6.29	6.32
Cysteine	5.45	5.49	5.52	5.41	5.47
Glutamine	5.58	5.22	5.36	5.73	5.81
Methionine	5.39	5.43	5.02	5.06	5.21
Phenylalanine	5.74	5.73	5.73	5.93	5.67
Lysine	5.37	5.47	5.99	6.13	5.97

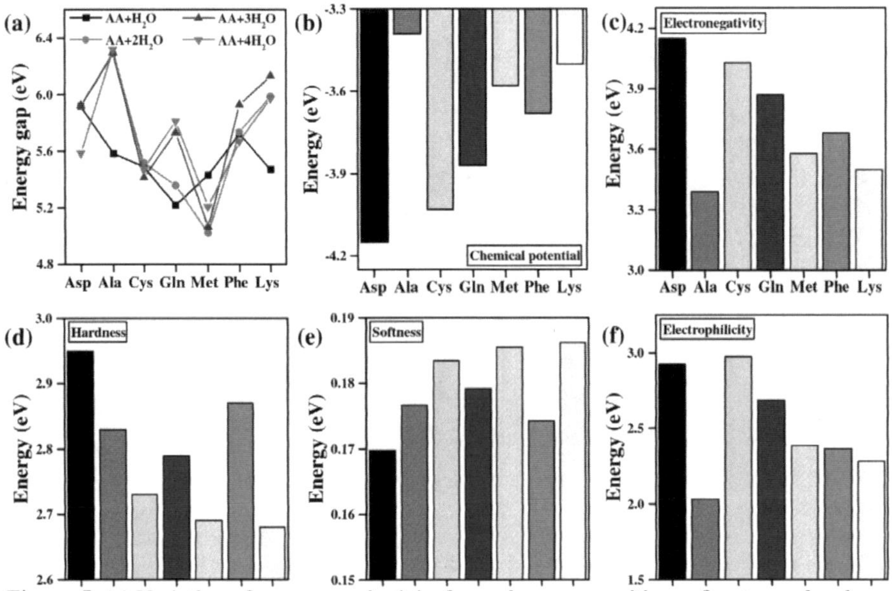

Figure 5. (a) Variation of energy gap in AAs for various compositions of water molecules, (b) Chemical potential, (c) Electronegativity, (d) Hardness, (e) Softness, and (f) Electrophilicity index of pure AAs.

Among hydrophobic AAs, alanine shows gradual increment in energy gap with the addition of water molecules. In a way, alanine forms the most stable complex with four water molecules. Meanwhile, the energy gap decreases in methionine+4H_2O complex making it as a least stable complex. In the case of cysteine, the magnitude of change in energy gap in the presence of water molecules is not so high when compared to the pristine case. The trend is almost constant. This proves the strong hydrophobic nature of cysteine.

Table 3. Ionization potential (I) and electron affinity (A) of AAs.

AAs	**I (eV)**	**A (eV)**
Aspartic acid	7.10	1.21
Alanine	6.22	0.56
Cysteine	6.75	1.30
Glutamine	6.66	1.08
Methionine	6.28	0.89
Phenylalanine	6.55	0.82
Lysine	6.18	0.82

Table 3 shows the values of I and A. Lysine and alanine have the lowest ionization potential and hence they are best electron donors. Cysteine and aspartic acid have the highest electron affinity and they can accept electrons, easily. The values of μ, χ, η, S, and ω are calculated for allAAs and they are represented graphically in Figs. 5(b)-5(f).Lysine and alanine have lower chemical potential and hence they are least stable (Fig. 5b).Eventually, alanine and lysine have least electronegativity, whereas aspartic acid and cysteine have more electronegativity (Fig. 5c). Again, Hardness (η) and softness (S) are useful parameters in understanding the stability of AAs. Aspartic acid and phenylalanine are the most stable molecules among the hydrophilic and hydrophobic AAs (Fig. 5d), respectively. On the other hand, lysine, methionine and cysteine are the least stable compounds (Fig. 5e). Kaya *et al.*have calculated some chemical descriptors such as energy gap and electronegativity for alanine,methionine,lysine with the basis set 6-31G** and those values agree with our results**[11]**. Eddy *et al.*have calculated the electronegativity for aspartic acid and glutamine using MOPAC 2008 program and it is in agreement with our values**[12].**

The electrophilicity index gives information about the stabilization energy when the system gets saturated by number of electrons. Alanine has low electrophilicity index and therefore, it can donate electrons. Meanwhile, cysteine, aspartic acid and glutamine have high electrophilicity index and hence, they can accept electrons easily (Fig. 5f). Hence, these four AAs can interact with water through donor-acceptor interactions. Interestingly, these AAs show higher interaction with water molecules as reported in Table 1.

Conclusions

We have carried out first-principlesDFT calculations on certain hydrophilic and hydrophobic AAs to understand their interaction with finite number of water molecules. The gas phase calculations have shown that cysteine, aspartic acid and methionine interacts well with water molecules followed by other AAs. The energy gap analysis has projected that alanine forms the most stable complex with water molecules. On the other hand, the methionine+water complex is the least stable structure. The chemical descriptor calculations have revealed that aspartic acid and phenylalanine are the most stable AAs. More importantly, electrophilicity index analysis results match with the gas phase interaction energy results. Both the calculationshave proved that cysteine, alanine, aspartic acid and glutamine have higher interaction with water molecules when compared to other AAs. Similar computational studies on AAs help us to understand their behavior in solvent moderated chemical reactions, and other biological media/surfaces.

Acknowledgement

Y.S. thanks the Department of Science and Technology (DST) - Science and Engineering Research Board (SERB), Government of India for financial support (ECR/2017/001218). The authors thank High Performance Computing Center-HPCC, SRM IST for providing the computational facility.

References

[1] Ahadian, S.,Estili, M., Surya, V. J., Ramón-Azcón, J., Liang,X.,Shiku,H., Ramalingam,M., Matsue, T.,Sakka, Y., Bae, H., Nakajima, K., Kawazoe, Y., A.Khademhosseini, "Facile and green production of aqueous graphene dispersions for biomedical applications", Nanoscale 7 (2015) :6436-6443.

[2] Ahadian, S., Naito, U., Surya, V. J.,Darvishi, S.,Estili,M., Liang, X., Nakajima, K., Shiku, H., Kawazoe, Y.,&Matsue, T.,"Fabrication of poly(ethylene glycol) hydrogels containing vertically and horizontally aligned graphene using dielectrophoresis: An experimental and modeling study", Carbon 123 (2017): 460-470.

[3] Arakawa, T., K. Tsumoto, Y. Kita, B. Chang, and D. Ejima. "Biotechnology applications of amino acids in protein purification and formulations." Amino acids 33, no. 4 (2007): 587-605.

[4] Xue, Z. J., and K. C. Chao. "Solubility of four amino acids in their water solution." J ChemEng Data 37.2 (1992): 199-203.

[5] Zhou, X., Fan, J., Li, N., Du, Z., Ying, H., Wu, J., ...& Bai, J.. Solubility of lphenylalanine in water and different binary mixtures from 288.15 to 318.15 K. Fluid phase equilibria, 316(2012): 26-33.

[6] Xu, Shoujun, J. Michael Nilles, and Kit H. Bowen Jr. "Zwitterion formation in hydrated amino acid, dipole bound anions: how many water molecules are required?." The Journal of chemical physics 119.20 (2003): 10696-10701.

[7] Choi, K. E., Chae, E., Balupuri, A., Yoon, H. R., & Kang, N. S. Topological water network analysis around amino acids. Molecules, 24(2019):2653.

[8] Gaussian 16, Revision C.01, Frisch, M. J.; Trucks, G. W.; Schlegel, H. B.; Scuseria, G. E.; Robb, M. A.; Cheeseman, J. R.; Scalmani, G.; Barone, V.; Petersson, G. A.; Nakatsuji, H.; Li, X.; Caricato, M.; Marenich, A. V.; Bloino, J.; Janesko, B. G.; Gomperts, R.; Mennucci, B.; Hratchian, H. P.; Ortiz, J. V.; Izmaylov, A. F.; Sonnenberg, J. L.; Williams-Young, D.; Ding, F.; Lipparini, F.; Egidi, F.; Goings, J.; Peng, B.; Petrone, A.; Henderson, T.; Ranasinghe, D.; Zakrzewski, V. G.; Gao, J.; Rega, N.; Zheng, G.; Liang, W.; Hada, M.; Ehara, M.; Toyota, K.; Fukuda, R.; Hasegawa, J.; Ishida, M.; Nakajima, T.; Honda, Y.; Kitao, O.; Nakai, H.; Vreven, T.; Throssell, K.; Montgomery, J. A., Jr.; Peralta, J. E.; Ogliaro, F.; Bearpark, M. J.; Heyd, J. J.; Brothers, E. N.; Kudin, K. N.; Staroverov, V. N.; Keith, T. A.; Kobayashi, R.; Normand, J.; Raghavachari, K.; Rendell, A. P.; Burant, J. C.; Iyengar, S. S.; Tomasi, J.; Cossi, M.; Millam, J. M.; Klene, M.; Adamo, C.; Cammi, R.; Ochterski, J. W.; Martin, R. L.; Morokuma, K.; Farkas, O.; Foresman, J. B.; Fox, D. J. Gaussian, Inc., Wallingford CT, 2016.

[9] Yuvaraj, S. V., Subbotin, O. S., Belosludov, R. V., Belosludov, V. R., Kanie, K., Funaki, K., & Kawazoe, Y. Theoretical evaluation on solubility of synthesized task specific ionic liquids in water. Journal of Molecular Liquids, 200, (2014): 232-237.

[10] Bendjeddou, A., Abbaz, T., Gouasmia, A., &Villemin, D.Studies on chemical reactivity of p-aminophenyl benzene-fused bistetrathiafulvalenes through quantum chemical approaches. American Journal of Applied Chemistry, 4(3)(2016): 104-110.

[11] Kaya, S., Tüzün, B., Kaya, C., &Obot, I. B., Determination of corrosion inhibition effects of amino acids: quantum chemical and molecular dynamic simulation study. Journal of the Taiwan Institute of Chemical Engineers, 58, (2016): 528-535.

[12] Eddy, NnabukOkon. "Part 3. Theoretical study on some amino acids and their potential activity as corrosion inhibitors for mild steel in HCl." Molecular Simulation 36.5 (2010): 354-363.

Magnetic Properties of Ultrathin As-deposited and Annealed Ta/CoFeB/TaO_x Heterostructures

Jaydeb dey[1], Soumya Ranjan Das[1], Jaivardhan Sinha[1*]

[1]Department of Physics and Nanotechnology, SRM Institute of Science and Technology, Kattankulathur 603203, Tamil Nadu, India

E-mail.: jaivardr@srmist.edu.in

Abstract. In order to meet the ever-increasing demand of magnetic recording industry, it is important to developferromagnetic thin film heterostructure compatible for magnetic memory device. Here, we have developed ultrathin ferromagnetic film of transition metal borides(CoFeB) which has huge potential to be integrated in magnetic memory devices. In particular, we have studied the surface roughness and magnetic properties of sputter deposited Substrate/1 nm Ta/1.5 nm CoFeB/1nm TaO_x heterostructures. Magnetic properties investigation of as-deposited and 300 ^{0}C annealed Ta/CoFeB/TaOx heterostructure using vibrating sample magnetometer indicates the presence of in-plane anisotropy in both the film stacks and a reasonable increase in the saturation magnetization of annealed film stack. Importantly, possible boron diffusion as well as partial crystallization of CoFeB layer due to annealing play crucial roles in governing the magnetic properties in these film stacks. These results provide in-depth insight into the factors affecting saturation magnetization of such ultrathin film heterostructures.

Keywords: Ultrathin ferromagnetic film, Saturation Magnetization, Thin film heterostructures, Magnetic Properties, Surface roughness

1. Introduction.

The detailed understanding of ultrathin transition metal boride film placed adjacent to heavy metal layer with strong spin orbit coupling issignificantly important due to its application potential in magnetic recording industry [1, 2]. The presence of heavy metal layer next to the ferromagnetic layer substantially influences the magnetic properties such as saturation magnetization and magnetic anisotropy. It may assist to achieve perpendicular magnetic anisotropy (PMA) and interfacial Dzyaloshinskii-Moriya interaction(IDMI) etc which are useful to implement in modern spintronics based devices for improving the device performance [2, 3, 4].An integral component of magnetic tunnel junction (MTJ) which is used in magnetic random-access memory is ferromagnetic thin film of CoFeB. In order to improve the tunnelmagnetoresistance (TMR) in MTJ, it is crucial to anneal it at optimum temperature [5, 6]. In general,the as-deposited thin films of CoFeB shows amorphous nature.Ferromagnetism in amorphous transition metal borides has been studied with a great attention due to its smooth surface properties and ease of integrating in devices [7].On annealing, CoFeB shows an interesting change as the boron from CoFeB layerdiffuses out at certain temperature leaving crystallized bcc-phase CoFe [8]. It has been reported that thecrystallization leads to higher TMR ratio in such annealed MTJ [5]. Particularly in Ta/CoFeB basedheterostructures, annealing results in interlayer mixing which markedlyaffect the interfacial roughness thereby modifying the magnetic properties. Annealing can furthercause partial oxidation of Ta and boron which results in suitable modification in properties neededto achieve large TMR [9]. In order to improve the performance of magnetic memory devices it is crucial to fabricateMTJ with added functionalities (PMA, IDMI etc) originating from the interface. Furthermore, ultrathin ferromagnetic film sandwiched between heavy metal and oxide layer can be suitably integrated in modern spintronics devicesthat aim utilize the interface effects. Amongst various combinations of heavy metal/ferromagnet/oxide based heterostructures, Ta/CoFeB/TaO_x has drawn significant attentionin recent time due to the possibility of stabilizing Skyrmions in them [10, 11]. Additionally, it is preferred to use β-Ta in this heterostructure as the spin Hall angle, which is measure of charge current to spin current conversion efficiency, is reasonably large for β phase [12].It is worth mentioning here that the role ofboron diffusion in ultrathin Ta/CoFeB/TaO_x heterostructures in governing the surface roughness and magnetic properties is notwell understood and calls for a systematic investigation. Here, we attempt to address the effect of annealing on Ta/CoFeB/TaO_x thin film stack by comparing the results of surface roughness and in-plane magnetization hysteresis loop for both as-deposited and annealed thin films.

2. Experimental Methods:

The ferromagnetic thin film heterostructure 1nm Ta/1.5nm CoFeB/1nm TaO_x was deposited using dc-magnetron sputteringon Si substrate coated with 300nm SiO_2. Commercially available Ta and$Co_{20}Fe_{60}B_{20}$ sputtering target of 2inch diameter was used for deposition. The base pressure ofsputtering unit was better than 2× 10^{-7}torr whereas working pressure for the Ta and CoFeBdeposition was 4.3 mtorr and 3 mtorr respectively. Before the deposition of actual film stack,deposition rate of Ta and CoFeB was estimated by depositing thick films of them. DC power 30W (V = 218V and I = 131 mA) for Ta and 50 W (V = 352 V and I = 151 mA) for CoFeB was usedfor deposition. The deposition rate for Ta was estimated as 0.17 A°/sec whereas for CoFeBit was 0.15 A°/sec. Such optimization is crucial for the development of ultrathin film heterostrucures with small interfacial roughness. Thin film stackswere exposed to atmosphere after deposition due to which the capping layer Ta partially oxidizes and converts to TaO_x.Subsequently the film stack was annealed at 300°C for one hour in a vacuum chamber withpressure better than 10^{-5} torr. Schematic diagram of as-deposited and annealed film stack is shown in figure 1 (a) and (b) respectively. Annealing such heterostructures can result in interestingmodification in structure and magnetic properties due to boron diffusion out of CoFeB and likely intermixing at the interface. The magnetic properties of the ultra-thin films were characterized usingVibrating Sample Magnetometer with field applied along the in-plane direction.The nominaldimension of the substrate taken for the VSM measurement was 10 mm × 10 mm. Initially,the VSM setup was calibrated using a nickel foil of the same dimension to minimize the errorinvolved in measuring the saturation magnetization. Before performing the VSM measurement ofactualfilm stack, Si/SiO_2 substrate of same dimension was measured to estimate the background signal. Subsequently, the magnetic moment versus magnetic field (MH)data is measured for the film stacks. The surface roughness analysis of the thin films was done using non-contact mode Atomic force Microscopy.

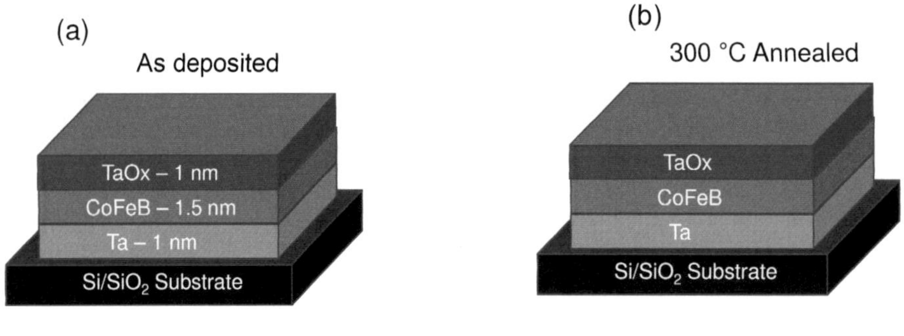

Figure1: Schematic of as-deposited (a) and annealed (b)thin film stack Si-SiO_2 (Sub)/(1 nm) Ta/(1.5 nm)CoFeB/(1 nm)TaO_x

Results and Discussions:

In order to know the phase of the Ta under the sputter deposition condition used in the experiment, a thick layer of tantalum (thickness - 47 nm) wasdeposited. The X-ray diffraction plot for this Ta film is shown in fig.2. The peak at 34.1° correspond to (002) plane of β-Ta.The full width at halfmaximum of this peak is 0.43°. A sharp peak adjacentto Ta appears due to the crystalline lattice structure of silicon substrate.In general, for variousspintronics device application it is desirable to achieve β-Ta phase as it exhibits large spin-orbit coupling strength and significantly modify the magnetic property of ferromagnetic thin filmplaced adjacent to it.

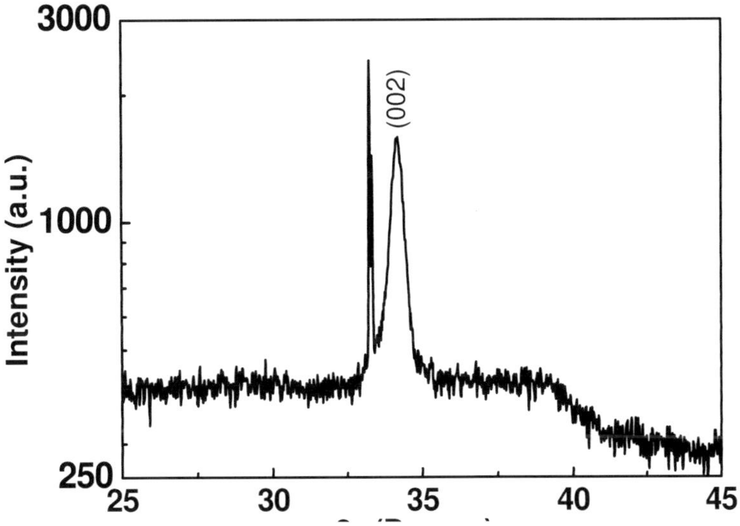

Figure 2:X-ray diffraction plot of thick Ta film revealing the β-phase of Ta in the used sputter deposition condition

In figure 3(a) and (b), scanning electron micrograph of as-deposited and annealed film stacks are shown. In figure 3(a) we observe a few dark spots on the film surface which disappear once the thin film is annealed. It is likely that these dark spots are due to possible localized Ta rich TaO_x formation over the surface of the film. Due to ultrathin nature of each layers, it is difficult to verify our premise using energy dispersive x-ray spectroscopy (EDAX) of these regions. Interestingly, the surface morphology undergoes suitable modification as a result of annealing and in turn lead to improved uniformity of the surface. Nevertheless, from these images we infer that overall, the surface of both the films are smooth and there is no apparent defect present on the surface over the large scale.In order to investigate the surface topography at fine scale weuse non-contact mode of AFM. Figure 4 (a) and (b) showAFM imagesfor both the thin films overscan area5μm × 5μm.The gradient color bar on the left shows the variationof topographical height. The average surface roughness obtained from theAFM image for as-deposited thin film is ~ 0.3 nm while the average surface roughness of annealed thin film is ~0.24 nm. Similar roughness values in Ta/CoFeB/TaOx has been reported in previous study [3]. It is worth mentioning here that the surface roughness observed insuch ultrathin film will have contribution from the interfacial roughness as the overallthickness of the film is significantly small (less than 5 nm). Thesmall surfaceroughness indicates that the interfacial roughness present at Ta/CoFeB and CoFeB/TaOx interfaceis significantly small and moreover annealing leads to some extent sharpening of the interface. It is worth to mentionhere that it is quite challenging to achieve such small surface roughness in the annealedTa/CoFeB/TaO_x heterostructures due to possible intermixing that may happen due to annealing [13].

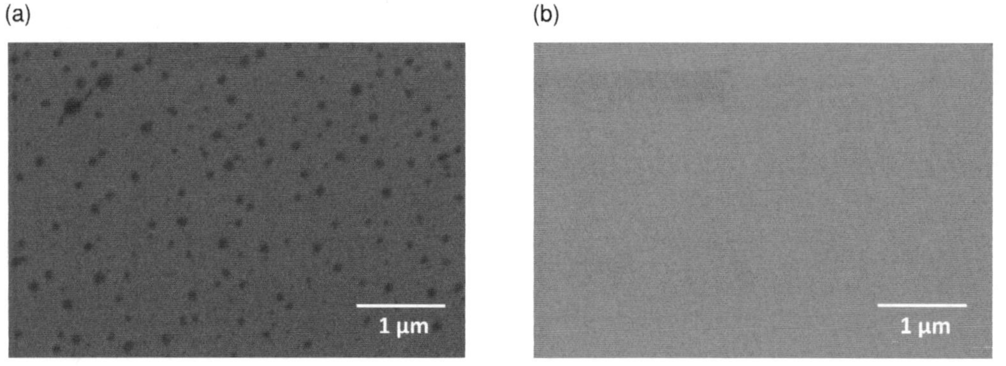

Figure 3: Scanning electron microscopy image of (a) as-deposited and (b) 300 °C annealed film stack

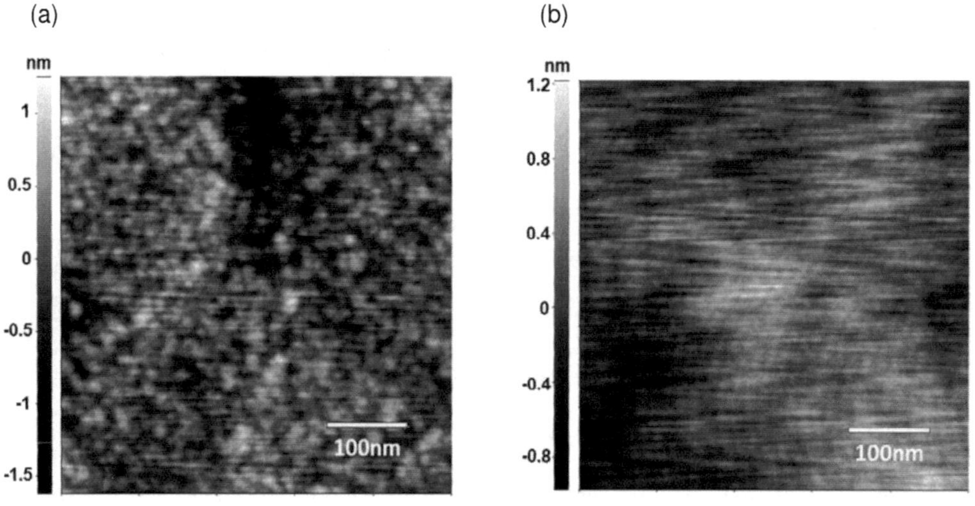

Figure 4: Atomic force microscopy image of (a) as-deposited and (b) 300 °C annealed film stack.

In figure 5(a) and (b) we show the magnetization versus magnetic field (MH) response for the as-deposited and annealed film stack when the field is swept along in-plane direction of the film. For the as-deposited thin film stack the hysteresis loop shows the saturationmagnetization to be 650 emu/cc.The coercive field ~ 5Oe for the as-depositedthin film indicates the soft ferromagnetic nature of the film. For the annealed film stack the MH hysteresis loop shows M_s ~ 850emu/cc and H_c ~ 40 Oe.From these measurements we infer the in-planedirection is the easy axis direction in both these films. It is likely that the boron diffuses out of CoFeB due to annealing which in turn results in partial crystallization. Interestingly, in the process of partial crystallization a suitable CoFe-phase get stabilized which following Slater-Pauling rule may exhibit enhanced magnetization [14]. Achieving reasonably large saturation magnetization in annealed Ta/CoFeB/TaOxheterostructurein-plane magnetic anisotropy compared with as-deposited thin film in such ultrathin CoFeB layer can havesignificant application potential in various spintronics based devices.It is worth to mention here that we have not been able to estimate the effective anisotropy in the studied thin film due to relatively large signal to noise ratio in the MH hysteresis loop measured for the field applied along the out of plane direction of the film.With proper understanding of in-plane magnetic anisotropy and saturation magnetization of thesethin film heterostructure,in future it will be interesting to investigate the magnetic domain and corelate it withsurface roughness. Additionally, x-ray photoelectron

IOP Conf. Series: Materials Science and Engineering 1219 (2022) 012007 doi:10.1088/1757-899X/1219/1/012007

spectroscopy study may provide further in-depth insight into the mechanism of magnetization enhancement in these heterostructures [15].

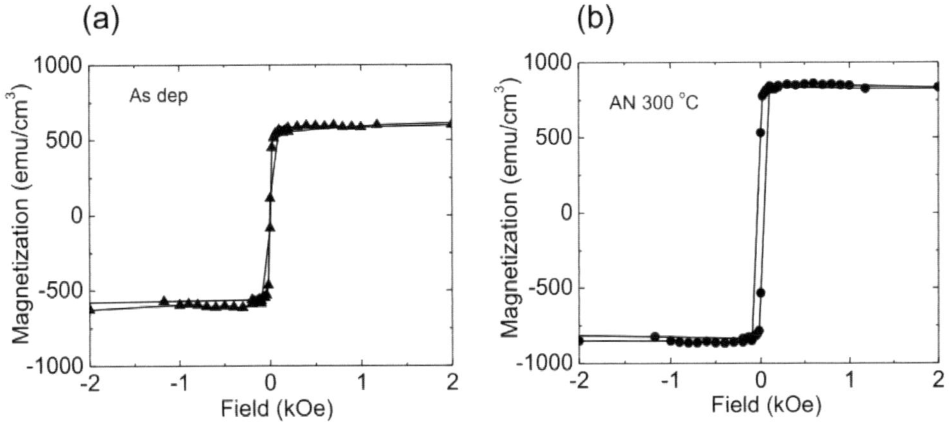

Figure 5: Magnetization versus field hysteresis loop for as-deposited (a) and annealed (b) film stack for the in-plane direction of field sweep

Conclusion:

In conclusion, we have studied the magnetic and structural properties of as-deposited and annealed Ta/ CoFeB /TaO_x thin film stack. Comparing both the cases we found the Saturation magnetization for as-deposited thin film as ~ 650 emu/cc and for the annealed thin film as ~850 emu/cc. The estimated average surface roughness in as-deposited case is 0.3 nm and for the annealed case is 0.24nm. This indicates that the interfacialroughness present in the annealed film stack is relatively smaller than the as-deposited stack. Observation of small surface roughness and reasonably large saturation magnetization in annealed thin film heterostructures hints at the possibility of interface sharpening in Ta/CoFeB based annealed heterostructures. We believe, achievingsmall interfacial roughness and reasonable large saturation magnetization in ultrathin film ofCoFeB with Ta as underlayer and TaO_x as overlayer is crucial for achieving pronounced interface effects such as PMA and IDMI. In future, it will be important to investigate the effective magnetic anisotropy along with separately estimating the bulk and interface contribution, magnetic domain and Boron diffusion in these heterostructures.Furthermore, such thin film heterostructures are envisioned to have application potentialin the spintronic based devices where the interface related effects are utilized for improving thedevice performance.

Acknowledgement:

We acknowledge UGC-DAE Consortium for scientific research, Indore for providing us the opportunity to use the sputter deposition facility. We gratefully thank Mukul Gupta, Dileep Kumar, SaileshKalal, Syamlal S K and Hariprasanth P for help in experiments. We acknowledgeNanotechnology Research Centre (NRC), SRM Institute of Science and Technology for providingVSM measurement facility.JS acknowledges DST-SERB, India for Ramanujan Fellowship (SB/S2/RJN-093/2014) and CRG/2018/004175, for providing financial support.

Reference:

[1] Vaz C. A. F., Bland J. A. C., and Lauhoff G., 2008 Rep. Prog. Phys. 71 056501.
[2] 2. Hellman F., Hoffmann A., Tserkovnyak Y., Beach G. S. D., Fullerton E. E., Leighton C., MacDonald A. H., Ralph D. C., Arena D. A., Dürr H. A., Fischer P., Grollier J., Heremans J. P., Jungwirth T., Kimel A. V., Koopmans B., Krivorotov I. N., May S. J., Petford-Long A. K.,

IOP Conf. Series: Materials Science and Engineering 1219 (2022) 012007 doi:10.1088/1757-899X/1219/1/012007

Rondinelli J. M., Samarth N., Schuller I. K., Slavin A. N., Stiles M. D., Tchernyshyov O., Thiaville A., and Zink B. L., 2017 Rev. Mod. Phys. 89 025006.
[3] Chaurasiya A. K., Choudhury S., Sinha J., and Barman A., 2018 Phys. Rev. Appl. 9 014008.
[4] Sinha J., Hayashi M., Kellock A. J., Fukami S., Yamanouchi M., Sato H., Ikeda S., Mitani S., Yang S.-h., Parkin S. S. P., and Ohno H., 2013 Appl. Phys. Lett. 102 242405.
[5] Karthik S. V., Takahashi Y. K., Ohkubo T., Hono K., Ikeda S., and Ohno H., 2009 J. Appl. Phys. 106 023920.
[6] Ikeda S., Miura K., Yamamoto H., Mizunuma K., Gan H. D., Endo M., Kanai S., Hayakawa J., Matsukura F., and Ohno H., 2010 Nat. Mater. 9 721.
[7] Gupta R., Gupta A., Gupta M., Rajput P., and Wildes A., 2013 J. Appl. Phys. 114 063903.
[8] Sinha J., Gruber M., Kodzuka M., Ohkubo T., Mitani S., Hono K., and Hayashi M., 2015 J. Appl. Phys. 117 043913.
[9] Greer A. A., Gray A. X., Kanai S., Kaiser A. M., Ueda S., Yamashita Y., Bordel C., Palsson G., Maejima N., Yang S. H., Conti G., Kobayashi K., Ikeda S., Matsukura F., Ohno H., Schneider C. M., Kortright J. B., Hellman F., and Fadley C. S., 2012 Appl. Phys. Lett. 101 202402.
[10] Jiang W., Upadhyaya P., Zhang W., Yu G., Jungfleisch M. B., Fradin F. Y., Pearson J. E., Tserkovnyak Y., Wang K. L., Heinonen O., te Velthuis S. G. E., and Hoffmann A., 2015 Science 349 283.
[11] Je S.-G., Vallobra P., Srivastava T., Rojas-Sánchez J.-C., Pham T. H., Hehn M., Malinowski G., Baraduc C., Auffret S., Gaudin G., Mangin S., Béa H., and Boulle O., 2018 Nano Lett. 18 7362-7371.
[12] Hoffmann A., 2013 IEEE Trans. Magn. 49 5172-5193.
[13] Miyakawa N., Worledge D. C., and Kita K., 2013 IEEE Magn. Lett. 4 1000104-1000104.
[14] Bozorth R. M., *Ferromagnetism*. (Wiley-IEEE Press, New York, 1993).
[15] Kozina X., Ouardi S., Balke B., Stryganyuk G., Fecher G. H., Felser C., Ikeda S., Ohno H., and Ikenaga E., 2010 Appl. Phys. Lett. 96 072105.

Effect of annealing on the structural and optical properties of pristine and cerium doped calcium tungstate nanoparticles synthesized via hydrothermal method

C Nagarajan, S Monesh Kumar, R Annie Sujatha*, K. Mani Rahulan, Angeline Little Flower

Nanophotonics Research Laboratory, Department of Physics and Nanotechnology, SRM Institute of Science and Technology, Kattankulathur, Chennai 603 203, India.

E-mail: anniesujatha@gmail.com

Abstract. The work is an investigation of the effect of annealing temperature on pristine and rare earth doped $CaWO_4$ nanoparticles with hydrothermal route. The sample was annealed at two varied temperatures 800 °C and 1200 °C for 5 h. The structural analysis reveals the reduction in the broadening of the peaks as a result of annealing. Surface morphology shows the formation of agglomerated structures. In Ce^{3+}:$CaWO_4$ nanoparticles the morphology is modified to dumbbell like and flake like structures after annealing. Raman shift observed at 909 cm^{-1} A_g modes confirms the presence of $CaWO_4$ nanoparticles with its own rotational and vibrationalmodes. Increase in the absorption peak intensity and a double absorption edge appears as an intermediate energy level state after the inclusion of Ce^{3+} ion as dopant.

Key words: Raman shift, Double absorption, Intermediate energy levels, annealing

1. Introduction

Non-spherical nanoparticles have attracted a great attention in the recent years due to its exceptional properties when compared with its spherical nanostructures [1]. Divalent metal tungstatelike ($CaWO_4$) is noteworthy inorganic material due to its excellent luminescence properties. Still $CaWO_4$has adrawback of high absorption cross section with a broad band gap.The emissionof $CaWO_4$starts from 300-800 nm, a broad emission band very difficult to tune the color for display applications [2]. To solve this problem,a material with low absorption like lanthanide ionscan be incorporated via doping in a $CaWO_4$.The other advantage of doping rare earth ions in $CaWO_4$is the forbidden nature of f-f transitionsof lanthanide ions can be relaxed by the surrounding crystal field [3]. It can also act as a host for many rare earth elementslike Europium (Eu^{3+}), Cerium(Ce^{3+}), Terbium(Tb^{3+}), samarium(Sm^{3+}) and gadolinium(Gd^{3+})[4].Lanthanide doped divalent metal tungstate emerged as a new class of nanophosphors owing to its excellent chemical and thermal stability along with the ability to tune its optical properties [5]. Among the divalent metal tungstate groups $CaWO_4$ and $CaMoO_4$have the highest optical emission intensities. Comparing with the $CaWO_4$, $CaMoO_4$has a drawback of lower decomposition temperature therefore $CaWO_4$become widely studied luminescent material [1]. The incorporation of rare earth ions in the particular Ca^{2+}sites of $CaWO_4$changes its luminescence spectra due to the generation of emission centers which gives rise to red, yellow and orange emissions [6]. The bandgap tuning of $CaWO_4$can be done by adding different rare earth elementsas dopants [7].Several studies have been proposed with the incorporation of many rare earth ions in $CaWO_4$in connection with the application of white light-emitting diodes [8].Puneetkaur et al synthesized Eu, Dy, and Sm doped $CaWO_4$nanoparticlesannealed with 800 and 1150°C to study its changes in the photoluminescence properties [9]. Sharma et al synthesized Eu^{3+} doped

$CaWO_4$nanoparticlesannealed at 500 and 900°C to study the behavior of thermoluminescence and photoluminescence properties [10]. Parchur et al synthized Tb^{3+} doped $CaMoO_4$ nanoparticles and annealed at 500 and 900°C for studying its luminescence properties [2].
In this work, we have reported pristine $CaWO_4$ and 2 and 5 wt.% of Ce^{3+}:$CaWO_4$synthesized by hydrothermal method followed by annealingat 800 and 1200°C. The structural, morphological, vibrational and optical analysis of the sample is been studied. The changes in the morphology after the inclusion of doping Ce^{3+}ions is been observed clearly using SEM images. The changes in the optical absorption after annealing the samples at1200 °C gives rise to a change in the bandgap of the material. The lowering of bandgap is been studied using tauc-plot method for the synthesis of nanoparticles.

2. Materials and Methods

The chemicals used, $CaWO_4$ and Ce^{3+}:$CaWO_4$ nanoparticles were prepared via hydrothermal method at 150 °C for 12 h following the same procedure as reported in literature [7]. The prepared samples remained subjected for annealing at 800 and 1200 °C for 300 minutes.

3. Results and Discussion

Structural Analysis

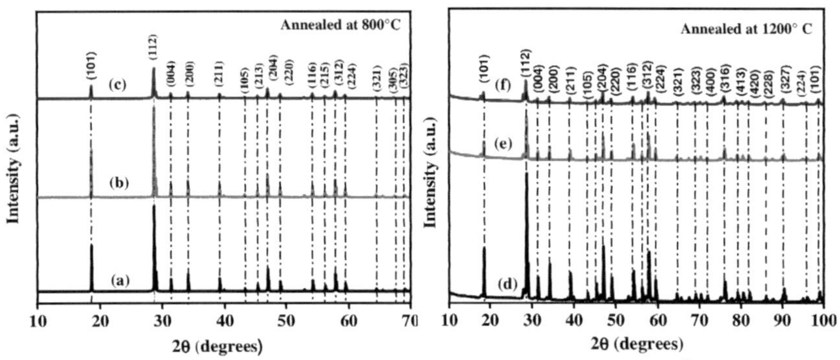

Figure. 1 XRD pattern of pristine $CaWO_4$, 2 and 5 wt. % of Ce^{3+}:$CaWO_4$ annealed at 800 °C and 1200 °C.

The Fig. 1a, 1b and 1c illustrates the XRD pattern of $CaWO_4$ and Ce^{3+}:$CaWO_4$ at 2 and 5 wt. % annealed at 800 °C. The diffracted pattern (1d, e and f) depicts the XRD pattern of pristine $CaWO_4$ and Ce^{3+}:$CaWO_4$ at 2 and 5 wt. % annealed at 1200 °C. The XRD pattern fits with $CaWO_4$ a Scheelite with $I4_1/a$ belonging to a space group of non symmorphic exhibiting a tetragonal structural formation with the values of a=b= 5.242 Å and c= 11.373 Å which is in confirmation with (JCPDS No. 41-1431) [11]. The dominant 2θ values 18.43°, 28.41° and 48.58° respectively matches with the crystal planes of (1 0 1), (1 1 2) and (2 2 0). An appreciable decrease in the intensity of the peaks is observed after the inclusion of Ce^{3+} dopant.Some new peaks are visible after increasing the annealing temperatures from 800 to 1200 °C.

IOP Conf. Series: Materials Science and Engineering 1219 (2022) 012008 doi:10.1088/1757-899X/1219/1/012008

Morphological analysis

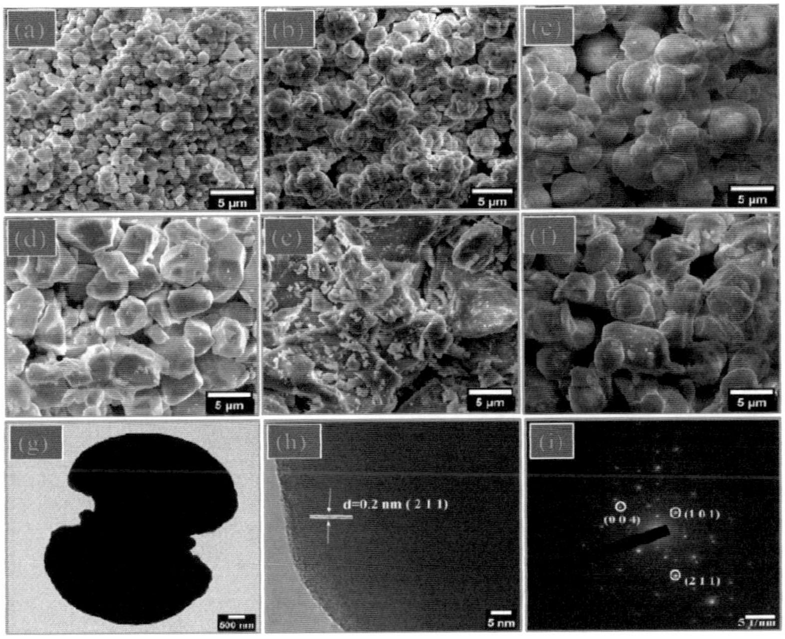

Figure. 2 SEM images of pristine $CaWO_4$, 2 and 5 wt. % of Ce^{3+}:$CaWO_4$ annealed at 800 °C and 1200 °C.

Figure. 2g, h and i TEM images of 5wt. % of Ce^{3+}:$CaWO_4$ annealed at 800 °C.

The morphology of the samplesannealed at 800 and 1200°C has been studied using scanning electron microscopy technique.The Figure (2a, b and c) depicts the FESEM images of pristine $CaWO_4$and Ce^{3+}:$CaWO_4$ nanoparticles 2 and 5 wt.%annealed at 800°C. The FESEM micrograph (2d, e and f) represents the annealing temperature of1200°C. Figure (2a)reveals anon-spherical structures with evenlydistributed particles.FESEM image (2b) clearly shows the agglomerated spherical structures. The micrograph (2c)showsthe agglomerated nanoparticles forms a dumbbell like structures. Figure (2e)forms flake like agglomerated structures whereas in the FESEM image (2d) and (2f)shows flake like structures disappear to give a particle nature of the sample and it shows an non uniform particle distribution. The significant changes in the functional properties of nanoparticles can be attained using these type of nanostructures due to an enhancement of electron transfer between these structures[12]. 5 wt.% of Ce^{3+}:$CaWO_4$ nanoparticles annealed at 800 °C is subjected to TEM analysis the image Fig. 2g depicts the agglomerated dumbbell shaped microparticles is witnessed clearly with the bottom breadth and center breadth is to be around 4.5 and 3.3 μm. The HRTEM image with perfect lattice fringes corresponding to d space of 0.2 nm belongs to (2 1 1) plane as shown in Fig. 2h. Fig. 2i shows the SAED pattern that matches with (1 0 1), (0 0 4) and (2 1 1) planes.

IOP Conf. Series: Materials Science and Engineering 1219 (2022) 012008 doi:10.1088/1757-899X/1219/1/012008

Raman analysis

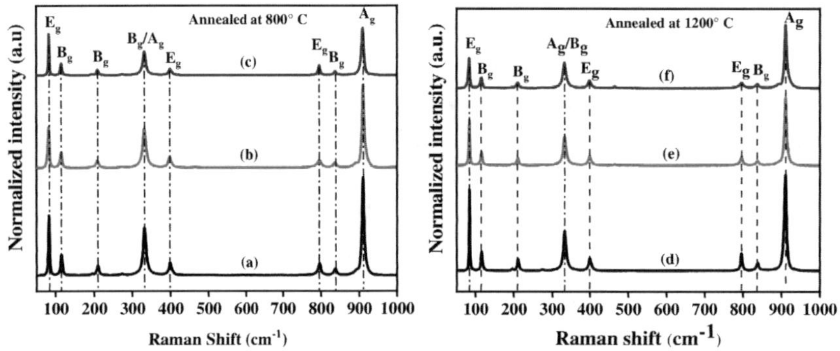

Figure. 3a-f Raman spectra of pristine $CaWO_4$, 2 and 5 wt. % of Ce^{3+}:$CaWO_4$ annealed at 800 and 1200 °C.

The Raman spectra is excited with a 532 nm of continuous wavelaserat room temperature. Increase in the intensities of Raman peaks are observed after increasing the annealing temperature from 800 to 1200°C.The vibrational modes of the Raman spectra belong to the vibrations of WO_4^{2-} against the Ca^{2+} are called as external modes. The vibrations occur only in WO_4^{2-} whereas the cation remains stationary.The peaks present in the lower Raman shift regions are 83, 116 and 207 $cm^{-1}$are called as external modes of $CaWO_4$[13]. The other peaks at 330, 396, 794, 839 and 909 $cm^{-1}$are called as internal modes of $CaWO_4$. Among the 13 Raman active modes for $CaWO_4$among them only 8 modes are active other modes are invisible. The peak at 330 cm^{-1}ascribed to the bending of O-W-O vibrations.The 396 cm^{-1}peak belongs to the antisymmetric bending of W-O bonds vibration. The peaks present at 794 and 839 $cm^{-1}$is due to the antisymmetric stretchingvibrations of O-W-O bonds. The major peak at 909 cm^{-1}ascribed to the symmetric stretching vibrations of W-O bonds [14].

UV- Visiblediffuse reflectance spectra

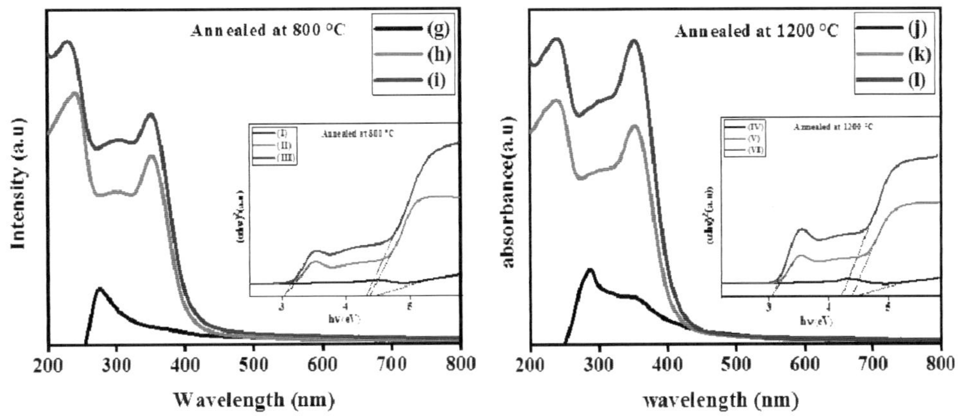

Figure 4a-f UV Visible absorption spectra of pristine $CaWO_4$, 2 and 5wt. % of Ce^{3+}:$CaWO_4$ annealed at 800 and 1200 °C.**Figure inset** I-VI, Tauc plot of pristine $CaWO_4$, 2 and 5wt. % of Ce^{3+}:$CaWO_4$ annealed at 800 and 1200 °C.

UV- Visible absorption spectra of all the samples has been recorded in the range from 200-800 nm. The Figure (4a, b and c) represents the UV- Visible absorption spectra of pristine $CaWO_4$and Ce^{3+}:$CaWO_4$ nanoparticles at 2 and 5 wt.%annealed at 800°C. UV- Visible absorption spectra images (4d, e and f) belongs to the annealing temperature of 1200°C. Increase in the intensity of the absorption peak for samples annealed at1200°Cis clearly marked as a retraction of increasing annealing temperature. Absorption intensity increases in 4b, c, e and f is due to the inclusion of dopant ions. There is a red shift in the absorption after increasing the dopant concentration is due to the increase in the particle size. The presence of two absorption band edgesof 2 and 5 wt.% of cerium doped $CaWO_4$is encountered in both the annealing temperatures. The first absorption edge is around 430 and 424 nm in the Ce^{3+}doped 4b,4c,4e and 4f samples. The second absorption edge is starting at 272 and 268nm.The absorption at $WO_4$is due the transition of electrons from filled O^{2-}states tohalffilledW 4f state. The formation of two absorption band edge is due to the creation of additional energy levelsbetween the conduction and valence band due to the inclusion of Ce^{3+} ions [15]. The optical bandgap of the material depends on the absorption edge of the sample.

The energy gap values are calculated using tauc plot method [15]. Bandgap values for samples subjected to a temperature of 800 °C is shown in Fig. inset I, II and III. The energy gap values for pristine $CaWO_4$ is calculated to be around 4.46 eV. 2 wt.% of Ce^{3+} doping has two energy levels of 3.0 and 4.38 eV and for 5 wt.% the values of the bandgaps are 3.0 and 4.32 eV whereas for the samples annealed at 1200 °C is around 4.49 eV is shown in Fig. inset IV, and Fig V and VI shows the bandgap of Ce^{3+}:$CaWO_4$ nanoparticles annealed at 1200 °C to be 4.39 eV and 4.22 eV respectively.

Conclusions

Pristine $CaWO_4$and Ce^{3+}:$CaWO_4$nanoparticles synthesized by hydrothermal method and further annealed with 800 and 1200°C. The changes observed in the structural, morphological and optical properties are analyzed as a function of annealing temperature. 5 wt.% of Ce^{3+}:$CaWO_4$annealed at 800°C exhibits dumbbell like morphology. 2 wt.% of Ce^{3+}:$CaWO_4$annealed at 1200°Cexhibits a flake like structures. The optical absorption spectra show two absorption edges asan effect of doping which gives rise to an additional intermediate energy level.

References

[1] Boltaev, G. S., Ganeev, R. A., Krishnendu, P. S., Zhang, K., &Guo, C. (2019). Sci. Rep, 9(1), 1-8.
[2] Parchur, A. K., Prasad, A. I., Ansari, A. A., Rai, S. B., &Ningthoujam, R. S. (2012). DaltonTransactions, 41(36), 11032-11045.
[3]Ningthoujam, R. S. (2012).Nova Science Publishers, 145-84.
[4] Cavalli, E., Boutinaud, P., Mahiou, R., Bettinelli, M., &Dorenbos, P. (2010). J. Phys. Chem. C, 49(11), 4916-4921.
[5] Sotiriou, G. A., Schneider, M., &Pratsinis, S. E. (2011). J. Phys. Chem. C, 115(4), 1084-1089.
[6] Singh, B. P., & Singh, R. A. (2015). New J. Chem, 39(6), 4494-4507.
[7] Nagarajan, C., Sujatha, R. A., Rahulan, K. M., Flower, N. A. L., &Vinitha, G. (2020). Opt. Mater, 110, 110512.
[8] Barbosa, H. P., Silva, I. G., Felinto, M. C. F., Teotonio, E. E., Malta, O. L., & Brito, H. F. (2017),J. Alloys Compd.696, 820-827.
[9] Kaur, P., Khanna, A. and Fábián, M., 2020. Ceramics International, *46*(17), pp.27262-27274.
[10] Sharma, K. G., Singh, N. S., Devi, Y. R., Singh, N. R., & Singh, S. D. (2013). J. AlloysCompd, 556, 94-101.
[11] Tian, B., Wang, M., Liu, C., Liu, S., Jin, L., Yan, X., ... & Wang, Z. (2015).RSC Adv, 5(126), 104172-104178.
[12] Chen, Y., Moon, B. K., Choi, B. C., Jeong, J. H., & Yang, H. K. (2013).J. Am. Ceram, 96(11), 3596-3602.
[13] Singh, B. P., & Singh, R. A. (2015).New J. Chem, 39(6), 4494-4507.
[14] Pereira, P. F. S., De Moura, A. P., Nogueira, I. C., Lima, M. V. S., Longo, E., de Sousa Filho, P. C., ... & Rosa, I. L. V. (2012).J. AlloysCompd, 526, 11-21.
[15] Jubu, P. R., Yam, F. K., Igba, V. M., &Beh, K. P. (2020).J. Solid State Chem, 290, 121576.

Acknowledgements

We wish to acknowledge the HRTEM FACILITY at SRMIST set up with support from MNRE (Project No. 31/03/2014-15/PVSE-R&D), Government of India, and also place our acknowledgement to SRMIST for high resolution scanning electron microscope (HR-SEM) facility, likely We acknowledge the XRD FACILITY at SRMIST set up with support from MNRE (Project No. 31/03/2014-15/PVSE-R&D), Government of India and We acknowledge the support from Nanotechnology Research Centre (NRC), SRMIST for offering various research facilities.

IOP Conf. Series: Materials Science and Engineering 1219 (2022) 012009 doi:10.1088/1757-899X/1219/1/012009

Mechanical and Morphological study of Copper slag admixtured cement mortar under different curing media

R. Gopalakrishnan, M.Akash, S.Subashini

Department of Physics and Nanotechnology, SRM Institute of Science and Technology, Kattankulathur 603203, India

E-mail: gopalakr@srmist.edu.in

Abstract: The hydration of ordinary Portland cement was measured on the respect of the setting time, compressive strength and morphology measurements in the presence of 0%, 10%, 20% and 30 wt% copper slag. These samples were cured in distilled water, seawater and groundwater. The results show that the replacement rates of copper slag (0%, 10%, 20% and 30 wt.%) raises the compressive strength and brings down the setting time. Both these findings suggest that at an early stage, seawater accelerates cement hydration but slows the later stage of hydration.

Key words: Copper slag,setting time, compressivestrength, Microstructure

1. Introduction

Copper slag (CS) is one of the great reactive pozzolanic materials [1]. To use high-performance concrete and mortars, CS is considered as one of the best supplementary cementing materials. Using these reduces the costs, CO_2 emissions and global warming and increases the properties of cement such as strength, toughness and workability. If calcium hydroxide is removed from the system, the object of CS carries out the root problem of efflorescence.

In the chemical composition of cement, the consciousness of setting behaviour is important. While determining the different phases of the construction process, such as transporting, positioning, condensing and appearance, the information that if the admixture is set accelerator or retarder is important. The cement hydration phase and CS pozzolanic reactivity were identified using DTA, XRD, SEM, FTIR, and electrical conductivity measurements [2-6].These different instrumentation techniques illustrate that the pozzolanic materials have a strong influence and enhance the mechanical and durability properties. The present study explains the setting time, compressive strength and morphology of CS-admixtured cement cured with distilled water (DW), groundwater (GW) and seawater (SW).

2. Materials and Methods

Vicat apparatus by standard method (ASTM C125) was used to measure the setting time of OPC and CS-admixtured cement paste [7]. The compressive strength of CS-admixtured cement (0%, 10%, 20% and 30%) was measured with a water-to-cement ratio of 0.4. Distilled water (DW), groundwater (GW) and seawater(SW)were used to cure the samples 1 day at room temperature.The compressive strength of CS-admixtured mortar cubes was measured by Compression Testing Machine (Techno-Science Trading Company, New Delhi, India) at different hydration times (3 days, 2weeks, 4 weeks) and by standard method (ASTM C780) [7]. The microstructure analysis of samples was carried out using scanning electron microscopy (SEM) on JEOL JSM-6510LV. Before this analysis, the humidity was extracted from the samples and then gold-coated conductive samples were carried out. To research the specimens at an accelerating voltage range of 15 kV to 20–22 kV with a working distance of 12 mm, the secondary

Cement nomenclature:CSH-Calcium silicate hydrate, CH-Calcium hydroxide electron detector was worked in a high vacuum. Post 28 days of curing, the analysis of specimens was performed. The chemical composition of the powder of cement and copper slag (CS) is studies using XRF and the results are presented in Table 1, whereas Table 2 provides Content of solvents (μg/g) of DW, GW and SW.

Table 1. Chemical composition of Portland cement and CS (wt. %)

Samples	CaO	SiO_2	Al_2O_3	Fe_2O_3	MgO	SO_3	MnO	Loss on ignition	Insoluble residue
Portland Cement	63.32	21.70	5.40	3.40	2.09	2.10	0.12	0.79	1.08
CS	0.45	47.24	23.25	0.14	0.22	1.01	0.02	1.55	0.23

Table 2. Content of solvents of different waters (μg/g).

Water	Total dissolved solvents	Total hardness	Chlorine	Magnesium	Calcium	Sulphur	Sodium
Distilled water	45	4	-	1	2	-	-
Sea water	30500	1060	16400	1080	272	5.80	9100
Ground water	1600	170	70	20	68	2.50	155

3. Results and Discussion

Setting time measurements

Table 3 demonstrates the initial and final setting times of the cement paste and CS-admixtured cement paste. In the table 3,it is shown that the incorporation rate of CS increases with the decrease in setting time. As the amount of silicon dioxide and aluminium oxide content is high in CS, they can react with cement phases and produce large volumes of calcium sulphoaluminate. This hydration phase speeds up the reaction rate in the early hydration state and hence decreases the setting time. As the substitution ratio is further increased to 20% and 30% when related to CS10% admixtured cement paste, there is a slight decrease in setting times.It is due to the higher demand for water from the higher CS material that could have created a denser binder process and the atmosphere could have run up. This implies that CS acts as a gentle admixture to continue it when OPC is followed by up to 30%, which may be due to the presence of a higher amount of chloride and sulphate ions in the seawater. It reduces the setting time for SW-cured samples. These ions promote the hydration of cement and serve as higher amounts for hydration product enhancement (ettringiteand C–S–H) [8].The setting time is reduced at all times by these procedures. Groundwater has few ion combinations when compared with seawater, though stronger than distilled water.

IOP Conf. Series: Materials Science and Engineering 1219 (2022) 012009 doi:10.1088/1757-899X/1219/1/012009

Table. 3. Setting time and Compressive strength of CSadmixturedcement treated with different water.

Water	CS%	Setting time (h.min)		Compressive strength (MPa)		
		Initial	Final	3 days	2 weeks	4 weeks
DW	0	5.05	6.50	11.7	40.4	47.7
	10	4.45	6.25	20.3	47.2	55.3
	20	3.25	5.40	27.2	61.5	69.3
	30	2.30	4.40	32.5	69.2	74.2
SW	0	4.10	6.20	14.1	43.0	33.1
	10	3.55	6.00	24.3	50.7	41.7
	20	2.40	5.05	32.5	64.1	52.2
	30	1.55	4.15	37.5	72.6	57.5
GW	0	4.45	6.35	12.9	42.4	48.5
	10	4.15	6.15	22.5	49.1	58.2
	20	3.15	5.30	29.4	62.3	71.5
	30	2.10	4.30	34.2	71.2	78.5

Compressive Strength Measurement

For the various CS replacement rates and three hydration intervals (3 days, 2 weeks, 4 weeks), the compressive strengths obtained are shown in Table 3. CS cement samples hydrated with distilled water demonstrated greater compressive strength than those of the control samples at all ages and all replacement rates (OPC). According to Wild and Khatib[9], a wide range of particle size distribution, greater quantity of Al_2O_3, filler effect and acceleration of cement hydration where CS particles occupy the space between cement particles and CS and hence increase the compressive strength.The pozzolanicreaction of CS with CH is explained with the following equation [8].

Copper slag+ CH + H →Additional cementitious Materials(C-S-H)(1)

Calcium hydroxide Water

$$C_3A + 3C\bar{S}H_2 + 26H \rightarrow C_6A\bar{S}_3H_{32} \quad (2)$$

TricalciumAluminate Gypsum Water Ettringite

$2(3CaO.SiO_2) + 6H_2O$ $3CaO.2SiO_2.3H_2O + 3Ca(OH)_2$(3)

Tricalcium silicate water C-S-H Calcium hydroxide

$2(2CaO.SiO_2) + 4H_2O$ $3CaO.2SiO_2.3H_2O + Ca(OH)_2$(4)

IOP Conf. Series: Materials Science and Engineering 1219 (2022) 012009 doi:10.1088/1757-899X/1219/1/012009

Dicalcium silicate water C-S-H Calciumhydroxide

The compressive strength improved the formation of a higher amount of C–S–H gel at 28 days of hydration. Moreover,it increases when the seawater mixes with CS-admixtured cement at an early stage (3 days). The acceleration of the early stage compressive strength increase may be due to the presence of a high amount of chloride and sulphate ions in the seawater. In the hydration reaction and crystallisation salt, ions fill the pores and hence increase the strength at an early stage (7 days). But in the later stage the thickening time decreases, which therefore creates voids and subsequently decreases the strength at 4 weeks compared to DW and GW. This can be attributed to the decrease in chloride, sulphate ions, and the earlier attitude of C_2S to seawater has been identified, therefore falling of rate in the subsequent process.

Microstructural analysis

The SEM analysis was conducted within 28 days of collecting samples. The specimen of 0% CS in the DW mix involves several cracks, and the presence of a small number of needle-shaped structures, i.e. ettringite, with the rough C–S–H gel pattern, concerns the porous internal structure (Fig.1a). The blend of 30% CS with DW occurred in a better homogenous and denseC–S–H gel structure, although some microcracks can be seen in the micrograph shown in Fig. 1(b). This C–S–H gel induces interfacial transition zone (ITZ) densification that may be associated with the use of calcium hydroxide (CH) by acting to increase C–S–H with pozzolanic silica shown in the CS. The 30% CS replacement has overlapping amalgamation sheets of C–S–H gel to fill the gaps (Fig. 1c) and reduces the thickness of ITZ treated with GW.However, with 30% CS replacement and SW-treated sample, less C–S–H gel production is seen with some microcracks and water holes (referto Fig. 1d). This could be due to SW and CS impurities that accumulate in the fresh state, further leading to the presence of excess water content in the CS neighbourhood. In free water, the low concentration components and polished surface of CS also emerged, causing microbleeding and ITZ thickness production leading to microcracks and capillary channels.

Fig.1a

IOP Conf. Series: Materials Science and Engineering 1219 (2022) 012009 doi:10.1088/1757-899X/1219/1/012009

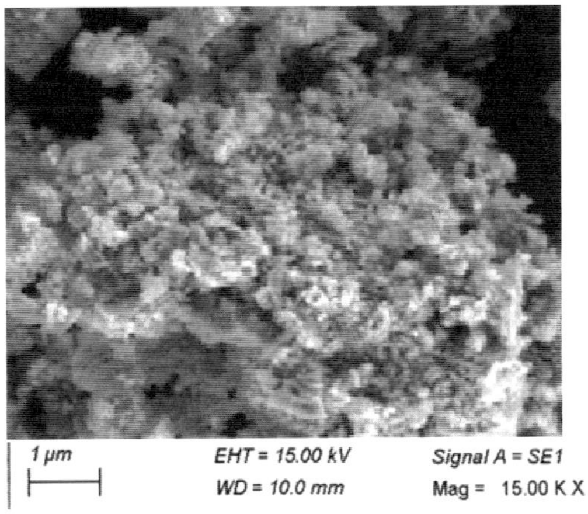

Fig.1b

[redacted]

Fig.1c

Fig.1d

Fig 1. Microstructure of (a) control cement mortar treated with DW at 28days (b) 30% CS mortar treated with DW at 28 days (C)30% CS mortar treated with GWat 28 days (d)30% CS mortar treated with SW at 28 days

Conclusions

1. The effect of rising the CS percentages to have quicker setting times.
2. When compared to OPC cement pastes, CS improves the early strength. From these systems, a relationship between intensity and setting time has been established to examine all hydration periods.
3. All experimental investigations show that when combined with seawater, hydration is accelerated in the early age and retarded in the later stage. It is observed that the sample hydrated with groundwater has an intermediate chemical reaction compared to DW and SW.
4. Compressive strength and setting time experiments check a well-accelerating effect of CS cured with DW, SW and GW.
5. The SEM images indicate that, compared to the control mortar, the CS mortar produces a denser pore structure – for the improved strength and impact resistance of the mix.

References

[1] SaikaN J, SenguptaP, Gogoi P K, BorthakurP C2002*Cem. Concr. Res.* **32** 297-302
[2] Moises F R 2006*Cem. Concr. Res.***36**827-831
[3] LoveC A, Richardson I G BroughA R2007*Cem. Concr. Res.***37(2)** 109-117
[4] WengL, Sagoe-crentsilK, Brown T, SongS 2005*Mater. Sci.Engg B.***117** 163-168
[5] Alonso Sand PalomoA 2001*Cem. Concr. Res.***31**125-128
[6] SaadMorsyM, Abo El-EneinS A,HannaG.B 1997 *Cem. Concr. Res.***27(9)** 1307-1312
[7] M.S. Shetty, Concrete Technology (S.Chand and Co.Ltd. New Delhi, 2004).
[8] CultroneG, Dela Torre M J, SebastianE M, CazallaO,NavarroC R *Water. Air. Soil Pollution.* **119**191-207
[9] WildS, KhatibJ M 1998*Cem. Concr. Res.***28(1)**83-92
[10] Badogiannis E, PapadakisV G, Chaniotakis E 2004*Cem.Concr.Res***34**1035-104

Synthesis, growth, spectral, optical and thermal studies of phthalate strontium nitrate hydrochloric acid (PASNHCL) single crystal

Surya. S[1],Dheenadhayalan. S[1], Surya.L[1], Ramesh V[1]and Gunasekaran. B[1]*

[1]Department of Physics, SRM Institute of Science and Technology, Kattankulathur, Chennai 603203, Tamil Nadu India

E-mail.:gunasekb@srmist.edu.in

Abstract. In present investigation, the nonlinear optical single crystal of phthalic acid strontium nitrate hydrochloric acid (PASNHCL) was grown by the solution evaporation method from a mixture of an aqueous solution at room temperature. The grown PASNHCL crystals crystalline nature, cell parameters, functional groups, and elemental compositions were characterized by PXRD, UV-Vis-NIR, FTIR, FT-Raman, EDX, and Thermal analysis.The PXRD analysis shows the PASNHCL crystal has good crystalline nature. Single crystal XRD confirmed that the grown crystal belonged to the orthorhombic system with space group of $P_{222}$with alattice parameters a = 6.785 Å, b = 7.420 Å, c = 11.637 Åand volume 585.86 $Å^3$. UV-Vis-NIR spectrum study reveals that the lower cut off wavelength was found to be 333 nm and the crystal has good optical transparency. The presence of functional groups was identified by FTIR spectrum. The Thermogravimetric analysis curves confirms melting point of the grown PASNHCL is 233.17℃. Itis interesting to note that PASNHCL $[Sr^+(C_8H_7N_2O_{10})Cl^-]$ single crystal has a greatly resourceful second harmonic generation (SHG) material for Optoelectronics applications.

Keywords: PASNHCL, single crystal, slow evaporation solution,Thermal stability, and SHG

1. Introduction:

In recent years, nonlinear optics is a new area of research and development of organic, inorganic,and semi-organic NLO materials. NLO materials arebeneficialowing to their commercialized applications in the fields of optical switches,optical storage device, telecommunication,optoelectronics, frequency conversion,photonics etc.,[1]. Semi-organic crystals possess numerous properties like high laser damage threshold, high optical nonlinearity, broad transparency range,high nonlinear coefficient, temperature stability and good mechanical properties.[2].Phthalic acid, strontium nitrate, and hydrochloride acid is a semi-organic materials which are the important NLO compounds in the alkali metal phthalate acid family [3]. Strontium nitrate is a white granular material can be used for the creation of high-purity crystal. Hence, in this present work, the combination of Phthalic acid, strontium nitrate, and hydrochloride acid single crystal was synthesized by gradual evaporation method at room temperatureand analyzed by PXRD, UV-visible absorption, (EAX), FT-Raman and TG/DTA.

2. Experimental section

2.1 Materials and crystal growth

Commercially available Phthalic Acid, strontium nitrate, and hydrochloric acid were obtained from LOBA CHEM are taking in a 1:1:3 stoichiometric ratio. All the used reagents were analytically pure and as received without any further purification. The essential materials for the preparation of PASNHCL crystal using the following equation:

$$C_8H_6O_4 + Sr(NO_3)_2 + 3HCl \rightarrow [Sr^+ [C_8H_7N_2O_{10}] CL^-$$

PASNHCL crystal was grown from (1.66 g/100 mL) of phthalic acid, strontium nitrate (2.11 g/mol), and hydrochloric acid (1.08 ml)was taken in 1:1:3 stoichiometric ratio, which dissolved with a mixed solvent of water. The mixed solution allowed long time evaporation processes at normal room temperature (28°C). Owing to the development of monomers at the single crystals with Bulk size were achieved from the mother solution, subsequently the conclusion of the nucleation and growth processes. The solvent slowly evaporatesin a time period of 28 days which results in the formation of PASNHCL crystal with dimension ~8 x 4 x 3 mm^3 was obtained. The image of PASNHCL single crystal is presented in Figure 1.

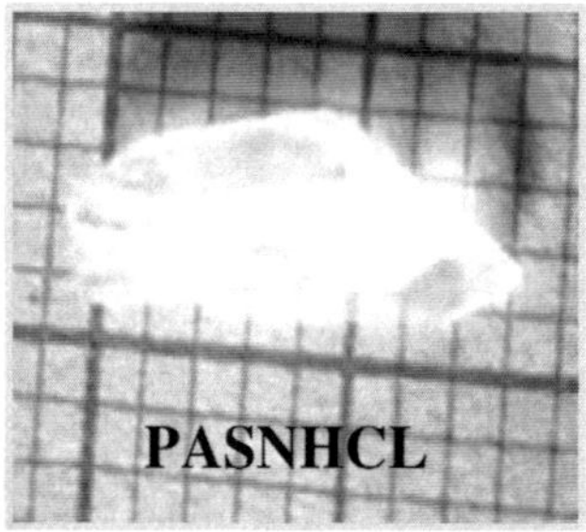

Figure: 1 PASNHCL single crystal

3. Results and discussion

PXRD analysis

The powder X-ray diffraction analysis was done byRich model diffractometer having Seifert with monochromatic radiation of Cu Kα (λ = 0.15406 nm). PXRD has beenperformed between 2θ values 10 - 80°.Figure 2 Shown the PXRD result of PASNHCL crystal, (0 3 0) crystal plane is shown the highest peak intensity. Additionally, the unit cell values were measured by the TERROR program, and the values calculated are a= 6.785 Å, b = 7.420 Å, c= 11.634 Å, and V = 585.86 Å. The analysed prominent peaks confirm the crystalline property of the PASNHCL crystal.The selected PXRD data of the PASNHCL crystal tabulated (Table1). This investigationconfirms the PASNHCL crystal showsorthorhombic crystal system with non-centrosymmetric space groupP_{222}.

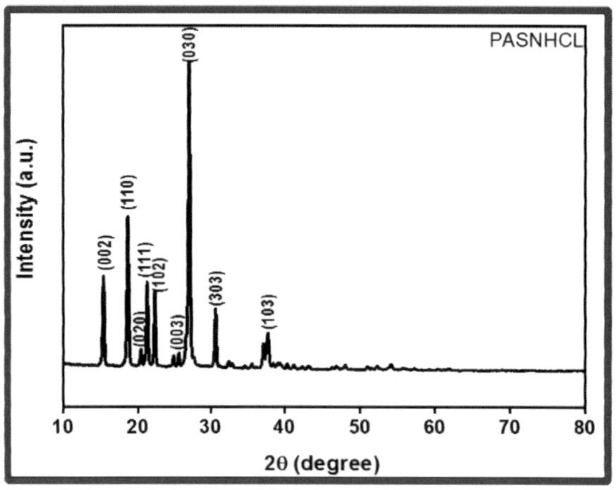

Figure: 2 PXRD pattern of PASNHCL single crystal

Table 1: PXRD date of PASNHCL single crystal

Parameters	PASNHCL structural data
Diffractometer	ENRAF NONIUS CAD - 4
Refinement method	Full matrix least square method
Crystal System	Orthorhombic
Space group	P_{222}
Unit cell parameters	a = 6.785 Å, b = 7.420 Å, c = 11.637 Å
Volume (V)	V = 585.86 Å ^{3}
Z, D (calc) [g/cm ^{3}]	4, 1.278
Crystal Size [mm]	8 mm x 4 mm x 3 mm
Temperature (K)	295
Radiation wavelength	Cu Kα (λ = 0.15406 nm)

FTIR Spectral Analysis

FTIR spectrum is animportant tool for the identification of differentelemental groups in the grown crystal.The bands which appear in the spectral region 4000-400 cm^{-1} arise from the vibration of PASNHCL crystal.The observed wavelength 2879cm^{-1} is assigned to NH_2 and N-H bending.Peak at 2650cm^{-1} and 2524 cm^{-1} indicates the NH_3^+ bending vibration. C=O stretching is confirmed with the help of peak 1650cm^{-1}.C=C stretching vibration is detected at at 1498cm-1and Asymmetric stretching of CO_2 is attributed by peak at 1585 cm^{-1}.The medium band appears in the region 1400cm^{-1} and 1066cm^{1} due to COO$^-$ symmetric stretching and OH bending of the carboxylic group in Phthalic acid. The peak at 1147 cm^{-1}confirmed C-N stretching vibration. C-OH out-of-plane deformations are confirmed at 974 cm−1.The imidazole ring asymmetric stretching mode vibrations are confirmed by peaks obtained 833cm^{-1}.The peak at 669 cm^{-1} confirm the COO$^-$ symmetric bending mode. The peak attained ~418 cm^{-1} indicate C-C plane bending corresponds to the presence of aromatic ring [4,5] and complete FT-IR assignments to the vibrational frequencies PASNHCL crystal are assigned in Table 2.

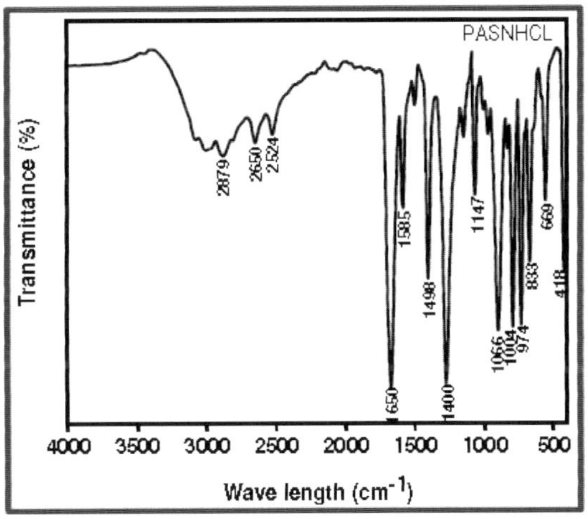

Figure: 3 FT-IR spectra of PASNHCL single crystal

Table: 2 vibrational Assignments for PASNHCL single crystal

Frequency Range (cm^{-1})	Band Assignments
2879	$NH_3$and NH bending
2650,2524	NH_3^+ bending vibration
1650	C = O stretching
1585	CO_2 Asymmetric stretching
1498	C=C stretching vibrations
1400	COO^- symmetric stretching
1066	OH bending
1147	C-N stretching vibration
974	C-OH out-of-plane deformation
669	COO^- symmetric bending
418	C-C plane bending

EDX analysis

EDX analysis helps to gain information of elemental composition of the grown PASNHCL crystal. The weight percentage of the C, O, Sr, Cl and N was achieved from EDX studies. The wt% of different composition in the PASNHCL crystal as shown in figure 4.

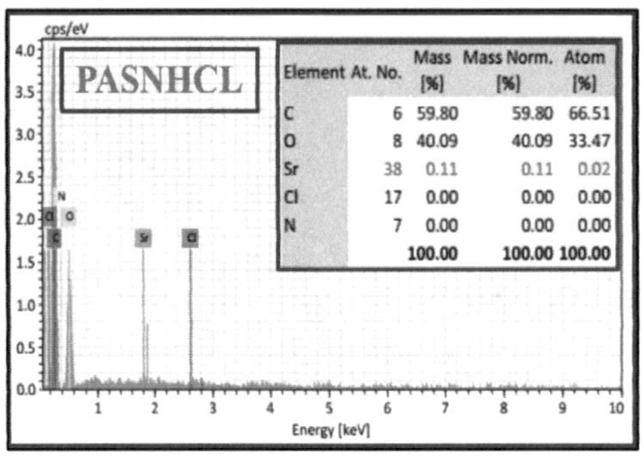

Figure: 4 EDX spectrum of PASNHCL single crystal

Linear Optical Properties:

UV–Vis spectra have been performed to find the absorption and transmittance of PASNHCL crystals. UV–Visible spectrum of the PASNHCL crystal is recorded between 200 to 1000 nm at room temperature and shown in Figure 5a. The broad absorbance in the UV–vis-NIRregion delocalization of electrons and strong intramolecular charge transfer interactions. The 333 nm cut-off wavelength is observedand moreover the PASNHCL crystal exhibits 30% of transmission. In the visible and NIR regions the absorption of grown PASNHCLcrystal is decreased. The high transmittance of the PASNHCL crystalleads to application of second harmonic generation frequencies with suitable optoelectronic applications. The absorption coefficient (α) is given by

$$\alpha = \frac{2.303 * \log(1/T)}{d} \quad \text{..........................(1)}$$

Where d is the thickness of the crystal, T transmittance .The band gap(E_{gap}) of PASNHCL crystal was calculated bytauc plot (hvvs $(\alpha h\nu)^2$) (figure 5b)and band gap was found by the extrapolation of the linear part. The optical band gap of PASNHCL is 3.88eVis comparable to the similar compound [6]. The long rangeband gap of PASNHCL crystal authenticates the enormous transmittance in the visible region and the crystal is applicable for optoelectronic applications.

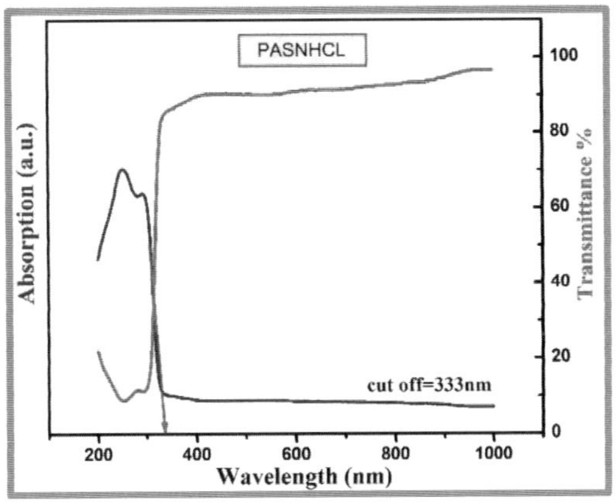

Figure: 5a Absorption and Transmittance spectra of PASNHCL single crystal

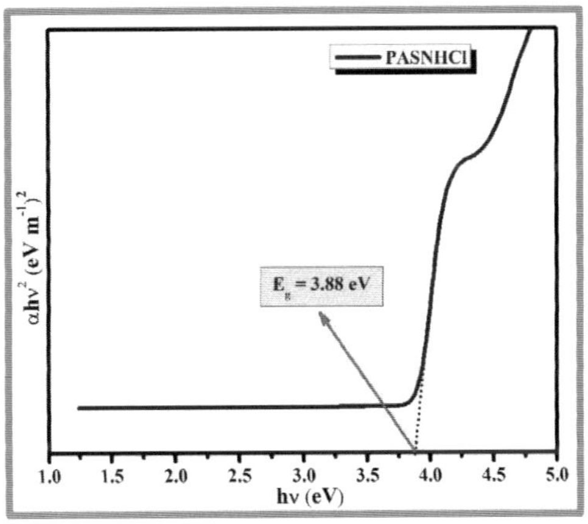

Figure:5b Tauc plot of PASNHCL single crystal

Raman spectrum

The RAMAN spectrum analysis was used to examine the basis characteristic vibrations of various elemental groups, viz.,hydroxyl, carbonyl, and amino groups. Inthe RAMAN spectrum of the PASNHCL crystal sample (figure 7), The C=O and C-H in-plane bending vibration and was record at 771cm^{-1},778cm^{-1} and 892 cm^{-1}. The OH deformation vibrations usually occurin between 1380 – 1280cm^{-1}. A peak with high intensity ~1263cm^{-1}corresponds the Phthalic acid compound [7]. The alcohol C–O group attributable to the stretching vibration at the Raman counterparts was observed at 1268cm^{-1}[8]. C–H deformation arises at 1142 cm^{-1}. C=O out of plane bending vibration was perceived at the band ~547cm^{-1}[9]. The band Ring deformation was observed at 643cm^{-1}. The peak at 1172 cm^{1}indicates C—H in-plane bending vibration[10]. The complete FT-Raman assignments to the vibrational frequencies of PASNHCL single crystal are assigned in Table 3.

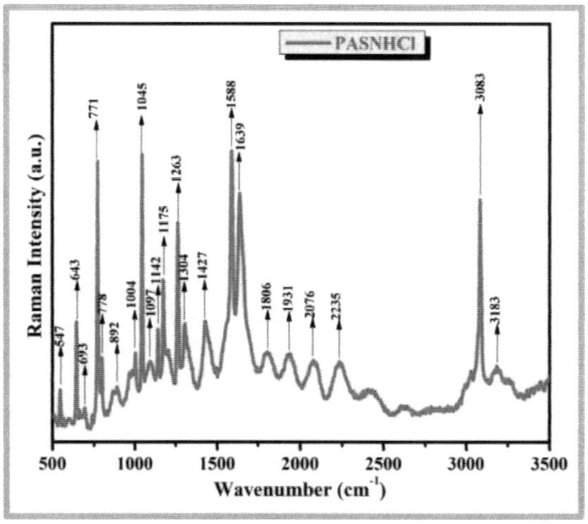

Figure: 6 FT-Raman spectra of PASNHCL single crystal

Table: 3 Ramanvibrational Assignments for PASNHCL single crystal

Wavenumber (cm^{-1})	**Functional group**
547	C=O out of plane bending vibration.
643	Ring deformation
693	NO2 wagging
771	C=O bending
778	C-H bending
892	C-H out of plane bending
1045	C-C stretchingmoede.
1142	C–H deformation
1175	C-H in plane bending
1263	OH deformation
1639	C=O vibration
3080	C-H stretching

Thermal Studies:

Thermal stability of PASNHCL single crystal was analysed byTGA/DTA. The measured sample with 2.75 mg weight was used for the thermal test. The grown sample was heated at 20°C/min using argon gas for temperature ranging from 35°C to 500°C(figure 8). The TGanalysis proves that thereoccurs no weight loss upto190 C. Major weight losses appear in the temperature range of 190 with 99.16% indicates the decomposition of all the materials present in the crystal. A sharp endothermic peak for melting point at 233 °C was observed using DTA spectra and implies good degree of crystallinitywas observed[11]. The study concludes that the grown compound is stable up to 190°C and issuitablefor

IOP Conf. Series: Materials Science and Engineering 1219 (2022) 012010 doi:10.1088/1757-899X/1219/1/012010

Non-linearoptical applications.

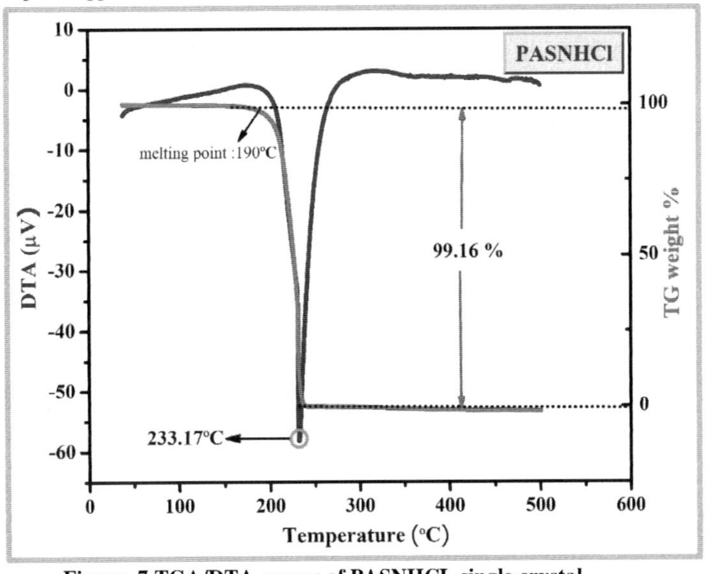

Figure: 7 TGA/DTA curves of PASNHCL single crystal

Conclusion

Single crystal of PASNHCL were successfully synthesized and grown from water using a solvent evaporation method at ambient temperature. PXRD analysis confirms the high crystallinity and the grown PASNHCL crystal of space group is P_{222} which belongs to the orthorhombic crystal family with non-centrosymmetric. FT-IR and RAMAN spectral study is used to confirm the diverse modes of frequencies and elementalgroups appear in the PASNHCL single crystals. The lower cut-off wavelength and band gap energy (E_{gap}) evaluated from the UV-Vis-NIRspectrum was endowed to be 333nm and 3.83eV. The TGA/DTA curveshows the melting point of the PASNHCL single crystal is 233.17℃. PASNHCL is a second order Nonlinear optics crystal which can find applications in photonics and optoelectronics.

Referernce:

[1] Nirmala Sri Devi S, Arun Kumar R, &Girija E. K. 2016 Materials Research Innovations 20(4) 293– 299.
[2] SenthilA, RamasamyP, Sunil Verma B. 2011 Journal of Crystal Growth 318 757–761.
[3] Miniewicz A, and BartkiewiczS 1993Adv Mater Opt Electr 2 157–63.
[4] Hemalatha V A,Arulmani S2019 Materials Today: Proceedings 8 142–147.
[5] Suresh Kumar M, Vijayaraghavan G V, Rajesh K and Krishnan S 2020Materials Research Innovations 422-432.
[6] Krishnan S and Justin Raj C 2008 Cryst. Res. Technol 43(6) 670 – 673.
[7] Tamil Elakkiya, M, PremKumar, S, Sathiyendran, M, Suresh, P, Shanmugaiah, V, andAnitha, K. 2018. Journal of Molecular Structure 1173 635–646.
[8] Krishnan, P, Gayathri, K, Rajakumar, P. R, Jayaramakrishnan, V, Gunasekaran, S, and Anbalagan, G. 2014 SpectrochimicaActa Part A: Molecular and Biomolecular Spectroscopy 131 114–124.
[9] Sangeetha V, Kanagathara N,Mariusz K. M, Gunasekaran S and Anbalagan G2014Journal of Physical Science 25(1) 93–111.
[10] Rejeena V. R, Merin G, Javeesh A, Sajana D andVinitha G 2018 Optical materilas86 198-212.
[11]Theras J. E. M, KalaivaniD ,Jayaraman D and Joseph V 2015 Journal of Crystal Growth 427 29– 35.

Design and Analyses of Broadband Perfect Light Absorber for Visible and Near IR Frequencies using Plasmonic Metamaterials

GurukaelaiarasuTamilarasi Mani[1], Renilkumar Mudachathi[1,2*]

[1]Department of Physics and Nanotechnology, SRM Institute of Science and Technology, Kattankulathur, Tamilnadu, India, 603203
[2]Metamaterials Laboratory, RIKEN center for advanced photonics, RIKEN, Wako, Japan, 351-0198

E-mail: renilkum@srmist.edu.in

Abstract. Metal-insulator-metal nanostructures based plasmonic metamaterials have been extensively explored for the design of broadband perfect light absorption in all the technologically relevant windows of electromagnetic spectrum. We have computationally studied a unique metal-insulator-metal nanostructure for the design of a broadband perfect light absorber in the visible and near infrared frequencies. We explored different metal combinations for the same metal-insulator-metal geometry for the absorber design. The study shows that the dependence of spectral response is dominated by the geometrical shape and arrangement, rather than the nature of the metal combinations. We report an average light absorption of more than 80% from 400nm to 1100nm wavelength range.

Key words: Metal-insulator-metal nanostructures, plasmonic metamaterials, perfect light absorber

1. Introduction

Structured photonic materials such as plasmonic nanostructures and metamaterials (MMs) have been explored for manipulating the electromagnetic propagation owing to their tailored electric and magnetic responses. The bianisotropic response of metamaterials to the electromagnetic waves has provided unusual optical properties such as negative/zero refractive index and selectively tunable magnetic and electric responses to the incident electromagnetic radiation, which resulted the development of devices like electromagnetic cloak and flat super lenses [1,2]. Metamaterials are often designed using sub wavelength metal and dielectric inclusions, which introduce undesired optical losses to the MM, based devices. These inherent losses in such structures can be exploited for the design of perfect light absorbers and Landyet. al. has reported the first MM based light absorber in the microwave region [3]. Followed to this,several narrow and broadband light absorbers at specific wavelength ranges havebeen demonstrated for applications such as sensing [4, 5], structural colouartion [6, 7], selective emitters [8, 9], IR camouflage etc. [10]. Usually MMs have two structural elements to produce net electric and magnetic responses simultaneously. Smith et. al. has demonstrated negative refractive index using metamaterial consisting of a planar split ring in combination with a cut wire [11], in which the split ring excites the magnetic response and the cut wire excites the electric response.
Three dimensional metal-insulator-metal (MIM) structures can be used as the structural elements for the design of MMs, which provides both responses and several such designs have been reported for narrow and broadband perfect light absorption [12-16]. We present the design of a MIM based broadband metamaterial perfect light absorber in the visible and near IR wavelengths. We have computationally investigated different noble metal combinations for the design of MIM structures and found that the metal combinations have least influence on the electromagnetic response of the resultant MM. We report a broadband absorption of more than 80% for the entire visible and near IR wavelengths using our MM based light absorber, which finds applications in

IOP Conf. Series: Materials Science and Engineering 1219 (2022) 012011 doi:10.1088/1757-899X/1219/1/012011

thermophotovoltaics as broadband absorber and narrowband emitter and anti-reflection surfaces.The design of the MIM structures and its analyses with different metal combinationsfollowed by results and discussion form the content of the succeeding sections of this paper.

2. Design of the broadband absorber

The schematics of the proposed MIM structure is shown in the Figure 1, which consists of top, middle and bottom layers in the form of discs of diameter 200nm each. The MIM structures are arranged in a square lattice of periodicity 400nm to form the MM for broadband light absorption.

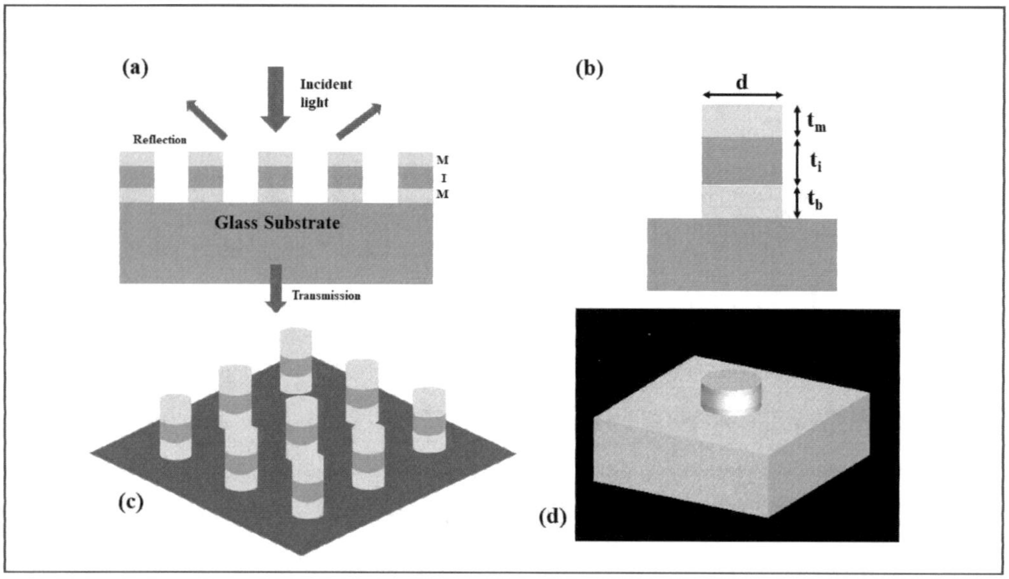

Fig. 1. Schematics of the proposed MIM structure shows (a) the cross section of the MM, (b) cross section of a meta atom, (c) perspective view of the proposed structure and (d) the perspective view of the computational model

The top and bottom metal layers are designed with a thickness of 50nm and the middle dielectric layer with a thickness of 100nm in the computer model. We have used gold, silver and aluminum in different combinations to form homo and hetero MIM structures with SiO_2 as middle dielectric layer. The geometries of the MIM structure and the MM are kept constant for both the homo and hetero MIM based MMs.

3. Results and discussion

We have first studied the spectral characteristics of MM based on homo MIM structures using gold and silver as functional layers and depicted in the figures 2 (a) and (b) respectively. In both figures the blue, black and red lines represent absorption, reflection and transmission respectively.

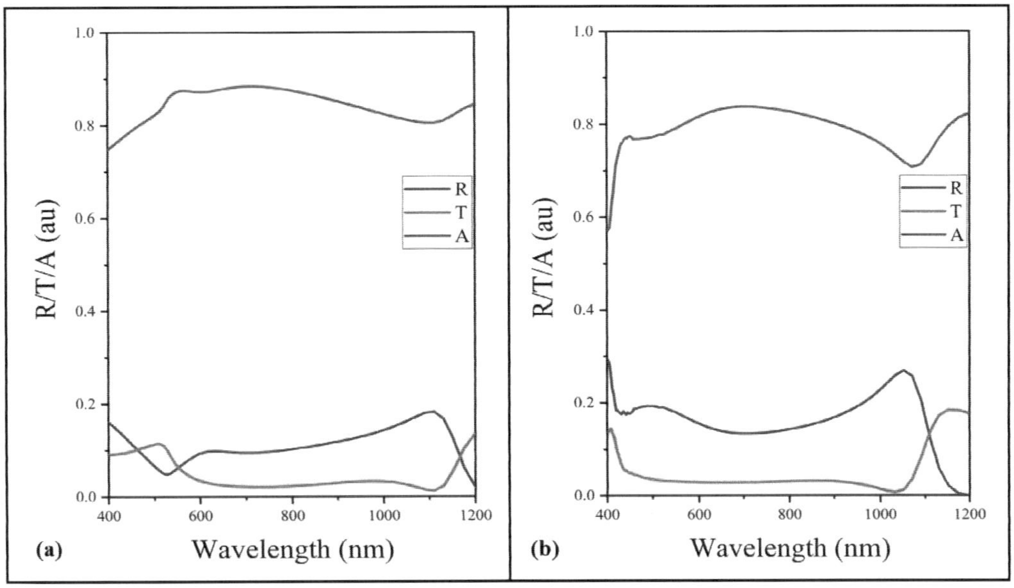

Fig. 2. Optical spectra of homo MIM based MMs using (a) gold and (b) silver a functional layers.

Irrespective of the metals, both the MMs shows broadband absorption of more than 75% over a broad wavelength range of 800nm in the visible and near IR regions. As shown in figure 2 (a) the MM using gold functional layer provides better absorption of more than 80% when compared with that of the MM using silver functional layer. This can be attributed to the inherent optical absorption of gold in the visible wavelength region. In all the cases the absorption is calculated using the equation $\boldsymbol{A = 1 - (R + T)}$, where $\boldsymbol{A}$, $\boldsymbol{R}$ and $\boldsymbol{T}$are absorbance, reflectance and transmittance respectively.

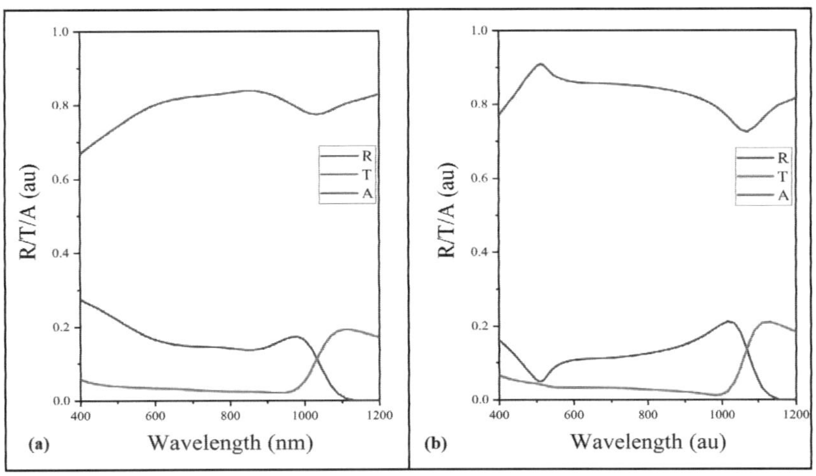

Fig. 3. (a) Optical spectral measurement of MM based on aluminum homo MIM structures and (b) that of the MM based on gold and aluminum based hetero MIM structures

A similar trend is observed in the case of MM based on aluminum homo MIM structures as shown in the figure 3(a), for which the absorption in the visible region is comparatively weaker.

IOP Conf. Series: Materials Science and Engineering 1219 (2022) 012011 doi:10.1088/1757-899X/1219/1/012011

Further we analyzed the optical absorption in MM based on hetero MIM structures. First we studied the gold/aluminum combination in which the top layer is gold and bottom layer is aluminum and the results are shown in the figure 3(b). In this case the MM shows better absorption in the visible region but has significant transmission and reflection in the 1000nm- 1200nm wavelength range. Similar trend is also observed in the cases of other metal combinations. The optical spectrum of the silver/aluminum combination is shown in the figure 4(a), in which top layer is silver and bottom layer is aluminum.

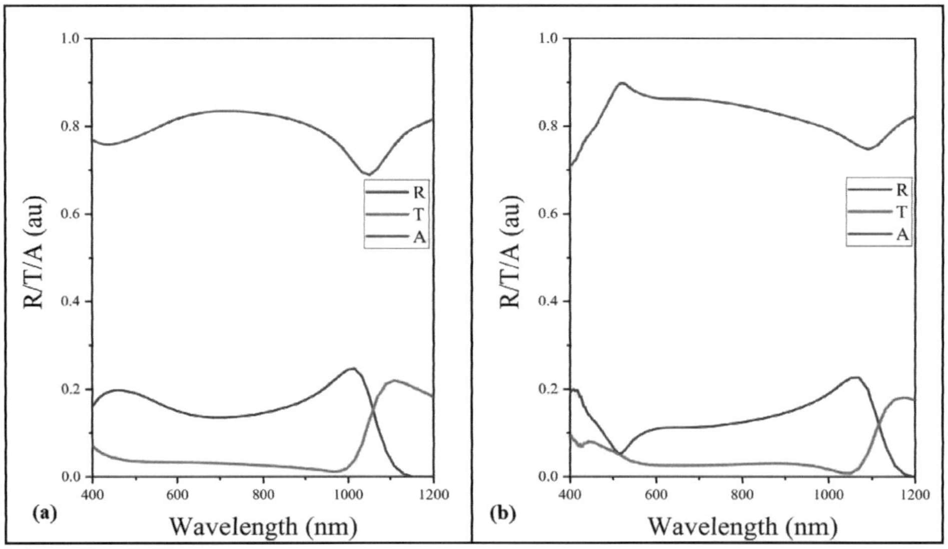

Fig. 4. Optical spectra of MM based on hetero MIM structures (a) silver/aluminum and (b) gold/silver combinations.

Figure 4(b) depicts the optical spectrum of the gold/silver combination in which the top layer is gold and the bottom layer is silver. In this case the MM shows better absorption in the visible region, which can be attributed to the inherent optical absorption of gold. In all the calculations we have used the same MIM geometry and we found that from these analyses the different material combinations have least influence on the absorption spectrum. A comparative plot of the absorption for all MMs is shown in the figure 5.

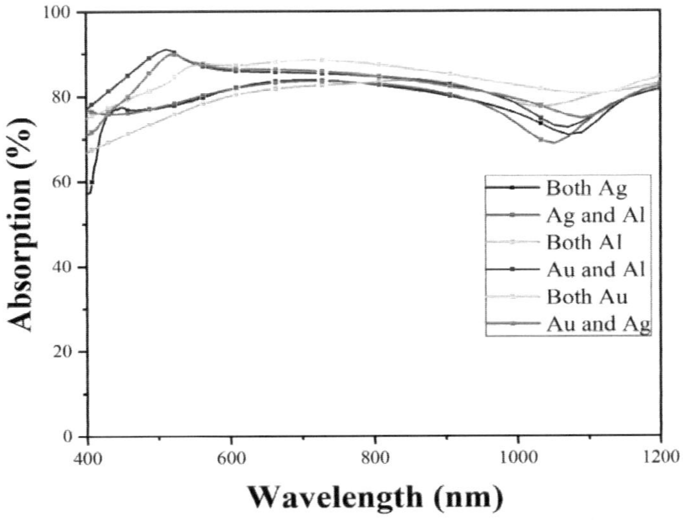

Fig. 5. Absorption spectra of all MM geometries

Figure 5 shows the comparison of spectra of all the proposed MMs where it is evident that all the structures provide similar performance with slight variations. The MM based on homo Metal MIM structure made of gold metal layer shows the highest average absorption in the given wavelength range. Table 1 shows the calculated average broadband absorption of all the proposed structures in the 400nm – 1200nm wavelength range.

Table 1. Average broadband absorption of different geometry

Structure	Absorption (%)
Au/SiO2/Au	84.5
Ag/SiO2/Ag	78.7
Al/SiO2/Al	79.3
Au/SiO2/Al	82
Ag/SiO2/Al	79
Au/SiO2/Ag	84

Conclusion

In conclusion, we have studied the broadband optical absorption in metamaterials based on MIM structures of homo and hetero metal layers. We found that the optical absorption is mainly determined by the geometry of the metamaterial systems and it is independent of the different metal combinations used for defining the meta atoms. We have demonstrated a broadband absorption of more than 80% using the MIM based MM for the visible and near IR wavelength ranges.

References

[1] Schurig D, Mock J J, Justice B J, Cummer S A, Pendry J B, Starr A F and Smith D R 2006 Metamaterial Electromagnetic Cloak at Microwave Frequencies *Science* **314** 977–80

[2] Pendry J B 2000 Negative Refraction Makes a Perfect Lens *Physical Review Letters* **85** 3966–9

[3] Zhou W, Wang P, Wang N, Jiang W, Dong X and Hu S 2015 Microwave metamaterial absorber based on multiple square ring structures *AIP Advances***5** 117109

[4] Lu X, Zhang L and Zhang T 2015 Nanoslit-microcavity-based narrow band absorber for sensing applications *Optics Express***23** 20715–20

[5] Huang H, Xia H, Xie W, Guo Z, Li H and Xie D 2018 Design of broadband graphene-metamaterial absorbers for permittivity sensing at mid-infrared regions *Scientific Reports***8** 4183

[6] Galinski H, Favraud G, Dong H, Gongora J S T, Favaro G, Döbeli M, Spolenak R, Fratalocchi A and Capasso F 2017 Scalable, ultra-resistant structural colors based on network metamaterials *Light: Science & Applications***6** e16233–3

[7] Cheng F, Gao J, Stan L, Rosenmann D, Czaplewski D and Yang X 2015 Aluminum plasmonic metamaterials for structural color printing *Optics Express***23** 14552–60

[8] Chirumamilla M, Krishnamurthy G V, Knopp K, Krekeler T, Graf M, Jalas D, Ritter M, Störmer M, Petrov A Y and Eich M 2019 Metamaterial emitter for thermophotovoltaics stable up to 1400 °C *Scientific Reports***9** 7241

[9] DeMeo D F, Pfeister N A, Shemelya C M and Vandervelde T 2014 Metamaterial selective emitters for photodiodes ed M J F Digonnet and S Jiang *Optical Components and Materials XI***8984**

[10] Lee N, Kim T, Lim J-S, Chang I and Cho H H 2019 Metamaterial-Selective Emitter for Maximizing Infrared Camouflage Performance with Energy Dissipation *ACS Applied Materials & Interfaces***11** 21250–7

[11] Landy N I, Sajuyigbe S, Mock J J, Smith D R and Padilla W J 2008 Perfect Metamaterial Absorber *Physical Review Letters***100**

[12] Fann C-H, Fann C-H, Zhang J, Zhang J, ElKabbash M, Donaldson W R, Campbell E M and Guo C 2019 Broadband infrared plasmonic metamaterial absorber with multipronged absorption mechanisms *Optics Express***27** 27917–26

[13] Segovia, Ni, Wong Z, Mrejen M, Wang Y, Zhang ; 6 X, Wen, Yue F, Ardron M, Chen ; 7 X, Cong, Tan S, Yahiaoui R, Yan F, Zhang W, Singh ; 8 R, Singh, Cao W, Al-Naib I and Cong L 2012 Experimental demonstration of ultrasensitive sensing with terahertz metamaterial absorbers: A comparison with the metasurfaces*IEEE Trans. Antenn. Propag***60** 20347

[14] Yan M 2013 Metal–insulator–metal light absorber: a continuous structure *Journal of Optics***15** 025006

[15] Aalizadeh M, Khavasi A, Butun B and Ozbay E 2018 Large-Area, Cost-Effective, Ultra-Broadband Perfect Absorber Utilizing Manganese in Metal-Insulator-Metal Structure *Scientific Reports***8** 9162

[16] Mudachathi R and Tanaka T 2018 Broadband plasmonic perfect light absorber in the visible spectrum for solar cell applications *Advances in Natural Sciences: Nanoscience and Nanotechnology***9** 015010

Selective Defect Metal Sulfide for Localized Surface Plasmon Resonance

Pooja P. Sarngan,[†]Manju C P,[†]Pranjali Jadhao, and Debabrata Sarkar[*]

Applied NanoPhysics Laboratory, Department of Physics and Nanotechnology, SRM Institute of Science and Technology, Kattankulathur-603203, India
† Authors with equal contribution

E-mail. deboo_phy@yahoo.com

Abstract. As a p-type semiconductor with metal vacancies in the lattice,coppersulfide $Cu_{2-x}S$ has been proved asan effective photocatalyst with its unique semiconducting character, non-toxic nature andexcellent optoelectronic properties. In this work, we have synthesized copper sulphide nanostructure via hydrothermal method by using cetyltrimethylammonium bromide (CTAB) as a morphology controlling agent. Due to plasmonic nature, results from copper vacancy $Cu_{2-x}S$ nanocrystals possess a property called localized surface plasmon resonance which makes them capable of utilizing the NIR region of solar spectrum. The aim of this study is to explore copper sulphide catalyst which is capable of absorbing NIR region of the solar spectrum. It is found that copper sulphide has significantly enhanced light absorption over the entire wavelength range (from ultra-violet to near-infrared).

Keywords: Crystal defect, copper sulfide, localized surface plasmon resonance

1. Introduction:

As global warming has become a major issue now a days, we reached to apoint where fossil fuel utilization should be reduced or it should be replaced with more cleaner fuels[1]. Hydrogen has gained much importance as an ideal clean fuel because of the fact that it is environmental friendly, abundant from various sources (biomass or water) and its high energy yield[2][3]. Among various techniques used, photocatalytic water splitting was found an effective method by making use of solar energy for Hydrogen production as it produces pollution free by products[4][5][6].With the incident light on a semiconductor (SC), electrons and holes generate in the CB and VB simultaneously[7][8][9]. For this the incident light energy should be greater than the SCs band gap. The generated e_{CB}^- and h_{VB}^+ due to leads to water splitting through reduction of H_2O molecules and oxidation to form O_2[10][11].

For efficient hydrogen production the catalyst plays an important role. A good photocatalystshould have properties of high Solar to hydrogen conversion efficiency (STH). The solar spectrum is composed of ultraviolet, visible and infrared radiation. More than half of the spectrum is occupied by infrared radiation whereas a very less percentage by ultraviolet and the rest is occupied by visible light[12][13]. Hence a catalyst which is capable ofabsorbing visible to infrared radiation of solar spectrum is a proper choice for photocatalysis as we need high solar to hydrogen conversion efficiency. In order to achieve visible to NIR solar energy absorption, substances called plasmonic nanocrystals including noble metals, self-doped and heavily doped semiconductors exhibit unique property called localized surface plasmon resonance (LSPR) which offers to effectively absorb and utilize the low energy photons, that is infrared radiation of solar spectrum[14]. Due to remarkable light scattering and absorption properties plasmonic nanocrystals has gained much attention in application such as photocatalysis. Among plasmonic nanocrystals (NCs) self-doped NCs are more suitable as a photocatalyst. By self-doping or vacancy doping their carrier density can be increased giving rise to LSPR corresponding to near infrared (NIR) region of the spectrum. Interesting candidates for vacancy doping are turned out to be Copper chalcogenide NCs, mainly copper sulfides ($Cu_{2-x}S$), copper selenide ($Cu_{2-x}Se$), and copper telluride ($Cu_{2-x}Te$). $Cu_{(2-x)}S$is a significant metal chalcogenide[15][16]. Less attentions have been paid to the NIR absorption of $Cu_{2-x}S$ nanostructures. Due to its unique p-type semiconducting nature and its non-toxic nature, it has found application in photocatalysis[8][17].

In this study, we demonstrate the synthesis of self-doped copper sulfides at gram scale and investigated their plasmonic properties. $Cu_{2-x}S$ nanostructure were synthesized via solvothermal technique by changing the amount of cationic surfactant Cetyltrimethyl Ammonium Bromide (CTAB)and the nanostructure and plasmonic properties were controlled under different synthesis conditions.

IOP Conf. Series: Materials Science and Engineering 1219 (2022) 012012 doi:10.1088/1757-899X/1219/1/012012

2. Materials and Methods:

Herein,all the chemicals were used without any further purification. Copper(II) nitrate tri hydrate ($Cu(NO_3)_2.3H_2O$) and Glutathione was purchased from Sigma Aldrich. Cetyltrimethyl Ammonium Bromide (CTAB, assay 99%) was obtained from SRL.

Material Synthesis:

For the synthesize of diversified copper sulfide $Cu_{2-x}S$ 0.2 gram of Copper nitrate tri hydrate [$Cu(NO_3)_2.3H_2O$] is used as copper precursor and 0.307 gram of Glutathione as sulphur precursor. CTAB is used as a surfactant and 30 ml of DI water was used to dissolve all this component. The solvent and precursors weremixed well in order to form homogeneous solution by making use of a magnetic stirrer with rpm is about 400-700. The prepared solution was transferred to a Teflon lined stainless steel auto clave, maintained at 120°C for 24 hours. Later, the as-prepared samples after cooling down wascollectedand washed several times using ethanol and distilled water to remove all the unreacted and other products from the sample.The washed sample was then kept for drying at 60for 6-7 hours. Hence the effect of CTAB concentration was studied. The final samples denoted as CS-x (x= 0, 0.25,0.5, 1 represent CTAB amount in gm).

Characterisations:

For the structural and morphological conformation of the sample Field emission Scanning electron microscope (FESEM, FEI QUANTA 200, USA) was used with an accelerating voltage of 15 kV. The crystallinity of the sample was recorded using XRD technique(D8 Advanced Bruker Davinci). Thediffused reflectance UV-Vis-NIR spectrum (DRS) of the sample was obtained by an Agilent Cary 5000 UV-Vis-NIR spectrometer and thereby the band gap of the material was calculated.

3. Results and Discussion:

Morphological Analysis

Morphological analysis was done using FESEM. The following Fig.1 shows SEM images of different samples synthesized with different CTAB conditions. With varying CTAB condition various surface morphologies and size were obtained. The goal was to obtain the sample in the nanoscale dimension in order to get high surface to volume ratio which results in good photocatalytic activity. But some of the sample were not in the nanoscale dimension and did not show the expected structure. CS-0.5 synthesized with 0.5gram CTAB condition produced a nanoflake

structure.

***Fig.1**Represents the FESEM images of the sample: a) and b) CS-0 and CS-1, b)and e)Lower and higher magnification of CS-0.25, c) and f) represents the lower and higher magnification of CS-0.5 respectively.*

From the Fig.1 SEM analysis, it is clear that formed structure was flakes. As per the calculations with the scale bar we can conclude that size of the flakes is not uniform. It spreads over the range of 90-200 nm. This structure will enhance the photocatalytic activity as it is having a surface area to carry out the surface reactions. As the amount of CTAB is increased to 1 gram, the nanoflakes gathered and form micron size crystals.

Crystallinity and Phase Analysis

The crystallinity and phase of the samples were studied with XRD analysis. With different CTAB conditions crystallinity of the samples were changed. The Fig.2 shows the XRD peaks obtained for CS-0 sample which was synthesized without CTAB. The sample was unable to give sharp peaks as crystalline property of the material was not reached.For CS-0.5 sample, the sharp peaks were obtained due to the presence of crystalline phases. Each diffraction peak is attributed from a specific set of parallel planes of atoms. These planes are represented using Miller indices. The position of diffraction peaks is compared against library of known stochiometric compositions of Copper Sulfide. The peak position corresponding to CS-0.5 sample was found to match with $Cu_{1.9375}S$. The crystal phases obtained are (2 1 1), (3 1 0), (4 1 0), (2 3 0), (1 3 2), (2 4 1), (1 1 2) which corresponds to 2theta values of 10.819, 11.077, 14.423, 18.254, 21.661, 24.507, 27.287 and 46.44 (JCPDS card no. 83-1462, 65-3817 and 24-0061). The extra peaks in the XRD pattern are due to presence of Aluminium in the sample. From XRD spectra it wasconfirmed that it was able to create copper vacancy to obtain LSPR which results in good photocatalytic activity.

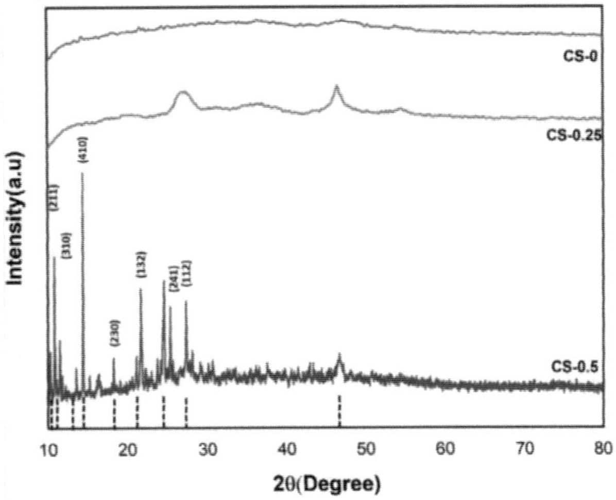

Fig.2 *X-ray diffraction spectra of the sample (i) CS-0, (ii) CS-0.25 and (iii) CS-0.5*

Evolution of NIR absorption

For the application of photocatalysis the samples optical properties really have an important role. The sample should be capable of absorbing most part of the sunlight. To study the absorbing ability of the sample UV-Vis-NIR spectroscopy was performed (Fig.3).There observed absorbance peak in both UV and NIR region. Sample CS-1 showed absorbance only in the UV region. Apart from CS-1 all the other sample showed a broad peak starting from NIR region. The noise in the spectra after 800 nm is due to the changing of lamp from visible to IR. Among these samples CS-0.5has showed good absorbance peak. CS-0.5 was able give the nanoflake structure also. For the good photocatalytic activity high surface to volume ratio and good absorbance of sunlight is important and the sample CS0.5 have both the factors. CS-0.5 sample was synthesized with 0.5 gram of CTAB hence the good condition for CTAB was found to be 0.5gram.Here the absorbance in the NIR region is due to the plasmonic resonance resulting from the copper vacancy and the absorbance in the UV region results from the band gap of the samples.

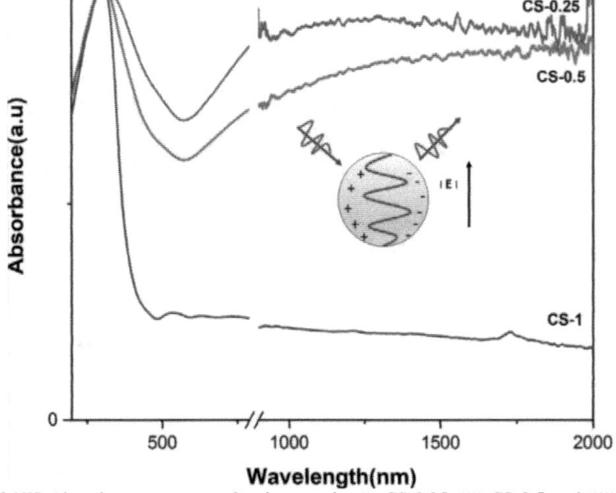

Fig.3 NIR Absorbance spectrum for the samples (i) CS-0.25, (ii) CS-0.5 and (iii) CS-1

NIR absorption sensitive to morphologies, different compositions, size as-well-as crystalline-phase feature.Copper sulfides are p-type semiconductor due to copper vacancies in their lattice and hole generation affects the valance band structure which has a major contribution from sulfur 3p orbitals. In the absorption spectra, the first peak associated with the photon energy required to promote an electronic transition from the top of the valence band to the bottom of the conduction band. The second broad absorption band corresponds to the LSPR in cation vacant $Cu_{2-x}S$, located at 650 to 1800 nm. It is reported that the LSPR frequency is related to the carrier density in the nanocrystal.

$Cu_{2-x}S$ represents both direct and indirect optical transition. The band gap of the samples was calculated using Tauc plot. For $Cu_{2-x}S$ NC both direct and indirect optical transitions are possible with direct bandgap in the range of 1.5eV and indirect bandgap in the range of 2.6 eV. For direct transition the Tauc plot will be hνv/s $(\alpha h\nu)^2$where hν is the photon energy, α is the absorption coefficient which can be obtained from absorbance peak. And for indirect transition Tauc plot will be hνv/s $(\alpha h\nu)^{1/2}$. Then the bandgap is calculated from the Tauc plot by fitting of a linear region in the plot.Selected region shows the fitted straight line along the plot. The X intercept will give the bandgap. Using Tauc plot both the direct and indirect band gap of the sample CS-0.5 was found out. Indirect bandgap ofthe sample was found out to be 2.3 eV.

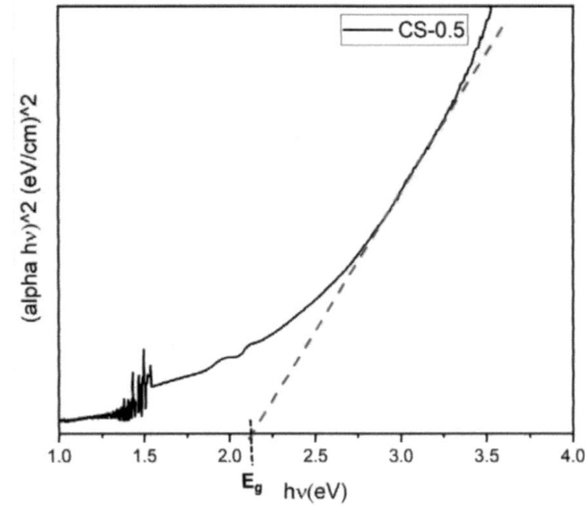

Fig4. *EnergyBand gap estimation of CS-0.5 by linearly extrapolating towards energy axis.*

Localized surface plasmon resonance is collective oscillation of free charge carries in resonance with the incident light. Solar radiation being part of an electromagnetic spectrum consist of simultaneously varying electric and magnetic field. When light incidents on a plasmonic catalyst the free carries are perturbed by the electromagnetic field present in the light (polarization). If the natural frequency of charge carrier is in resonance with the frequency of incident light then there is a collective oscillation of free carriers coupled with the light. The oscillation of the electrons absorbs part of electromagnetic spectrum. LSPR is observed in objects which are much smaller than wavelength of the incident light.

To better describe LSPR in NCs concepts related to optical response in substances were introduced. Drude model is one of them [18][16]. The model assumes that the optical response in plasmonic NCs is determined by the collective oscillation of carriers via plasmon frequency ω_p, the damping parameter γ and a background dielectric constant. The dielectric permittivity of metals is given by

$$= {}_{\infty} - /(\quad +)$$

The damping parameter represents the scattering events of carriers. In case of bulk materials, the damping parameter is given by

$$= /^{*}$$

Where, is the carrier mobility and * is the effective mass. In nanoscale dimension the damping parameter becomes bigger than in the bulk dimension because now scattering of carriers with surface become relevant as size of the NC is comparable to mean free path of the carriers. This effect is accounted as

$$= + \qquad /$$

Where, is the damping parameter in the bulk, is the Fermi velocity and is the diameter of the C. The plasmon frequency[19],

$$= \qquad / \qquad ^{*}$$

Where, is the free carrier density. From the above equation the LSPR frequency depends on the carrier density. In case of metal nano crystals, the carrier density is very high compared to self-doped or heavily doped traditional semiconductors resulting in a high LSPR frequency corresponding to a short wavelength. Hence for good photocatalytic activity either self- doped or heavily doped metal chalcogenide or metal oxide plasmonic nano structures should be chosen as effective photocatalyst.

Conclusions:

Diversified Copper Sulfide was synthesized using $Cu(NO_3)_2.3H_20$ (0.2 gram) , Glutathione (0.307 gram) and CTAB by following Hydrothermal process. By varying CTAB concentration different samples were synthesized. Among these sample CS-0.5 with 0.5 gram of CTAB condition was able to give nanoflakes structure and good absorbance of solar radiation.

Here CTAB works as a surfactant to reduce the surface tension between the precursors and help mixing them well, due to which concentration of CTAB plays an important role in this synthesis. Conditions for obtaining a better sample with high photocatalytic activity is analyzed by varying CTAB concentration.

Acknowledgements:

Authors would also like to thank SRM Institute of Science and Technology (SRMIST)for providing the scanning electron microscope (SEM), XRD and UV-Vis research facilities.

References:

[1] Abbasi T and Abbasi S A 2011 Decarbonization of fossil fuels as a strategy to control global warming *Renew. Sustain. Energy Rev.* **15** 1828–34

[2] Winter C J 2009 Hydrogen energy - Abundant, efficient, clean: A debate over the energy-system-of-change *Int. J. Hydrogen Energy* **34** S1–52

[3] Navarro R M, Peña M A and Fierro J L G 2007 Hydrogen production reactions from carbon feedstocks: Fossil fuels and biomass *Chem. Rev.* **107** 3952–91

[4] Acar C, Dincer I and Naterer G F 2016 Review of photocatalytic water-splitting methods for sustainable hydrogen production *Int. J. Energy Res.* **40** 1449–73

[5] Liao C-H, Huang C-W and Wu J C S 2012 Hydrogen Production from Semiconductor-based Photocatalysis via Water Splitting *Catalysts* **2** 490–516

[6] Fajrina N and Tahir M 2019 A critical review in strategies to improve photocatalytic water splitting towards hydrogen production *Int. J. Hydrogen Energy* **44** 540–77

[7] Tang L, Deng Y, Zeng G, Hu W, Wang J, Zhou Y, Wang J, Tang J and Fang W 2016 CdS/Cu_2S co-sensitized TiO_2 branched nanorod arrays of enhanced photoelectrochemical properties by forming nanoscale heterostructure *J. Alloys Compd.* **662** 516–27

[8] Tang Y, Zhang D, Pu X, Ge B, Li Y and Huang Y 2019 Snowflake-like Cu 2 S/Zn 0.5 Cd 0.5 S p–n heterojunction photocatalyst for enhanced visible light photocatalytic H 2 evolution activity *J. Taiwan Inst. Chem. Eng.* **96** 487–95

[9] Joy J, Mathew J and George S C 2018 Nanomaterials for photoelectrochemical water splitting – review *Int. J. Hydrogen Energy* **43** 4804–17

[10] Osterloh F E and Parkinson B A 2011 Recent developments in solar water-splitting photocatalysis *MRS Bull.* **36** 17–22

[11] Barber J 2017 A mechanism for water splitting and oxygen production in photosynthesis *Nat. Plants* **3** 1–5

[12] Guan R, Zhai H, Li J, Qi Y, Li M, Song M, Zhao Z, Zhang J, Wang D and Tan H 2020 Reduced mesoporous TiO_2 with Cu_2S heterojunction and enhanced hydrogen production without noble metal cocatalyst *Appl. Surf. Sci.* **507** 144772

[13] Lakhotiya G, Belsare N, Rana A and Gupta V 2019 Cu_2S nanocrystals incorporated highly efficient nonfullerene ternary organic solar cells *Curr. Appl. Phys.* **19** 394–9

[14] Zada A, Muhammad P, Ahmad W, Hussain Z, Ali S, Khan M, Khan Q and Maqbool M 2020 Surface Plasmonic-Assisted Photocatalysis and Optoelectronic Devices with Noble Metal Nanocrystals: Design, Synthesis, and Applications *Adv. Funct. Mater.* **30** 1906744

[15] Zhao Y, Pan H, Lou Y, Qiu X, Zhu J and Burda C 2009 Plasmonic $Cu_{2-x}S$ nanocrystals: Optical and structural properties of copper-deficient copper(I) sulfides *J. Am. Chem. Soc.* **131** 4253–61

[16] Liu X, Reviews M S-C S and 2014 U 2011 Heavily-Doped Colloidal Semiconductor and Metal Oxide Nanocrystals: An Emerging New Class of Plasmonic Nanomaterials Article Type: Tutorial Review *Chem. Soc. Rev.* **43** 3908–20

[17] Yue Y, Zhang P, Wang W, Cai Y, Tan F, Wang X, Qiao X and Wong P K 2020 Enhanced dark adsorption and visible-light-driven photocatalytic properties of narrower-band-gap Cu_2S decorated Cu_2O nanocomposites for efficient removal of organic pollutants *J. Hazard. Mater.* **384** 121302

[18] Comin A and Manna L 2014 New materials for tunable plasmonic colloidal nanocrystals *Chem. Soc. Rev.* **43** 3957–75

[19] Müller H 1996 Optical Properties of Metal Clusters *Zeitschrift für Phys. Chemie* **194** 278–9

STRUCTURAL, OPTICAL AND ELECTRICAL PROPERTIES OF NANOCRYSTALLINE NICKEL OXIDE THIN FILMS PREPARED BY HYDROTHERMAL AND SOL-GEL METHODS

Ajaijawahar Kaliyaperumal, Nivin T S, Karthikeyan B, Gokuladeepan Periyasamy*

Hydrogen Storage Materials and Nanosensors Laboratory, Department of Physics and Nanotechnology, College of Engineering and Technology, SRM Institute of Science and Technology, Kattankulathur, Tamil Nadu, India, 603203

E-mail. gokuladp@srmist.edu.in

Abstract. The present work focuses on the comparison of the structural, optical and electrical properties nickel oxide thin films synthesized by hydrothermal and sol-gel methods and their characterizations. The films obtained from both methods are having a cubic crystal lattice. The morphological studies have shown that the films are having a uniform coverage with nanocrystalline grains. The NiO film obtained through the sol-gel method possess an optical band gap of 2.86 eV and its electrical resistance tends to decrease from 2.64GΩ to 11MΩ with an increase in temperature (RT to 150°C). The hydrothermally obtained film has an optical band gap of 3.14 eV and its electrical resistance change from 6.83GΩ to 3.86GΩ with an increase in temperature.

1. Introduction

Nanostructured thin films are very attractive for many applications in the field of energy, health, electronics, and environmental monitoring. Because they possess a high surface area to volume ratio to exhibit quantum effects. Nanomaterials exhibit different physical, chemical, and biological properties corresponding to material in bulk form. Fabrication of these materials into thin films using solutionbased preparation techniques spin coating, dip coating, hydrothermal growth, spray coating, ink-jet printing make the process less complex and cheap [1].

Nickel Oxide nanostructures were chosen owing to their useful properties [2] and make them a good candidate in different applications as solar absorber, as a catalytic material in fuel cells, as an electrochromic material in display devices and a gas sensitive material. The important properties of NiO films can be tailored for different applications by optimizing synthesis conditions [3].

As compared to study of electric properties of NiO thin films fabricated by complex methods such as sputtering, pulsed layer deposition, oxidation electrochemical and spray coating [4-5], reports on electrical performance of films derived from hydrothermal and sol-gel are less in number comparatively [6-8]. In the present work, we report the nanowall like NiO thin films prepared by a facile hydrothermal synthesis method and the nanoparticle-like NiO thin films were prepared by the sol-gel growth.

2. Experimental section

Materials

All chemicals and solvents purchased were of reagent quality and used without further purification. The nickel nitrate, ammonium fluoride, urea, isopropanol, ammonium hydroxide, polyethylene glycol 400, and triton X-100 were obtained from Sisco Research Laboratories Pvt. Ltd.

Methods NiO thin film synthesis by sol-gel process

For coating, a solution was prepared by mixing 0.1 mol/nickel nitrite hexahydrate [$Ni(NO_3)_2 \cdot 6H_2O$] into 20 ml of isopropanol and polyethylene glycol 400 [$H(OCH_2CH_2)_nOH$] taken in 1:1 ratio. The mixture was stirred at 40°C for 90 minutes. A colloidal solution of Nickel hydroxide was prepared by adding dilute ammonium hydroxide [NH_4OH] dropwise under continuous magnetic stirring. Triton X-100 [$C_{14}H_{22}O\ (C_2H_4O)_n$] was finally added in the solution for smooth spin coating. Before coating, the substrates were soaked into chromic acid for making hydrophilic nature followed by cleaning in distilled water. Then, the substrates baked at 100°C for 10 minutes. The final mixed solution was used to coat on the microscopic glass substrate (1×1 cm^2) using a spin-coater at a rotation speed of 1500 rpm for 20 s and 3000 rpm for 60 s. After coating each layer, the sample was heated at 350°C for 5 min to evaporate solvents. The resultant films were annealed at 350°C for 3 hours in the air [9-10]. **NiO thin film synthesis by hydrothermal growth**

In this work, 0.15 mmol of $Ni(NO_3)_2{\cdot}6H_2O$, 0.3 mmol of NH_4F, and 0.75 mmol of $CO(NH_2)_2$ were dissolved in 35 mL of distilled water under continuous mixing by a magnetic stirrer. After obtaining a homogeneous solution, it was poured into a Teflon-lined stainless-steel autoclave with well cleaned glass substrates (1×1 cm^2) immersed into the reaction solution. One side of the substrate was protected from solution contamination using polytetrafluorethylene (Teflon) tape. Finally, the autoclave was sealed and maintained at 110°C for 5 hours and allowed to cool down to the room temperature naturally. Finally, the substrates from autoclave were thoroughly cleaned with distilled water and dried in hot air oven subsequently annealed at 350°C in the air for 3 hours [11].

3. Result and discussion

Structural properties

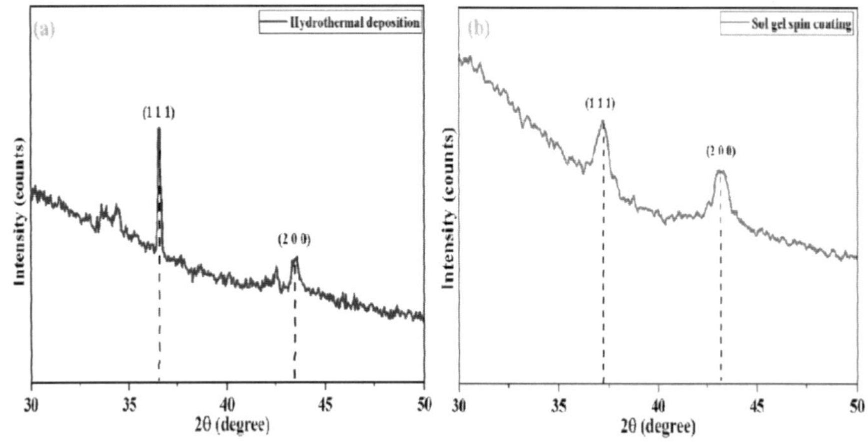

Fig. 1. XRD pattern of NiO thin films by (a) hydrothermal growth (b) sol-gel process

The following results were observed from the X' pert powder XRD Panalytical system which was indexed to the standard NiO data of JCPDS card number 01-078-0423 and 01-073-1519 for the sol-gel process [910] and hydrothermal deposition respectively [11]. As shown in Fig.1. (a) and (b), it was observed that the both NiO films, XRD patterns show two peaks at 2θ values of 37° and 43° corresponding to (111) and (200) planes respectively. From the XRD data, it was confirmed that the NiO thin films possessed a wellcrystallized cubic structure (a (Å) = 4.17). The average crystal size calculated using Scherrer's formula was found to be 13.44 nm and 3.145 nm for the films prepared by sol-gel process and hydrothermal growth respectively with similar d-spacing.

IOP Conf. Series: Materials Science and Engineering 1219 (2022) 012013 doi:10.1088/1757-899X/1219/1/012013

Morphological properties

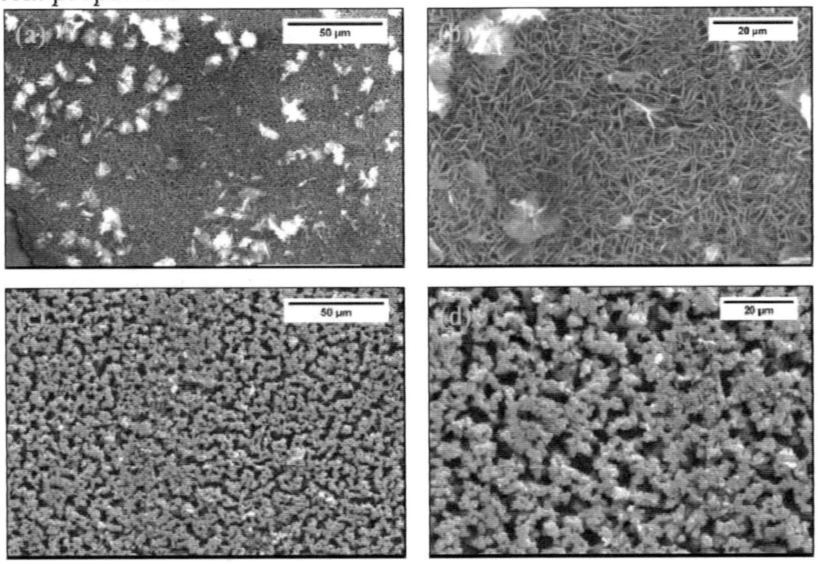

Fig. 2. (a) and (b) SEM images of NiO thinfilm by hydrothermal growth (c) and (d) NiO thinfilm by sol-gel process

The surface morphology and crystallographic information of the films were examined using DESKTOP Mini-SEM with EDS (SNE- 3200M, SEC) system. As depicted in Fig.2. (a) and (b) hydrothermal deposition produced NiO thin films with a porous nanowall structure composed of nanoflakes and the solgel derived film as shown in Fig.2. (c) and (d) exhibit cluster or circular grains like shapes [9-11]. The products obtained through the spin coating method and hydrothermal growth method were annealed at 350 ºC for 3 hours. Similar reports [9-11] also confirm the obtained results. As compared to sol-gel coated film, hydrothermal grown film showed continuous surface structure.

Elemental Analysis

The film composition was qualitatively analyzed for finding the elements present with the Energy-Dispersive Spectroscopy (EDS) during the SEM imaging.

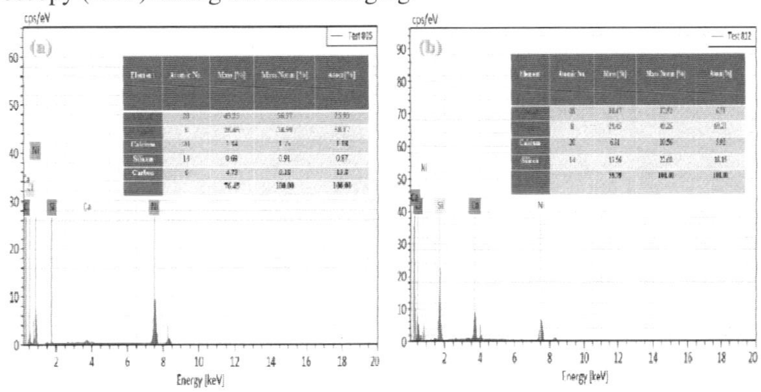

Fig. 3. (a) EDS images of NiO thinfilm by hydrothermal growth (b) NiO thinfilm by sol-gel process

The contribution from the microscopic slides in the form of Si and Ca in the EDS spectrum along with Carbon contamination as presented in Fig.3. (a) and (b) EDS spectrum. The elemental composition for both films exhibits oxygen-rich stoichiometry. And the atomic ratio O/Ni approximately at 58.17/25.93 and 69.21/6.71 for sol-gel process and hydrothermal deposition respectively. Both films show Ni/O mass normality ratio as 56.57/34.59 and 17.52/49.25 for sol-gel process and hydrothermal deposition respectively [10-11].

Optical properties

UV–visible measurements were carried out with JASCO UV-V-750 spectrophotometer for both the NiO thin films. The optical properties of the NiO films prepared by the sol-gel synthesis and hydrothermal growth were analyzed. The room temperature optical transmittance of the samples is shown in Fig.4. (a) and (b). By the results, the sol-gel spin-coated sample having transmittance around 95% at the visible region, and it gradually decreases into around 80% transmittance.

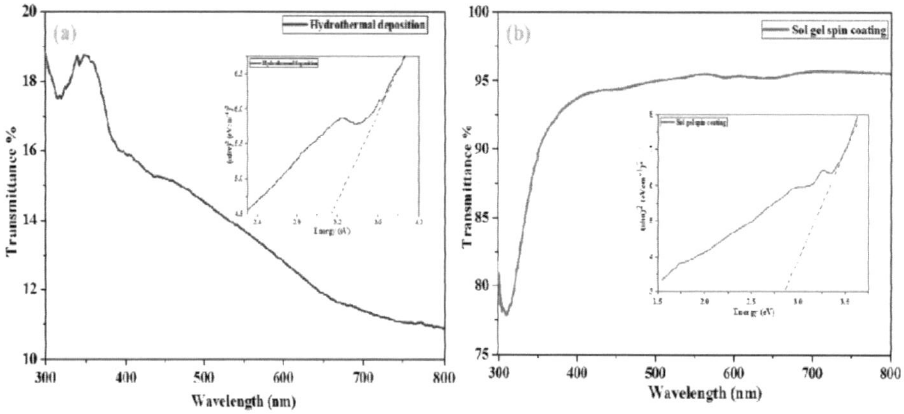

Fig. 4.UV Visible transmittance spectra of NiO thin films by (a) hydrothermal growth (b) sol-gel process

While the hydrothermally coated sample having 11-16% transmittance only at the visible region and gradually increases to 18% transmittance. In both the samples, the transition starts at 390 nm. It is generally known that Oxygen Nickel ratio bound to affect transmittance of the NiO thinfilms [12]. Amit Kumar Srivastava et al.earlier reported similar transmittance for the NiO films [13-14].

The tauc plot of $\alpha h\upsilon^2$ vs $h\upsilon$ is used to determine the energy bandgap of the samples as shown in Fig.4. (a) and (b) that the linear portion of the curve when extrapolated intersects the x-axis and gives the energy bandgap Eg. The Eg value of the films will be higher than the bulk samples [14]. The energy bandgap of the hydrothermal deposited sample and sol-gel deposited samples are found to be 3.14 eV and 2.86 eV respectively. So varying the NiO films synthesis method can be a potential method to finely tune the energy bandgap based on the bandgap engineering requirements.

Electrical properties

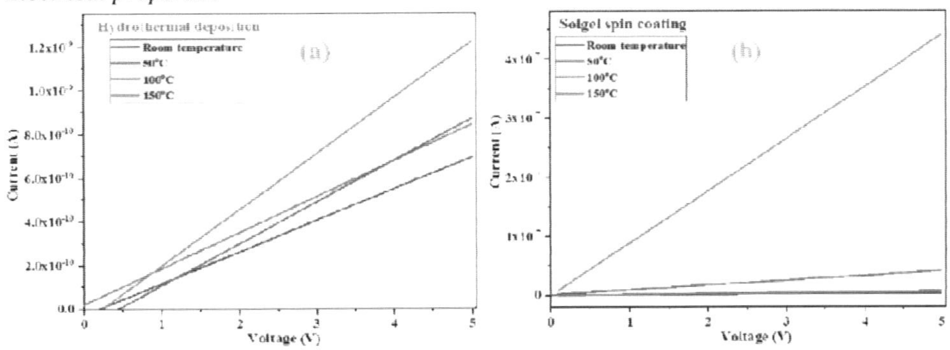

Fig. 5. I-V characteristics of NiO thinfilm by (a) hydrothermal growth b) Sol gel growth

The temperature dependence of the films are studied using Keithley Source meter Model 2636-B and the IV characteristics were plotted in Fig. 5 (a) and (b) after linear fitting. Since the NiO is a p-type semiconductor, it shows an increase in its conductivity with respect to increased temperature. The sol-gel spin-coated film with an energy band gap of 2.76 eV shows a decrease in resistance from 2.64×10^9 to 11.33×10^6 ohms once 150 °C is reached while the hydrothermally deposited film with an optical energy bandgap of 3.14 eV shows changes from 6.93×10^9 to 3.89×10^9 ohms on the same temperature range. The ohmic behavior has been shown by the films. The low resistance value achieved by the sol-gel spin coated film possibly due to multiple coatings perfomed. A. Hakim et.al. also reported a similar decrease in the resistivity of the NiO films on the increase in temperature for undoped NiO thin films [15-16].

Table 1 Comparison of resistance for hydrothermal growth and sol gel process films

Fabrication method	Room temperature	50 °C	100 °C	150 °C
Sol gel	2.64 GΩ	0.95 GΩ	128.62 MΩ	11.33 MΩ
Hydrothermal	6.93 GΩ	6.08 GΩ	5.24 GΩ	3.89 GΩ

Conclusion

NiO thin film was prepared by spin coating technique and hydrothermal deposition. The structural, morphological, optical and electrical properties were characterized by X-Ray Diffraction, Scanning Electron Microscope with EDS and UV Visible transmittance and I-V characteristics respectively. XRD studies indicated that prepared product well crystallized with cubic structure and the optical transmittance propertied showed different behavior possibly owing to Nickel oxygen ratio. The drastic decrease in optical transmittance in hydrothermally deposited NiO film required further investigation.

Acknowledgement

The authors thank the DST-SERB (DST-SERB-SB/S2/CMP/-073/2013) and Department of Physics and Nanotechnology, SRMIST for their experimental and characterization support. The authors also thank Prof. A. Karthigeyan for his support and guidance.

References

1) Ion N. Mihailesc, 2019 Nanomaterials 9 10 1427.
2) Teboho P. Mokoena, 2019 Journal of Alloys and Compounds 805, 267-294.
3) Mari Napari, Tahmida N. Huq, Robert L. Z. Hoye, 2020 Info.Mat. 1–41.
4) Lalithambika, K.C., Thayumanavan, A, 2016 Der Pharma Chemica, Volume 8 2 415-421.
5) Michelle, P. B., Hugo, N., Nina, C. B, 2016 Int. J. Electrochem. Sci., 11, 6636–6647.
6) Mohammad, K. and Leila, M, 2014 Research and Reviews: J. of Mat. Sci. Vol. 2 1 37.
7) Sajjadi, Seyed Pooyan, 2005 J. of Poly. Eng. and Tech. 13 38-41.
8) Pilban-Jahromi S., Huang N.M., 2012 J. Nanomaterials 173825.
9) Amir M, Soleimanpour, Ahalapitiya H, 2013 App. Surf. Sci. 276 291– 297.
10) Amir M, Soleimanpour, Yue Hou, 2013 Sensors and Actuators B 182 125-133.
11) Z. Chen, A. Xiao, Y. Chen, 2013 J. of Phy. and Chem. of Solids, 74 11 1522-1526.
12) Maria Isabel Pintor-Monroy, 2019 ACS Appl. Mater. Int., 11, 27048–27056.
13) Srivastava, Amit Kumar; Thota, 2008 J. of Nanosci. & Nanotech. 8 8 4111–4115.
14) M. Jlassi, I. Sta, M. Hajji, 2014 Mat. Sci. in Semicond. Process. 21 7-13.
15) G.Sinha, K.Adhikary, and S.Chaudhuri, 2006 J. Phys.: Condens. Matter 18 2409.
16) A. Hakim, J., 2009 Hossain Renewable Energy 34 2625–2629.

Solution based process of ZnO nanostructured thin films: A review

Sourav M, Leena B and Bakiyaraj G
Department of Physics and Nanotechnology, SRM Institute of Science and Technology, Kattankulathur, Tamil Nadu, India, 603203.

Authors correspondence: bakiyarg@srmist.edu.in

Abstract. The solution-based technique is recently gaining momentum for the fabrication of optoelectronic materials due to its cost-effectiveness, fabrication of large-scale thin films, processing at low temperature, low cost and simplicity of the process requirements. The properties of thin films depend on the diverse parameters like substrate, deposition time, source materials, doping agent, pH and temperature at which the process is done. ZnO nanostructured thin film has been widely used in organic and hybrid solar cells due to its precise characters such as reduced cost, simple processing, non-toxicity, better optoelectronic properties and good stability. In this review article, we have focused on different solution-based methods which are capable of yielding good quality thin films of ZnO with different morphology for various applications.

Keywords: ZnO thinfilms, SILAR, CBD, spin coating.

1. Introduction

Zinc Oxide (ZnO) Nanostructured materials have fascinated great impact in the last few decades because of interesting structural, morphological, optical, electrical properties, when compared to bulk counterparts and also their size and shape-dependent optical and electrical characteristics that have really motivated many researchers to investigate ZnO thin films. The qualities of the thin films were highly dependent on material structure such as crystallite size, morphology, orientation, density and roughness. Nanostructured ZnO is an applications in several fields, particularly photovoltaic cells, electro-chemical cells, UV lasers, LED (light emitting diodes), display device, chemical-sensors, bio-sensors, photo-catalysis, etc and its investigation owing to interest in materials research [1-2]. Because of its numerous interesting properties and potential uses, different morphological nanostructured ZnO thin films were prepared so far. These techniques also offer a low price, large-scale production, reliability, repeatability and simplicity [3, 4]. In the current review paper, a survey of nanostructured ZnO thin films deposited by chemical bath deposition, dip coating, spin coating, spray coating and SILAR method are presented. The morphological background for nanosheets, nanoflakes, nanobundles, nanowalls, nanoflowers, nanorods of ZnO thin films is discussed. In addition, the different applications of ZnO nanostructured thin films are summarized.

2. Materials and Methods

Fig. 1 shows the graphical representation for solution processing of thin films techniques. Chemical Both Deposition method, it has two step processes: (i) nucleation and (ii) particle growth. In this technique a cleaned substrate was placed inside the solution containing the precursors of anions and cations. The temperature of solution, pH, deposition time and concentration of solution is playing important role in the case of formation of films. After the film growth, the substrate has to be removed from the solution and thoroughly washed with de-ionized water. Then it was packed in tight container. In Spray pyrolysis, the chemical compound prepared as solution and it was sprayed on the substrates with very tiny nozzle. During the deposition process the substrate was placed on hot surface, on substrate the reaction has been occurred to form thin films. In this process the thickness of the thin films mainly depends on the deposition time, nozzle size and the growth temperature of substrate. SILAR cycle contains 4 steps: step (i) the cleaned substrate is immersed into cationic precursor, step (ii) washed with deionized water to eliminate the loosely bound species, step (iii) immersed into anionic precursor and step (iv) again washed with de-ionized water.

IOP Conf. Series: Materials Science and Engineering 1219 (2022) 012014 doi:10.1088/1757-899X/1219/1/012014

The thickness of the films is depends on the number of SILAR cycle processing, deposition time and pH of the precursor. In dip coating, the substrate was dipped within the solution for the depositing material at a regular speed. The substrate stayed inside the solution for deposition of films and began to withdraw. The thickness of the coated films is depends on dipping speed, withdrawal speed, dipping duration, solution temperature and number of cycles. Spin coating is a method used to disperse uniform thin films by centrifugal force on flat substrates. A material solution is dispensed onto the middle of a wafer, and is then rotated at a high speed. Rotation occurs until the layer is spun off by the excess solution and the desired film thickness is left on the substrate. The solvent added is normally volatile and evaporates during deposition.

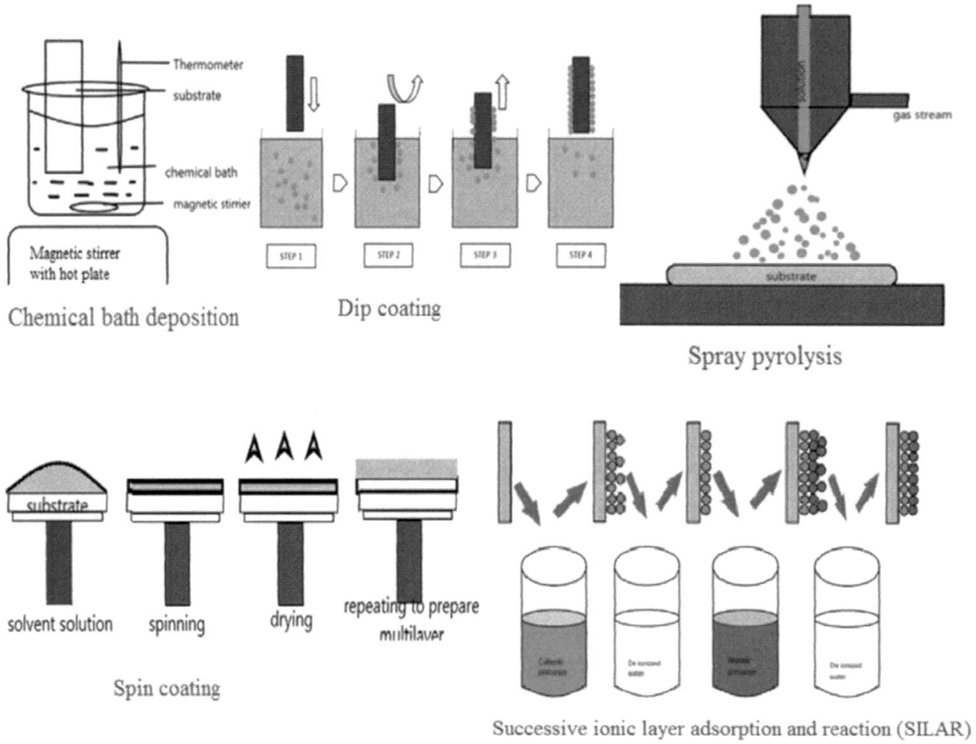

Fig. 1 Solution processing of thin films techniques

3. ZnO nanostructured thin films by solution process

This section deals with solution process of different morphological ZnO nanostructured thin films and its remarkable results. The detailed growth conditions and the important properties are discussed in table 1-5.

IOP Conf. Series: Materials Science and Engineering 1219 (2022) 012014 doi:10.1088/1757-899X/1219/1/012014

Table 1 Different Morphological ZnO thin films prepared by CBD and remarkable results

Morphological type	Growth condition	Remarks	Ref.
Nanorods	ITO substrate Temperature 90 °C Reaction time 2 hrs	The pure ZnO has Schottky barrier of 0.55 eV and Samarium doped nanorods 0.72 eV.	[5]
Nano-flowers	Glass Substrate Temperature 80 °C Time 2 hrs	Pure ZnO has flower like shape and it has been changed to hexagonal form after adding the Eu^{3+} dopant.	[6]
Nanowire	Glass substrate Temperature 70 °C Reaction time 10 hrs	The fabricated Ag/ZnO/Ag photoconductive detector illustrates high photocurrent (113.83 µA) with 5 volts applied bias.	[7]
Nanorod (Well aligned)	Si substrate Temperature 95 °C Time 5 hrs	The room temperature photoluminescence shows only strong near band edge emission and it indicate the grown ZnO nanorods are high optical quality.	[8]
Nanorods to porous thin films	Glass substrate Temperature 60 °C Time 24 hrs	Influences of citric acid adding reagent with the bath solution the structure only form as microspheres.	[9]
Nano-flower	Temperature 80 °C	The obtained estimated value of grain size and energy band gap values dependents on the zinc acetate concentration.	[10]
Nanorods	Si substrate Temperature 110-210 °C	High intensity with low FWHM of (002) peak indicated the high crystallinity of the material.	[11]
Nanorods	Glass Substrate Temperature 60 °C Time 45 mints	Optical studies revealed that the potassium doped ZnO thin films are preferable for various optoelctronics and sensing applications.	[12]

Table 2 Different Morphological ZnO thin-films prepared by dip coating and remarkable results

Morphological type	Growth condition	Remarks	Ref.
Nano-flakes	Aluminium Substrate Temperature 90 °C Reaction time 4 h	Honeycomb structured ZnO nanoflake coated on boro-silicate glass capillary and it has been used as bio sensor to study intracellular glucose concentrations for living cells. It has a very good reproducibility, selectivity and reliable stability.	[13]
Nano-sheets	Glass substrate Temperature 80 °C Reaction time 24 h	photocatalytic degradation of 4-chlorophenol on nanosheet-based films is ~ 2 times greater than those on sol-gel technique prepared nano-columnar ZnO films.	[14]
Rose like Nano flowers	Bamboo slice Dipped for 5 min Dried at 80 °C for 5 hrs	In order to increase surface roughness, the large-scale rose-like ZnO nanoflowers were first planted on the bamboo surface and then the surface was subsequently treated with low surface energy fluoroalkylsilane content.	[15]
Nano rods	Dipped for 30 s Dried at 180 °C for 30 min	The alcohol, acetone and benzene vapor concentration also affects their gas sensitivities, but as the vapor concentration increases, the sensitivities do not rise linearly. But the response time is approximately 10 s for acetone and alcohol, but the time for benzene is comparatively longer.	[16]
Nano rods	Glass substrate Room Temperature	PVP-modified precursor solutions, the pure and Al doped thin films to exhibit highly preferential orientation along the c-axis.	[17]
Nano rods	Glass substrate 20 mm/min Dried for 1hr at 400 °C	In comparison with the undoped ZnO films, the Al doped ZnO thin films display better photocatalytic behaviour.	[18]
Nanorods	Glass substrate Room Temperature Time 3 min	The response and recovery time for acetone is in the order of seconds. The lowest detection limit for ZnO thin film is usually at room temperature was observed at 2 ppm of acetone with a response of 1.08.	[19]

Table 3 Different Morphological ZnO thin films prepared by Spin coating and remarkable results

Morphological Type	Growth condition (Disc Speed in rpm & Growth time in second)	Remarks	Ref.
Nano rods	Glass substrate 700-1600 rpm for 10 s & dried at 100 °C	Spinning speed, number of deposition cycles and annealing temperature may influence the film properties.	[20]
Micro and Nano particles	Glass substrate 2400 rpm for 30 s & dried at 100 °C	The electrical measurement was measured by four point probes methods. It is apparent that by increasing the number of ZnO layers, the average resistivity decreases.	[21]
Spherical porous structure	Glass substrate 2000 rpm for 40 s & dried on a hot plate at 200 °C for 5 min	Gas sensing efficiency was investigated for different gases (NO_2, NH_3, CH_3OH, Cl_2 and H_2S) at temperatures. There was a high response for100 ppm concentration of NO_2 gas at 200 °C	[22]
granular structure	Glass substrate 3000 rpm & dried at 300 °C	The results of FESEM and AFM indicated the improvements in surface morphology and decreases in grain size as the degree of doping (Al and Ga) was increased.	[23]
Spherical porous	Glass substrate 2000 rpm for 30 s & dried at 200 °C	Synthesized ZnO thin film gas sensing output was initially tested for H_2S gas at various operating temperatures as well as concentrations. In contrast to other gases such as CH3OH, Cl2, NH3, LPG, CH3COCH3, CH3COCH3, the optimum gas response to H2S gas is achieved at 300 °C operating temperature, at 100 ppm gas concentration	[24]

IOP Conf. Series: Materials Science and Engineering 1219 (2022) 012014 doi:10.1088/1757-899X/1219/1/012014

Table 4 Different Morphological ZnO thin films prepared by Spray coating and remarkable results

Morphological type	Growth condition	Remarks	Ref.
Nanowall	Glass substrate Temperature 95 °C	3D ZnO nano-wall architecture developed on a transparent conductive substrate used for dye-sensitized (DSSC) and perovskite solar cell electron transport (PSCs)	[25]
Nano sheets	FTO glass substrate Temp. 170 to 200 °C 30 min	The morphology of the nanosheets is significantly impacted by growth temperature. The PCE_{max} of 15.91%, is the highest efficiency firstly ever reported for perovskite solar cells based on ZnO Nanosheets using single-step electrospray deposition.	[26]
Nano grains	Glass substrate Temperature 450 °C	Super-hydrophobic behaviour with contact angle (CA) value 154 ° is seen in the film. PL confirmed the lower defect density in the samples and optimized ZnO thin film exhibited high sensing response to C_3H_8O as compared to C_2H_5OH and C_3H_6O.	[27]
Nano grains	Glass substrate time 7 to 25 min	Above 600 K, the individual crystals grow preferentially perpendicular to the substrate with their crystallographic c-axis.	[28]
Spherical shaped	Glass substrate 40 to 120 min	Substrate temperature, solution flow rate (droplet size), substrate-to-nozzle distance, applied voltage and deposition time are important factors of electro-spraying process that can affect the morphology of ZnO thin films. Al-doped ZnO thin film obtained the power conversion efficiency (PCE) of 12.0% in perovskite solar cells.	[29]

IOP Conf. Series: Materials Science and Engineering 1219 (2022) 012014 doi:10.1088/1757-899X/1219/1/012014

Table 5 Different Morphological ZnO thin films prepared by SILAR method and remarkable results

Morphological type	Growth condition	Remarks	Ref.
Nano rods	60-120 deposition cycle Temperature 70 °C Glass substrate	The deposition using SILAR of well-oriented ZnO nanorods are free of seed layers has a huge advantage over the other multi-step deposition methods. The orientation and alignment variations in ZnO nanorods with increasing the number SILAR cycles. Better PEC output obtained at 1 V with photo-current of 0.19 mA/cm^2 and maximum photoconversion efficiency of 0.084 percent at 0.45 V	[30]
Nano rods	Glass substrate	The obtained films have best homogeneity, compactness, highest c-axis orientation and least crystallite size.	[31]
Rice like and flower like structure	Glass, Quartz and Si-SiO$_2$ substrate 100 cycle	The flower like architecture was created from rice-like structure and do not alter considerably with the type of substrate. Even though there are several variations in grain sizes and the shapes were same.	[32]
flower and spindle-like	Glass substrate 15-25 cycle Temperature 80 °C	SILAR (25 cycles) deposited seed layered substrate was shown higher orientation along the directions of (0 0 2), which seed layer used for the growth of ZnO Nanorods by placing them in a chemical bath deposition.	[33]
Nanowires	Temperature 80 °C 80 cycles	Sr doped ZnO showed a change in the orientation of crystals from (002) to (101). 5 wt% Sr-doped ZnO film provided an excellent response to the detection of ammonia gas at room temperature	[34]
Nanorod like structure	Temperature 85 °C 12 cycles	SEM Images showed that the silica substrates are well covered by homogeneous hexagonal ZnO nanorod-like structure.	[35]
Rod to Nanosheet	30 cycles	Morphological changes from rod-like structures to 3D nanosheet hierarchical structures were succeeded by Al doping. It has resulted in an increase in SHG and THG	[36]
spherical grains	30 cycles	ZnO film surface morphology was significantly influenced by the dipping time period. The correlation among the surface morphology and liquid petroleum gas (LPG) sensing properties has been identified.	[37]

IOP Conf. Series: Materials Science and Engineering 1219 (2022) 012014 doi:10.1088/1757-899X/1219/1/012014

4. Conclusion

This article enlighten various solution based thin films processing methods to produce ZnO nanostructured thin films with different surface morphology like nanograins, nanosheets, nanoflakes, nanobundles, nanowalls, rose like flowers, rise form flower, nanorods, nanowires and so on. These methods are reliable, ease to produce large area thin films, economical and possible to manufacture good quality. The information summarized in the table pointed out that, the different morphological formation could be dependent on various parameters such us growth temperature, time, deposition cycles, processing dopent atom, substrates chosen for growth and so on. In addition, various devices such as photoconductive detectors, solar cell, photocatalysis, chemical sensors and bio sensors were discussed in detail. Such solution processing multifunctional ZnO thin films are one of the promising candidates for the diverse applications.

Reference

[1] Özgür Ü, Alivov YI, Liu C, Teke A, Reshchikov MA, Doğan S, Avrutin V, Cho S-J, and Morkoç H 2005 J. Appl. Phys. 98 041301

[2] Ong CB , Ng LY and Mohammad AW 2018 Renew. Sustain. Energy Rev. 81 536-551

[3] Wahab R, Ansari SG, Kim YS, Khang G and Shin HS 2008 Appl. Surf. Sci. 254 2037.

[4] Wahab R, Ansari SG, Kim YS, Seo HK, Kim GS, Khang G and Shin HS 2007 Mater. Res.Bull. 42 1640.

[5] Ahmed MAM, Mwankemwa BS, Carleschi E, Doyle BP, Meyer WE and Nel JM 2018 Mater. Sci. Semicond. Process. *79* 53-60

[6] Koao LF, Dejene BF, Swart HC, Motloung SV and Motaung TE 2016 Opt. Mater. *60* 294-304

[7] Shaikh SK, Inamdar SI, Ganbavle VV and Rajpure KY 2016 J. Alloys. Compd. *664* 242-249

[8] Li Q, Bian J, Sun J , Wang J, Luo Y, Sun K and Yu D 2010 Appl. Surf. Sci. *256*(6) 1698-1702

[9] Wang H and Xie C 2006 J. Cryst. Growth *291*(1) 187-195

[10] Koao LF, Dejene FB and Swart HC 2014 Mater. Sci.Semicond. Process. *27* 33-40

[11] Lee TH, Ryu H and Lee WJ 2014J. Alloys. Compd.*597* 85-90

[12] Shanmuganathan G, Banu I S, Krishnan S and Ranganathan B 2013 J. Alloys. Compd. *562* 187-193

[13] Fulati A, Ali SMU, Asif MH, Alvi NH, Willander M, Brännmark C, Strålfors P, Börjessonb SI, Elinder F and Danielsson B 2010 Sens. Actuators. B.Chem.*150* (2) 673-680

[14] Hynek J, Kalousek V, Žouželka R, Bezdička P, Dzik, P, Rathouský J, Demel Jand and Lang K 2014 Langmuir. *30* (1), 380-386

[15] Jin C, Li J, Han S, Wang J and Sun Q 2014 Appl. Surf. Sci. *320* 322-327

[16] Ge C, Xie C and Cai S 2007 Mater. Sci. Eng. B*137* 53-58

[17] Ratana T, Amornpitoksuk P, Ratana T and Suwanboon S 2009 J. Alloys. Compd. *470* 408-412

[18] Islam MR, Rahman M, Farhad SFU and Podder J 2019 Surf. Interfaces. *16* 120-126

[19] Muthukrishnan K, Vanaraja M, Boomadevi S, Karn R K, Singh V, Singh PK. and Pandiyan K 2016 J. Alloys. Compd.*673*,138-143

[20] Smirnov M, Baban C and Rusu GI 2010 Appl. Surf. Sci. *256*(8) 2405-2408

[21] Khan MI, Bhatti KA, Qindeel R, Alonizan N and Althobaiti HS 2017 Results. Phys. *7* 651-655

[22] Patil NP, Nimbalkar AR and Patil MG 2018 Mater. Sci. Eng. B. 227 53-60

[23] Ng ZN, Chan KY, Low CY, Kamaruddin SA and Sahdan MZ 2015 Ceram. Int. *41* S254-S258

[24] Nimbalkar AR and Patil M G 2017 Physica. B.Condens. Matter. *527* 7-15

[25] Islavath N, Das D, Joshi SV and Ramasamy E 2017 Mater. Des. *116* 219-226

[26] Mahmood K and Khalid A 2018 Mater. Lett. *224* 78-81

[27] Tarwal NL, AVRajgure, AI Inamdar, RS Devan, IY Kim, SS Suryavanshi, YR Ma, JH Kim and PS Patil 2013 Sens. Actuator. A. Phys.199 67-73

[28] Andrade E and Miki-Yoshida M 1999 Thin Solid Films *350*(1-2) 192-202

[29] Mahmood K, Swain BS and Jung HS 2014 Nanoscale *6*(15) 9127-9138

[30] Desai MA, Sharma V, Prasad M, Jadkar S, Saratale GD and Sartale SD 2020 Int. J. Hydrog. Energy. *45*(10) 5783-5792
[31] Mitra P and Mondal S 2013 Prog. Theor. Exp. Phys.1 *1*17–31
[32] Jiménez-García FN, Londoño-Calderón CL, Espinosa-Arbeláez DG, Del Real A and Rodríguez-García ME 2014 Bull. Mater. Sci. *37*(6) 1283-1291
[33] Kumar PS, Raj AD, Mangalaraj D and Nataraj D 2008 Appl. Surf. Sci. *255*(5) 2382-2387
[34] Radhi Devi K, Selvan G, Karunakaran, Poul Raj IL, Ganesh V and Alfaify S 2020 Mater. Sci. Semidond. Process. 119 105117
[35] Şahin B, Soylu S, Kara M, Türkmen, Aydin R and Çetin H 2021 Ceram. Int. 47 341-350.
[36] Jellal I , Nouneh K, Jedryka J, Chaumont D and Naja J 2020 Opt. Laser. Technol. 130 106348
[37] Shinde VR, Gujar TP, Lokhande CD, Mane RS and Han SH 2007 Sens. Actuators B. 123 882-887

IOP Conf. Series: Materials Science and Engineering 1219 (2022) 012015 doi:10.1088/1757-899X/1219/1/012015

Thermal and Layer Thickness Dependence of Defect and Band Edge Transmissions in 1D PCs Based on ZnS/CaF_2 Multilayers

Shilpa R[1], Abhinaya[1], Yamuna[1], Renilkumar Mudachathi[1,2*]

[1]Stuctured Electromagnetic Materials and Devices Lab, Department of Physics and Nanotechnology, SRM Institute of Science and Technology, Kattankulathur, Tamilnadu, India, 603203

[2]Metamaterials Laboratory, RIKEN center for advanced photonics, RIKEN, Wako, Japan, 351-0198

**corresponding author email:* renilkum@srmist.edu.in

Abstract. Alternating layers of different materials with sub wavelength thickness in one dimension are used for controlling the propagation of electromagnetic waves owing to their ability to selectively transfer and confine photons of specific wavelengths, called photonic band gap (PBG). Temperature independent large PBG in the visible wavelength range is highly desired for many applications and we have theoretically investigated the thermal and layer thickness dependence on PBG and proposed a novel strategy to choose materials based on their coefficients of thermal expansion and thermo optic effect. The thermal dependence of PBG can be eliminated by making the spectral shift due to thermo optic effect equal and opposite to that of the thermal expansion effect. We have chosen ZnS/CaF_2 multilayers for the design of 1D photonic crystal (PC) and our calculations have shown that the spectral shift due to thermal fluctuations could be significantly reduced by the negative thermo optic coefficient of these materials compensated by the positive thermal expansion coefficient. The proposed 1D PC shows a temperature dependent spectral shift of ~14 pm/°C in the visible range.

Key words: One dimensional photonic crystal, visible range, temperature independent PBG, thermal drift, passive thermal compensation

1. Introduction

Multiple layers of alternatingly arranged high and low refractive index (RI) medium have the capacity to impede the propagation of a specific spectrum of light. Such structures, known as one dimensional Photonic Crystals (PC) [1,2] give rise to a unique Photonic band Gap (PBG) owing to their geometrical structure, periodicity as well as their RI contrast. The structure can be further manipulated by breaking the periodicity with a defect to produce singular or multiple transmission modes within the forbidden region [3]. Over the years various studies on the dynamics of the transmission modes [4–6] within the band gap have been exploited to give way to several tunable optoelectronic applications [7–11]. External factors such as temperature, pressure and strain affect the mechanical and material properties of the layers constituting the PC and enable the tuning of the PBG [12–14]. On the downside, these external factors can become a drawback in applications that require stable transmission modes. Thermal factor in particular has a two-fold effect on the PBG in the form of structural and RI variation in the PC layers arising from thermal expansion and thermo optic effects respectively. Temperature compensating techniques [15–17] are then a crucial aspect to consider in the design of applications such as wavelength division multiplexing (WDM) [18], narrow bandwidth filters [19] or sensors requiring precise error free measurements [20,21].

Most studies on temperature dependence of PBG are in the communication wavelength range. However Sepehr Razi and Fatemeh Ghasemi [22] report a ~5nm red shift of the PBG in a temperature range from 50°C to 300°C. This gives a thermal sensitivity S = $\Delta\lambda / \Delta T$~20pm/°C.

In this paper we propose a passive thermal compensation technique to significantly reduce the thermal drift of PBG and transmission modes in a 1D PC designed in the visible spectrum. Transfer Matrix Method (TMM) [23] was implemented to theoretically investigate the thermal dependence on PBG and we propose a novel strategy to choose materials based on their coefficients of thermal expansion and thermo optic effect. We have chosen ZnS/CaF_2 multilayers for the design of 1D photonic crystal (PC) and our calculations have shown that the thermal drift could be significantly reduced by the compensation between the negative thermo optic coefficient and positive thermal expansion coefficient of these materials.

2. Theoretical Methodology and Simulation Parameters:

The proposed 1D PC consists of 8 bilayers of alternatingly stacked high and low RI medium as shown in the figure (1). The layer thicknesses $d_H$and d_H represent the thickness of higher and lower RI layers respectively and n_H and n_L represent the RI of denser and rarer medium respectively. The period of the PC is then given as$\Lambda = d_H + d_L$.

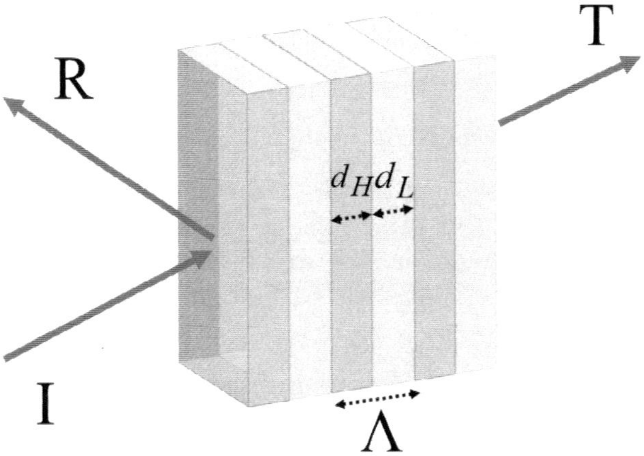

Figure 1: One dimensional Photonic crystal with alternating layers of high (blue) and low (yellow) refractive index layers. The thickness of the layers are given as d_H, d_L and the periodicity $\Lambda = d_H + d_L$. Incident ray (I) , Reflected ray (R) and Transmitted Ray (T) are shown as ray lines in red.

Successive reflection from the multiple interfaces of the 1D PC constructively interfere to produce the PBG. Theoretically the PBG of the PC can be calculated using TMM which relates the electromagnetic fields on either side of the PC using a 2 × 2matrix called as the Transfer Matrix (M) [23].

$$\begin{bmatrix} a_{i+1} \\ b_{i+1} \end{bmatrix} = \begin{bmatrix} M_{11} & M_{12} \\ M_{21} & M_{22} \end{bmatrix} \begin{bmatrix} a_i \\ b_i \end{bmatrix} \qquad (1)$$

The transfer matrix is calculated from the transmission matrix $D_j$and the propagation matrix P_j of each layer where$j = 0, H, L$represents the surrounding medium, high index medium and low index medium respectively. $D_j$and $P_j$are 2 × 2matrices given as

$$D_j = \begin{bmatrix} 1 & n_j \\ -n_j & 1 \end{bmatrix} \qquad (2)$$

$$P_j = \begin{bmatrix} e^{ikn_j d_j} & 0 \\ 0 & e^{-ikn_j d_j} \end{bmatrix} \qquad (3)$$

Here $k = 2\prod \square \lambda$ is the propagation vector for the wavelength $\lambda$and $n, d$are the RI and thickness of the corresponding layers. M is then given by,

$$M = D_0^{-1} ¿ \quad (4)$$

where m is the number of bilayers of the 1D PC. The reflection coefficient for any wavelength is then obtained from the elements of matrix M as follows,

$$r = \frac{M_{21}}{M_{11}} \quad (5)$$

from which the reflection intensity $R = |r|^2$ is calculated for that particular wavelength.

The RI variation arising from thermal parameter is calculated from the temperature dependent Sellmeier equations for ZnS and CaF_2 [24,25] and incorporated in equations (2) and (3). Layer thickness variation due to thermal expansion is given by the expression,

$$d_H = d_0 + \alpha \cdot d_0 \quad (6)$$

Where α is the coefficient of thermal expansion and d_0 is the thickness of the layer without the influence of temperature. The α values for ZnS and CaF_2 [26,27] at any given temperature is used to calculate the d_H values after expansion from equation (6) and incorporated in equation (3). Both effects contribute to the thermal sensitivity S = $\Delta\lambda / \Delta T$ of the structure which can be obtained from the spectral shift due to temperature calculated using TMM.

3. Result and Discussion

Theoretical calculations using equation (2) through (4) was performed for a 1D PC comprising 8 bilayers of ZnS and CaF_2. The PC was designed to have a central wavelength $\lambda_c = 550\,nm$ and the quarter wavelength layer thickness of the two mediums are $d_H 57.6\,nm$ and $d_L 95.8\,nm$. The calculations were performed for temperature ranging from 30°C to 300°C.

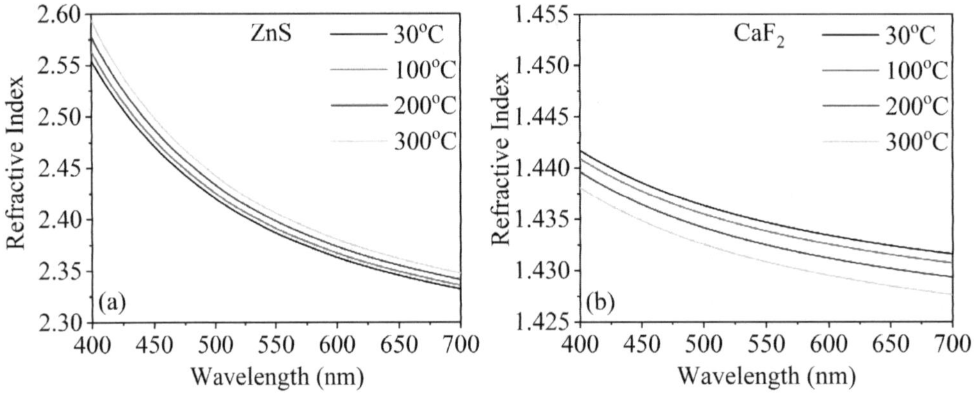

Figure 2: Refractive index variation with wavelength at different temperatures for (a) ZnS and (b) CaF_2

Figure 1 shows the value of refractive index for wavelengths in the visible region (400nm-700nm) at 30, 100, 200 and 300 degree Celsius depicted in black, red, blue and green solid lines respectively. A positive thermo optic trend is observed in figure 2 (a) where the RI values of ZnS increases with an increase in temperature while a negative trend is evident in figure 4 (b) where the RI values of CaF_2 decreases with an increase in temperature. The negative thermo optic effect ultimately increases the overall RI contrast of the PC as temperature increases which contributes to the slight enlargement of the absolute PBG.

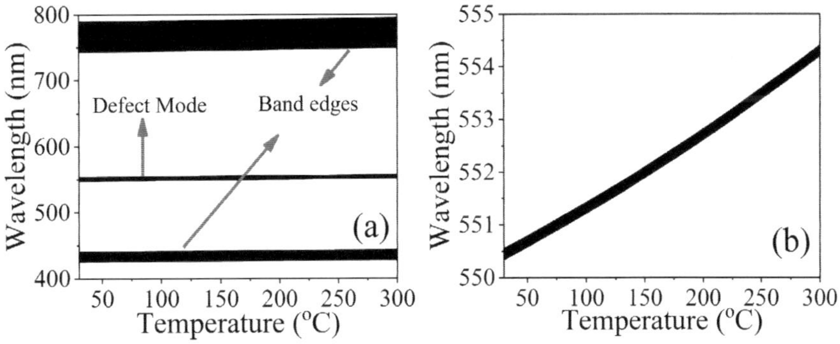

Figure 3:(a) Photonic Band Gap map of the transmission modes for varying temperature and (b) the close up of the defect mode varying with temperature.

The thermal compensation can be observed in the PBG map shown in figure 3 (a) of the PC having a CaF_2 defect layer thickness of 275nm corresponding to a defect mode transmission at 550nm. The thermal drift is almost insignificant with a spectral variation of $\Delta\lambda$=3.87nm from 30°C-300°C with ΔT= 270°C as seen in figure 3(b) giving us a thermal sensitivity $S = \Delta\lambda / \Delta T \sim$ 14pm/°C. More insights can be obtained from wavelength-reflection intensity plots on the band edge and transmission mode dynamics with temperature.

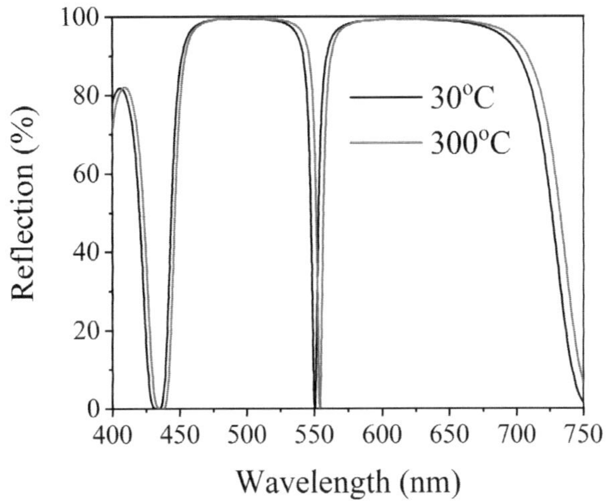

Figure 4: Photonic Band Gap of ZnS/CaF2 PC with defect mode transmission at 550nm at two temperature values. Black solid line is the PBG calculated for 30°C and red solid line is the PBG calculated for 300°C.

The PBG from wavelength vs reflection intensity is plotted in figure 4. The solid black line is the PBG calculated at 30°C and solid red line is that at 300°C. It is clear from the plot that the longer wavelength band edges are broader in comparison to the shorter wavelength band edges. This can be corresponded to the lower RI contrast at the longer wavelengths as observed from figure 2. With this observation we further calculated the thermal sensitivity of the defect mode transmission for every 50 degree rise in temperature. The position of the defect mode for every 1 degree change in temperature at 50 degree temperature interval is recorded and presented in the following table.

Table1 : Spectral shift and thermal sensitivity at every 50 degree rise in temperature

Temperature range $\Delta T = 1^{\circ}C$	Defect mode position at lower temperature (nm)	Defect mode position at higher temperature (nm)	Thermal Sensitivity $S = \Delta\lambda/\Delta T$ (pm/°C)
49°C - 50°C	550.6721	550.6843	12.2
99°C - 100°C	551.3023	551.3154	13.1
149°C - 150°C	551.9777	551.9916	13.9
199°C - 200°C	552.6993	552.7142	14.9
249°C - 250°C	553.4668	553.4826	15.8
299°C - 300°C	554.2784	554.2951	16.7

Every 50°C rise in temperature shows S increasing by ~0.9 pm/°C. The approximate value of S calculated earlier from figure 3 is an average of the above data. The thermal sensitivity increases with temperature due to the increase in RI contrast from the thermal effects but the variation in the thermal sensitivity is negligible due to the compensation of the positive thermal expansion with the negative thermo optic response of the rarer medium.. We can then say that S varies from 12.2 pm/°C to 16.7 pm/°C with an average value of ~14 pm/°C.

4. Conclusion

Thermal dependence of band edge and defect mode transmission of a 1D PC having PBG in visible wavelength range was studied. Defect mode with negligible thermal drift was obtained by compensating the positive thermal expansion effect with negative thermo optic property of the rarer medium. Thermal sensitivity for the PC was determined with S increasing by 0.9 pm/°C at every 50°C rise in temperature varying from 12.2 pm/°C to 16.7 pm/°C and having an overall average value of 14 pm/°C.

References

[1] E. Yablonovitch 1987 Phys. Rev. Lett. 58 2059–2062.
[2] J.D. Joannopoulos, P.R. Villeneuve, S. Fan 1997 Nature 386 143–149.
[3] J.D. Joannopoulos, S.G. Johnson, J.N. Winn, R.D. Meade, 2011 Photonic Crystals: Molding the Flow of Light - Second Edition, Princeton University Press.
[4] M. Renilkumar, P. Nair 2011 Opt. Mater. 33 853–858.
[5] F. Segovia-Chaves, H. Vinck-Posada 2018 Optik 154 467–472.
[6] F. Segovia-Chaves, H. Vinck-Posada 2018 Optik 163 16–21.
[7] H. Cai, B. Liu, X.M. Zhang, A.Q. Liu, J. Tamil, T. Bourouina, Q.X. Zhang 2008 Opt. Express 16 16670–16679.
[8] M. Djavid, M.S. Abrishamian 2007 Opt. Quantum Electron. 39 1183–1190.
[9] R. Mudachathi, P. Nair 2012 J. Microelectromech. Syst. 21 190–197.
[10] G. Lammel, S. Schweizer, P. Renaud 2001 14th IEEE International Conference on Micro Electro Mechanical Systems Technical Digest. MEMS 2001 pp. 578–581.
[11] M.J. Thorpe, D. Balslev-Clausen, M.S. Kirchner, J. Ye 2008 Opt. Express 16 2387–2397.
[12] X. Dai, Y. Xiang, S. Wen, H. He 2011 Journal of Applied Physics 109 053104.
[13] A. Baseri, A. Keshavarz, A. Hatef 2020 Journal of Applied Physics 127 214304.
[14] F. Segovia-Chaves, H. Vinck-Posada 2018 Physica E: Low-Dimensional Systems and Nanostructures 104 49–57.

[15] Y. Huang, J. Li, G. Kai, S. Yuan, X. Dong 2003 Microwave and Optical Technology Letters 39 70–72.
[16] M.C.P. Huy, G. Laffont, V. Dewynter, P. Ferdinand, D. Pagnoux, B. Dussardier, W. Blanc 2008 IEEE Sensors Journal 8 1073–1078.
[17] N. Mothe, D. Pagnoux, M.C.P. Huy, V. Dewinter, G. Laffont, P. Ferdinand 2008 Opt. Express 16 19018–19033.
[18] A. Doyle, C. Juignet, Y. Painchaud, M. Brown, N. Chummun-Courbet, E. Pelletier, M. Guy 2002 Electronics Letters 38 1561.
[19] F. Wang, Y.Z. Cheng, X. Wang, Y.N. Zhang, Y. Nie, R.Z. Gong 2018 Materials 11.
[20] L. Liu, H. Zhang, Q. Zhao, Y. Liu, F. Li 2007 Optical Fiber Technology 13 78–80.
[21] K. Srimannarayana, P. Vengal Rao, M. Sai Shankar, P. Kishore 2014 Optical Sensing and Detection III.
[22] S. Razi, F. Ghasemi 2019 Laser Physics 29 046204.
[23] P. Yeh 2005 Optical Waves in Layered Media, Wiley.
[24] H.H. Li 1984 Journal of Physical and Chemical Reference Data 13 103–150.
[25] W.J. Tropf 1995 Optical Engineering 34 1369.
[26] D.C. Harris 2008 Optical Engineering 47 114001.
[27] B. Schumann, H. Neumann 1984 Crystal Research and Technology 19 K13–K14.

Synthesis and characterization of electrical properties of nanostructured zinc oxide film prepared by spray coated seed layer based hydrothermal method

Bhuvaneshwari P, George Vasanthan S, Ajaijawahar Kaliyaperumal, Gokuladeepan Periyasamy*

Hydrogen Storage Materials and Nanosensors Laboratory, Department of Physics and Nanotechnology, College of Engineering and Technology, SRM Institute of Science and Technology, Kattankulathur, Tamil Nadu, India, 603203.

Corresponding author: gokuladp@srmist.edu.in

Abstract. The aim of study was to investigate the carrier transport properties of seed layer assisted nanostructured ZnO thin film. The structural analysis by XRD showed the predominant peak at 34.63° (2θ) corresponding to <002> plane. From the SEM image, it) corresponding to <002> plane. From the SEM image, it was found that the particles grown vertical nanoparticles on the glass substrate. The optical bandgap of the material was calculated from UV-Visible absorption spectrum which was found as 3.28 eV. The I-V characteristics showed that the film possess semiconducting nature. The electron mobility was determined as $1.60\times10^2 cm^2/Vs$ and the resistivity was around $1.99\times10^{-4}\,\Omega$.cm and the carrier concentration of the film was $7.54\times10^{20} cm^{-3}$ by Hall effect measurement. The seed assisted nanostructured film exhibited good electron transport properties which can be applied in many electrical applications like a solar cells, sensors and optoelectronic devices.

1. Introduction

Zinc oxide (ZnO) is one of the best wide bandgap semiconductor materials because of its numerous properties that makes it suitable in different fields. Zinc oxide (ZnO) is the n-type semiconductor metal oxide with wide band gap energy (3.37 eV), large exciton binding energy (60 meV), higher electron mobility (~100 $cm^2V^{-1}s^{-1}$) , optical transparency in the visible region (80% transmittance), and piezoelectric properties [1]. It is also easy to fabricate n-type ZnO material than a ZnO p-type layer. [2]. It is also an important semiconductor material used as a sensing element in metal oxide semiconducting sensors. Defects can also influence the carrier transport properties [3]. Material preparation by hydrothermal method has the advantages of low cost regulating chemical growth of the ZnO and also it can be implemented in a standard laboratory [4]. ZnO has been used in many applications such as gas sensors, solar cells, optoelectronic devices, and piezo-electronic devices [5-8]. ZnO based homojunction and heterojunction materials have been reported for LED applications [9]. In this report, we have prepared the ZnO nanostructures by combining the spray coating and hydrothermal deposition to study the carrier transport properties.

2. Experimental section

2.1 Preparation of seed layer

The procedure for the preparation of planar ZnO seed layer coating using the spray pyrolysis followed by hydrothermal growth of ZnO can be found elsewhere [10,11]. 4.2 g of Zinc acetate dihydrate ($Z_nC_4H_6O_4$) was added to 30 ml IPA under vigorous stirring at 70 °C for an hour where 1.2 ml of Monoethanolamine (MEA) (C_2H_7NO) was added drop wise to dissolve zinc salt in the solution. Once the solution became transparent the addition of MEA was stopped. The above solution was aged for 24 hours to obtain a homogenous solution. Before the coating, borosilicate glass slide was used as a substrate which was treated with chromic acid and deionized water thoroughly after it was baked at 120 °C for 10 min. Later the

IOP Conf. Series: Materials Science and Engineering 1219 (2022) 012016 doi:10.1088/1757-899X/1219/1/012016

prepared solution was coated on the glass substrate by spray pyrolysis method. The solution sprayed via nozzle for 3 sec and dried at 350 °C using substrate heater for 30 sec. This process was repeated until the solution gets over. The coated samples were kept for 5 hrs after reaching room temperature.

2.2 Preparation of hydrothermal process

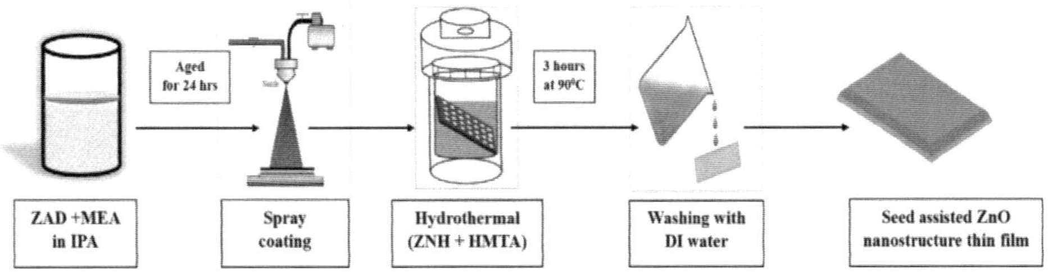

Fig.1. Schematic process of seed layer assist with nanostructured ZnO thin film

25 mM of Zinc nitrate hexahydrate ($Zn(NO_3)_2 \cdot 6H_2O$) and 25 mM of HMTA ($C_6H_{12}N_4$) was added to 70 ml DI water and stirred for half an hour and immediately transferred into Teflon lined stainless steel autoclave where the seed layered glass substrate was kept inclined at 45 ^{0}C touching the walls of the autoclave. The autoclave was kept at 90 °C for 3 hrs in hot air oven. After 3 hrs the autoclave was cooled down to room temperature and the substrate was taken out and washed thoroughly with DI water several times and dried. The films were calcined at 500 °C for 1 hour to remove the organic residues and to obtain good crystalline films.

2.3 Device fabrication

Interdigitiated electrode was coated on the nanostructured ZnO thin film grown on the seed layer with gold using thermal evaporation system.

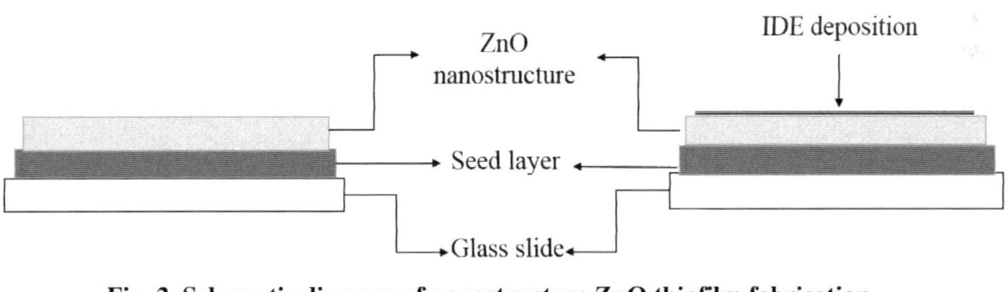

Fig. 2. Schematic diagram of nanostructure ZnO thinfilm fabrication

3 . Results and discussion

3.1 Structural properties

The following results were observed from the X' pert powder XRD Panalytical system which were shown in Fig.3. (a) and (b). As per the results, the observed XRD patterns were well matched with standard ZnO data which JCPDS card number is 01-089-0511 and 01-089-051 for seed layer and nanostructured ZnO thin film respectively. In that Fig.3. (a) the patterns were observed for the seed layer ZnO film that corresponding 2θ) corresponding to <002> plane. From the SEM image, it values related to the planes
(100) , (002), (101), (102), (110), (103) and (201) respectively. In the particular (002) plane 2 θ) corresponding to <002> plane. From the SEM image, it peak value (34.63°) was observed as predominant. The average crystal size was calculated by using Scherrer's formula found as

IOP Conf. Series: Materials Science and Engineering 1219 (2022) 012016 doi:10.1088/1757-899X/1219/1/012016

25 nm for the seed layer and 28 nm for nanostructured ZnO thin films prepared by spray coating process and hydrothermal process respectively. Thus, the particle size was increased by 3 nm after the hydrothermal process [12-15].

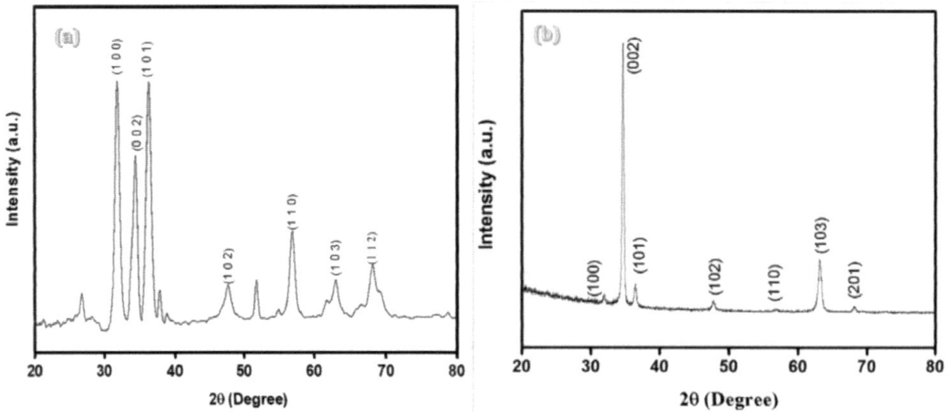

Fig. 3. XRD pattern of ZnO thin films by (a) Seed layer and (b) nanostructured ZnO thin film

3.2 Morphological properties

The surface morphology of the ZnO films were examined by using SEM with EDS (SNE3200 M, SEC) system. From the Fig.4. (a), (b) and (c), nanoparticles were observed at 10 μm, 5 μm and 2 μm scale respectively, which were grown by spray coating process and after hydrothermal process. The nanoparticles are irregular but it is in uniform dimension and both images were taken as top view only. These nanoparticles were grown from the seed layer whose morphology is similar to flakes arranged randomly over the film. The average particle size of the film was determined by using 'Image J' software which showed the seed layer flakes having 214 nm and nanostructure ZnO thin film around 253 nm [11].

IOP Conf. Series: Materials Science and Engineering 1219 (2022) 012016 doi:10.1088/1757-899X/1219/1/012016

Fig. 4. SEM images of nanostructure ZnO thinfilm

3.3 Optical properties

UV–visible measurements were carried out with UV-Vis-750 spectrophotometer for ZnO nanostructured film. Fig. 5. shows that the room temperature optical absorption spectrum of nanostructured ZnO thin film taken at UV-visible region from 200 nm to 800 nm.

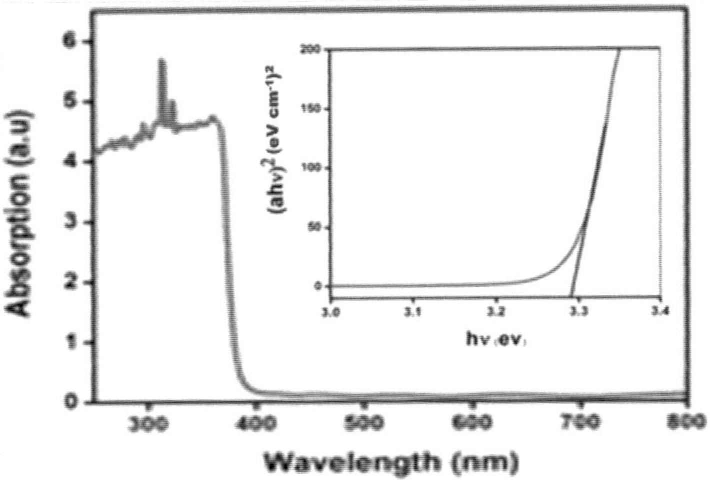

Fig. 5. UV Vis absorption spectra of nanostructure ZnO thinfilm

From the results, nanostructured ZnO thin film absorbed the light in the UV region and the optical band gap of ZnO thin films were estimated by extrapolation of the linear relationship between (α)^2.(eV cm^{-1})^2 and . From the Tauc's Plot, it was estimated that the nanostructured ZnO thin film after hydrothermal growth found to be 3.3 eV bandgap [14-15].

3.4 Hall measurement analysis

The synthesized ZnO films were characterized by Hall effect measurement system HMS5300 at room temperature as shown in Table 1. The charge carrier transport properties of ZnO thin films generally depend on their structure like grain size and shape, defects and the purity (concentration of impurities like absorbed and adsorbed gases, moisture, etc) and on the preparation conditions. The Hall measurement revealed that all the films are n-type exhibiting a resistivity of ~1.99×10^{-4} Ω.cm and bulk carrier concentration of 7.54×10^{20}/cm^3. The measurement was recorded after the ZnO thin film annealed at 500 °C with Silver (Ag) paste as contact. The maximum bulk carrier concentration and Hall mobility achieved is 7.54×10^{20} cm^{-3} and 1.60×10^2 cm^2/V. s respectively [16-17].

Table 1 Hall effect measurements of nanostructure ZnO thinfilm

Average hall coefficient	-1.07×10^{2}cm/C
Mobility	1.60×10^{2} cm^2/Vs
Bulk Concentration	7.54×10^{20} $/cm^3$
Resistivity	1.99×10^{-4} Ωcm
Conductivity	1.71×10^{4}/Ωcm

3.5 I-V characteristics

I-V Characterization was taken for the nanostructured ZnO thin film (annealed at 500 °C) using Keithley source meter model 2636-B, for which the voltage was swept from -5V to +5V. All samples demonstrated n-type conductivity. The I-V measurement was recorded for the different temperatures starting from 30°C (near to room temperature) to 150°C. As shown in Fig. 6. as the temperature increases, then the conductivity is always increasing. The increase of carrier mobility may be related to the grain boundary scattering of the free electrons. The mobility of a thin film may be affected due to scattering centers such as ionized and neutral impurities, grain boundaries, lattice vibrations and dislocations with semiconducting nature of the film [18-19].

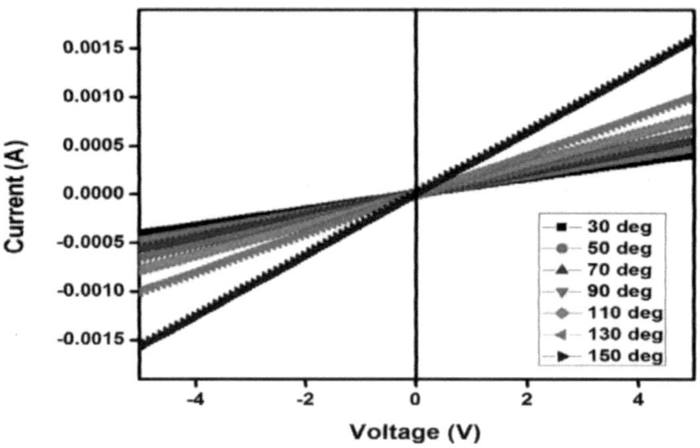

Fig. 6. I-V characteristics of nanostructure ZnO thinfilm

4. Conclusion

Nanostructured ZnO thin films were prepared successfully by spray coating technique followed by hydrothermal growth technique. The structural, morphological, optical and electrical properties were characterized by using X-Ray Diffraction, Scanning Electron Microscope, UV Visible analysis, Hall effect measurement and I-V characteristics respectively. XRD studies estimated that the prepared samples having hexagonal structure. The electron mobility, resistivity and carrier concentration were determined as $1.60\times10^{2}cm^{2}/Vs$, $1.99\times10^{-4}\,\Omega.cm$ and $7.54\times10^{20}cm^{-3}$ by hall measurement respectively.

5. Acknowledgement

The authors thank the DST-SERB (DST-SERB-SB/S2/CMP/-073/2013) and Department of Physics and Nanotechnology, SRMIST for their experimental and characterization support. The authors also thank Prof. A. Karthigeyan for his support and guidance.

References

1) N. Singh, P. Pandey, F.Z. Haque, 2012 Optik 123 1340–1342

2) Van de Walle, C.G., 2000 Phys. Rev. Lett., 85, 1012–1015.

3) S.J. Chang, W.Y. Weng, C.L. Hsu, T.J. Hsueh, 2010 Nano Com. Net. 1, 283–288.

4) S Agarwal, P Rai, E Navarrete Gatell, E Llobet, F Güelle, M Kumar, K Awasthi, 2019 Sens. Actu. B Chem., 292, 24–31.

5) Y Zhang, M.K Ram, E.K Stefanakos, Y.D J Goswami, 2012 Nanomater., 624520.

6) M. Yano, K. Koike, K. Ogata, T. Nogami, 2012 Phys. Status Solidi C, 9, 1570–1573. 7) W. Zhang, R. Zhu, X. Liu, B. Liu, S. Ramakrishna, 2009 Appl. Phys. Let., 95, 043304.

8) C.B. Ong, L.Y. Ng, A.W. Mohammad, 2018 Renew. Sustain. Ene. Rev., 81, 536–551.

9) Ali Baltakesmez, Su¨leyman T, and Sebahattin T, 2011 Jour. of App.Phy.110, 054502. 10) Y.C.Ching, N.Kalyani, K.S.Mohammad, W.Yhoong, 2014 Cera. Inter., 40, 9997–10004.

11) Jianguo Lv, Fengjiao S, Guangcai P, Feng W, Zhitao Z, 2014 J. Mater Sci: Mater Elect., 25, 882–887.

12) Shaila T, Rahul B, Swati N, & Sudarsan N, 2015 Ind. Chem. Eng. Taylor & Francis.

13) Sonalika A, Prabhakar R, Eric N G, Eduard L, Frank G, Manoj K, Kamlendra A, 2019 Sens.& Act.: B. Chem. 292 24–31.

14) Deepu T, Sunil C. V, Sunny M, and Simon A, 2014 Hind. Pub. Corp. Vol., 549019. 15) Mohammad Reza Arefi, Saeed Rezaei-Zarch, 2012 Int. J. Mol. Sci., 13, 4340-4350. 16) Seunghak S, Changhun K, Jung-A. L, Young-W. H, Joon-H. L, Jeong-J. K, 2017 Ceram. Inter., 43, 11163–11169.

17) Miran C, Hsing-Chao Chen, Miin-Jang Chen, 2010 Mat. Trans., Vol. 51, 2, 219-226.

18) E.K. Kim, H.Y. Lee, S.E. Moon, 2008 Journ. of Nanosci. & Nanotech., 8, 4698–4701. 19) Q. Humayun, M. Kashif, and U. Hashim, 2013 Hind. Pub. Corp., Journ. of Nanomat., 792930.

Synthesis and Structural studies of lightweight CNT/Cu composites: An overview

Shivani Dhall[a*], Srijin Pushkarn[a], Poonam Kharangarh[b]

[a]Department of Physics, DAV College, Jalandhar, Punjab, 144008, India.

[b]Department of Physics and Astro Physics, University of Delhi, Delhi 110007, India

*Email: shivani.dhall24@gmail.com

Abstract. An overview on structural and electrical properties of carbon nanotubes (CNTs) and Copper (Cu) based composite is presented keeping in mind their applications today and in the near future. CNTs are emerging as the most promising conducting wires in interconnection technology due to their higher electrical conductivity and extraordinary current-carrying capacity. In this work, we have found that, metal sputtering technique proved useful for the deposition of Cu nanoparticles on the surface of nanotubes. The structural and morphological studies of Cu deposited nanotubes were investigated by Raman spectroscopy and scanning electron microscopy (SEM). It was found that, the presence of prominent Raman bands in Cu deposited nanotubes is clear indication of the nanocompoites formation without destruction of carbon tubular structure. CNTs/Cu composites showed high conductivity and ampacity, making it uniquely suited for applications in microscale electronics and inverters.

Keywords: CNTs, Cu, FE-SEM and Raman.

Introduction

In recent years, there is huge advancement in electronic devices in their performance and simultaneously decrease in their dimensions. In the present era, the electronic devices of small dimensions are proved more portable, efficient and easy **to handle**. The carbon materials including graphene, CNTs and carbon dots are fascinating the researchers because of their unique structural, electronic, optical, thermal, sensing and chemical properties [1-2]. Among these carbon materials, CNTs has tremendous mechanical and thermal properties **such as** young modulus ~1TPa, tensile strength ~63GPa and thermal conductivity of 6600 WmK^{-1}. Furthermore, the tubular and nano dimensional structure of CNTs is responsible for anisotropic behavior and high current carrying capacity of the order of 10^9 A cm^{-2}. Moreover, CNTs in powder form are used in different commercial products such as automotive parts and rechargeable batteries and so on. These CNTs finds application **in** different areas such as interconnected electronics, H_2 production, photo catalysts as sensing layer, electrode in batteries, super capactiors and many more [2-3].

Keeping in mind aforesaid CNTs properties and applications, Industry and researchers are developing a new class of composites materials. Various CNT composites for different applications have been reported in literature and type of these composites are alumina-based composites [4], polymer-based composites [5-6], hybrid structured composites [7-8] and metal-CNT composites. Metal-CNT composites have attracted wide attention due their applications in various fields of science and technology. The incorporation of CNTs in different metal matrices have resulted in enhanced thermal, electrical, mechanical and sensing properties. Selective CNT incorporation on Metal matrix can enhance the stiffness, toughness and strength of the materials [9]. Among various metal CNT composites, Cu-CNT based composites have capabilities to surpass the current geometrical integration of semiconductor industry [9-10]. Also, the optimum ratio of Cu-CNT and synthesis optimization can achieve their room temperature conductivity much higher than currently existing materials [10]. Xing *et al.* [11] reported high thermal (2.83 W m^{-1} K^{-1}) and electrical conductivity (1500 S cm^{-1}) of CNT-copper nanowires. These nanowires synthesized by self-fuelled electro-deposition have resulted in high electrical conductivities (~3 X 10^5 Scm^{-1}) which are of the order of metallic conductivity and mechanically robust [11]. Subramaniam *et al.* [12] reported a CNT/Cu composite material with a conductivity near to copper and ampacity is 100 times higher than copper.

There are various methods such as hydrothermal, electrode-position, microwave plasma-assisted chemical vapor deposition (MPCVD), arc discharge method and many more have been reported for the synthesis of CNTs/Cu nanocomposites. Setlu *et al.* [13] initially reported the simple method for the synthesis of bulk amount of CNTs/Cu composites. They reported hydrogen arc method for the filling of Cu in CNTs shells. In 2007, Hassel *et al.* [14] reported simple and single step route for the synthesis of CNTs/Cu based nanostructures. In 2010, Lim *et al.* [15] have investigated the mechanical properties of CNTs/Cu composites. They reported controlled oxidation and molecular level mixing approach for the synthesis of CNTs/Cu powder based composite [15]. They used hot compression technique for the study of electrical and mechanical properties of CNTs/Cu powder. They observed that hot compression enhanced the Cu and CNTs grain alignment, which further reduce the scattering of electrons and strength of CNTs/Cu composites. Sundaram *et al.* [16] reported lightweight MWCNTs/Cu composites based electrical wires using electro-deposition method. They investigated that Cu filling in CNTs shows low temperature coefficient of resistance (TCR) and claimed that it is alternatives to Cu wire for weight reduced device applications. In 2016, Sun *et al.* [17] synthesized CNTs/Cu composite using electroplating technique to decorate vertically aligned CNT bundles with Cu nanoparticles. They reported that the deposition of Cu content on CNTs fiber affect their electrical and thermal properties significantly. However, the presence of Cu continuous form, either as outer-coating and as internal filling in CNTs, offering a lower resistive path for electron transport, which leads to reduced ρRT values for the full, partial, and no internal Cu filling composite wires, in comparison to the Cu-seeded wires. Therefore, it is important to control the Cu content on CNTs during synthesis of CNTs/Cu composites. In summary, most of researchers adopted electrodeposition/electroplating method for the synthesis of CNTs/Cu composites based electrical wires. However, with best of our knowledge, there is hardly any publication on the synthesis of CNTs/Cu using metal sputtering technique.

By keeping this in mind, we have synthesized the MWCNTs/Cu composites using metal sputtering method and found that control deposition of Cu on CNTs is possible by this route without the destruction of tubular structure of nanotubes. We have investigated that, the adopted synthesis methods is simple and found that it significantly improves their structural properties. In addition, we have discussed overview of CNTs/Cu composites. However, CNT agglomeration in Cu matrix increases porosity and defects, which degrade their electrical

IOP Conf. Series: Materials Science and Engineering 1219 (2022) 012017 doi:10.1088/1757-899X/1219/1/012017

conductivity. Therefore, it is very important to select appropriate method for the synthesis of CNTs/Cu composites for applications point of view.

Materials and Methods

Pristine MWCNT were purchased from JK-impex (Mumbai, India) with a purity level >95%.These nanotubes have been synthesized by chemical vapor deposition (CVD) method. The average outer diameter of as procured MWCNTs were ~10–20 nm.

We have synthesized MWCNT/Cu by metal-sputtering route. In this, firstly we have functionalized MWCNT by simple chemical route. The details of adopted functionalization process have been reported in our previous work [7]. The functionalized MWCNT were dissolved in the ethanol and then micro drop of well dispersed solution was drop-casted on desired cleaned substrate. Afterwards, substrate kept on hot plate for few min to remove the solvent. The deposition of Cu by metal sputtering has been optimized for different times and RF power. The deposition of Cu metal was done at 20 W for 1 min with pressure of the order 10^{3} bar. The annealing of MWCNTs/Cu was performed using rapid thermal processing (RTP) system. This sample is named as CNTs/Cu composite. The schematic of adopted method as shown in Figure 1.

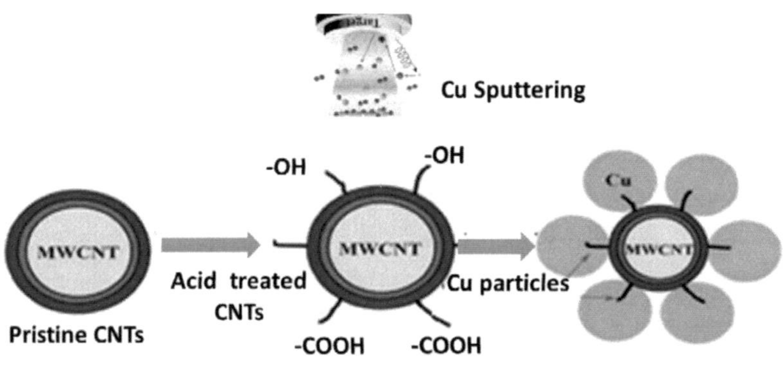

Figure 1 : Schematic illustration of the process used for the synthesis of MWCNT/Cu composite.

Results and Discussion

The Raman spectra **of** pristine and MWCNT/Cu are shown in Fig. 2. As shown, it contains three prominent standard bands namely D, G and G bands. There is an increment in theintensity D band, which indicates the attachment of function groups [7].The presence of G band in MWCNT/Cu indicate the enhancement in metallic character of nanotubes after the attachments of Cu nanoparticle on the surface of nanotubes [3,7]. After attachment of the Cu nanoparticles, it is found that there is upshifting in shifting D and G band in Raman spectrum. It is well known that, the surface structure ofnanoparticles plays an important role in shifting of the raman bands related to CNTs [3]. The presence of Raman characteristic bands in the Raman spectrum of MWCNTs/Cu clearly indicates that the graphitic tubular structure of the nanotubes is maintained as reported in previous work [3,7].

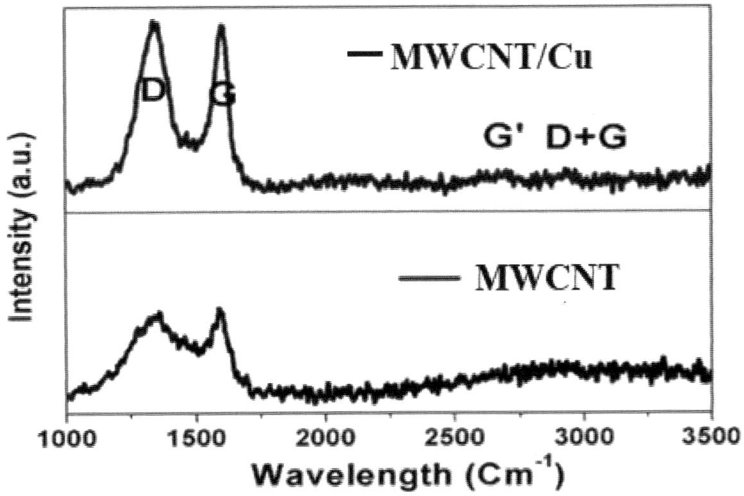

Figure 2 : Raman spectra of MWCNT and MWCNT/Cu composite.

The FE-SEM image of pristine nanotubes and MWCNT/Cu **are** shown in Fig. 3. The blue ovals in Fig. 3(a) indicate the well dispersion on the nanotubes after acid functionalization. Whereas, the red ovals in Fig. 3(b), indicate the deposition of Cu nanoparticles on the nanotubes. As reported in previous work [3,18], the presence of functional groups on carbon materials proved helpful for the nucleation of metal nanoparticles. This nucleation process follows a diffusion limited aggregation (DLA) model, in which defect sites create the initial nucleation centers for the growth of particles [19]. However, the difference between cohesive enthalpy and adsorption, along with the mobility of Cu-metal atoms on the graphitic surface is cause of the formation of clusters as shown in Fig. 3(b).

IOP Conf. Series: Materials Science and Engineering 1219 (2022) 012017 doi:10.1088/1757-899X/1219/1/012017

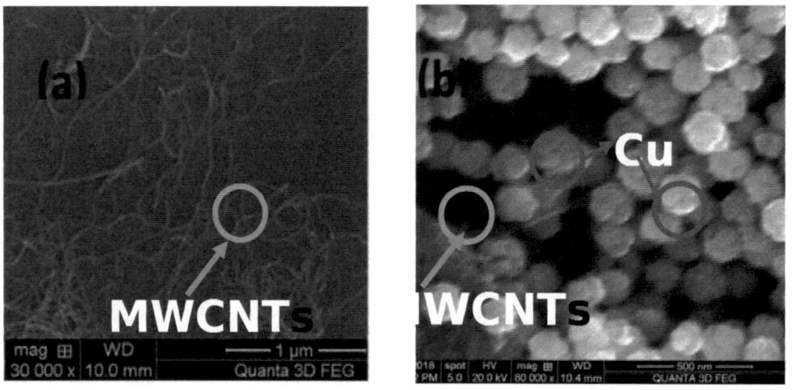

Figure 3: The FE-SEM image of pristine MWCNTss and MWCNT/Cu composites .

Electrical Study

The electrical conductivity of the CNTs are reported **of** the order 10^4 S cm^{-1} with the ampacity ranges between $10^7 - 10^{13}$ A cm^{-2}. On the other hand, if we consider the pure copper, which is very good conductor of electricity has the conductivity 5.9×10^5 S cm^{-1}, an order greater than the CNT, with the ampacity $\sim 10^6$ A cm^{-2} [10]. After the fabrication of CNT-Cu composite, where both of these materials have an important regime and contributes to the overall conductivity of the composite material, the electrical conductivity of CNT-Cu composite is recorded in the range 2.3-4.7×105 S cm^{-1} approximately near to the value of copper but the ampacity of CNT-Cu is reported as 6×10^8 A cm^{-2}, which is 100 times higher than Cu and is the highest till date [12]. The composite CNT-Cu is the only material with high conductivity and high ampacity. The electrical resistivity of the composite CNT-Cu is in the range $1.7\text{-}1.9\times10^{-6}$ Ω cm for the film thickness of about 6.8μm. This conductivity of CNT-Cu increases after the alignment of CNTs in the Cu matrix and reached to the value 5.25×10^5 S cm^{-1} [20]. The electrical conductivity of CNTS/Cu nano-composites depends on the volume fraction and dispersion of CNTs and the method adapted to synthesize CNTs/Cu composites always play major role. So, control of Cu deposition on nanotube for better interface should be important. The electrical investigation of reported composites is underway.

Conclusions

From the present study, it is found that metal sputtering is simple and effective method for the synthesis of CNT/Cu composites, which are emerging as promising materials in conducting wire technology and it shows remarkable electrical properties that **are** important for future electronics. Also, CNT is strong candidate for the formation of new class of composites, which further shows excellent chemical, electrical, thermal, mechanical and structural properties. However, control on CNTs agglomeration during composites formation is still a challenge. Therefore, the present study needs to be further explored for advances in electronic devices.

Acknowledgements

The author S. Dhall acknowledge to SERB (DST, Govt. of India) for providing TARE funding (letter no. TAR/2019/000113) to support the research work.

References

[1] Iijima S 1991 Nature 354 56.

[2] Jariwala D, Sangwan VK, Lauhon LJ and Marks T J 2013 Chem. Soc. Rev. 42 2824.

[3] Dhall S, Sood K and Nathawat R, 2017Int J Hydrogen Energy 42 839.

[4] Coleman J N, Khan U, Blau WJ and Gun'ko Y K 2006Carbon44 1624.

[5] Spitalsky Z, Tasis D, Papagelis K and Galiotis C 2010 Prog. Polym. Sci. 35 357.

[6] Ma P C, Siddiqui N A, Marom G and Kim J K 2010 Composites Part A 41 1345.

[7] Dhall S, Jaggi N, Nathawat R 2013 Sens Actuators A 201 321.

[8] Estili M and Kawasaki A Scr. 2008 Mater. 58 906.

[9] Bakshi S R, Lahiri D and Agarwal A 2010 Int. Mater. Rev. 55 41.

[10] Chinnappan A,Baskar C, Kimdand H, and Ramakrishna S 2016 J.Mater.Chem.A, 4 9347.

[11] Xing Y, Cao W, Li W, Chen H, Wang M, Wei H, Hu D, Chen M and Q. Li, 2015 J. Nanosci. Nanotechnol. 15 3265.

[12] Subramaniam C, YamadaT, K. Kobashi, A. Sekiguchi, D. N. Futaba, M. Yumura , K. Hata, Nature communication, 4 (2013) 2202.

[13] Setlur A A, Lauerhaas J M, Dai J Y and Chang R P H 1996, Appl. Phys. Lett., 69 345.

[14] Haase D, Hampel S, Leonhardt A, Thomas J, Matternand A, B ̈uchner B 2007 Surf. Coat. Tech. 201 9184.

[15] Lim B K, Mo C B, Nam D H and Hong S H, 2010 J. Nanosci. Nanotechnol. 10 78 .

[16] Sundaram R, Yamada T, Hata K and Sekiguchi A 2017 Scientific Reports 79267.

[17] Sun S, Mu W, *et al* 2016 Nanotechnology 27 335705.

[18] Dhall S, Jaggi N 2015 Sensors and Actuators A 224 50.

[19] Bittencourt C, Hecq M, *et al* 2008, Chem. Phy. Lett. 462 260.

[20] Jayathilaka W A D M, Chinnappan A, Ramakrishna S, 2017, J. Mater. Chem. C 5 9209.

Thermodynamic study of high thermal stability Carbon Dots

Vishal Rimal, P.K. Srivastava

Birla Institute of Technology, Mesra, Ranchi, India, 835215

vishalrimal19@gmail.com

Abstract. Carbon Dots (CDs) in the recent years have been found to exhibit exceptional thermal stability. The superior nature CDs demonstrate in their physical, chemical, optical and conductive properties together with their low cytotoxicity make them a true contender to conventional inorganic counterparts. In this study the traditional thermodynamic properties that support the high thermal stability of CDs have been investigated. The thermal stability of CDs has been found to have a superior relationship with numerous parameters thereby paving a strong evidence for further research and advancement in the field. This result will further help in the development of organic nanoparticles which can be employed in environments where high-temperature resistance is a crucial requirement.

Keywords: Carbon Dots, Stability.

1. Introduction

Ever since the accidental discovery of Carbon Dots (CDs) in 2004; carbon in the nano dimension has always fascinated scientists. In the recent year's CDs have turned out to be an exceptional substitute to metal-based semiconductor quantum dots. They generally exist with sp^2 conjugated core and have suitable oxygen content in the forms of multiple oxygen-containing species signified by carboxyl, hydroxyl, and aldehyde groups, complimented by a size below 10 nm. Convenient synthesis, brilliant chemical and photophysical properties, prominent biocompatibility, colorful photoluminescence along with factors like low-cost and toxicity have had a strong role in making CDs a favorable choice. On the application end CDs have gained prominent significance in bio-imaging, drug delivery, chemo/biosensors, catalysis, optoelectronic devices and white light-emitting diodes, etc. [1, 2]
With the development in the field of nanotechnology the applications of CDs for diverse applications in the industry have quadrupled over the years. The thermodynamic properties of nanoscale particles in the recent years have received notable attention due to the difference in properties exhibited from their bulk counterparts. Particles with moderately low melting points were mostly used for experiments till date [3]. Sufficient studies can be found on the thermal stability of carbon nanotubes and nanomaterials that are metal based but limited research has been done on the high-thermal stability of CDs.
The investigations and studies on the thermal and thermodynamic characteristics of CDs have often been set aside by the research community. In this manuscript the thermodynamic parameters that strongly influence the stability of CDs have been studied and elaborated.

2. Thermodynamic Consideration

The thermal stability of materials holds a prominent position for real-world applications at diverse temperatures. Molar heat capacities of materials at different temperatures have been key data sets significant for further research and development in the field and industry. An important level of interest has been gained by researchers on the heat capacity determinations of compounds. Thermodynamic properties including enthalpy and entropy can be measured with the heat capacity data [4, 5]. For any possible change in nature, the thermodynamic stability arises from the Clausius inequality [6] i.e.,

$$dS \geq \frac{đQ}{T} \qquad (1)$$

Conversely,

$$dS < \frac{đQ}{T} \qquad (2)$$

IOP Conf. Series: Materials Science and Engineering 1219 (2022) 012018 doi:10.1088/1757-899X/1219/1/012018

Here our symbols shall follow their usual meanings i.e., S = entropy, Q = heat transfer, and T = temperature.
We can see that for all permissible change in state, the system cannot spontaneously leave the current state irrespective of what the conditions are. The system is thus in what is called the stable equilibrium. Theoretically a stable equilibrium is said to occur if a body after slight displacement returns to its original position when the displacing force has been removed; while a metastable equilibrium exists when additional energy must be introduced before the body can reach its true stability. An unstable equilibrium on the other hand occurs if a small disturbance will produce a large change, the same has been illustrated in Figure 1.

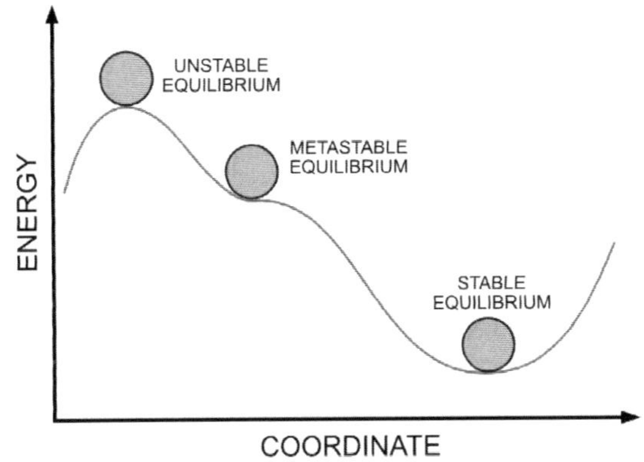

Figure 1: Graphical representation of unstable, metastable and stable equilibrium states respectively

If the system is not adiabatically isolated, where P = pressure, V = volume, E = energy, N = number of molecules and μ = chemical potential; we will have,

$$đQ = dE + PdV - \mu dN > TdS \qquad (3)$$

Consequently, for all permitted changes if the system is in equilibrium we will have,

$$dE + PdV - \mu dN - TdS > 0 \quad (4)$$

The above equation is the main stability result. If E, V, and N are held fixed $đQ = 0$ and the stability condition will become the same as that before i.e., $dS < 0$. If S, V, N is held constant, from equation 4 we see that if $dE > 0$ the system will be stable; i.e., the energy is therefore a minimum.
If N, T, V are fixed, equation 4 becomes

$$d\,(E - TS)_{N,\,T,\,V} > 0 \qquad (5)$$

Equation 5 is the Helmholtz free energy ($A \equiv E - TS$) which is a minimum for stable equilibrium when (N, T, V) are forbidden to fluctuate.
Let us consider (S, P, N) to be fixed analogous to the case with enthalpy. Equation 4 thus becomes

$$d\,(E + PV)_{S,\,P,\,N} = (dH)_{S,\,P,\,N} > 0 \qquad (6)$$

which implies that for stable equilibrium the enthalpy is a minimum.
If T, P, and N are held fixed. Equation 4 becomes,

$$d\,(E + PV - TS)_{T,\,P,\,N} > 0 \qquad (7)$$

This gives rise to Gibbs free energy i.e., $G \equiv E + PV - TS = H - TS$ which is a minimum for a stable equilibrium when T, P, and N are fixed.

Table 1: Conditions for Stable Equilibrium

Held Fixed	State Function	Definition	Differential	Stable Equilibrium Criterion
E, V, N	S	-	$dS = \frac{dQ_{rev}}{T}$	maximum
S, V, N	E	-	$dE = TdS - PdV + \mu dN$	minimum
S, P, N	H	$H \equiv E + PV$	$dH = TdS + V\,dP + \mu dN$	minimum
T, V, N	A	$A \equiv E - TS$	$dA = -SdT - PdV + \mu dN$	minimum
T, P, N	G	$G \equiv H - TS$	$dG = -SdT + V\,dP + \mu dN$	minimum

Chemical reactions generally occur when they are thermodynamically favorable i.e., when ΔG is negative. Table 1 enlists the necessary criteria's to be observed for a stable equilibrium. It is a well-known that lower the potential energy of the system, the more stable it is. The thermodynamic favorability implies a change from a state of high energy to low energy or in other words from a less stable state to more stable state.

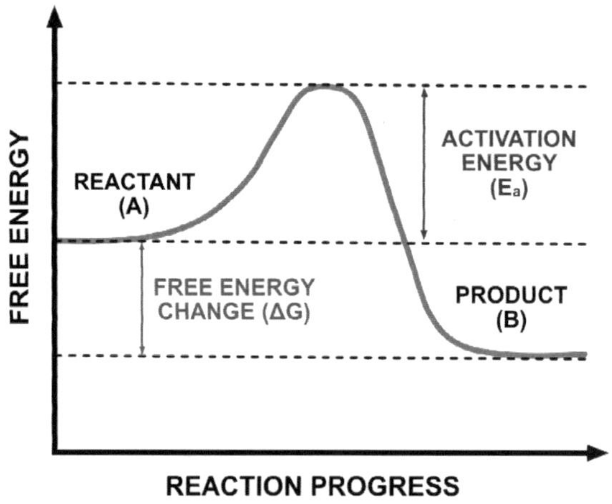

Figure 2: Free Energy vs Reaction Progress

In Figure 2 we can see that B is at lower energy than A and as discussed ΔG is negative. The reaction should hence proceed spontaneously and B is the more thermodynamically stable species. The reaction however must surpass the barrier called the Activation Energy (E_a). Thermodynamically we can predict the starting and ending position of the reaction but however the intermediate or transition states cannot be estimated.

Studies on the thermal stability of carbon nanotubes have been done in the past and considerable inferences were observed for the samples. The stability was concluded in terms of the heat capacities and TG-DTG-DSC curves for CNTs with different sizes. It was found that the value of C_p of CNTs are independent of their lengths when the diameters are below 10 nm. TG-DTG results also showed that the thermal stabilities of the CNTs were influenced with the diameter. When the length increases the thermal stabilities of CNTs increase when the diameters are 60–100 nm but decrease when the diameter is less than

60 nm. DSC analysis concluded that heat release increases when the diameter of CNT increases but is unaffected by the lengths of CNTs. *Xu et al* have elucidated detailed studies in the foresaid inferences [7]. The results underlined in the former paragraph point to a perspective where suitable implications can be assumed to be obvious for CDs with size playing a dominant role in its stability. The TGA and DSC studies as reported by our group in past strongly align with the results for carbon nanotubes. A stable curve was observed for the TGA and DSC observations on CDs derived from oleic acid. This observation maybe a result of similar stability of CDs in a particular size range. It can be thought about that a variation of particle size may potentially alter the stability of CDs. Since we have a single criterion to consider, the diameter shall be the sole factor for stability followed by the raw material, synthesis methodology etc. considered in the experiment. This theoretical study can be followed with suitable experimentation to conclude and potentially get verified results on the dependence of stability on the basis of size.

3. Conclusion

The main motive of this paper has been to discuss the concept and physical meaning of stability in Carbon Dots. The thermodynamic study of CDs is something that has been long neglected by the scientific community. With limited resources in the field to usher the importance in learning the behaviour of the nanoscale particles and their uniqueness compared to the bulk scale counterparts' further studies should be prioritised by researchers in the domain. Following adequate research and consideration in the topic in the years to come, we will be able to understand the science of stability in these nano-dimensional particles better.

References

[1] Paola Lagonegro, Umberto Giovanella, and Mariacecilia Pasini. Carbon dots as a sustainable new platform for organic light emitting diode. *Coatings*, 11(1):5, 2021.

[2] Vishal Rimal, Shubham Shishodia, and PK Srivastava. Novel synthesis of highthermal stability carbon dots and nanocomposites from oleic acid as an organic substrate. *Applied Nanoscience*, 10(2):455–464, 2020.

[3] Xianlong Wei, Ming-Sheng Wang, Yoshio Bando, and Dmitri Golberg. Thermal stability of carbon nanotubes probed by anchored tungsten nanoparticles. *Science and Technology of advanced Materials*, 2011.

[4] Y Arita, T Ogawa, B Tsuchiya, and T Matsui. Heat capacity measurement and dsc study of hafnium hydrides. *Journal of thermal analysis and calorimetry*, 92(2):403–406, 2008.

[5] J Leitner, K Ruzicka, D Sedmidubsky`, and P Svoboda. Heat capacity, enthalpy and entropy of calcium niobates. *Journal of thermal analysis and calorimetry*, 95(2):397–402, 2009.

[6] JR Brownson. Solar energy conversion systems. *Academic Press*; 2013 Nov 9.

[7] F Xu, L Sun, J Zhang, Y Qi, L Yang, H Ru, C Wang, X Meng, X Lan, Q Jiao, et al. Thermal stability of carbon nanotubes. *Journal of thermal analysis and calorimetry*, 102(2):785–791, 2010.

Synthesis and characterization ZnO nanoparticles using sol-gel method and their antibacterial study

[1]Vignesh K, Anju S Nair, C. Udhayakeerthana & T..Kalaivani ,

Department of Physics & Nanotechnology· SRM Institute of Science and Technology, Kattankulathur, Tamilnadu, India, 603203

Corresponding Author: kalaivat@srmist.edu.in

Abstract. Zinc oxide nanoparticles were prepared using the sol-gel method. Zinc is one of the most important elements in living organisms and similarly zinc derived compounds have attracted much attention because of their applications in various fields like agricultural, medicinal, industrial, etc. These nanoparticles have piezoelectric, semiconducting, photocatalytic activities, and very high surface area. Also, they possess the unique property of being biodegradable and biocompatible. To synthesize the pure ZnO nanoparticles, zinc nitrate hexahydrate [$Zn(NO_3)_2.6H_20$] is used as the precursor. Sol-gel can be done either in an aqueous or non-aqueous medium. But in this work, the ZnO nanoparticles are prepared in the aqueous medium (i.e in water) at room temperature. The prepared zinc oxide nanoparticles are characterized using X-ray diffraction to determine their crystallinity and scanning electron microscopy (SEM) is used to study the morphology of the sample.

Keywords: Zinc oxide, nanoparticles, sol-gel, biosynthesis, biocompatible.

Introduction

Zinc oxide (ZnO) is an n-type semiconducting material with bandgap energy, Eg of 3.37eV and has high recombination speed. Also, it is an essential part of many enzymes, sunscreens, ointments for itching and pain. If ZnO is in the form of microcrystals they tend to be very good light absorbers in the UV-A and UV-B region of the ultraviolet spectrum due to its high bandgap. Zinc oxide in the nanoscale gives very good antibacterial activity and is also more stable at high temperatures and pressure. Many other such inorganic materials are antibacterial and at the same time are metallic nanoparticles and metal oxide nanoparticles like gold, silver, titanium oxide, copper, etc. Hence we can say that ZnO nanoparticles can be used in various fields due to its astonishing properties (1-5). Zinc oxide is non-toxic and can be synthesized at low costs. As it can be used as a good photocatalyst and to increase its effectiveness the electron-hole recombination speed and bandgap energy needs to be reduced. The recombination speed is reduced by doping of the zinc oxide material. Also, these are incorporated in packaged food to prevent spoilage of food due to contamination. The impact of ZnO depends on the particle size, morphology, pH and biocompatibility. And the morphology of the ZnO depends on the synthesis method. The method which we are going to use for the synthesis of ZnO and mesoporous ZnO nanoparticles is the Sol-gel method, which is a bottom-up approach technique. During this process, the mixture containing the precursor and surfactant are stirred over a magnetic stirrer till a gel formation takes place. The white precipitate thus formed is removed using the centrifugation method and thus we will finally obtain the gel. This is then kept for drying and is then heated (calcination) to favor further polycondensation or to improve its mechanical properties.

Hydrolysis: $\text{M-OR} + \text{HOH} \rightarrow \text{M-OH} + \text{R-OH}$

Condensation: $\text{M-OH} + \text{OH-M} \rightarrow \text{M-O-M} + H_2O$

$\text{M-OR} + \text{M-OH} \rightarrow \text{M-O-M} + \text{M-OH}$

One of the biggest advantages of the sol-gel method is that it doesn't require high temperatures for synthesis. The other advantage of this method is it can easily shape materials into complex geometries in a gel state. It can produce high purity products as the organometallic precursor of the desired ceramic oxides can be mixed, dissolved in a specified solvent, and hydrolyzed into a sol and then into a gel. The composition can be highly controllable. Thus, in our research, we will be synthesizing ZnO nanoparticles. We will make use of various characterization methods to study the prepared pure ZnO nanoparticles. To characterize the prepared materials we will be using XRD (x-ray diffraction) which will give us information about the crystallinity of the material and SEM (scanning electron microscopy) to study the morphology.

2. Materials and methods

All chemical reagents were purchased from EMERCK(A.R. grade) such as zinc nitrate hexahydrate [$Zn(NO_3)_2.6H_20$] and sodium hydroxide (NaOH) are used without further purification. The crystal structure of the synthesized nanoparticles were characterized by X-ray diffraction (X'pert PAN analytical) with Cu-K radiation (= 1.5405Å) in the 2 range 20◦–80◦. The experiment is done in an aqueous medium i.e water. Morphology analysis were performed using Field Emission-Scanning Electron Microscope FE-SEM(6).

2.1. Synthesis of mesoporous zinc oxide nanoparticles

Two beakers containing 50ml of deionized water in each of them. Let's mark these beakers as 1 and 2. In the next step, we will add 14.4g of zinc nitrate hexahydrate in beaker 1 and will place a magnetic pellet within the solution and the beaker 1 is then placed on the magnetic stirrer. Solution 1 in beaker 1 is thus stirred for around 5 minutes. Similarly, we will add sodium hydroxide, NaOH, to beaker 2 with a magnetic pellet in it and will place it on the magnetic stirrer. Solution 2 is also stirred for around 5 minutes. The solution placed in the beaker gets stirred fast due to the rotating magnetic field generated. Solution 2 consisting of sodium hydroxide is then added to solution 1 consisting of zinc nitrate hexahydrate. While adding solution 2 into solution 1, we should add it dropwise using a syringe that allowing the solutions to mix properly and hence initiate the mechanism or reaction. Also, the entire process of stirring takes place at normal room temperature (7-10).

On adding the solution 2 dropwise using a syringe, we will be able to see a slight difference in the color of the mixture. The mixture will be turning into white color slowly as we add the solution 2 dropwise. This transition of the solution from transparent to opaque shows the initiation of the chemical reaction. Thus, after the complete addition of solution 2 into solution 1 we can notice that the mixture appears cloudy. Thus, the resultant solution 3 is stirred for another 2 hours on the magnetic stirrer. After stirring it for 2 hours the solution 3 is transferred into the centrifuge tube. Solution 3 thus undergoes a process called centrifugation where the precipitate is removed or filtered out and we obtain the gel at the end of the process. Around 3 rounds of centrifugation is done. After the first round, the centrifuge tubes are filled with deionized water up to the marking and then is placed in the centrifuge machine again. After doing two rounds of centrifugation using deionized water, we will now use ethanol and centrifuge it. Finally, we will obtain the gel as the residue. Here also the entire process of centrifugation takes place at the normal room temperature i.e 28 °C. The gel obtained as the residue of the centrifugation process is then spinned by keeping on a spinner. This is done to displace the gel stuck at the bottom of the centrifuge tube. Also while doing this a small amount of ethanol is added. Thus, the gel is then finally taken into a petridish and is broken using a spatula. The petridish containing the gel is then covered using an aluminium foil and tiny perforations are made over this aluminium foil covering the petridish. The petridish containing the gel is placed overnight to dry. Once dried the petridish is placed in a hot air oven for 10 minutes and is then placed in the muffle furnace at a temperature of 500°C for 3 hours. This process of heating is called calcination. After undergoing calcination the particles on the petridish are transferred to a mortar and are finely grinded into a powder form. The finally obtained powdered particles are then transferred into tubes and stored. To determine if the synthesized material is ZnO nanoparticles we will have to use various characterization methods.

3. Results and discussion

3.1. X-ray diffraction analysis of mesoporous SiO_2nanoparticles

The pure ZnO nanoparticles were prepared using the sol-gel method. The powdered sample thus obtained were characterized using X-ray diffraction method to analyze the crystallinity of the sample. The Debye-Scherrer equation is used for determining the crystalline size:

$$\text{“ } \mathbf{D = K.\lambda / \beta.cos\theta} \text{”}$$

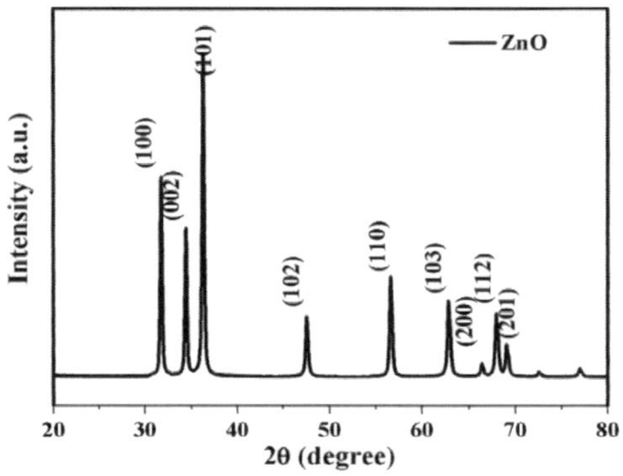

Figure 1: XRD pattern of pure ZnO nanoparticle

Thus, Fig. 1 shows the XRD pattern of the pure ZnO nanoparticles prepared during this research work. The average crystalline size of the zinc oxide nanoparticle was found to be 36nm using the Scherrer equation with the width of the peaks at (100) and (101).

3.2. Morphological Studies

The morphology of the prepared pure ZnO was obtained using SEM (scanning electron microscopy). Using the SEM image we can observe the surface of the sample. Fig. 2 gives us the SEM image of the prepared pure zinc oxide nanoparticle at a magnification of 100,000. From the SEM image, we can see that the nanoparticles are in the form of flocks. Also as a result of this, the nanoparticles have different shapes(11).

Figure 2: FESEM image of Zno

The analysis of the SEM image shows that the size of the ZnO nanoparticle is greater than that obtained from the XRD. This increase in size thus indicates that the size of the nanoparticle obtained using SEM is not of a single crystal but the entire flock of nanoparticles.

Antibacterial analysis:
The antibacterial examinationof ZnO was examined against infective bacteria *Escherichia coli* (*E.coli) by* well diffusion technique. [12]. A small drill tool is used to make a well. Bacteria were spread separately in the sterilized plates. At different concentrations (25, 50, 75 and 100 µg/ml),ZnOnanoparticleswereadded.Theµg/ ml), ZnOnanoparticles were added. The plates go into the incubation process for 24 h at 37°C. After 24 h, the inhibition zone was analysed [13]

ANTIBACTERIAL STUDIES:

The antibacterial activity of ZnO was examined with four different concentrations (25 µg/ml, 50 µg/ml,75 µg/ml,100 µg/ml). Figure shows the zone of inhibition after 24 h of incubation in both dark and visible light atmospheres.(*Escherichia coli*) it shows 5.4mm, 6.2mm, 6.7m, 7.2mm inhibition zone for the concentration of 25 µg/ml, 50 µg/ml, 75 µg/ml, and 100 µg/ml. the zone of inhibitonvolues are shown in table 1.

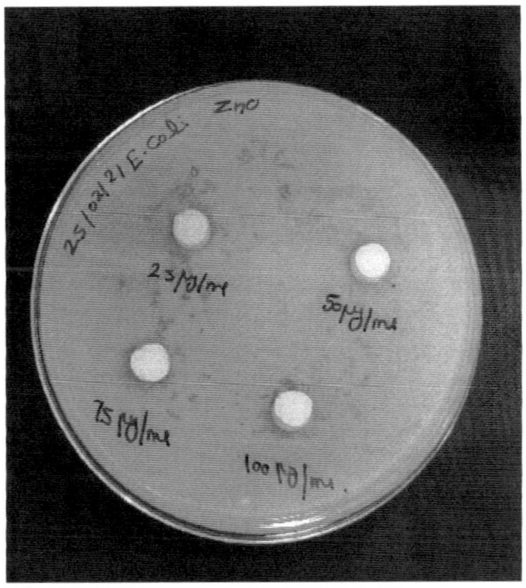

Figure 3: Antibacterial activity ZnO nanoparticles

4. Conclusion:

ZnO nanoparticles were prepared using the simple sol-gel method. The prepared zinc oxide nanoparticles were characterised using XRD and SEM. From XRD it was found that the crystalline size of the prepared pure zinc oxide nanoparticles is 36nm. From the SEM image it was found that the particles were present in the form of flocks. ZnO nanoparticles can be used in various fields due to its unique properties like low toxicity, biocompatibility and biodegradability. It has a vast range of future applications. It can be even applied in the field of electronics, cosmetics and biomedical devices. Zinc oxide nanoparticles can be used in medical fields to even treat cancer through drug delivery or gene delivery methods. It has also shown its potential in treatments of antibacterial and antidiabetic fields. ZnO is now also being used for bioimaging because of its luminescent properties. There are many new fields in which ZnO nanoparticles are being used like the textile industry as adding this material to fabrics has been found to increase the resistance against absorption of UV and visible light, bacteria, and deodorant. This method of synthesis can be used for the large scale production of ZnO nanoparticles at room temperature.

Table 1: Antibacterial activity ZnO nanoparticles

Sample	**Escherichia coli**			
Cons.	**25 µg/ml**	**50 µg/ml**	**75 µg/ml**	**100 µg/ml**
ZnO	**5.4 mm**	**6.2 mm**	**6.7 mm**	**7.2 mm**

REFERENCES

1. Nikhil Chauhan, Virender Singh, Suresh Kumar, Monika Kumari & Kapil Sirohi, 2019 Preparation of silver and nitrogen co-doped mesoporous zinc oxide nanoparticles by evaporation induced self-assembly process to study their photocatalytic activity", Journal of Sol-Gel Science and Technology 90 2 390-403.

2. Zhang Y L, Yang Y J, Zhao H, Tan R Q, Cui P & Song W J, 2009 Preparation of ZnO nanoparticles by a surfactant-assisted complex sol-gel method using zinc nitrate." Journal of sol-gel science and technology 51 2 198-203.

3. Baodian Yao1, Huazhong Shi, Huijuan Bi and Lide Zhang, 2000 Optical properties of ZnO loaded in mesoporous silica Journal of Physics: Condensed Matter12 28 6265.

4. Nikhil Chauhan, Virender Singh, Suresh Kumar, Monika Kumari and Kapil Sirohi. 2019 Synthesis of sulfur & indium co-doped mesoporous zinc oxide nanoparticles via hydrothermal method to study their photocatalytic activity Optik 185 838-846.

5. Hasnidawani J N, Azlina H N, Norita H, Bonnia N N, Ratim S, and Ali E S, 2016 Synthesis of ZnO nanostructures using sol-gel method Procedia Chemistry 19 Pages: 211-216.

6. Madhubala V, Kalaivani T 2018 Phyto and hydrothermal synthesis of Fe3O4@ZnO core-shell nanoparticles using Azadirachta indica and its cytotoxicity studies Applied surface science 449 584-590.

7. Sebastian Polarz, Andrey V. Orlov, Ferdi Schüth and An Hui Lu 2007 Preparation of high surface area zinc oxide with ordered porosity, different pore sizes, and nanocrystalline walls Chemistry–A European Journal 13.2 592-597.

8. Sunandan Baruah and Joydeep Dutta.2009 pH-dependent growth of zinc oxide nanorods Journal of Crystal Growth 311 8 2549-2554.

9. Mostafa Zamani, Mojtaba Rostami, Mozhgan Aghajanzadeh, Hamidreza Kheiri Manjili, Kobra Rostamizadeh & Hossein Danafar 2018 Mesoporous titanium dioxide@ zinc oxide–graphene oxide nanocarriers for colon-specific drug delivery Journal of materials science 53 3 1634-1645.

10. Riyadh M Alwan, Quraish A Kadhim, Kassim M Sahan, Rawaa A Ali, Roaa J Mahdi, Noor A Kassim, Alwan N Jassim 2015 Synthesis of zinc oxide nanoparticles via sol–gel route and their characterization Nanoscience and Nanotechnology 5 1 1-6.

11. Lihitkar P B, Samuel Violet, Mandar Shirolkar, Jai Singh, Srivastava O N, Naik R H, Kulkarn S K, 2012 Confinement of zinc oxide nanoparticles in ordered mesoporous silica MCM-41 Materials Chemistry and Physics 133 2 3 850-856.

12. Venkataprasanna K.S., Prakash, J. Vignesh S, Bharath, G ManigandanVenkatesan, Fawzi Banat, Sahabudeen, S. Saravanan Ramachandran, G. DevanandVenkatasubbu. 202019.Fabrication of Chitosan/PVA/ GO/CuO patch for potential wound healing application. Journal of biological macromolecules, 143, 744-762.

13. J. Prakash , T. Sushil Kumar , K.S. Venkataprasanna , R. Niranjan , M. Kaushik , DebashreeBanitaSamal , G. DevanandVenkatasubbu. 2019 PVA/alginate/hydroxyapatite films for controlled release of amoxicillin for the treatment of periodontal defects. Applied Surface Science.495, 143543.

IOP Conf. Series: Materials Science and Engineering 1219 (2022) 012020 doi:10.1088/1757-899X/1219/1/012020

Exploitation of Versatile Naturally Derived Blue Colorant Pigment 3-[2-hydroxy-5-(5-hydroxy-1H-indol-3-yl)-1H-pyrrol-3-yl] indol-one

Ishani[1], **Isita**[1], **T.Vijayakumar**[2]

[1]Department of Biotechnology, Faculty of Engineering and Technology, SRM Institute of Science and Technology, Kattankulathur,Tamilnadu,India,603203
[2]Futuristic Materials Research Centre for Planetary Exploration, Department of Physics and Nanotechnology, College of Engineering and Technology, Faculty of Engineering and Technology, SRM Institute of Science and Technology, Kattankulathur - 603 203, Kancheepuram, Tamil Nadu, India.

*Corresponding Author email: vijayakt1@srmist.edu.in

Abstract. 3-[2-hydroxy-5-(5-hydroxy-1H-indol-3-yl)-1H-pyrrol-3-yl] indol-one depicts a multifaceted pigment which was discovered in distinct bacteria species like *Chromobacterium violaceum, Duganella sp. , Collimonas sp. , D. violaceinigra , Pseudoalteromonas sp. , J. svalbardensis , Janthinobacteriumlividum,* etc. 3-[2-hydroxy-5-(5-hydroxy-1H-indol-3-yl)-1H-pyrrol-3-yl] indol-one is liable for the stunning purple pastels which demonstrates a variety of biological behaviors and is reportedly garnering growing popularity in industrial- market segments, for instance, textiles, medicine, cosmetics etc. It is perhaps a novel therapeutic category of drug which is L-tryptophan originated displays anti-cancer along with antibiotic action. The intention of this review was to identify the numerous applications of 3-[2-hydroxy-5-(5-hydroxy-1H-indol-3-yl)-1H-pyrrol-3-yl] indol-one and evaluate its varied biochemical properties in a pharmacological sense, including its antioxidant, antitumor, anti-trypanocidal, antibacterial, anti-ulcerogenic, anticancer and immunomodulatory activities. Additionally, its usage in cosmetics, food, textiles, toys along with insecticides has stemmed as peculiar emerging segments of applications of 3-[2-hydroxy-5-(5-hydroxy-1H-indol-3-yl)-1H-pyrrol-3-yl] indol-one have been reviewed here.

Keywords: Violacein; Industrial Applications; Biological properties

1. Introduction

3-[2-hydroxy-5-(5-hydroxy-1H-indol-3-yl)-1H-pyrrol-3-yl] indol-one, Violacein, a purple indole derivative pigment comprises of three units in its structure: 5-hydroxyindole, oxindole, and 2-pyrrolidone has been researched for their biological significance. This pigment possesses prospective medical implementations along with distinct immunomodulatory activities inculcating antifungal activity, anti-tumoral activity, antiparasitic activity, antiprotozoal, antiviral, antioxidant, antidiarrheal, antiulcerogenic, antipyretic, analgesic, antibacterial, antitrypanocidal along with anticancer activity [1]. Violacein is procured via distinct bacterial strains, for instance, *Janthinobacterium lividum, Pseudoalteromonas sp., Duganella violaceinigra, Alteromonas luteoviolacea, Collimonas, D.violaceinigra* etc. Violacein, for its robust usage in food, pharmaceutical, cosmetic along with textile industries, is a bacterial quorum-sensing chromophore, also known as 'purple treasure'. It also offers a wide range of biological processes that is the object of comprehensive scientific study [2]. They are attractive factor for human consumption because of biodegradability, non-toxicity alongwith non-carcinogenicity of natural pigments, colours and dye. Violacein being a marker molecule for quorum sensing (QS) studies alongwith inhibition of QS regulated production of violacein in C. Violaceum [3]. The introduction of violacein pigment into the nylon membranes as a natural antibacterial agent would give the wound wrapping procedure added benefit [4].

Fig.1. shows the molecular structure of 3-[2-hydroxy-5-(5-hydroxy-1H-indol-3-yl)-1H-pyrrol-3-yl]indol-one (Violacein) that belongs to the class of hydroxy indole developed by the oxidative coupling of the third position of 1,3-dihydro-2H-indol-2-one with the third position of 1,3-dihydro-2H-pyrrol-2-one, which is substituted in the fifth position through 5-hydroxy-1H-indol-3-yl group subsequently formed double - bond having E configuration. Antimicrobial, antiprotozoal, antileishmanial, antimycotic, antiparasitic, antibacterial activity ,antitumor and antiviral properties have all been discovered in violacein. The antimicrobial activity of violacein via Duganella violaceinigra (NI28) against S. aureus has recently been demonstrated [5].

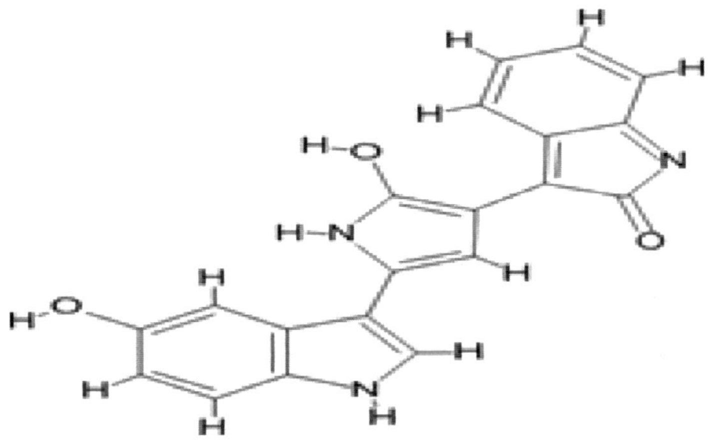

Fig.1. 3-[2-hydroxy-5-(5-hydroxy-1H-indol-3-yl)-1H-pyrrol-3-yl] indol-one molecular structure

Moreover, violacein inhibits the growth of phytopathogenic fungi like Rosellinia necatrix. It also depicts antiprotozoal properties against nanoflagellates. In Ehrlich ascites tumors and MOLT(4) leukemia, the proapoptotic action of violacein derived through C. violaceum (CCT 3496) was reported in vitro and in vitro [6]. In vitro and in vivo studies have shown that violacein has antimalarial activity. Violacein micromolar concentrations were found to be effective in in-vitro condition for killing chloroquine sensitive along with resistant Plasmodium falciparum varieties, while in in-vivo condition it inhibited parasitemia although after the parasite had established itself, and Plasmodium infected mice was protected from undergoing any lethal challenge [2]

Table 1: Various applications of Violacein

Microorganism	Location	Industry	Property exploited	Application	References
Janthinobacterium lividum 1522 strain ATCC 12473	Amsterdam (Netherlands)	Textile	antimicrobial	Optimization for dyeing polyamide fabrics	[7]
Chromobacterium sp. NIIST (MTCC 5522)	Kerala, India	NA	Antifungal activity	economic methodology for recovering pure crystalline violacein from crude-violacein	[8]
Iodobacter sp. 7MAnt	Antarctic Peninsula.	NA	Antibacterial and antifungal	n is a wide-spectrum abiotic, effective against bacteria and fungi	[9]

2. Material and Methods

Placing emphasis on the promising techniques for acquiring violacein pigment, distinct extraction approaches of pigment through cell pellets/exhausted culture – broth of microbes have been reported [10]. Chemical solvents, water alongwith boiling-buffered ethanol solution for cell lysis were utilized for extraction of the violacein pigment accompanied by washing alongwith drying/dehumidification. Several enzymatic alongwith mechanical approaches have also been reported [10] for distinct organisms. Separated components were further characterized utilizing the blend of distinct technologies, for instance, NMR,

UV/VIS, IR, FTIR and ESR. Various other strategies were also used for determining the key functional groups along with bonds existing within the pigment [10]. Spectra acquired through characterization was analyzed and contrasted with recognized spectra of pigments. In the analysis, every pigment was found distinct from other types of extraction along with character which indicates pigment diversity developed via microorganisms [10].

3. Results and Discussion

Violacein offers numerous fascinating biological activities of therapeutic significance and plenty of these have indeed been extensively analyzed. Preliminary research has documented the different violacein functions within bacterium serving the role of an antioxidant agent which is followed by the elaboration of its significance in two certain bacterial violacein producing species. The results of such research focus the fact that they are still to be investigated in the realm of violacein activities and to reveal the awareness of its position in the community. Then, on the other hand, further study has emphasized the medicinal significance of violacein, like trypanocidal, leishmanicidal, antiviral, antifungal, antibacterial, antiprotozoal, antioxidant and antitumor alongwith its modes of action [11].

Anti-bacterial activity

Free alongwith nanoparticles-loaded violacein demonstrated in vitro antibacterial activity against ATCC 29213 alongwith ATCC 25923 strains of *Staphylococcus aureus* .They demonstrate approximately two to five fold lower minimum inhibitory concentration (MIC) then free Violacein [12]. Janthinobacterium lividum, the bacterial species that commonly infests amphibian skin, is noteworthy for its synthesis of violacein, a dark purple pigment that demonstrates multitude of inhibitory activities against growth of various fungi, bacteria, parasites alongwith tumor tissues [13].

Anti-tumor activity

Violacein natural indolic pigment extracted from Chromobacterium violaceum was assessed in-vivo alongwith in-vitro for its antitumor efficacy against the Ehrlich ascites tumor (EAT) [14].

Anti-oxidant

Furthermore, an imbalance in homeostasis alongwith production of reactive oxygen species can trigger several pathologies. Both reactive oxygen species (ROS) alongwith reactive nitrogen species (RNS), specifically peroxynitrite, can significantly increase the affected organism, when mammal is attacked by an antigen like bacterium. Their natural role in overcoming the foreign species and their inactivation alongwith degradation of residues generated by reactive species, enzymatic and/or non-enzymatic antioxidant defenses protect invading of organisms from oxidative stress [15].

Anti-fungal activity

Pure violacein extracted from current isolate was extremely highly efficient against all pathogenic fungi of humans and plants verified with yeast like Candida albicans, Aspergillus flavus, Rhizoctonia solani, Phytopathogenic Fusarium oxysporum, Penicillium expansum, compared with common antifungal drugs agents like bavistin and amphotericin B. The zone of inhibition induced by the minimum concentration of violacein inhibitors and the antimicrobial action of pure violacein against medically significant fungi have yet been identified [16].

Anti-cancer potential of bacterial pigments

The probable cytotoxic effect of this material has been identified on various cell lines like V79, Vero, Hep2, NCI-H460 , HT29, SW480, KM12.. It has also been documented in tumor-cultivated cell line in leukemia patients and its important activities toward cancer cells originating from B cell chronic lymphocytic in leukemia patients. Violacein isolated from C.violaceum demonstrates the cytotoxic effects along with apoptosis of multiple cancer cells comprising colorectal cancer, leukemia, uveal melanoma and lymphoma cells [17]. Multi-target cytotoxicity processes characterize violacein`s anti-tumoral activity. Caspase-8, MAP kinase- p38, tumoral necrosis factor-1 (TNF) activation are responsible for violacein property of serving as a promising apoptosis inducer in the leukemic HL 60- lineage. In colorectal tumoral of the HT-2922 lineage, violacein can also cause cell death through generalized oxidative stress and in lymphocytes violacein serves the role of direct-enzyme modulation [4].

Anti-parasitic activity

Evidently, the efficient concentration of pigment violacein (LC50-N 30 μmol.dm^{-3}) against Caenorhabditis elegans was investigated in a series of potential drug leads alongwith other established agents of chemotherapy including flavopiridol, (LC50 b 48.3. dm^{-3}). The analysis of mechanisms engaging in this activity showed an Escherichia coli accumulation which is non-pathogenic in the worms' intestine in violacein presence alongwith apoptosis induction. Studies of the operation of violazole anti-nematode not only have assessment of its therapeutic ability but also an appreciation of its toxicity with potential targets for multicellular eukaryotic drugs [18].

Table 2: An overview of Violacein

Microorganism	Medium	Incubation	Analysis technique	Pigment produced	Comments	References
Chromobacterium violaceum	sugarcane bagasse, L-tryptophan	30 °C, 200 rpm for 24 h	UV-Vis, FTIR, Thin-layer chromatography, NMR and mass spectrometry analysis	Violacein	Produced from agricultural waste	[19]
Pseudoduganella violaceinigra sp. NI28	Lysogeny Broth (LB) broth (Difco BD, USA)	30 °C and 250 rpm	Antibiotic resistance determination , Spot viability assay, Statistical analysis	Violacein	Violacein utilized for killing polymicrobial pathogenic communities	[20]
violaceum UTM5	liquid pineapple waste, Nutrient agar, nutrient broth	200 rpm, 30 C for 24 h	FT-IR, TLC	Violacein	Produced from liquid pineapple wasted	[21]

A range of review papers have appeared in literature and this review inculcates the biological actions of violacein, for instance, anti-fungal activity, anti-tumoral activity, anti-parasitic activity, anti-protozoal, anti-viral, anti-oxidant, anti-diarrheal, anti-ulcerogenic, antipyretic, analgesic, antibacterial, anti-trypanocidal alongwith anti-cancer activity. This review also highlights the significance of violacein pigment, it`s properties alongwith their pharmacological potential applications. The presence of purple pigment Violacein, which is potent marker feature during QS studies, is single greatest factor contributing to the interest of the research community to this bacterium [22]. It has been possible to further explore bioactive behaviors identified in this research in order to produce added-value goods for the food industry, pesticide industry, fabric industries and to enhance food quality alongwith protection during storage [23].

4. Conclusion

Violacein tends to be a vital compound and possesses significant pharmacological activities like anti-biotic, anti-viral, immunomodulatory, anti-pyretic, analgesic, anti-cancer effects, etc. This analysis highlights Violacein reported to be of economic value due to its biological activities.

References

1. Durán, N., et al., Advances in Chromobacterium violaceum and properties of violacein-Its main secondary metabolite: A review. Biotechnology advances, 2016. **34**(5): p. 1030-1045.
2. Durán, M., et al., Potential applications of violacein: a microbial pigment. Medicinal Chemistry Research, 2012. **21**(7): p. 1524-1532.
3. Vijay Kothari, S.S., Divya Padia, Recent research advances on Chromobacterium violaceum. Asian Pacific Journal of Tropical Medicine, 2017.
4. Siti Hajar Osman, S.A.S., Fabrication of Nylon-66 Membranes Coated with Violacein Pigment for Wound Dressing Application. eProceedings Chemistry, 2018.
5. Masuelli, L., Pantanella, F., La Regina, G., Benvenuto, M., Fantini, M., Mattera, R., ... & Bei, R., Violacein, an indole-derived purple-colored natural pigment produced by Janthinobacterium lividum, inhibits the growth of head and neck carcinoma cell lines both in vitro and in vivo. Tumor Biology, 2016. **37**(3): p. 3705-3717.
6. Choi, S.Y., et al., High-level production of violacein by the newly isolated Duganella violaceinigra str. NI28 and its impact on Staphylococcus aureus. Scientific reports, 2015. **5**(1): p. 1-12.

7. Maria Kanelli, M.M., Margarita Kalakona, Sozon Vasilakos, Dimitris Kekos, Jasmina Nikodinovic-Runic and Evangelos Topakas, Microbial Production of Violacein and Process Optimization for Dyeing Polyamide Fabrics With Acquired Antimicrobial Properties. Front. Microbiol, 2018.
8. Anju Sasidharan, N.K.S., Dileepkumar Bhaskaran Nair Saraswathy Amma, Radhakrishnan Kokkuvayil Vasu, Anupama Vijaya Nataraja & Krishnakumar Bhaskaran, Antifungal activity of violacein purified from a novel strain of Chromobacterium sp. NIIST (MTCC 5522). Journal of Microbiology 2015.
9. Atalah, J., et al., Isolation and characterization of violacein from an Antarctic Iodobacter: a non-pathogenic psychrotolerant microorganism. Extremophiles, 2020. **24**(1): p. 43-52.
10. SUDHA, C. and S. Aggarwal, Extraction, Purification and Characterization of Microbial Pigments–A Review. Biosciences, 2014: p. 2070.
11. Ramesh, C., N.V. Vinithkumar, and R. Kirubagaran, Marine pigmented bacteria: A prospective source of antibacterial compounds. Journal of Natural Science, Biology and Medicine, 2019. **10**(2): p. 104.
12. Martins, D., et al., Evaluation of the antibacterial activity of poly-(D, L-lactide-co-glycolide) nanoparticles containing violacein. Journal of Nanoparticle Research, 2011. **13**(1): p. 355-363.
13. Durán, N., et al., Violacein: properties and biological activities. Biotechnology and applied biochemistry, 2007. **48**(3): p. 127-133.
14. B., N.L.D.V.R.V.P., Growth inhibition and pro-apoptotic activity of violacein in Ehrlich ascites tumor. Chemico-Biological Interactions, 2010: p. 43-52.
15. Marlon Konzen, D.D.M., Clarissa A. S. Cordova,Tiago O. Vieira, and R.V.A.n.a.T.n.B. Creczynski-Pasa, Antioxidant properties of violacein: Possible relation on itsbiological function. Science Direct, 2006.
16. Sasidharan, A., et al., Antifungal activity of violacein purified from a novel strain of Chromobacterium sp. NIIST (MTCC 5522). Journal of Microbiology, 2015.
17. Ahmad, W.A., et al., Production and characterization of violacein by locally isolated Chromobacterium violaceum grown in agricultural wastes. Applied biochemistry and biotechnology, 2012. **167**(5): p. 1220-1234.
18. Hansol Im, S.Y.C., Sangmo Son & Robert J. Mitchell, Combined Application of Bacterial Predation and Violacein to Kill Polymicrobial Pathogenic Communities. scientific reports, 2017. **7**.
19. Kanelli, M., et al., Microbial production of violacein and process optimization for dyeing polyamide fabrics with acquired antimicrobial properties. Frontiers in microbiology, 2018. **9**: p. 1495.
20. Sasidharan, A., et al., Antifungal activity of violacein purified from a novel strain of Chromobacterium sp. NIIST (MTCC 5522). Journal of Microbiology, 2015. **53**(10): p. 694-701.
21. Ahmad, W.A., C.K. Venil, and C.A. Aruldass, Production of violacein by chromobacterium violaceum grown in liquid pineapple waste: current scenario, in Beneficial microorganisms in agriculture, aquaculture and other areas. 2015, Springer. p. 45-58.
22. Rodrigues, A.C., et al., Quorum quenching and microbial control through phenolic extract of Eugenia uniflora fruits. Journal of food science, 2016. **81**(10): p. M2538-M2544.
23. Lozano, G.L., et al., A Chemical Counterpunch: Chromobacterium violaceum ATCC 31532 Produces Violacein in Response to Translation-Inhibiting Antibiotics. Mbio, 2020. **11**(3).

IOP Conf. Series: Materials Science and Engineering 1219 (2022) 012021 doi:10.1088/1757-899X/1219/1/012021

Spectral Identification of Constituents of Rock Minerals

S.Joshua Samuel[1], M.R.Kannan[1], Y.Sheeba Sherlin[2],T.Vijayakumar[1,*]

[1]Futuristic Materials Research Centre for Planetary Exploration, Department of Physics and Nanotechnology, College of Engineering and Technology, Faculty of Engineering and Technology, SRM Institute of Science and Technology, SRM Nagar, Kattankulathur-603203, Kanchipuram, Chennai, Tamil Nadu, India.
[2]Department of Physics, Scott Christian College (Autonomous), Nagercoil-629003, Tamil Nadu, India.

**E-mail: vijayakt1@srmist.edu.in*

AbstractThe investigations of various planetary materials like mineral, organic, and biogenic materials, pre-solar grains, primitive meteoroids and soils from the moon and asteroids have been an effective research interest in planetary exploration. Spectroscopic analyses of terrestrial analog materials can give a best of understanding geology structure, chemical elements and processes on surfaces *viz,* valuable elements present in the earth and mars. The present work aims at understanding the identification of different types of rock minerals, its structure and chemical compositions, structural and micro-structural characterization. The rock sample is taken from the hills of Yelagiri located in the Eastern Ghats of Tamil Nadu. The samples were analyzed both as whole rocks and as fine powders as prepared similar to the ExoMars SPDS. The rock minerals are characterized using X-ray Diffraction, Scanning Electron Microscopy, FT Raman, X-ray Fluorescence.

Keywords: Minerals; Spectroscopy; XRD; SEM; XRF

1. Introduction

Mineralogy is an area of science which is used to evaluate various geological features that are revealed and still a lot to be explored. It has been the area of interest due to its close connection with the development of human society [1-3]. The rapid growth of science and technology demands a lot from mineralogy. A mineral is an element mostly found on the earth surface which is a normally crystalline structure and forms as a result of geological process. It must be a homogenous substance with defined composition. Chemistry and crystal structure defines a mineral. Each mineral species has its own name, while its chemical composition is expressed by means of an idealized formula [4-7]. According to their definition minerals of same composition have different structures pertain to different species. Two minerals have same composition but with different crystal structure are known as polymorphs. Crystal structure generally identifies a mineral whose physical and chemical properties are characterized by the presence of additional elements with certain natural boundaries. Some specific features of the crystal structure distinguish the structural varieties of the minerals with the chemical composition remaining constant. The demand for mineral is increasing and mineralogical studies have great deal of importance [8-10]. This work aims to understand the composition and structural features of rock forming minerals present in Eastern Ghats of Tamil Nadu using various spectroscopic techniques.

2. Materials and Methods

Spectroscopy is the scientific technique and interpretation of electromagnetic source for radiation absorbed or emitted, when molecules, atoms of a sample move from ground state to excited state. Spectroscopy can be described based on two categories Atomic and molecular spectroscopy. Atomic spectroscopy is connected with the interaction of the electromagnetic radiation along with atoms in the lowest energy state. Molecular spectroscopy tells about with the interaction of radiation with molecules, result in transition between rotational and vibrational energy levels. Most of the experimental methods of spectroscopy depend upon the study of interaction between electromagnetic radiation and the molecular system the nature of information required decides the selection of particular spectral technique. The rock mineral has been analyzed using X-ray Diffraction, X-ray Fluorescence FTIR, FT Raman and Energy Dispersive Spectroscopy.

3. Results and Discussion

X-ray Diffraction (XRD)

X-rays diffraction study leads to very much clear understand the physics and chemical property of the metal, polymer materials, solids, minerals. It is a convenient method for the qualitative identification of the crystal compound. X-Ray Diffraction is a most useful technique for elemental analysis and mineral analysis and characterizations and the BRUCKER X-ray diffractometer was used to measure XRD pattern. The prominent peaks were observed at 26.6, 35.6, and 41.6 corresponding to the miller indices (0 1 1), (0 0 2) and (3 1 3) which indicates the presence of graphite, quartz and fullerite respectively. XRD analysis confirms that the rock mineral contain of silicon in high content.

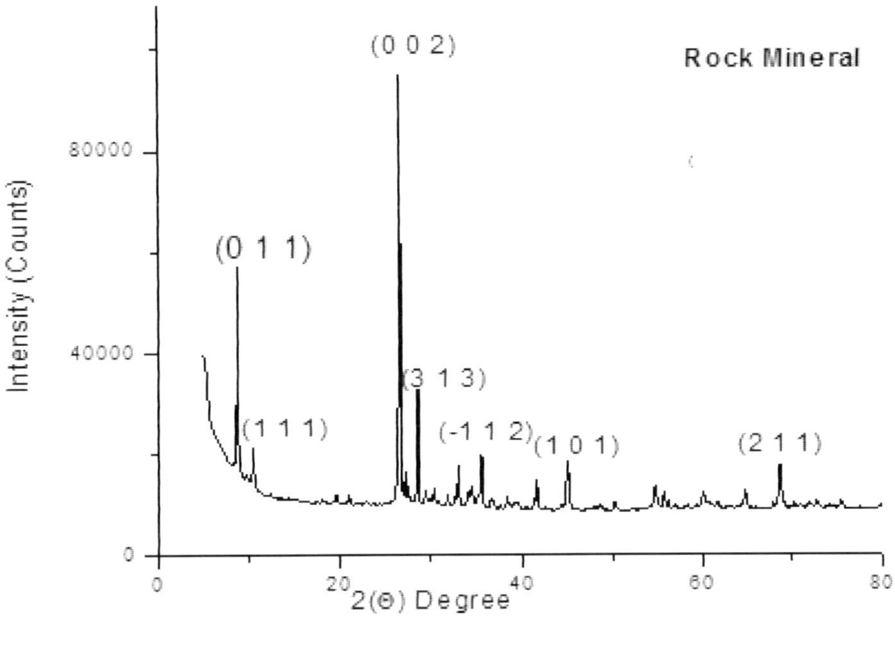

Fig.1 XRD Pattern of Rock mineral

X-ray Fluorescence (XRF)

XRF is an analytical technique which can analyze the sample directly with little sample preparation and the intensity of peaks in the XRF spectra is proportional to the percentage concentration. The elemental analysis shows that the mineral contains Fe, Al, Mg, Ca, Si, Ti, Zn, K and it is observed that the silicon (Si) and the dypsorpium (Dy) have high level of concentration.

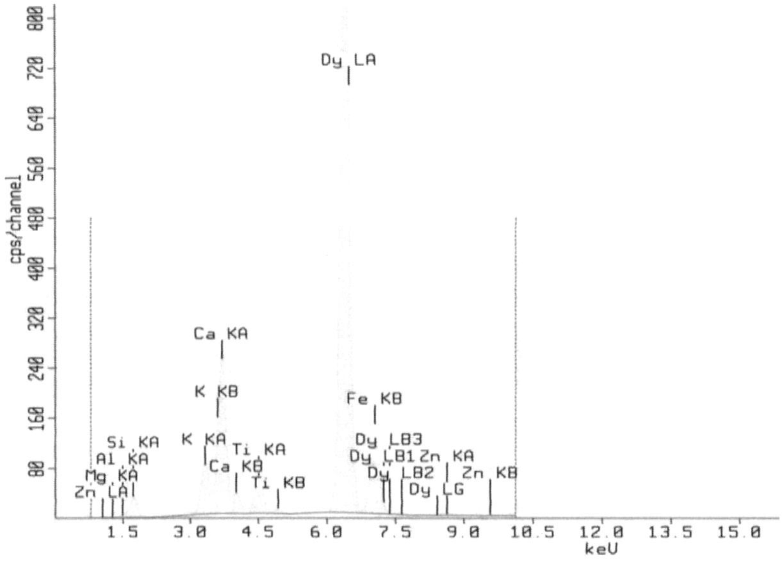

Fig.2 XRF Spectra of Rock mineral

Table 1 XRF Peak values of Mineral

ELEMENT	ATOMIC NUMBER	CONCENTRATION
Mg	12	15.17
Si	14	718.9
Al	13	78.81
K	19	1893.4
Ca	20	5870.1
Ti	22	1294.3
Fe	26	24632.7
Zn	30	58.0
Dy	66	47.8

Vibrational Spectroscopy

FT-Raman (Fig. 3) and FT-IR (Fig. 4) and are used to indicate the vibrational modes present in the material. The observed vibrational modes are analyzed for their frequencies, visual intensities and the tentative assignments are provided in the Table 2. The vibrational spectraexplain the presence of hydroxyl group (O-H), alkyl group (C-H) and carbonyl (C-O) group in the rock mineral.

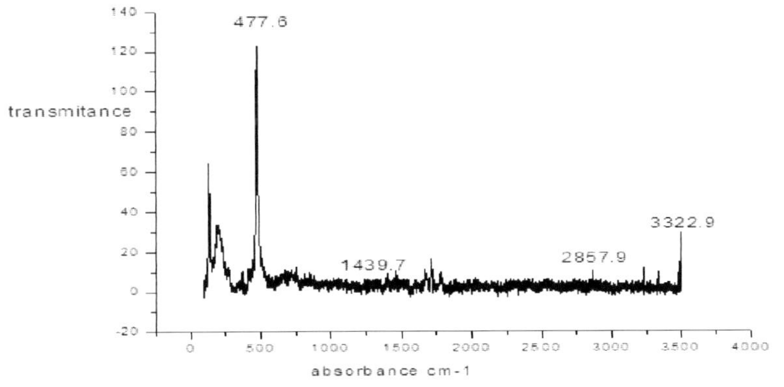

Fig. 3FT-Raman Spectrum of Rock Mineral

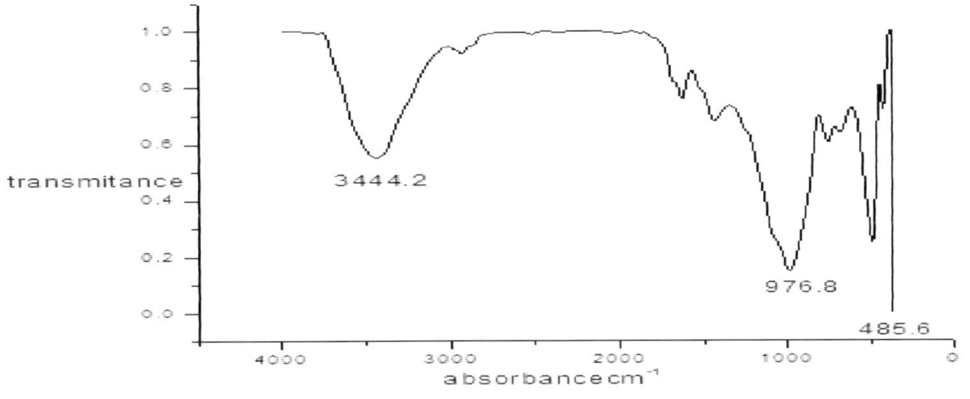

Fig. 4 FT-IR Spectrum of Rock Mineral

Table 2 Tentative vibrational assignments of the rock mineral

□IR	□RAMAN	BONDING
3495.8		O-H
	3444.2	O-H
2016.7		Si-O
1676.2		C-O
	1617.1	C-O
1468.3		C-O
	976.8	C-H
	745.3	C-H
	485.3	C-H

IOP Conf. Series: Materials Science and Engineering 1219 (2022) 012021 doi:10.1088/1757-899X/1219/1/012021

Energy Dispersive Spectroscopy (EDS)

Energy Dispersive Spectroscopy is used as an alternate tool in the Elemental identifications and the morphological structure analysis. Energy Dispersive Spectroscopy (EDS) analyzes the quantitative and qualitative analysis of the mineral sample. EDS micrograph (Fig.5) of mineral shows the presence of Oxygen (63.8%), Carbon (18.7%) and Dysprosium (2%). EDS confirms (Table 3) that the rock mineral has the high level content of silicon and oxygen and the Dysprosium is in low content.

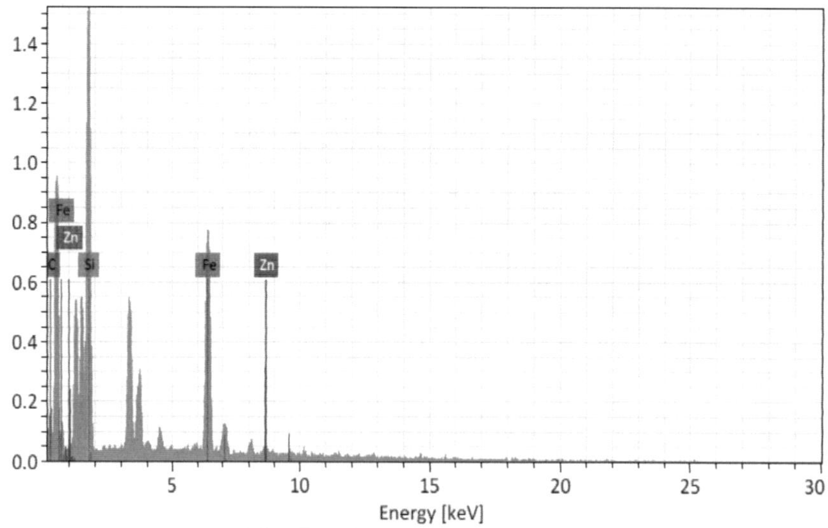

Fig. 5 EDS micrograph of Rock Mineral

Table 3 Mass and atomic percentages of Rock Mineral

ELEMENT	ATOMIC NUMBER	MASS(%)
O	8	63.8
C	6	20.7
Si	14	18.2
Fe	26	8.3
Mg	12	6.9
Al	13	6.4
K	19	5.5
Ca	20	3.0
Ti	22	2.3
Dy	66	1.0

4. Conclusion

XRD analysis confirms that the rock mineral contains silicon in high content and the XRF analysis reveals that the silicon and the dysprosium have high level of intensity. Vibrational spectra explain the presence of hydroxyl group (O-H), alkyl group (C-H) and carbonyl (C-O) group in the rock mineral. AAS confirms that the silicon has the high concentration in the rock mineral. EDS shows that the rock mineral has high content of silicon and low content of dysprosium.

Acknowledgements
Authors thank High Performance Computing Center, SRM Institute of Science and Technology for providing the computational facility.The author T.Vijayakumar acknowledges the financial support obtained from the Department of Space, Government of India [Grant no.B.19012/57/2016-II] through RESPOND project and the selective excellence initiative award received from SRM Institute of Science and Technology.

References

[1] Ramasamy V et al 2009Ind. Jour. pure and app. Phy. 47586-591
[2] Brown, G., Cry. . Clay Mine. X-ray Iden (eds Brindley, G. W. and Brown G.), 1980Miner.Soc, London, 361–410
[3] Borg, I. Y. and Smith, D. K. 1968 Am. Miner. 53 1709-1723
[4] PalanivelR.,VelrajG.,FTIR and FT-Raman spectroscopic studies of fired clay artifacts recently excavated in Tamil Nadu 2007Indian Journal of Pure and Applied Physics 45 501-508
[5] VelrajG., PrabakaranK., Mohamed Musthafa A. and HemamaliniR. FT-IR and Micro-Raman Spectroscopic studies of archaeological potteries recently excavated in Tamilnadu2010 India Recent Researches in Science and Technology 2(10)94-99
[6] Sathya P. and VelrajG. FTIR spectroscopic and X-ray diffraction analysis of archaeological grey potteries excavated in Alagankulam, Tamil Nadu 2011 Indian Journal of Experimental Sciences2(5) 04-06
[7] Viruthagiri, G, Ponnarasi, K and Rajamannan. B, Minerological study of clay samples and clay mines at different Depths in cuddalore district, Tamilnadu2010 International Journal of Current Research2035-038
[8] Mineral powder Diffraction file, Joint Committee of Powder Diffraction Standards (JCPDS) 2002
[9] RamasamyV.,SureshV., Characterization of Minerals and Naturally occurring Radionuclides in River Sediments2009 Research Journal of Applied Sciences, Engineering and Technology 1(3) 140-144
[10] Preethi Sagar Nayak and B.K SinghInstrumental Characterization of clay by XRF, XRDand FTIR2007Bulletin of Materials Science 30(3) 235-238

IOP Conf. Series: Materials Science and Engineering 1219 (2022) 012022 doi:10.1088/1757-899X/1219/1/012022

Investigation on Spinel Co_3O_4 Nanoparticles through Mechanochemical Technique

Bharath Nivas[1], Aleena Rose[1], B. Shunmugapriya[1], T. Vijayakumar[1*]
[1]Futuristic Materials Research Centre for Planetary Exploration, Department of Physics and Nanotechnology, College of Engineering and Technology, Faculty of Engineering and Technology, SRM Institute of Science and Technology, SRM Nagar, Kattankulathur, Chengalpattu, Chennai, Tamil Nadu 603203, India

[*]Corresponding author mail id: vijayakt1@srmist.edu.in

Abstract. The spinel type Cobalt Oxide (Co_3O_4) nanoparticles were synthesized through simple mechanochemical method. The synthesized nanoparticles were characterized through X-Ray Diffraction (XRD), Field Effect Scanning Electron Microscope (FE-SEM), Energy Dispersive spectroscopy (EDS) and Raman spectroscopy respectively. The Crystallite structure of the prepared sample was analyzed through X-Ray diffractograms. The surface morphology and the presence of elements were investigated through energy dispersive spectrum. Raman spectroscopy was utilized for recording the vibrational pattern of the sample. The cyclic voltammetry and electrochemical impedance spectroscopy reveals the good electrochemical performance of the Co_3O_4 nanomaterials.

Key words: Cobalt Oxide, Nano particles, Raman, Micrograph, crystalline size

1. Introduction

Nanoparticles are tiny materials, with size between 1 to 100 nm. To define, one nanometre is one billionth of a metre. Synthetic methodologies of nanomaterials include two types; bottom-up approach and top-down approach. The process of assembling the nanostructures with small constituents such as atoms or molecule is known as bottom-up approach (chemical vapour deposition, sol gel process). The building of nanomaterials starting with larger structures is known as top-down approaches (ex: lithography, ball milling). Nanoparticles can have enhanced properties than their bulk counterparts because the smaller particles have the larger surface area. Also, nanoparticles hold the ability to move around more in solution than the bulk particles [1-4]. Cobalt oxide (Co_3O_4) is a mineral compound which is insolvable in water and solvable in alkalis. It is present as black dense powder. Nanoparticles possess a cubic crystal structure with a molecular weight of 240.80 g/mol and density of 6.11 g/cm^3. Boiling point and melting point of this material is 900°C and 895°C respectively. Magnetic susceptibility of Co_3O_4 is $+7380.10^{-6} cm^3/mol$. The magnetic behaviour of Co_3O_4 plays an important role in the field of storage devices and in optoelectronics for flat-panel display. Agilandeswari et al., reported the antiferromagnetic property of Co_3O_4 nanoparticles and it mainly depends on its size and shape of the materials. Similarly, Saeed Farhadi et al., studied the optical absorption band gap of Co_3O_4 nanoparticles and estimated the values at 2.20 eV and 33.55 eV [5, 6]. Generally, Transition metals can be used to construct spinel structures (AB_2O_4) because of its multiple oxidization states which lead to improve additional pseudo-capacitance. In this category, Co_3O_4 spinel occupies a prominent role in energy storage because of its high theoretical specific capacitance, electrical conductivity, and higher electrochemical activity and more redox reactions [7,8]. Co_3O_4 nanomaterials can also be used in vast applications including Lithium ion battery, electronic gas sensors, medicine, thin films electrode, supercapacitor and magnetic material. The properties of Co_3O_4 in energy storage applications are highly related to its particle size. All the previous reports reveal that Nanocrystalline Co_3O_4 is good for promoting the electrochemical properties [9, 10]. It is hard to obtain Nanocrystalline Co_3O_4 through methods comprising pulsed laser deposition, sol–gel route, homogeneous precipitation, cobalt salt decomposition etc. In this present article, spinel type Co_3O_4 was synthesized through simple mechanochemical technique. The obtained sample was characterized through XRD, SEM, Micro Raman, cyclic voltammetry and electrochemical impedance spectroscopy.

2. Materials and Methods

Cobalt (II) Nitrate Hexahydrate $Co(NO_3)_2.6H_2O$ (Loba chemie 97%), Ammonium hydrogen carbonate (NH_4HCO_3), (Loba chemie, 98.5%) were used as a precursor material without further purification. The deionised water was used throughout the reaction. In this synthesis technique, 2.978 g of Cobalt nitrate and 2.022g of Ammonium hydrogen carbonate were taken in the mortar/pestle and mechanically milled it for 2h. After that, the attained pink colour precursor was cleaned several times with deionised water and dried in oven at 100°C for 12 h. The resultant product was annealed under atmosphere at 300°C for 2h in muffle furnace. Finally, the obtained sample was labelled as Co_3O_4 **[2]**.

The X-Ray pattern of the synthesized sample was analyzed through X-Ray diffractometer (XRD) in the 2θ range from 30° to 80° by using BRUKER USA D8 Advance, Davinci diffractometer. The surface morphology and the presence of the element were examined through Field Effect Scanning Electron Microscope (FE-SEM), and Energy Dispersive Spectrum (EDS) respectively. The Vibrational pattern of the synthesized sample was investigated using Micro Raman Spectroscopy (HORIBA France, Labram HR Evolution). The electrochemical performance of the prepared sample was analyzed through three electrode system (Bio logic workstation SP-300).

3. Results and Discussion

3.1 Structural characterization

The Crystalline structure and phase purity of the prepared sample was confirmed through XRD which is illustrated in fig.1. The obtained result is identical with the ICDD file no: 01-074-1656. The XRD pattern reveals the cubic crystal structure of Co_3O_4 nanoparticles. The crystalline size (D) of Co_3O_4 was computed through the Debye's Scherer formula as,

$$D=\frac{k\lambda}{\beta cos\theta}$$

Where k be the shape factor, λ be the wavelength, β be the (FWHM) Full Width Half Maximum calculated from the peak and θ be the diffraction angle. The average crystallite size of the prepared nanoparticles is found as 18.4 nm [9-11].

Table 1 Crystalline size of Co_3O_4 nanoparticles

S.No	Synthesized material Name	2θ (˚)	FWHM (rad)	d-spacing (Å)	Crystallite size (nm)
1	Co_3O_4	37.020	0.5353	2.426	**18.4**

IOP Conf. Series: Materials Science and Engineering 1219 (2022) 012022 doi:10.1088/1757-899X/1219/1/012022

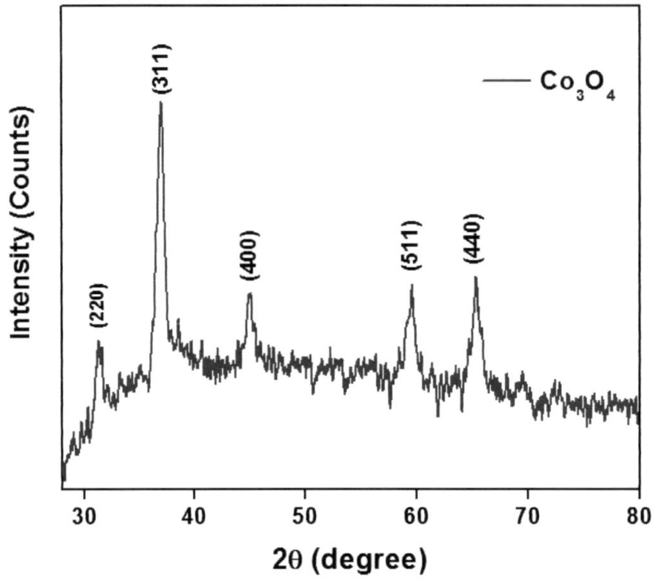

Fig.1 XRD Pattern for the prepared Co_3O_4

3.2 Scanning Electron Microscopy

Morphology of the nanoparticles was analyzed utilizing the FESEM. SEM micrographs shown in Fig.2 confirm the sphere morphology of Co_3O_4 nanoparticles. A small amount of agglomeration is visible in some parts of the micrographs. Fig.3 demonstrates the EDS spectrum of nanoparticles and the percentage of each element is shown in Table 2. It confirms the presence of elements Co and O in the sample. The pattern further proves the purity of the prepared compound [12,13].

Table 2. Mass and atomic percentages of Co_3O_4 nanoparticles

Element	Atomic Number	Mass (%)	Atom (%)
Co	27	73.89	43.98
O	8	25.55	56.02

Fig.2 SEM micrographs of Co_3O_4 nanoparticles

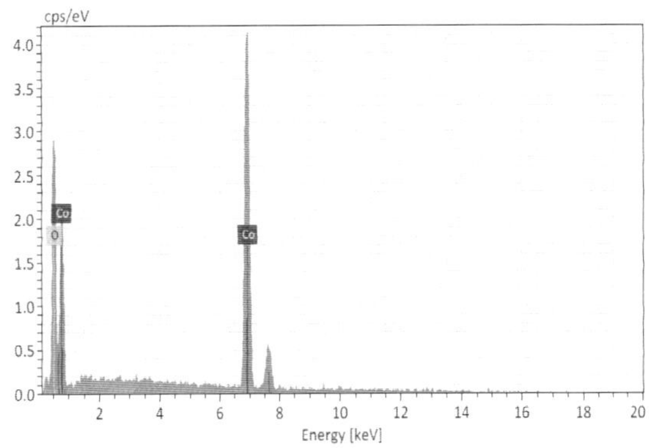

Fig.3 EDS spectrum of Co_3O_4 nanoparticles

3.3 Raman spectroscopy

The optical behavior of Co_3O_4 nanoparticles was analyzed through Raman spectroscopy. Four Raman active modes can be seen in the Raman spectrum as shown in Fig.4. Wavenumber 189cm^{-1}, 466cm^{-1}, 508cm^{-1}, and 670cm^{-1} indicate the phonon modes F_{2g}, E_g, F_{2g}, and A_{1g} respectively. Crystallization of Co_3O_4 is in the regular spinel structure where one Co (2+) is positioned at tetrahedral site and two Co (3+) are located ate octahedral site with four O (2-) [14,15].

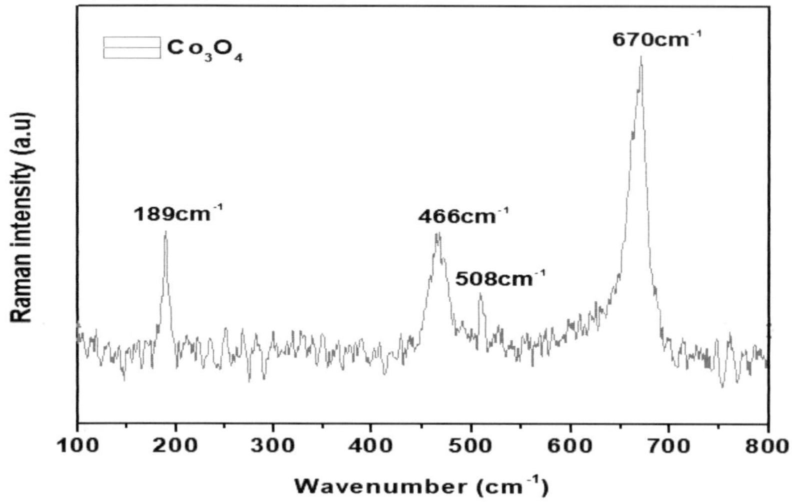

Fig.4 Raman spectrum of Co_3O_4 nanoparticles

3.4 Electrochemical characterization

In order to study the electrochemical behaviour of the prepared nanoparticles, the working electrode was mixed in the ratio of 80:12:8 with the help of NMP to form homogeneous slurry. The slurry was coated on the cleaned stainless steel mess and dried at 80°C for 12 hours. The coated electrode was analyzed through cyclic voltammetry (CV) and electrochemical impedance spectroscopy (EIS) in 1M Na_2SO_4 electrolyte. Fig.5 illustrates the cyclic voltammograms of the electrode at various scan rates ranges from 1 mV/s to 100 mV/s within the potential window as 0 to 1 Volt [11]. The quasi-rectangular shape of the curve clearly reveals the pseudo capacitive behaviour of the material.

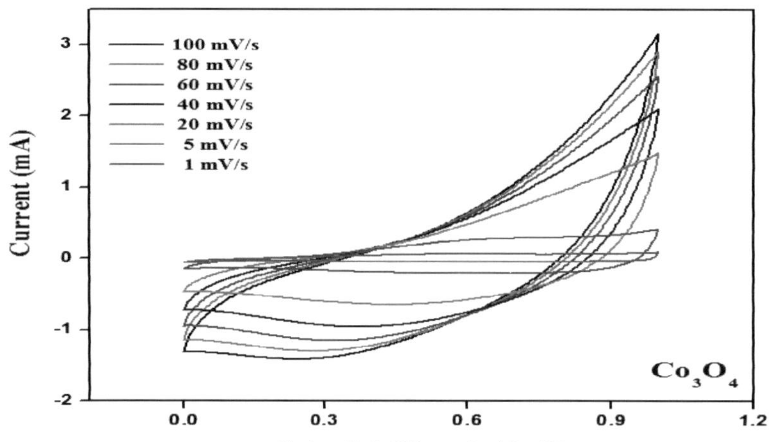

Fig.5 Cyclic voltammograms of Co_3O_4 nanoparticles

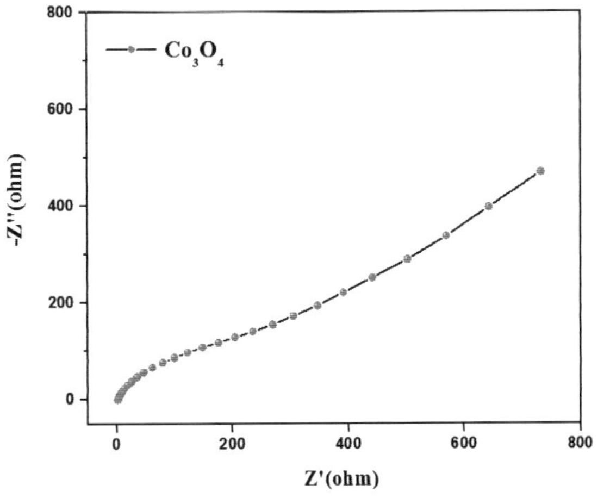

Fig.6 Electrochemical impedance spectroscopy of Co_3O_4 nanoparticles

Fig.6 shows the Nyquist plot of Co_3O_4 electrode which consists of a quasi-semicircle in higher frequency followed by the Warburg in lower frequency region. The shape of the EIS clearly confirms the characteristics behaviour of pseudocapacitor. Since, the intercept between the quasi-semicircle and the x-axis defines the value of solution resistance (Rs) as 2.04 ohm. Similarly, the straight line (Warburg) shows the diffusion resistance of the electrolyte into active material [14-16].

4. Conclusion

In summary, Cobalt Oxide nanoparticles were synthesized using a mechanochemical method. The synthesized nanoparticle is confirmed as Co_3O_4 by means of different analyses. Crystallographic parameters calculated from XRD revealed its cubic crystal structure. Raman spectrum displayed the vibrational modes of Co_3O_4. SEM micrographs showed the sphere like morphology exhibited by nanoparticles. EDS verified the presence of Co and O in the sample and their atomic percentages. The cyclic voltammetry and electrochemical impedance spectroscopy reveals the good electrochemical performance of the Co_3O_4 nanomaterials.

Acknowledgement
The authors would like to thank Nanotechnology Research Centre and SRM Central Instrumentation Facility of SRM Institute of Science and Technology, Kancheepuram, Tamil Nadu, India for extending support in characterizations.The author T.Vijayakumar acknowledges the financial support obtained from the Department of Space, Government of India [Grant no.B.19012/57/2016-II] through RESPOND project and the selective excellence initiative award received from SRM Institute of Science and Technology.

References

[1] Wei Wen, Jin-Ming, Jiang-ping Tu, 2012, J. Alloys Compd, 513.

[2] Yang, Huaming, Yuehua Hu, Xiangchao Zhang, and Guanzhou Qiu, 2004, Mater.Lett 58. 3-4.

[3] S.G. Kandalkar, D. S. Dhawale, Chang-Koo Kim, C.D. Lokhande, 2010, Synth.Met 160.

[4] KatalinSinko, Gezaszabo, MiklosZrinyi, 2011, J.NanoSci 11.

[5] K. Agilandeswari, A. Rubankumar, 2016, Synthesis and Reactivity in Inorganic Metal-Organic, and Nano-Metal Chemistry, 46 (4).

[6] Farhadi S., Safabakhsh J., and Zaringhadam P, 2013, Journal of Nanostructure in Chemistry, 3(1).

[7] Min-Sik Park, Jeonghun Kim, KiJae Kim, Jong-Won Lee, Jung Ho Kim and Yusuke Yamauchi, 2015, Physical Chemistry Chemical Physics 17 (46).

[8] Devi V.S, Athika M, Duraisamy E, Prasath A, Sharma A.S, and Elumalai, 2019 Journal of Energy storage 25.

[9] S.G. Kandalkar, J.L. Gunjakar, C.D. Lokhande, 2008, Appl.Surf.Sci 254.

[10] Juan Xu, LanGo, JianyuCau, Wenchang Wang, Zhidong Chen, 2010, Electrochim. Acta 56.

[11] Karthigeyan Krishnamurthy, Sang Jae Kim, 2015, J. Ind.Eng.Chem 32.

[12] Hailong Chen, Xinping Qiu, Wentao, Zhu, Paul Hagenmuller, 2002, Electrochem.commun 4.

[13] Guoxiu Wang, Xiaoping Shen, JosipHorvat, Bei Wang, Hao Liu, David Wexler and Jane Yao, 2009, J.Phys.Chem.C 133

[14] Hamid Emadi, Masoudsalvati-Niasari, Fatemech Davor, 2012, Polyhedron 31.

[15] Tao L, Shengjun L, Bowen Z, Bei W, Dayong N, Zeng C, Ying Y, Ning W and Weifeng Z, 2015, Nano scale research letters 10(1).

[16] Qian Xue and Qiang Zhang, 2019, Nanomaterials 9,503.

Quantum Chemical Computations of an Efficient Push-Pull NLO Chromophore 3-[4-Nitrophenyl Azo]- 9H- Carbazole-9-Ethanol

N.Balaji[1], M.R.Kannan[1], Y.Sheeba Sherlin[2],T.Vijayakumar[1,*]

[1]Futuristic Materials Research Centre for Planetary Exploration, Department of Physics and Nanotechnology, College of Engineering and Technology, Faculty of Engineering and Technology, SRM Institute of Science and Technology, SRM Nagar, Kattankulathur-603203, Kanchipuram, Chennai, Tamil Nadu, India.
[2]Department of Physics, Scott Christian College (Autonomous), Nagercoil-629003, Tamil Nadu, India.

email: vijayakt1@srmist.edu.in

Abstract. Electric-optic (EO) materials are being explored for applications ranging from fiber and satellite telecommunications, optical gyroscopes, to photonic detection of radar etc. Dipolar push-pull Organic chromophores that exhibit extended π-conjugation, in particular, show enhanced second order NLO properties. The present investigation reports the quantum chemical computations of the high efficiency push-pull NLO molecule 3-[(4-Nitrophenyl Azo)] - 9H-Carbazole-9-Ethanol (NPACE). The organic push-pull molecule is optimized in gaseous and in various solvent condition using Exchange correlation function (B3LYP/MP2). Molecular electrostatic potential, thermal properties and NBO analysis have also been performed in detail. The effective electron cloud moment in the molecule is mainly governed by the physical process termed intramolecular Charge Transfer (ICT).

Keywords: Push-pull NLO Chromophore;Carbazole; NBO; ICT; FMOs

1. Introduction

Over the past recent decades, the organic polymers have been a promising material for the large non-linear response to the applied electromagnetic field. The laser induced polymeric substances undergo a significant non-linear change results in the higher bandwidth potential giving an overview of the material response time [1, 2].Hence such materials show strong NLO response and have wide range of application [3, 4].The novel organic chromophore 3-[4-Nitrophenyl Azo]- 9H- Carbazole-9-Ethanol (NPACE), a dye substance consisting of azo, nitro phenyl and carbazole substituents has been identified as C_1 symmetry group. The molecule consists of electron-donor and electron acceptor character are interconnected by *Π*-bridge. The electron-donor group strongly displaces the electronic charge to acceptor moiety through conjugated bridge. Such molecules are known as 'Push-Pull molecule' [5, 6].In this work, the details on the molecules geometries, bandwidth and the NBO [6] results will provide good understanding on the properties of the material. Some chemical phenomenon such as back scattering, hyperconjugation and Intramolecular Charge Transfer (ICT) [7-9] are also important in determining the charge transfer. Along with this, electrostatic potential mapping and thermodynamic properties provides an additional support for the charge carriers.

2. Theoretical approach

The titled compound NPACE (Fig. 1) is geometrically optimized by using the DFT with Becke-3–Lee–Yang–Parr (B3LYP) combined with standard 6-311++G(d,p) basis set and GAUSSIAN 09W program package without any constraint on the geometry. The keyword POP=NBO is used to carry-out the natural bond orbital analysis.

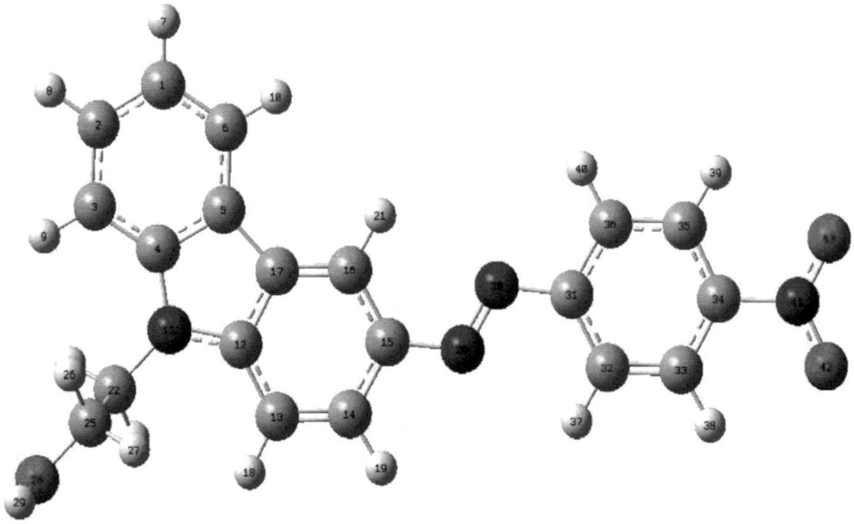

Fig. 1 Molecular Structureof NPACE

3. Results and Discussion

Table 1 Optimized Bond Angles (°), Bond Lengths (Å)and Dihedral Angles (°) of NPACE

Bond Angle (°)	Calc.	Bond Length	Calc.	Dihedral Angle (°)	Calc.
C_1-C_2-C_3	121.412	C_1-C_2	1.402	C_4-C_3-C_2-C_1	-0.1108
C_2-C_3-C_4	117.655	C_2-C_3	1.392	C_5-C_4-C_3-C_2	0.1706
C_3-C_4-C_5	121.613	C_3-C_4	1.394	C_6-C_5-C_4-C_3	0.0270
C_1-C_2-C_6	120.650	C_4-C_5	1.415	H_7-C_1-C_6-C_5	179.986
H_7-C_1-C_6	119.839	C_5-C_6	1.396	H_8-C_2-C_1-C_6	179.960
H_8-C_2-C_1	119.440	C_6-C_1	1.390	H_9-C_3-C_2-C_1	179.815
H_9-C_3-C_2	120.607	C_1-H_7	1.084	H_{10}-C_6-C_1-C_2	179.976
H_{10}-C_6-C_1	120.484	C_2-H_8	1.084	N_{11}-C_4-C_3-C_2	-179.356
N_{11}-C_4-C_3	129.330	C_3-H_9	1.083	C_{12}-N_{11}-C_4-C_3	-179.845
C_{12}-N_{11}-C_4	108.796	C_4-N_{11}	1.397	C_{13}-C_{12}-N_{11}-C_4	179.757
C_{13}-C_{12}-N_{11}	129.494	C_5-C_{17}	1.449	C_{14}-C_{13}-C_{12}-N_{11}	179.401
C_{15}-C_{14}-C_{13}	121.679	N_{11}-C_{22}	1.451	C_{16}-C_{15}-C_{14}-C_{13}	0.03216

IOP Conf. Series: Materials Science and Engineering 1219 (2022) 012023 doi:10.1088/1757-899X/1219/1/012023

Structural parameters:

Organic molecule NPACE is Optimized and the structural properties such as bond length; bond angle and dihedral angle are shown in Table 1.For our convenience the aromatic ring system, the fused carbazole ring is individually distinguished into CZ ring-I, 5- membered pyrrole ring attached strongly to the donor moiety and the CZ ring-III connected to azo-bridge(N=N) and the phenyl ring (Ph). Notable change in the internal angle C12-N11-C4, C4-C5-C17 induces elongation in the bond length N11-C4 and C5-C17 by 0.05 and 0.04Å.

In order to reduce steric repulsion fused ring (CZ1, CZ2) and Ph ring reveals increase in the endocyclic angle. Narrowing of bond angle C_{34}-N_{41} and C_{31}-N_{30} is balanced evidently by the endocyclic bond angle greater than 120°. Drastic increase in the exocyclic angle around the fused ring N_{11}-C_4-C_3 and N_{11}-C_{13}-C_{14} by 9.33° and 9.44°whereas the angle C_{33}-C_{34}-N_{41} and C_{35}-C_{34}-N_{41} falls down by 1 and 1.2°.

Additional negative charge on amino atom N_{11} also implies shrinking of the bond length C_{22}-N_{11} and C_{12}-N_{11}. At the junction of the acceptor moiety the two electronegative oxygen atom implies positive charge on the N_{41} atom. Lone pair O_{28} and N_{11} atom creates a back donation to the CH bond is evident from the NBO results. The intramolecular non-bounded hydrogen atom $H_{19}\cdots H_{37}$ and $H_{21}\ldots H_{40}$ are found to be at the distance 3.751 and 3.856 Å.
Lone pair nitrogen atom conjugating with ring system enhances the interatomic charge transfer (ICT) mechanism. The NC_3 moieties dihedral angle is deviated from the coplanarity C_{13}-C_{12}-N_{11}-C_{22}, C_{17}-C_{12}-N_{11}-C_{22}, C_3-C_4-N_{11}-C_{22} and C_5-C_4-N_{11}-C_{22} by 1.48, -178.87, -1.56 and -178.86is also evident by ICT state.

Electrostatic Potential Analysis

ESP plot (Fig. 2) is related to the density of electronic states indicating the potential distribution with different colour. It gives good agreement with localized electronic density of states. The red region indicates the electron-rich reflects most electronegative region in the centre and due to vicinity of oxygen atom.The electro-positive and the hydrogen bond formation are indicated in blue colour.

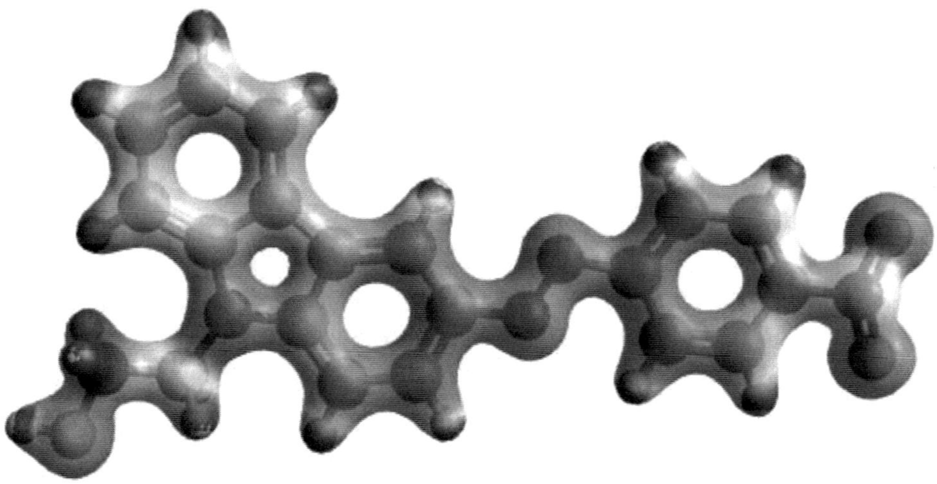

Fig.2 Molecular electrostatic potential surface for NPACE

FrontiersMolecular Orbital Analysis

HOMO is localized in donor, azo group and carbazole ring, while LUMO in contrast occurs in the *p*-substituted ring and nitro moiety. -electron overlap in the azo bridge is an inherent condition for successful electronic charge transition and the energy bandgap is found to be 1.9ev.The probability of higher order transition is low as it midst at 3.45ev. The complete HOMO-LUMO plot is shown in the Fig. 3

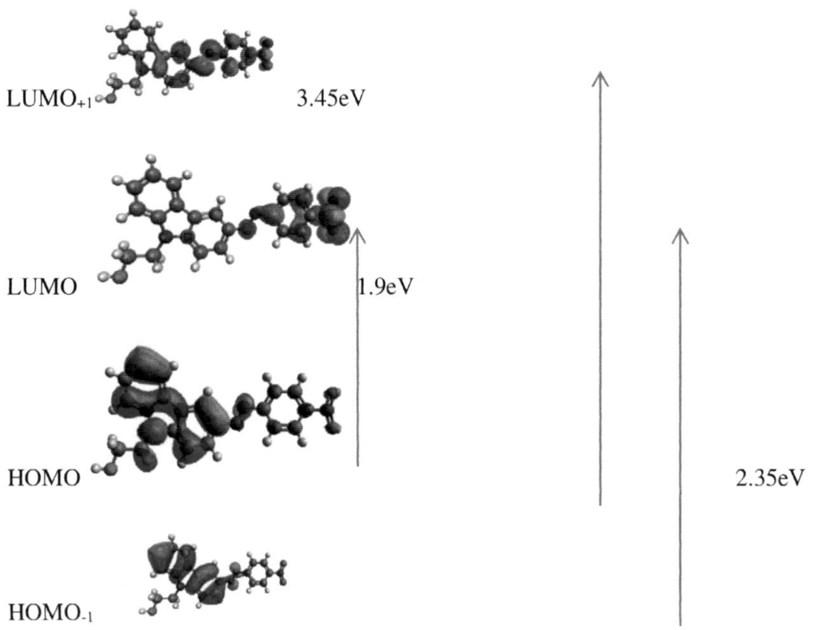

Fig 3 HOMO and LUMO plot of NPACE

Natural Bond Orbital Analysis

Natural bond orbital analysis helps greatly in assigning the vibrational bands. The hyperconjugation and the depolarization ratio from donor to acceptor NBO orbitals from second order perturbation theory are given in the Table 2. The most significant intramolecular charge transfer from lone pair nitrogen atom enhances the charge transfer very effectively. Therefore lone pair of N_{11} atom and the antibonding orbital of Π-bridge show high stabilization energy. The significant large stabilizing energy occurs from the lone O_{43} atom to the antibonding N_{41} - O_{42} state.

Solvent effect

On the influence of various solvent like water, DMSO, acetone the geometrical and structural properties of the NPACE molecule is notably deviated from the gaseous state. The alteration in bond angle and the bond length predicts the fast electron transfer. In addition, the HOMO-LUMO gap has been significantly reduced results in the large transition probability.

Table 2 Second-order perturbation theory analysis of Fock matrix in NBO basis

Donor NBO	Acceptor NBO	E(2) k J/mole[a]	E_j -E_I a.u[b]	F_{ij} a.u[c]
BD_2 ($C_1 - C_6$)	BD*($C_2 - C_3$)	97.28	0.27	0.072
BD_2 ($C_5 - C_3$)	BD*($C_1 - C_6$)	83.97	0.28	0.069
BD_2 ($C_{16} - C_{15}$)	BD*($C_{14} - C_{13}$)	96.23	0.28	0.072
BD_2 ($C_{13} - C_{14}$)	BD*($C_{17} - C_{12}$)	84.48	0.28	0.071
BD_2 ($C_{31} - C_{36}$)	BD*($C_{35} - C_{34}$)	93.30	0.27	0.069
BD_2 ($C_{35} - C_{34}$)	BD*($N_{41} - O_{42}$)	163.51	0.11	0.064
BD_2 ($C_{34} - C_{33}$)	BD*($C_{31} - C_{35}$)	84.47	0.28	0.067
LP N_{11}	BD*($C_5 - C_4$)	151.55	0.29	0.094
LP N_{11}	BD*($C_{17} - C_{12}$)	150.08	0.29	0.093
LP N_{11}	BD* (C_{22}–H_{22})	4.61	0.69	0.056
LP N_{11}	BD* (C_{22}-H_{23})	3.32	0.69	0.047
LP O_{43}	BD*($N_{41} - O_{42}$)	452.83	0.11	0.101
LP_2 O_{28}	BD^* ($C_{25} - H_{27}$)	6.15	0.74	0.060
BD*($N_{20} - N_{30}$)	BD*($C_{16} - C_{15}$)	146.44	0.04	0.066
BD*($N_{20} - N_{30}$)	BD*($C_{31} - C_{35}$)	121.71	0.04	0.057

[a]E_2 is the stabilization energy, [b]Energy difference between donor i and acceptor j NBO orbitals, [c]Fock matrix element

Thermodynamic properties

The thermodynamic functions listed in Table 3 are obtained from the theoretical harmonic frequencies. All thermodynamic calculations are done in gas phase can be related to estimate the direction of chemical reaction according to the second law of thermodynamics.

IOP Conf. Series: Materials Science and Engineering 1219 (2022) 012023 doi:10.1088/1757-899X/1219/1/012023

Table 3 Theoretically calculated energy: Zero-point energy, Enthalpy and Gibbs free energy (kJ·mol^{-1}), Rotational Constants (GHz)

Parameter	**NPACE Molecule B3LYP/6-311 G (d,p)**	**Parameter**	**NPACE Molecule B3LYP/6-311 G (d,p)**
Total	359.139	Rotational constant	0.48920, 0.05936, 0.05335
Translational	12.48	ZPE	86.583
Rotational	12.55	Enthalpy (H)	926.623
Vibrational	334.178	Gibbs free energy (G)	721.534

Conclusion

To summarize, the optimized molecular geometry and the electronic properties have been studied. The intramolecular hydrogen bonding and the depolarization ratios are calculated from NBO analysis. The result shows that the lone pair N_{11} atom is very significant for ICT and the charge transfer occurs across the molecule. Along with this ESP mapping, thermodynamic properties are also discussed in this paper. The large electron cloud movement indicates the charge transition and this novel material has a great potential in the electro-optic properties for the application in satellite telecommunication.

Acknowledgements

Authors thank High Performance Computing Center, SRM Institute of Science and Technology for providing the computational facility. The author T.Vijayakumar acknowledges the financial support obtained from the Department of Space, Government of India [Grant no.B.19012/57/2016-II] through RESPOND project and the selective excellence initiative award received from SRM Institute of Science and Technology.

References

[1] PrasadP.N., WilliamsD.J. Introduction to Nonlinear Optical Effects in Organic Molecules and Polymers 1991 John-Wiley & Sons Inc. New York

[2] SchoonveldW.A., WildemanJ., FichouD., BobbertP.A., Van WeesB.J., KlapwijkT.M., 2000Nature 404 977

[3] BosshardCh., SutterK., PretreP.h., HulligerJ., FlorsheimerM., KaatzP., GunterP. Organic Nonlinear Optical Materials, 1995Advances in Nonlinear Optics, vol. 1, Gordon and Breach, Amsterdam

[4] ShenY.R. The Principles of Nonlinear Optics, 1984Wiley, New York

[5] Larry R. Dalton, Aaron W. Harper, Bo Wu, Rima Ghosn, Joyce Laquindanum, Ziyong Liang, Andrea Hubbel, ChengzengXu, 1995Adv. Mater. 7 (6) 519

[6] JamesC., Amal RajA., ReghunathanR., JayakumarV. S. and Hubert JoeI. 2006 Journal of Raman Spectroscopy 37 1381

[7] ZyssJ., NicoudJ.F., CoquillayM., 1984J. Chem. Phys. 814160

[8] Glendening ED, Reed AE, Carpenter JE, Weinhold F. NBO Version 3.1. TCI 1998University of Wisconsin: Madison

[9] VijayakumarT., Hubert Joe I., Reghunadhan Nair C.P., JayakumarV.S.2008Chemical Physics 34383-99

IOP Conf. Series: Materials Science and Engineering 1219 (2022) 012024 doi:10.1088/1757-899X/1219/1/012024

TREMOR CLASSIFICATION USING WEARABLE IOT BASED SENSORS

A.Brindha, M.tech[1], K.A.Sunitha, PhD[1*], S. Robert Wilson, MBBS., MD, DM2

[1] EIE Department, SRM Institute of Science and Technology, Kattankulathur, Chengalpet, TamilNadu, India.

[2] Neurology Department, SRM Hospital and research centre, Kattankulathur, Chengalpet, TamilNadu, India.

Email id: sunithak@srmist.edu.in

Abstract. Tremors, a significant symptom of movement disorder, affects a part of the body ranging from slight to severe. These Tremors are symptoms of various neurological diseases such as Parkinson's disease (PD), Essential tremors (ET), Physiological tremors (PT), Cerebellar tremor, Dystonic tremor, Psychogenic tremor, and many more. Tremor features and types differ for various neurological disorders. During the early stages of the disease, clinical examination of tremors plays a significant role in diagnose management. This work aims to develop a wearable assistive system with an Inertial Measurement Unit (IMU) sensor to classify the tremor of three different neurological disorders based on the tremor position and frequency. This research has been carried out in SRM Medical college and Research Centre with 15 patients. The type of neurodegenerative disease of the subject with tremor is evaluated based on the tremor position and tremor frequency level. The data is collected, transmitted, and processed using the IMU sensor with Internet of things (IoT) and Node MCU board. The decision tree algorithm is used for the classification of tremors. ET, PD, and PT tremors are classified based on the tremor frequency and tremor position. A high rate of accuracy is achieved for the developed system when compared with the Neurologist results. The proposed device quantitatively classified the tremor based on the frequency and position among the three different neurological disorders, i.e., ET, PD, and PT tremors.

Keywords - Tremor classification· Parkinson tremor · Essential tremor · Physiological tremor · Decision tree algorithm

1. INTRODUCTION

Tremors, termed as a disorder experienced by a person while in rest or motion {1}. Though not life-threatening, tremors tend to extend difficulty in performing day-to-day tasks. Tremors can be of many forms as shown in table 1.

Table 1: Classification of tremors based on the occurrence

TREMOR TYPES	DEFINITION	FREQUENCY
Rest tremor	This occurs when a person is in a resting position. It appears when muscles are relaxed.	3-6Hz
Action tremor	Tremor occurs due to voluntary contraction of muscles, including postural, isometric, and kinetic tremors.	3-10Hz
Postural tremor	Occur when a part of the body is held against gravity, such as holding arms or outstretched legs.	5-9Hz
Kinematic tremor	Tremor occurs during any voluntary activities, such as the movement of the wrist.	4-6Hz

Neurological disorders are the primary cause of disability-adjusted life years (DALYs) and the second leading cause of death [2]. The pervasiveness of these disorders increases directly with age (considering an age group of 65). The global prevalence of ET is estimated to be at 4.6 % and PD at 1.8 % [3]. It is also observed that 7-10 million people live with PD. ET incidence increases with age, and more than 4% of people were affected [4,5]. PD is characterized by bradykinesia, rigidity, rest tremor, or asymmetric onset [6]. The National Institute of Neurological Disorders and Stroke (NINDS) states that it is possible to distinguish PD from ET if a subject is diagnosed to have any three characteristics of PD (i.e., bradykinesia, rigidity, rest tremor, or asymmetric onset) [7]. In contrast, tremors of Essential Tremor disease are primarily action tremors and contain bilateral and symmetric tremors [8]. There always exists a difficulty in distinguishing ET from PD in the early stages with subtle clinical signs. Indeed a study showed that one-third of PD subjects are misperceived to have ET [8]. The neural loss due to synuclein deposition in the substancia nigra is the pathological hallmark of PD, whereas the pathological hallmark of ET is challenges in the cerebellar and, in some cases, brainstem Lewy bodies. Also, the physiological tremor is caused due to mechanical and central components. Thus misdiagnosis of the tremor would lead to wrong medication errors, and that will worsen the disease [8].

The work is focused on differentiating three types of tremors, i.e., ET, PT, and PD tremors. The characteristic of these tremors is shown in the table 2 [9]. In this reference, thorough research was done to understand the underlying difference between the tremors for neurological disorders, say ET, PD, and others [10]. The study identified that the frequency of ET, PT, and PD have a narrow bandwidth and overlap [11].

The current diagnostic tool that is used by Neurologists is a UPDRS rating scale to categorize PD from other neurological disorders. Hand tremors of (3-12) Hz in Essential tremor are also a symptom of PD with (3-7) Hz [12]. Research reports that till today differentiating PD tremor from ET is always a challenge to avoid the misdiagnosis, thus resulting in a longer recovery rate [13]. Therefore, detection of the tremor, especially for PD, ET, and PT cases and further classification, is the need of the hour.

In few studies, sensors and classifiers are used to diagnose the PD tremor. Classification is a process of grouping similar data sets together. The Tremors can be classified by observing various features like frequency, phase difference, spectral density [14]. Author Luay Fraiwan, et al. used a mobile application to measure the acceleration of hand tremor and group them to differentiate PD tremors other tremors. This hand tremor was further classified using neural networks classifier. The results indicated that an accelerometer and classifier best suited the application [15]. Thus, the IMU sensor which is a combination of accelerometer and gyroscope is used to measure tremor in this study. Tremors of different types of disease that is considered and the specific characteristic is shown in Table 2 [2].

Neurological Disorder	**Clinical examination**	**Range of Frequency**	**Parts of the body that is affected**
Essential tremor	Concept of finger-to-nose test and S-band frequency	(4-12) Hz	Hand, head, and voice box
Parkinson's tremor	EMG signals and the concept of singular spectrum analysis	(3-7) Hz	Whole-body but mainly the limbs.
Physiological tremor	Clinical examination, physical and physiological exam.	(3-30) Hz	Only limbs when performing an activity

For the classification of hand tremors, different classifiers such as Naive Bayesian, KNN, ANN, SVM were used in the previous studies [16, 17]. In this study decision tree algorithm, a supervised learning-based algorithm, is used to classify the tremors. The algorithm works efficiently for both categorical and continuous values [18]. Also, this algorithm works for data set with fewer samples. This algorithm is proposed due to its ease of interpretation as the classification involves a data set that has many similar characteristics [19].

IOP Conf. Series: Materials Science and Engineering 1219 (2022) 012024 doi:10.1088/1757-899X/1219/1/012024

2. MATERIALS AND METHODS

2.1 Participants

Fifteen participants who were diagnosed with slight to severe tremors were involved in the study. The characteristics of all the participants are shown in table 3. In addition, each subject was asked to describe their symptoms and is recorded for further classification.

Table 3: Baseline characteristics of participants

	PD Subjects (n=5)	**ETsubjects (n=5)**	**PTsubjects (n=5)**
Age (years): mean ± SD (range)	45.8 ± 6.349	52.2 ± 6.763	57.3 ± 14.5
Sex(male/female)	3/2	2/3	4/1
BMI (kg/m2): mean ± SD	26.597 ± 11.773	25.464 ± 11.205	23.543 ± 11.432
Diabetes mellitus(yes/no)	¼	2/3	3/2

The patients were detailed about the procedure, and the consent was taken from the patients and their relatives. The patients are involved in the study with an Ethical certificate from SRM Medical College and Research Centre. Then, the participants were asked to take up the clinical trial with the wearable device under the supervision of neurologists from SRM Medical College and Research Centre.

2.2 Experimental Set Up and Procedure

The experimental setup consists of an IMU sensor and NodeMCU electronic board as shown in figure 1. This system has been set up to provide the raw data i.e, the displacement of the hand due to tremor. The IMU sensor is a device that is used to measure and report the orientation, angular rate, and specific force. It has been interfaced such that its input is an analog signal in the form of hand movements(tremors) and is converted into

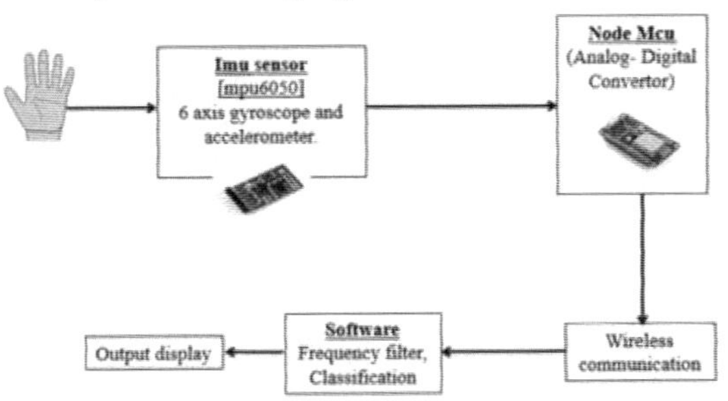

Figure 1: Process flow of the proposed system

a digital signal readable in terms of gravitational acceleration. The classification of signals based on their frequency requires this gravitational acceleration in terms of gms^2 to be converted into Hertz (Hz). The NodeMCU is used to communicate, convert, and ultimately display the acceleration output. The data then procured is in terms of mms^2 from the accelerometer. The acceleration readings have been stored, processed, filtered, and classified to predict the type of tremor.

PLX-DAQ is used to store the output(acceleration) in a spreadsheet. A PLX-DAQ is a parallax microcontroller data acquisition system used to import and store data from the serial port of the system to MS Excel. This helps in the easier conversion of recorded data into a database. Storing the data in the database is imperative for maintaining records of patients. The implementation for PLX-DAQ is coded into the Arduino IDE. While active detection there may be few readings that exceed the desired range. Thus, including the disease frequency range and using the band-pass filter concept, frequencies lesser than 1Hz and higher than 30Hz are discarded.

The classification of tremors is then performed using the probabilistic concept of the Decision tree, as shown in figure 2. The physiological tremor acquires a wide range for classification, ranging from 3Hz to 30Hz this prompts overlapping in frequencies. Since the frequencies overlap, the position of the tremor is considered. A patient experiences a wide variety of symptoms based on which the patient's tremor can be classified. The decision tree algorithm is a supervised learning-based algorithm, will update the data set of patients every time while hand tremor is recorded.

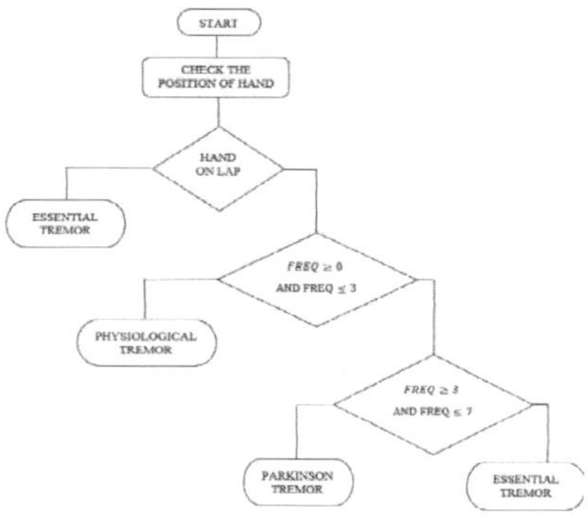

Fig. 2 Classification of tremor flow chart

Thus, when a patient experiences a tremor, the algorithm first checks for the position of the tremor. The three leading positions that is considered for examination are hands-on lap, while lying down, and hands stretched out. If a patient experiences tremor when the affected hand is in a stretched-out position, it can be due to ET, PD, and PT. So, the tremor frequency is then compared, and if it is greater than 7Hz, the tremor is categorized as a Physiological tremor. If the frequency is less than 7 Hz, then the UPDRS rating will be considered that is given by the Doctor. After considering the severity (occurrence and position) of the tremor, it will be categorized as a PD tremor.

2.3 Data collection and pre-processing

Before performing the test, the subject is asked to rest for 10 to 15 minutes, and then the procedure is detailed. Then the consent from both patient and relative is obtained. Then the patient is assisted to wear the hand glove with an IMU sensor. Now, based on the position of tremor that is experienced by the patient, the patient is made to either lie-down, sit or stretch their hand out as shown in figure 3 based on their convenience and environment. The reading is taken from the patient, subsequently the frequency and the possible tremors are displayed.

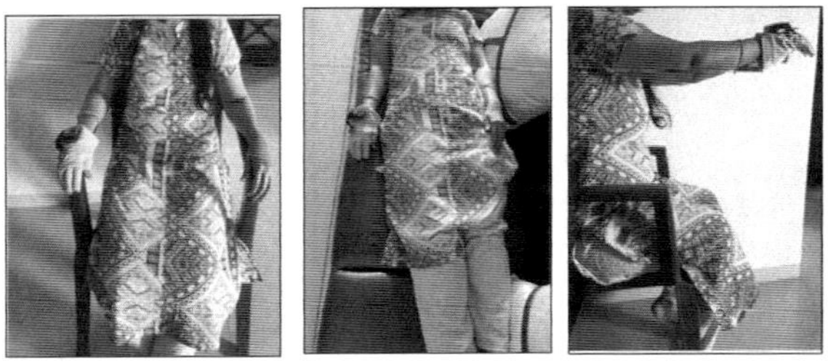

Fig. 3 Positioning of Patient (a) Patient Sitting with hand on the hand of a chair, (b) patient lying down, Patient with the hand stretched out

The data collected from the patient needs to be processed in a manner that can be read by the user. Once the acceleration is stored in the excel file from which average acceleration is found. Once the raw data is received, the acceleration values are converted using the following equation:

$a_{actual} = (a_{arduino} * 9812)/163$-----------------(1)

Further using the frequency equation, the frequency is found out and plotted as shown in figure 4(c). After this is done, topar get displacement values, the following equation (2) is used: The displacement Vs. time graph is shown in figure 4(b).

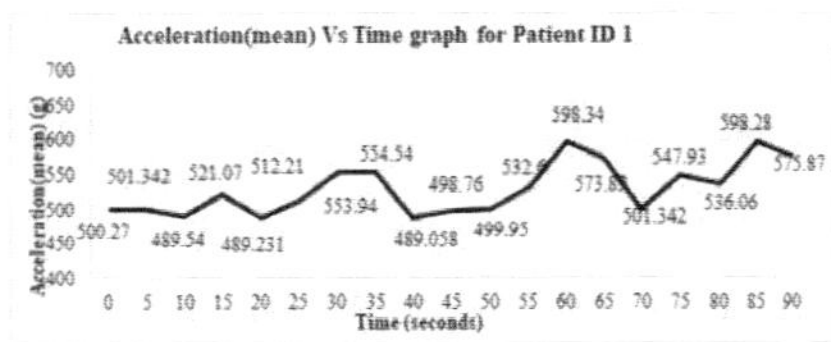

(a)

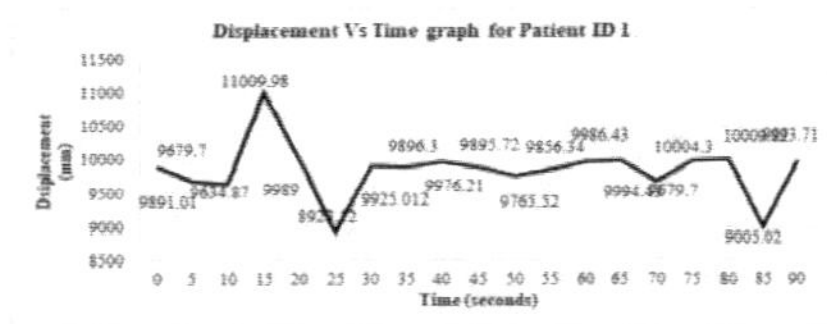

(b)

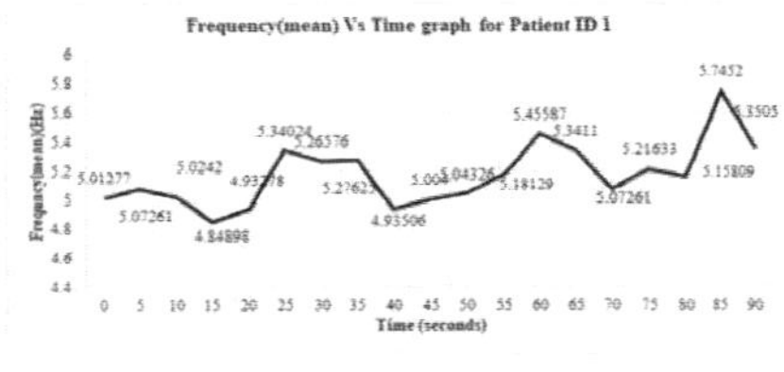

(c)

Fig. 4 (a) Acceleration Vs. Time graph for Patient with ID number 1,(b) Displacement Vs. Time graph for Patient with ID number 1, (c) Frequency Vs. Time graph for Patient with ID number 1

D = 0.5 ∗ a ∗ t^2 (4)-----------------(2)

Using the following formula in equation (3) frequency can be found out:

F = ((GA) 0 .5)/((2D) 0 .5) (5)-----------------(3)

The table 4 shows how the patient readings are stored in the database. Upon getting values of acceleration a plot was made as shown in the figure 4(a). After this stage the band pass filter is applied, and the frequency band is restricted to 1-30 Hz. This filter assists in obtaining the sample free of noise and unwanted signals. Using the values of frequency classification can be done.

3. RESULTS AND DISCUSSION

Table 4: The tremor response of subject with patient ID 1. The subject had hand tremor when the hand was placed on the lap and the readings are recorded using the IMU sensor. The table shows the sensor values in acceleration and the conversion to frequency.

Time (sec)	Mean acceleration (m/s^2)	Displacement (m)	Mean frequency (Hz)
0	500.27	9891.01	5.01277
5	501.342	9679.7	5.07261
10	489.54	9634.87	5.0242
15	521.07	11009.98	4.84898
20	489.231	9989	4.93278
25	512.21	8923.12	5.34024
30	553.94	9925.012	5.26576
35	554.54	9896.3	5.27625
40	489.058	9976.21	4.93506
45	498.76	9895.72	5.004
50	499.95	9765.52	5.04326
55	532.6	9856.34	5.18129
60	573.89	9994.43	5.45587
65	598.34	9986.43	5.45587
70	501.342	9679.7	5.07261
75	547.93	10004.3	5.21633
80	536.06	10009.82	5.15809
85	598.28	9005.02	5.7452
90	575.87	9993.71	5.3505

IOP Conf. Series: Materials Science and Engineering 1219 (2022) 012024 doi:10.1088/1757-899X/1219/1/012024

Table 5: Patient Sitting with hand on the hand of a chair (c) patient lying down (d) Patient with the hand stretched out Readings in the Excel File ZA

PATIENT ID	**ACCELERATION (Mean ± SD)** (m/s^2)	**DISPLACEMENT (Mean ± SD)** (m)	**FREQUENCY (Mean ± SD)** (Hz)
1	532.6±0.53	9856.34±0.15	5.181±0.034
2	443.82±0.32	17438.49±0.25	3.556±0.355
3	425.91±0.53	438.49±0.42	21.967±0.06
4	825.91±0.85	658.9±1.32	24.955±0.144
5	1025.64±1.6	9984.12±0.84	7.144±1.320
6	1452.32±0.45	4589.12±0.19	12.539±0.089
7	1892.4±0.38	8253.83±0.36	10.673±0.565
8	9426.95±0.74	10673.98±0.73	20.947±0.211
9	4756.25±0.35	35938.98±0.17	8.106±0.003
10	2463.54±0.75	84735.2±0.49	3.800±0.078
11	598.28±1.05	9005.02±2.09	5.745±0.885
12	4736.26±0.64	12647.3±0.41	13.640±1.880
13	1426.95±0.49	11946.31±0.86	7.703±1.515
14	489.058±0.94	9976.21±0.41	4.935±0.264
15	12464.3±1.28	46288.45±0.81	11.566±1.221

The patients involved in this study had hand tremors ranging from minimal to severe. The readings of 15 patients were taken and recorded in the manner shown below in table 4. Readings for each patient were taken over a period of 1.5 minutes. From this data, the mean and standard deviation is calculated. The hand tremor frequency and position of tremors are given as input to the classifier. With these readings, the patients were classified into the three categories, i.e., ET, PD, and PT tremor. The output is shown in table 5. The patient's hand tremor is analysed using three methods. In the first column, the interpretation obtained using an IMU device with classifier is listed. In the next column, interpretation is by IMU device without using the classifier. The tremor evaluation made by the Neurologist is given in the last column of Table 6. The validation of using a classifier is done by comparing the results with Doctor's observation. A GUI is developed that will help the Doctor diagnose the disease by just entering basic details and having a record of each subject. The device aids in easier data entry along with hassle-free data access, as shown in figure 5. Thus, when the inputs like Patient ID, Tremor frequency and position of the tremor is entered in the GUI and the Predicted tremor is given the type of disorder will be displayed in the GUI.

Table 6: The input parameters that is subject ID, age, position of tremor, frequency of tremor and the response after classification is shown in the table. The three types of analysis are compared in this table.

SUBJECT ID	AGE	POSITION OF TREMOR	FREQUENCY (Hz)	TYPE OF TREMOR		
				IMU DEVICE WITH CLASSIFIER	IMU DEVICE WITHOUT CLASSIFIER	OBSERVATIONS BY DOCTOR WITHOUT SENSOR
1	42	HANDS ON LAP	5.181	ESSENTIAL	ESSENTIAL	PARKINSON
2	48	WHILE LYING DOWN	3.556	PARKINSON	ESSENTIAL	PARKINSON
3	56	HANDS STRETCHED OUT	21.967	PHYSIOLOGICAL	ESSENTIAL	PHYSIOLOGICAL
4	62	HANDS STRECHED OUT	24.956	PHYSIOLOGICAL	PHYSIOLOGICAL	PHYSIOLOGICAL
5	78	HANDS ON LAP	7.144	ESSENTIAL	ESSENTIAL	ESSENTIAL
6	46	HANDS STRETCHED OUT	12.539	PHYSIOLOGICAL	ESSENTIAL	PHYSIOLOGICAL
7	51	WHILE LYING DOWN	10.673	ESSENTIAL	ESSENTIAL	ESSENTIAL
8	55	WHILE LYING DOWN	20.947	PHYSIOLOGICAL	PHYSIOLOGICAL	PHYSIOLOGICAL
9	45	HANDS ON LAP	8.109	ESSENTIAL	ESSENTIAL	ESSENTIAL
10	62	HANDS STRETCHED OUT	3.800	PARKINSON	PARKINSON	PARKINSON
11	48	WHILE LYING DOWN	5.745	PARKINSON	PARKINSON	PARKINSON
12	59	HANDS STRETCHED OUT	13.640	PHYSIOLOGICAL	ESSENTIAL	ESSENTIAL
13	63	HANDS ON LAP	7.703	ESSENTIAL	ESSENTIAL	ESSENTIAL
14	54	WHILE LYING DOWN	4.935	PARKINSON	PARKINSON	PARKINSON
15	67	HANDS ON LAP	11.566	ESSENTIAL	ESSENTIAL	ESSENTIAL

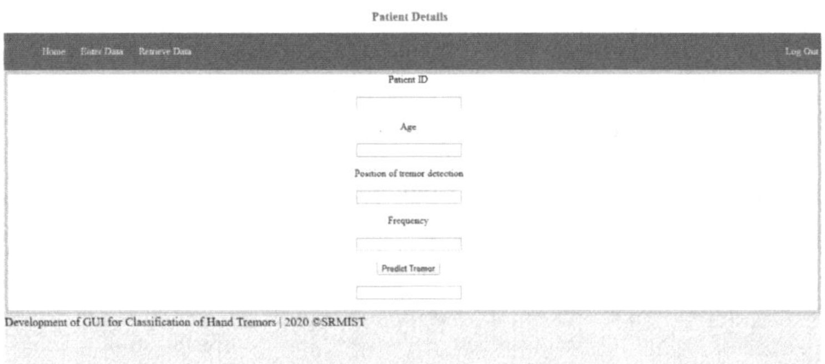

Figure 5: Outlook of the Interface

4. CONCLUSIONS

The proposed method serves as a lightweight, easy-to-use device for the detection and classification of tremors with high accuracy. The classification by the developed system helps in differentiating ET, PT and PD with accuracy of 86.67%, whereas a system without classifier had 73.33% as accuracy. This would help Doctors to actively detect tremor without prerequisite knowledge in handling the device. This device will serve as an assistive device for clinical assessment of patients with tremors. The device with the classifier will also help in remote monitoring of the patients without the need to move patients to clinic or hospital. Thus, helps in enhancing the healthcare services provided for the patients.

ACKNOWLEDGEMENT

This work was experimented in the Electronics and Instrumentation Engineering Department, SRM Institute of Science and technology. The authors thank the Department of Neurology, SRM Medical College, and Research Centre for the support to collect data and input to the study.

DISCLOSURES

Funding Sources and Conflict of Interest: No specific funding was received for this work, and we declare that there are no conflicts of interest relevant to this work.

Financial Disclosures for the previous 12 months: We declare that there are no additional disclosures to report.

Ethical compliance statement:

To perform the test for analysis, 15 patients were gathered and ethical approval was procured from SRM hospital and Research Centre. In this process, the right and privacy of the patients were maintained. The procedures are performed in compliance with the principles of declaration of Helsinki. Approval of ethical clearance ensures the accountability of the researcher to conduct the research. We confirm that we have read the Journal's position on issues involved in ethical publication and affirm that this work is consistent with those guidelines.

References

1. A. Puschmann and Z. K. Wszolek, "Diagnosis and treatment of common forms of tremor," in Seminars in neurology, vol. 31, no. 01. c Thieme Medical Publishers, 2011, pp. 065–077.

2. Feigin, Valery L., Emma Nichols, Tahiya Alam, Marlena S. Bannick, Ettore Beghi, Natacha Blake, William J. Culpepper et al. "Global, regional, and national burden of neurological disorders, 1990–2016: a systematic analysis for the Global Burden of Disease Study 2016." The Lancet Neurology 18, 459-480, no. 5 (2019).

3. E. D. Louis and J. J. Ferreira, "How common is the most common adult movement disorder? update on the worldwide prevalence of essential tremor," Movement Disorders, vol. 25, no. 5, pp. 534–541, 2010.

4. Louis, Elan D. "The roles of age and aging in essential tremor: an epidemiological perspective." Neuroepidemiology 52, no. 1-2, 111-118, 2019.

5. Benito-León, J., and E. D. Louis. "Update on essential tremor." Minerva medica 102, no. 6, 417-440, 2011.

6. K. B¨otzel, V. Tronnier, and T. Gasser, "The differential diagnosis and treatment of tremor," Deutsches Arzteblatt International ¨, vol. 111, no. 13, p. 225, 2014.

7. M. A. Thenganatt and E. D. Louis, "Distinguishing essential tremor from parkinson's disease: bedside tests and laboratory evaluations," Expert review of neurotherapeutics, vol. 12, no. 6, pp. 687–696, 2012.

8. S. Jain, S. E. Lo, and E. D. Louis, "Common misdiagnosis of a common neurological disorder: how are we misdiagnosing essential tremor?" Archives of neurology, vol. 63, no. 8, pp. 1100–1104, 2006.

9. K. Niazmand, K. Tonn, A. Kalaras, U. M. Fietzek, J.-H. Mehrkens, and T. C. Lueth, "Quantitative evaluation of parkinson's disease using sensor based smart glove," in 2011 24th International Symposium on Computer-Based Medical Systems (CBMS). IEEE, 2011, pp. 1–8.

10. M. Gourie-Devi et al., "Epidemiology of neurological disorders in India: Review of background, prevalence and incidence of epilepsy, stroke, Parkinson's disease and tremors", Neurology India, vol. 62, no. 6, p. 588, 2014.

11. C. W. Hess and S. L. Pullman, "Tremor: clinical phenomenology and assessment techniques", Tremor and other hyperkinetic movements, vol. 2, 2012.

IOP Conf. Series: Materials Science and Engineering 1219 (2022) 012024 doi:10.1088/1757-899X/1219/1/012024

12. O. Bazgir, S. A. H. Habibi, L. Palma, P. Pierleoni, and S. Nafees, "A classification system for assessment and home monitoring of tremor in patients with Parkinson's disease," Journal of medical signals and sensors, vol. 8, no. 2, p. 65, 2018.

13. G. Vivar, D.-L. Almanza-Ojeda, I. Cheng, J. C. Gomez, J. Andrade-Lucio, and M.- A. Ibarra-Manzano, "Contrast and homogeneity feature analysis for classifying tremor levels in Parkinson's disease patients," Sensors, vol. 19, no. 9, p. 2072, 2019.

14. A. M. Woods, M. Nowostawski, E. A. Franz, and M. Purvis, "Parkinson's disease and essential tremor classification on mobile device," Pervasive and Mobile Computing, vol. 13, pp. 1–12, 2014.

15. L. Fraiwan, R. Khnouf, and A. R. Mashagbeh, "Parkinson's disease hand tremor detection system for mobile application," Journal of medical engineering & technology, vol. 40, no. 3, pp. 127–134, 2016.

16. X. Yang, S. A. Shah, A. Ren, D. Fan, N. Zhao, D. Cao, F. Hu, M. U. Rehman, W. Wang, K. M. Von Deneen et al., "Detection of essential tremor at the s-band," IEEE journal of translational engineering in health and medicine, vol. 6, pp. 1–7, 2018.

17. L. Ai, J. Wang, and R. Yao, "Classification of parkinsonian and essential tremor using empirical mode decomposition and support vector machine," Digital Signal Processing, vol. 21, no. 4, pp. 543–550, 2011.

18. J. E. Alty and P. A. Kempster, "A practical guide to the differential diagnosis of tremor," Postgraduate medical Journal, vol. 87, no. 1031, pp. 623–629, 2011.

19. H. Sharma and S. Kumar, "A survey on decision tree algorithms of classification in data mining," International Journal of Science and Research (IJSR), vol. 5, no. 4, pp. 2094–2097, 2016.

20. Fahn, Stanley, Joseph Jankovic, and Mark Hallett. Principles and practice of movement disorders E-book. Elsevier Health Sciences, 2011.

Design and Fabrication of Knee Implant

S.K.Pradhan[1], Abhinav Mishra[1], Manoranjan Mishra[1*], Vedant Kedia[1], Satyam Kumar[1], P. Kour[2]

[1]Dept of Mechanical Eng., Birla Institute of Technology Mesra, Patna, Bihar, India, 800014

[2] Department of Physics, Birla Institute of Technology Mesra, Patna, Bihar, India 800014

Email: mishramanoranjan63@gmail.com.

Abstract. Stiffness, pain and impaired movement at various joints of the body are common with ageing and increase of obesity. The musculoskeletal chronic disorder causes such effect is named as rheumatoid osteoarthritis(OAs). Data shows that disease-Adjusted life years (DALYs) due to musculoskeletal disorder increases from 11.8% in 2000 to 13.5% in 2015.According to United Nations, by 2050, 20% of the world population will grow aged above 60 years of which 15% will account for symptomatic OAs, one-third will be severely disabled. Knee implants have been designed to replace the damaged part. Various designs and fabrications have been done to reduce its weight, flexibility and biocompatibility needs. This article is to highlight the use of new material or impart nanoparticle mixing with conventional ones for the design of knee implants so that it can be cost effective, light weight, to meet required balance strength, flexible and non-inflammatory. Hence, knee implant's computational and analytical design for different material is done to compare different parameters.

Keyword: musculoskeletal; rheumatoid osteoarthritis; implants; biocompatibility; non-inflammatory;

1. Introduction

Knee implant is an artificial assembly of components that was designed to replace the damaged bones of patellofemoral and tibiofemoral joints. Knee implant has many components such as femoral part, tibial part, base plate, patella plate[1]. The conventional materials used for the design of knee implant are titanium, ceramics, polyethylene, cobalt-chromium etc. These materials have certain merits like biocompatibility, flexibility, strength, but also have problems regarding various reactions taking place within the body like necrosis in periprosthetic tissue, poor frictional resistance[2]. This research was basically conducted to provide an alternative material for the fabrication of knee implant. To remove the problems raised by patients on using titanium based knee implant, magnesium was brought into consideration keeping in view of its presence within the body and its mechanical, thermal and physical strength. This was a theoretical based research where data was obtained from simulation and comparison was done with the data available from titanium based knee implant. The results obtained were promising. This article highlights the long and short term perspective of using magnesium as a material for fabrication of knee implant.

2. Materials and Methods:

The type of design considered for analysis is total condylar knee based design of size 3. This knee implant design is fixed and functional. These design provide high articulating surface area for femoral component over tibia surface and helps in rotation during gait if the bearing are revolving type. The conventional material used for comparison is Titanium (Ti). The new material taken into consideration is Magnesium (Mg). This is typically based on data collection, simulation and analysis of knee implant for different materials so experiments have not been performed only theoretical analysis is done based on other research work. ANSYS software has been apprehended with designing and simulation of the knee implant with specific geometric parameters based on the knee implant design taken i.e. TC knee based design. The assumptions taken into account while creating the design of knee implant are as follows:- The human body weight considered was 100 kgs which is equivalent to 1000N (approx.)[3]. The human body temperature was given as 37 degrees Celsius. The analysis is static in nature. Various other parameters like different sizes based on body size, gender, strength and biocompatibility have been not taken into consideration. The maximum number of element matrices after meshing was less than 0.5. The properties of the material taken into consider is extrapolated in the following table shown below.

Table 1: Property of two Different materials for designing knee implant [4]

Material name	Density(kg/m^3)	Young's Modulus(Pa)	Poisson's Ratio	Yield Strength(Pa)
Titanium alloy	4620	9.6E+10	0.36	9.30E+08
Magnesium	1740	4.5E+10	0.35	1.93E+08

3. Result and Discussion:

Table 2: Results obtained after simulation of knee implant, designed with two materials, separately displaying both components with its maximum and minimum values

Parameters		Magnesium(Mg)		Titanium(Ti)	
		Femoral	Tibial	Femoral	Tibial
Total Deformation (in mm)	Max Min	0.12 0	0.38 0	0.056 0	0.177 0
Directional deformation (mm)	Max Min	0.00652 -0.118	0.0117 -0.0588	0.00309 -0.0553	0.00545 -0.0027
Normal Stress(MPa)	Max Min	44.32 -14.747	132.96 -146.51	44.702 -14.76	134 -147.51
Normal Strain(mm/mm)	Max Min	0.00081 -0.00053	0.00192 -0.00238	0.000379 -0.00026	0.00089 -0.0011
Shear Stress (MPa)	Max Min	33.78 -11.762	33.509 -46.38	33.979 -12.184	33.314 -45.494
Shear Strain (mm/mm)	Max Min	0.00202 -0.000705	0.00201 -0.00278	0.000962 -0.00034	0.000943 -0.00128
Strain Energy (mJ)	Max Min	0.0251 7.46e-10	0.448 6.91e-08	0.0118 4.11e-10	0.20788 3.40e-08
Maximum Principal Stress(MPa)	Max Min	77.72 -8.931	161.07 -93.225	78.034 -9.367	161.79 -96.454
Minimum Principal Stress(MPa)	Max Min	14.255 -38.907	12.798 -333.2	14.642 -38.683	12.864 -336.08

The result is obtained in tabular form which clearly indicates different parameters like stress, strain, deformation etc. The data clearly depicts that normal stress caused on the tibial surface will be more than the femoral part. However, the shear stress on femoral part will be significantly more than tibial surface because of its sliding action over base plate. The normal and shear strain are very less in amount in spite of having higher values of stresses on the component. The deformation caused due to these stresses is in decimal places, taking the unit millimeter into consideration. On comparing the data, which is extrapolated from the simulation, for magnesium and Titanium it is quite evident that the difference in value is very minimal which shows that apart from titanium, magnesium can also be used as a material for manufacturing of knee implant. The simulation, done using Ansys software, was divided into various steps starting with geometry, followed by modelling, selection of material and defining of boundary condition, application of force on the component. The desired output where the component with various color code is obtained highlighting different values of parameters at different places on the component. This simulation helps us to identify the vulnerable points on the component which can easily be disrupted based on the number of gait cycles over a period of time. Starting with the initial steps, the figure stated below shows the geometry of the Knee implant of TC based design of size 3 which has a femoral component and a tibial surface. The data obtained from this table can be validated with various research papers. One of the papers applied a load of 2703.1 N where the deformation obtained was 0.150mm whereas in this the deformation is 0.05mm for 1000N which is nearly the same when if the load is acted on this magnitude[5]. Similar is the case for the maximum principal stress acting on the component. It is nearly 44.702 MPa for zero flex angle i.e. standing position whereas the data from the paper is 44.876 MPa and the material is titanium alloy which includes metals like aluminum(Al), vanadium (V), and titanium (Ti)[6].

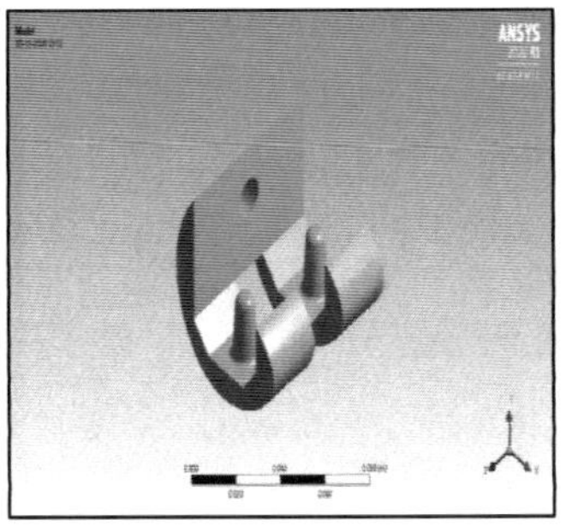

Figure 1: Anterior Side of Femoral Component geometry

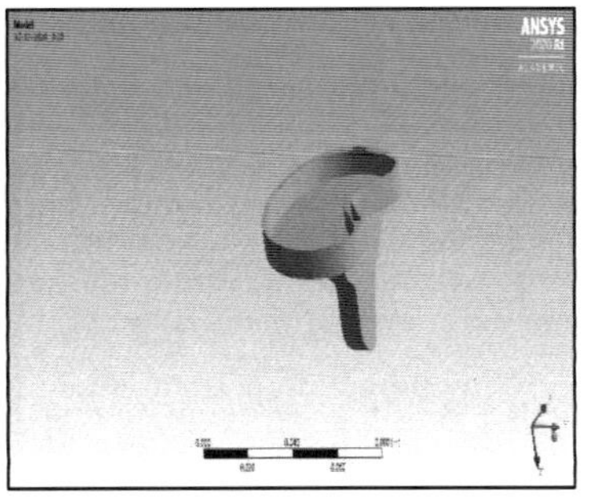

Figure 2: Geometry of Tibial Tray

The dimension of the geometry is as follows: -

Femoral component: Box height is 14.8 mm. The total width of the femoral component is 61 mm. The distance between two nails in femoral component is 17.8mm. The hole present in component is of dimension of 12.97mm* 11.94mm. This can allow a nail of 11 mm to be inserted into it. The height of the two flanges which will be sliding over base plate i.e. spine height is 31mm. Tibial Surface: The keel height is the height of the lower intruding portion which will be inserted in the bone is 38 mm. The width of this protruding part is defined as M/L keel is 37 mm. For size of 3, the offset component for tibia tray is 4mm. After completing all the steps, the final simulation, obtained with various color codes, is displayed in the figure below. The color displayed in the pictures highlights different values with red color showing the maximum value of the specific parameter and blue color showing the least. Deformation is measured in mm and normal strain is in mm/mm.

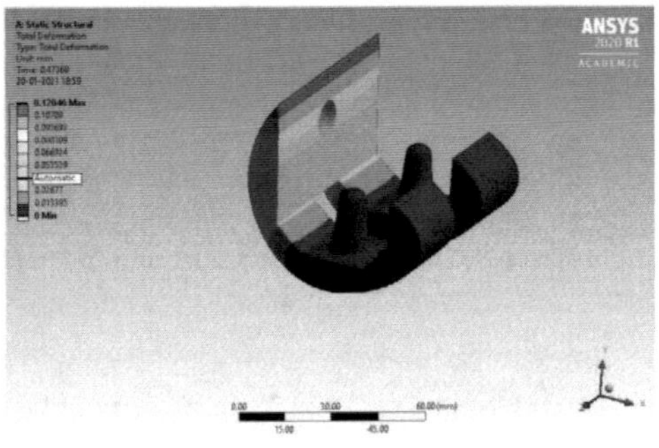

Figure 3: Deformation of femoral part

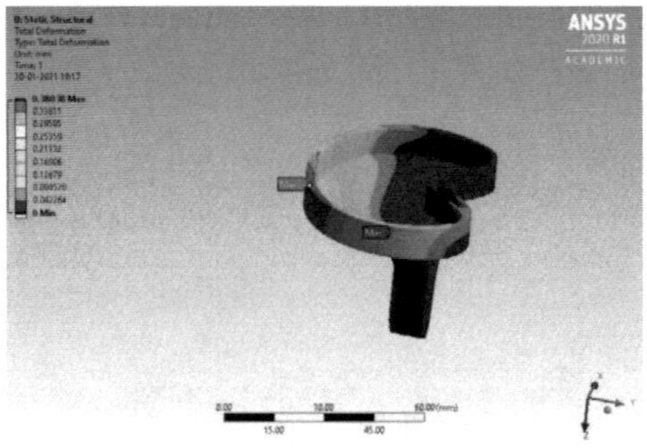

Figure 4: Deformation of Tibial part

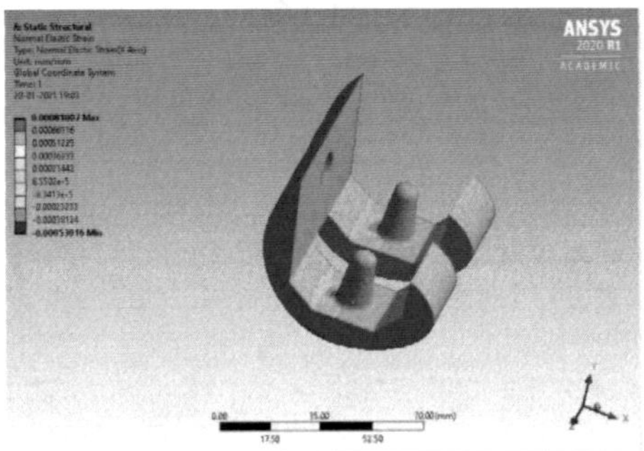

Figure 5: Normal Strain of Femoral part

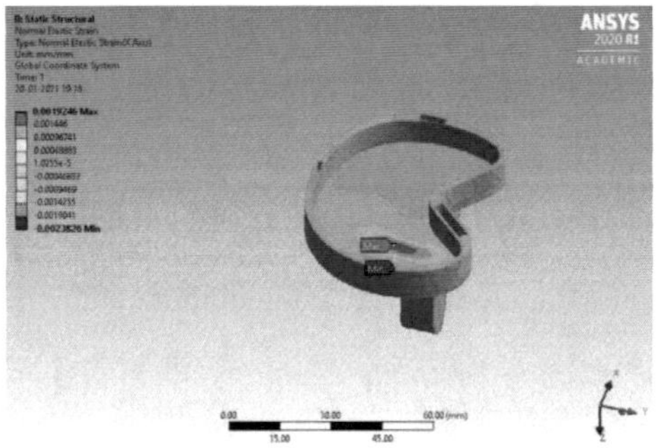

Figure 6: Normal strain of Tibial Surface

Most of the component is having zero deformation while the part which is vulnerable to deformation has a very gradual rise as observed from the color code. While in the normal strain simulation figure, most of the region is in complete green hence showing the presence of strain in almost every point of the component in spite of the fact the values are very minimal. Figure 3, figure 4, figure 5, figure 6 are the images of the component where the material is selected as Magnesium (Mg).The figure shown below are quite similar to the ones shown above. However, the material used here is Titanium(Ti). The color changing phenomenon is quite similar with the magnesium simulation however there is slight change in values depicting Ti to be a bit stronger and sustainable.

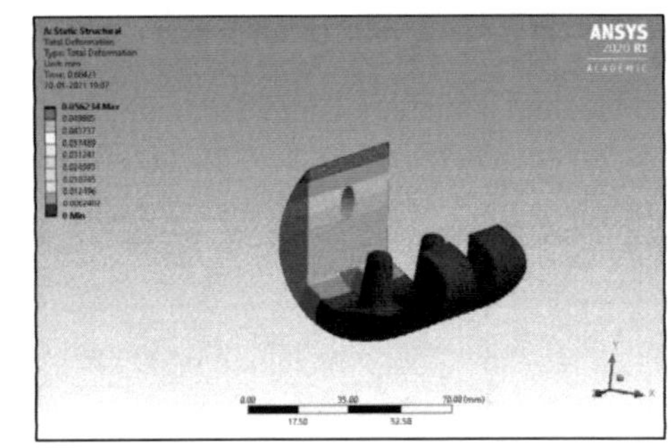

Figure 7: Deformation of Femoral Part

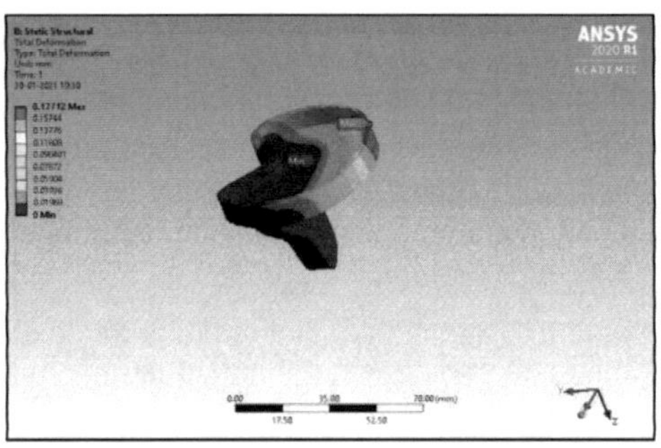

Figure 8: Deformation of tibial Part

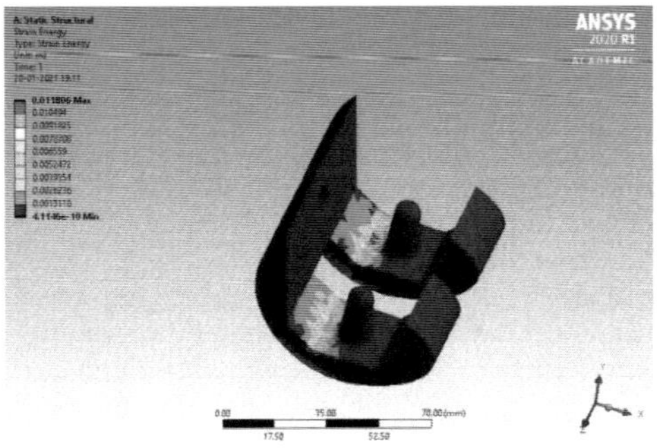

Figure 9: Strain Energy of Femoral Part.

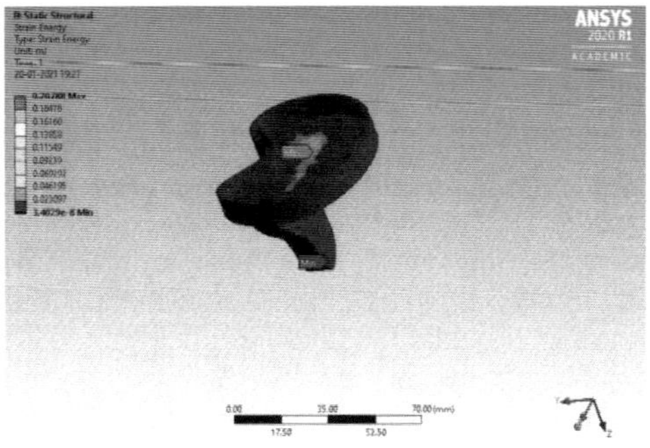

Figure 10: Strain energy of Tibial tray.

Deformation is measured in millimeters whereas strain energy in mJ. Different alloy material like calcium, zinc, manganese, strontium, silver etc. each can contribute differently as per their indiviual property to enhance magnesium and can give more prominent results. Magnesium (Mg) is the second most abundant element in human body. It is known for better mechanical compatibility, higher initial stability, initial support, reduces shielding of implants, light in weight and economical[7]. The density of Mg and its alloys are close to cortical bone[8]. It is known to have good biocompatibility with human philosophy. It is essential as a cofactor in enzyme and has significant impact in tissue healing. Extra amount of Mg is expelled via urine hence not causing excess trouble for its application as temporary implant. In spite of having so many advantages, it is essential to underline certain demerits light to clearly understand the slow speed evolution of this material in fabrication of knee implant. Mg is known for its significantly high its corrosion rate, which can cause loss of mechanical property and degradation of implant at a faster pace. Hydrogen evolution is also a major concern because it reduces the speed of the tissue healing process. The rate of evolution is reported for Magnesium and its alloys is 0.1 ml/cm^2/day[9]. Blood contains tremendous amount of alkalis, organic and inorganic compounds which can affect the corrosion rate at different speeds To resolve this issue, various energetic coating have proved to be promising to prevent corrosion. Different types of coating like biodegradable polymer coating includes poly lactic acid[10], polycaprolactone[11], etc. biodegradable silane coating[12], biodegradable calcium phosphate coating[13], graphene derivative based coating. These coating not only prevent the corrosion to a wider prospect but also enhance the biocompatibility by reducing the increase in pH and depriving the evolution of H_2 within the body fluids.

4. Conclusion:

Nowadays, knee implant is available in different types and designs as per requirement of different indiviual. The material should match with the present properties being used for fabrication of knee implant and should be economical as well. Hence, magnesium was taken into consideration. Corrosion was one of the key issue with this material but various coatings and mixing of other metals to improve strength has however brought promising results to use this material for the same. To eliminate the use of titanium, it is important to provide an alternative material for the manufacturing of the knee implant like ceramics and composite material which is being used nowadays.

5. References:

[1] Carr B C and Goswami T 2009 J. Materials and Design 30 398.
[2] Madeti B K and Rao C S 2018 Int. J. Experimental and Computational Biomechanics 4 113.
[3] Bergmann G, Bender A , Graichen F , Dymke J R ,Rohlmann A , Trepczynski A Heller M O and Kutzner I 2014 J.Plos one 9 1237.
[4] Kaivosoja E, Tianien V M , Takakubo Y , Rajchel B, Sobiecki J , Konttinen Y T and Takagi M 2013 Wear of Orthopedic Implants and Aritifcial Joints(Woodhead Publishing) p 178.
[5] Kustono D, Wulandari R, Ko A, Puspitasari P, Kurniawan G A and Putra A D 2018 Int. Conf. Mech. and Industrial Engg. 204 07015.

IOP Conf. Series: Materials Science and Engineering 1219 (2022) 012025 doi:10.1088/1757-899X/1219/1/012025

[6] Rajesh B, Radhakrishna L and Reddy K V 2017 Int. J. Mechanical Engineering and Technology 8 197.
[7] Radha R and Sreekanth D 2017 J. Magnesium and Alloys 5 286.
[8] Chen X 2017 J. of Orthopaedic Translation 10 1.
[9]Banerjee P C , Saadi S A , Choudhary L, Harandi S E and Singh R 2019 J. Materials 12 136.
[10]Alabbasi A, Liyanaarachchi S and Kannan M.B 2012 Thin Solid Films 520 6841.
[11] Ostrowski N.J, Lee B, Roy A, Ramanathan M and Kumta P.N 2013 J. Mater Sci Mater Med 24 85.
[12] Müller R, Abke J, Schnell E, Macionczyk F, Gbureck U, Mehrl R, Ruszczak Z, Kujat R, Englert C and Nerlich M 2005 J.Biomaterials 26 6962.
[13] Song Y, Zhang S, Li J, Zhao C and Zhang X 2010 J.Acta Biomater 6 1736.

IOP Conf. Series: Materials Science and Engineering 1219 (2022) 012026 doi:10.1088/1757-899X/1219/1/012026

An Improved double input dc converter for renewable energy application

A. Lavanya[1] J.Divya Navamani[1],K.Vijayakumar[1]

[1]SRM institute of Science and Technology, Kattankulathur, Tamilnadu, India, 603203

E-mail: lavanyaa@srmist.edu.in

Abstract.A hybrid renewable energy system with more than one source like solar photovoltaic, wind energy, battery, and fuel cell increase the system's reliability and flexibility. A suitable power electronic converter is required to integrate these sources and interface with the load simultaneously. Therefore, the paper's main objective is to recommend a novel double input dc converter for a hybrid renewable energy system. This power electronic converter efficiently integrates the sources and supplies the load demand. The critical waveforms for various modes of operation of the proposed converter are discussed in detail and the output voltage equation is derived. The proposed double input converter has a simple structure and reduced components. The performance comparison with the currently reported converters in the literature is presented, and the double input converter is simulated and analysed using MATLAB/Simulink software. The theoretical analysis is validated through the simulation results of the designed double input dc converter for renewable energy application.

Key words: Renewable energy system, Hybrid, Solar PV, double input, dc converter

1.Introduction

Fossil fuels have adverse effects on the environment, leading to the changeover to recently developing globally more sustainable and renewable sources that are environmentally friendly. Since the electric power demand is increasing day by day and there is a decrease in fossil fuel assets, inspired the energy sector to transfer towards solar PV based energy system. Hence, the world is moving towards renewable energy sources due to the reduced cost of solar photovoltaic (PV) panels have [1]-[2]. Rich solar energy reserves worldwide, its infinite potential, and low maintenance cost have supplemented the distinction of solar PV based standalone and grid-connected systems [3]-[4]. Numerous dc converter configurations are newly proposed with compact increased gain and efficiency to resolve the literature's issues [5], [6]. In this regard dual Input converters can broaden diverse vitality sources and expand the dependability of the system. DI converters can associate the benefits of various sources, such as photovoltaic cells, fuel cells, batteries etc. Spurred by expanding utilization of elective vitality sources, e.g., fuel cells, wind energy, sun-oriented energy etc., dual Input converters have been accepted as a better option with lower cost, effective operation for places with hybrid energy input sources. The conventional approach for associating two sources is to interface each source to an alternate Single Input Single Output (SISO) structure. This results in an increased complexity of the circuit. Such a system is less cost-effective and due to a more complex design, it leads to higher losses. Dual input converters carry out multiple power converters' operations with a reduced number of power components and switches, thus resulting in simple control, more stability, more versatility, and lower losses in the system. Electric hybrid vehicles require the most advanced power generation methods without compromising on their efficiency and life cycle. Existing topologies of buck converters used for electric vehicle control and charging batteries seemed to have variations in the desired output due to voltage ripples and switching losses. In contrast, the proposed system can be substituted as an adequate replacement for the techniques with more filter elements to achieve the desired output with lesser losses and more efficient and compact hardware used in electric vehicles.

A High-frequency transformer is employed in isolated converters to increase the voltage gain and control the input and output ports' voltage to the preferred values. But the transformer influences the performance of the converter by increasing the power losses and decreasing efficiency. Series connected dual input dc converter has extensive applications [7],[8]. Compared to single-input converters, dual input dc converters reduce the complexity, flexibility, and cost of PV based hybrid energy systems [9-12]. By connecting two PV panels to a

IOP Conf. Series: Materials Science and Engineering 1219 (2022) 012026 doi:10.1088/1757-899X/1219/1/012026

single dual-input dc converter, compared to the conventional structure as shown in Figure. 1a, 50% of the converter unit are reduced system architecture as shown in Figure.1(b) is simplified and the cost is also reduced.

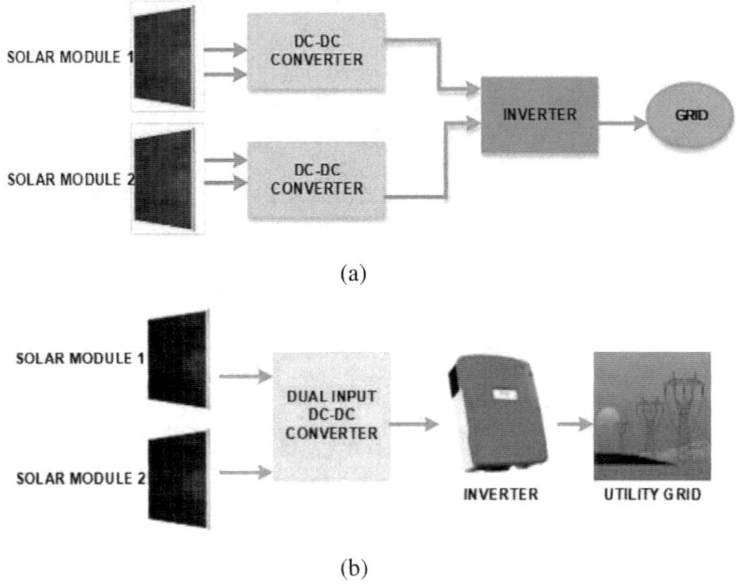

Figure 1. PV system (a) Conventional (b) Modern dual input dc-dc converter

Figure.2 illustrates the improved double input converter (IDIC) with input as two different DC voltage sources controlled by two switches (S_1, S_2), two diodes (D_1, D_2) and single output. The converter consists of an inductor and two capacitors along with two diodes and a boost switch. The boosting part of the design has the inductor L and capacitor C connected in parallel and series to the boost switch S and two diodes D and D_0 located in the circuit are complementary to each other in conducting at the time in the circuit while the capacitor C_0 is employed in series and parallel to the load side as an output capacitor to control the ripples in the output voltage.

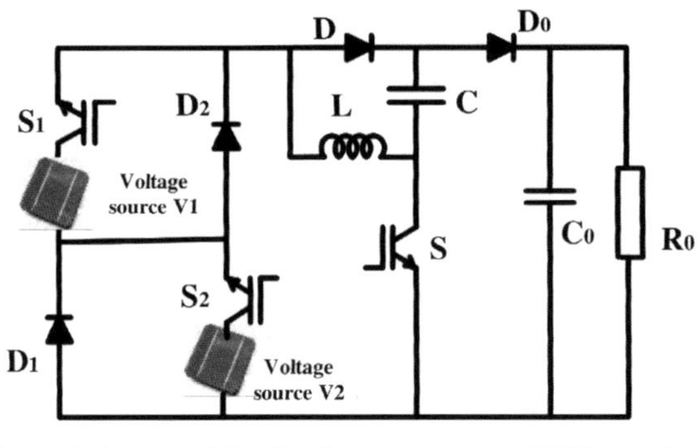

Figure 2. Improved Double input converter (IDIC) topology

2.Operating principle of IDIC

MODE I: In this mode voltage source V_1 is connected through switch S_1 while switches S_2 remain OFF state as shown in figure.3. The current path outlined by red colour travels in the conducting path with the components in which current flows in the circuit given below. Diode D is forward biased while D_0 is reverse biased. During ON time the inductor and the capacitor C is getting excited by the supply voltage V_1 while the output capacitor is feeding the load with the charge it held during its previous cycle.

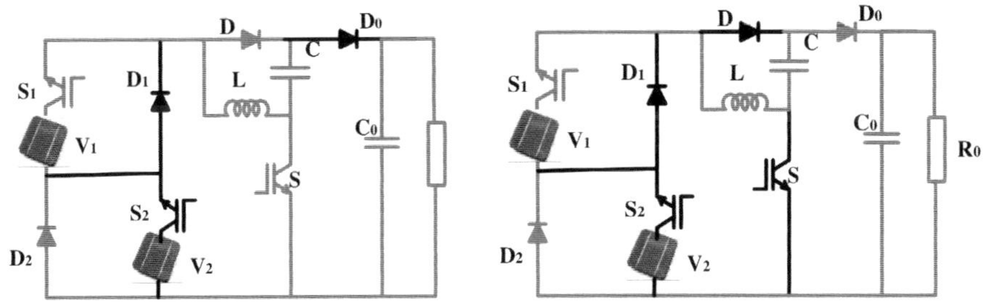

(a) (b)

Figure 3. IDIC topology in mode-I (a)charging(b) discharging

During OFF time the current is flows through the inductor, capacitor to the load through diode D_0 all connected in series. The diode D is reverse biased during this state of operation. While D_0 is forward biased and allows current flow through the load. The Kirchhoff's Current Law and Voltage Law equations for this state are given in equations (1)-(7)

ON: Input current, $I_g=I_L+I_C$ (1)

Inductor voltage, $V_L=V_1=V_C$ (2)

Output capacitor current, $I_{C0} = -I_0$ (3)

OFF: $I_g=I_L=I_C$ (4)

Output diode current, $I_{D0}=I_{C0}+I_0$ (5)

Output capacitor voltage, $V_{C0}=V_0$ (6)

Inductor voltage, $V_1 +V_C-V_{C0}= V_L$ (7)

MODE II:. In this mode as shown in figure.4 only input voltage source V_2 is connected to the converter through switch S_2, while S_1 remain OFF. Switch S is turned ON in this mode and remains conducting throughout this mode. Diode D_1 and D are forward biased while D_2 and D_0 are reverse biased. During ON time, the inductor and the capacitor C is getting excited by the source while the output capacitor is feeding the load with the charge it held during its previous cycle.

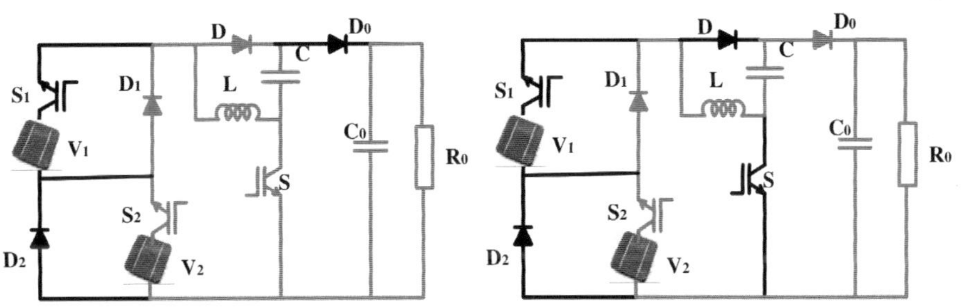

(a) (b)

Figure 4. IDIC topology in mode-II(a)charging(b) discharging

The Kirchhoff's Current Law and voltage Law equations for the above circuit are given in equations (8)-(10)

$I_g=I_L+I_C$ (8)

$V_2=V_L=V_C$ (9)

$I_{C0} = -I_0$ (10)

OFF Condition: In this state also input voltage source V_2 is connected to the converter through switch S_2, while Switch S_1 and S will be in OFF state. Diode D_1 still remains in forward biased while D_2 is reverse biased. During

OFF time, the inductor and the capacitor C is getting discharged while the output capacitor C_0 will be charging during this OFF state. The current direction is shown in red colour conducting path. The input current from the source voltage V_2 flows through the inductor L, capacitor C, diode D_0 to the load

The Kirchhoff's Current and Voltage Law equations for the above circuit are given in equations (11)-(14)

$$I_g=I_L=I_C \quad (11)$$

$$I_{D0}=I_{C0}+I_0 \quad (12)$$

$$V_{C0}=V_0 \quad (13)$$

$$V_2-V_L+V_C-V_{C0}=0 \quad (14)$$

MODE III: Figure.5 shows that in this mode, both V_1 and V_2 are linked together through switch S_1 and S_2. Mutually the switches S_1 and S_2 is turned ON throughout this mode of operation. The ON and OFF condition operation of other switches and diodes are similar to the previous two modes of operation but with two operating sources V_1 and V_2

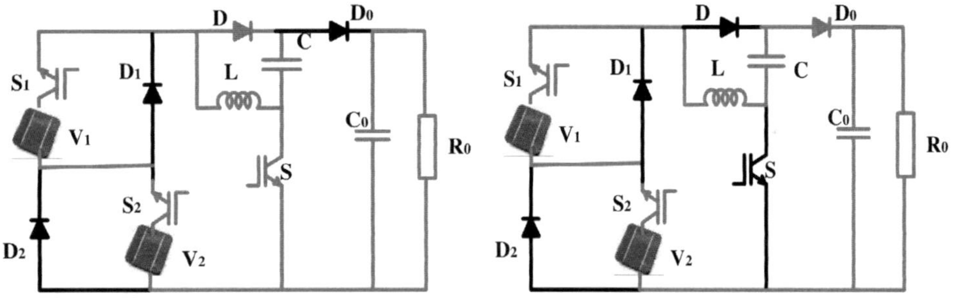

(a) (b)

Figure 5. IDIC topology in mode-III(a) charging(b) discharging

3.Simulation results and discussion

An improved double input dc converter with the voltage sources connected individually and in succession has been considered for simulation. The results have been analysed for continuous mode in steady-state conditions. Simulation results of output voltage and current, inductor voltage and current are shown. Figures 6 and 7 confirms that the proposed IDIC produces various output voltage of 90V,180Vand 300 V for different duty ratios. These voltages are in good agreement with the theoretical values as given in table 1.

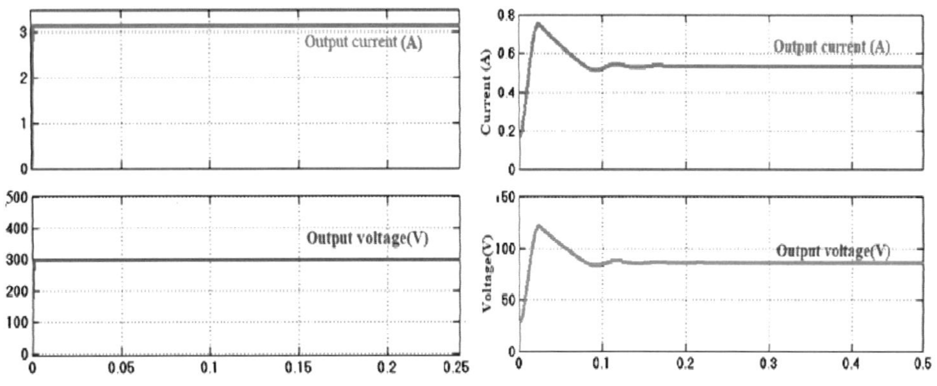

(a) (b)

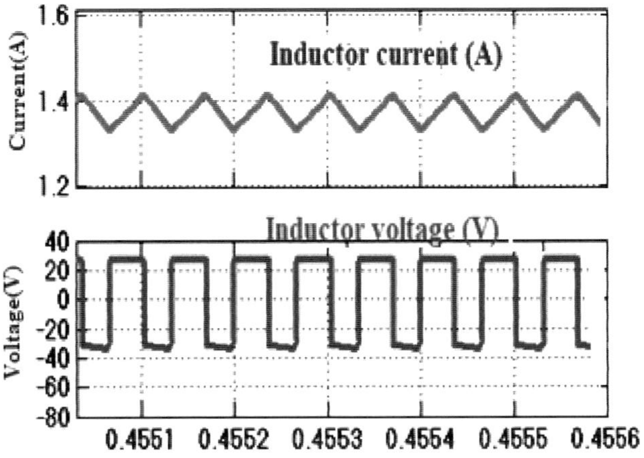

(c)

Figure 6. IDIC simulation results(a)Output voltage 300V and output current (b) Output voltage 90V and output current (c) Inductor voltage and current

Table 1: Various parameter relationship in proposed IDIC

Voltage gain(M)	V_L	I_L	V_0 Single input	V_0 Double input	I_0	Vsw	V_D
$\frac{(2-D)}{(1-D)}$	ON: V_g OFF: V_g+Vc-V_0	$\frac{I_0}{(1-D)}$	$V_0 = V_1 \frac{(2-D)}{(1-D)}$	$V_0 = (V_1 + V_2)\frac{(2-D)}{(1-D)}$	$I_g \frac{(1-D)}{(2-D)}$	$V_0 \frac{2(1-D)}{(2-D)}$	$V_0 \frac{2(1-D)}{(2-D)}$

To determine the IDIC Efficiency (η), η = Pout/Pin, the total converter loss is to be calculated, total converter power loss = (Pin–Pout), where Pin specifies the total input power and Pout specifies the total output power. IDIC converter efficiency is calculated as 91.6% at 50 W output power rating.

IOP Conf. Series: Materials Science and Engineering 1219 (2022) 012026 doi:10.1088/1757-899X/1219/1/012026

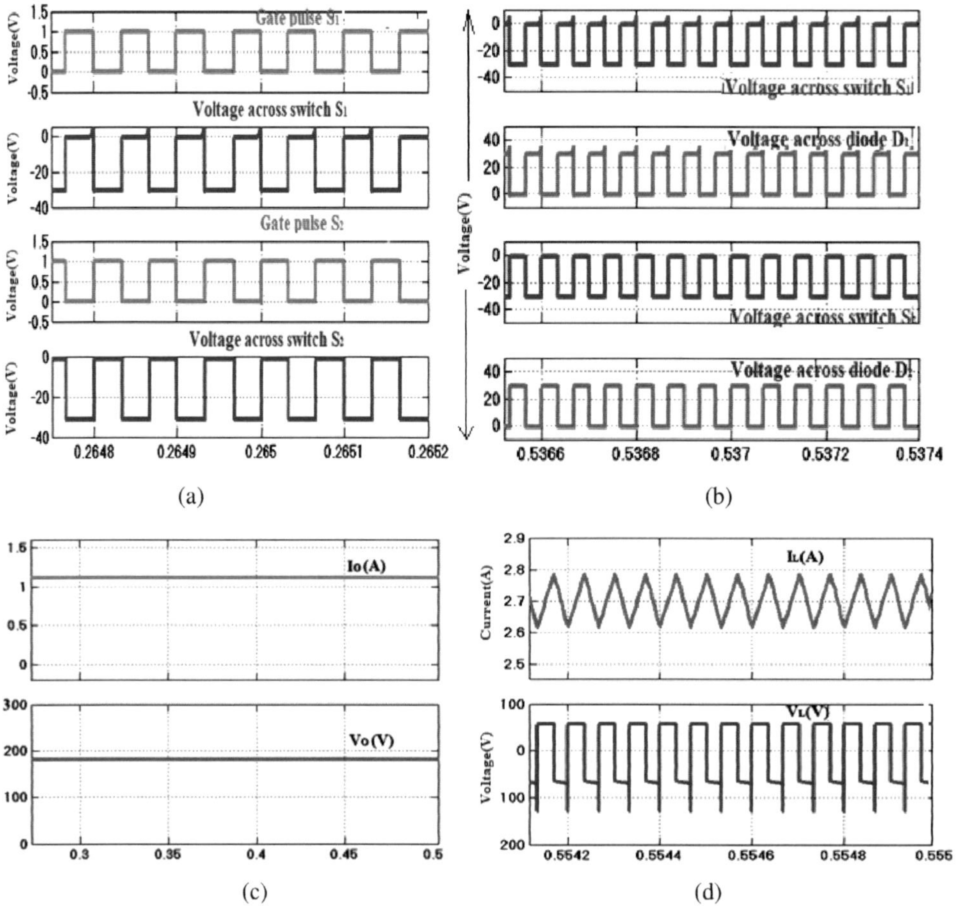

Figure 7. (a) switch waveforms (b) switch and diode waveforms (c) 180V output voltage and current (d) inductor current and voltage waveforms

4.Conclusion

An improved double-input converter has been proposed in this paper. Simplified structure, concurrent and autonomous power transferring ability of inputs, reduced components with reduced measurement devices are the key features of the proposed topology. Single inductor of the converter significances in low cost, compact simplified controller structure. The converter structure has the flexibility to interface two input sources. The proposed topology is suitable for hybrid renewable energy systems or hybrid electric vehicles HESs. The performance of IDIC topology and its theoretical concepts has been proved by simulation results.

References

[1] L. J, Chien., C. C, Chen., J, Chen., Y. P, Hsieh.: 'Novel Three-Port Converter with High- Voltage Gain', IEEE Trans.Power Electron., 2013, 29, (9), pp. 4693–4703.

[2] H, Behjati., A, Davoudi.: 'Single-stage multi-port DC–DC converter topology', IET Power Electron., 2013, 6, (2),pp. 392–403.

[3] F, Nejabatkhah., S, Danyali., S. H, Hosseini., M, Sabahi.,S. M, Niapour.: 'Modeling and control of a new three-input DC–DC boost converter for hybrid PV/FC/battery power system', IEEE Trans. Power Electron., 2011, 27, (5), pp.2309–2324.

[4] Zhijun Qian, Osama Abdel-Rahman, Issa Batarseh: 'An integrated four-port dc–dc converter for renewable energy applications', IEEE Trans. Power Electron., vol.25, no.7, pp. 1877-1887, Jul. 2010.

[5] Shireen W, Arefeen M.S., "An utility interactive power electronics interface for alternate/renewable energy systems", IEEE Trans. Energy Conversion, vol.11, no.3, pp.643-649, Mar. 1996.

[6] Dobbs B G, Chapman P L., "A multiple-input DC-DC converter topology", IEEE Power Electronics Letter, vol.1, no.1, pp.6-9, Jan. 2003.

[7] Yaow-Ming Chen, Yuan-Chuan Liu, Sheng-Hsien Lin., "Dual-input PWM DC/DC Converter for High-/Low-voltage sources", IEEE Trans. Ind. Electron., vol.53, no.5, pp.1538-1545, May 2006.

[8] Chen, Y.M., Liu, Y.C., Lin, S.H.: 'Double-input pwm dc/dc converter for high-/low-voltage sources', IEEE Trans. Ind. Electron., 2006, 53, (5), pp. 1538–1545.

[9] Divya Navamani.J ,Lavanya, A, Vijayakumar, "Modified Sepic converter with high boosting capability", IET Electronics letter, 55(13), pp. 759-761.

[10] Guo, F., Fu, L., Zhang, X., et al.: 'A family of quasi-switched-capacitor circuit-based dual-input dc/dc converters for photovoltaic systems integrated with battery energy storage', IEEE Trans. Power Electron., 2016, 31, (12), pp. 8237–8246

[11] Lavanya, A., Jegatheesan, R. & Vijayakumar, K. Design of Novel Dual Input DC–DC Converter for Energy Harvesting System in IoT Sensor Nodes. Wireless Pers Commun (2020). https://doi.org/10.1007/s11277-020-07048-0

[12] Ajami, A., Asadi Shayan, P.: 'Soft switching method for multiport dc/dc converters applicable in grid connected clean energy sources', IET Power Electron., 2015, 8, (7), pp. 1246–1254.

An Investigation of Mechanical Properties by Reinforcing Steel Mesh into Aluminium Alloy 6061

Harshit Tripathi[1*], Rajvir Singh[1], Sushil Kumar[1], J D Sharma[1], Sumit Bhardwaj[2]

[1]Department of Metallurgical and Materials Engineering, Punjab Engineering College (Deemed to be University), Sector-12, Vidya Path, Chandigarh-160012, India
[2]Department of Physics, University Institute of Sciences, Chandigarh University, Gharuan, Mohali, 140413, Punjab, India

Email - harshittripathi.phd19meta@pec.edu.in

Abstract. The application of aluminium-based composite is increasing day by day in structural and aerospace engineering sectors due its exceptional properties such as higher strength to weight ratio, excellent mechanical properties, and high corrosion resistance. In the current study the Metal Matrix Composite (MMC) with Aluminum alloy (AA) 6061 as matrix and steel mesh as reinforcement was fabricated by two routes. First, by hot rolling the steel mesh sandwiched between two 3 mm thick AA 6061 sheets and second by the pouring of AA 6061 over and below the steel mesh placed in horizontal plane in a sand mold such that the mesh gets completely incorporated at the centre of casting subsequently hot rolling them to obtain MMC. One specimen obtained by each route was given T6-heat treatment (HT). The microstructural study was done to characterize the composites formed and their mechanical properties such as hardness and tensile strength were compared with HT cast-rolled and HT rolled sheets of same thickness. The superior Ultimate tensile strength of 155.4MPa was shown by samples obtained by HT cast-rolled sheet with steel mesh sample than other samples.

Keywords: MMC, AA 6061, Steel Mesh, Mechanical Property

1. Introduction

In recent years, composite made by bimetal as well as laminated metal as sandwich panel of two or more dissimilar component materials have been used in many industrial applications, due to the combination of mechanical, thermal, corrosion resistant and different functional properties. There are several existing processing methods for the fabrication of metal composite, including diffusion bonding, explosive welding, casting and roll bonding. Different choices of fabrication technology and selections of component materials have a significant effect on the mechanical properties of composite. Since aluminium composite has high stiffness, high resistance to wear and seizure as compared to the aluminium alloy. For composite formation it is mixed with copper, zinc, magnesium or silicon and its reinforcement with other materials such as steel, epoxy etc. can be done [1]. AA are those in which the aluminium is the main constituent element. These AA carry both the elements qualities and thus make it better than the constituent elements in many aspects.

The AA 6061 is a precipitation-hardened aluminium alloy, containing magnesium and silicon as its major alloying elements. It has good mechanical properties, exhibits good weldability and is very commonly extruded. Due to general-purpose use of this alloy, it is frequently used. It is widely available in pre-tempered grades such as 6061-T6 (solutionized and artificially aged) [2]. The T6-HT represents a condition when aluminum's microstructure is solutionized, quenched and again heat treated (artificially aged).

The 6*xxx* series alloys containing magnesium and silicon provide moderate strengths and good corrosion resistance about other heat-treatable aluminium alloys. Because they are easily extruded, they are available in a wide range of structural shapes and sheet and plate products. These alloys' versatility is represented in 6061, which is one of the most commonly used aluminium alloys. Finally, the annealing is done by constant temperature in the furnace for some time. The high temperature is there to change the aluminum sheet's internal grain size to obtain the variety of mechanical properties and elongation [3]. Olexandr Grydin et al., in 2013 showed that aluminum–steel clad strips of a continuous, thin and uniform layer of intermetallic phases on the material having an adhesive strength over 70 MPa. Zhou Li et al., in 2018 showed analytical, numerical and experimental investigations were conducted on the bending characteristics of duplex stainless steel (2205) and low carbon steel (AH36) bimetal composite, this study to predict the deformation behaviour of each segment and particular planes [4-5].

IOP Conf. Series: Materials Science and Engineering 1219 (2022) 012027 doi:10.1088/1757-899X/1219/1/012027

In the present work, composites were developed by two routes, the samples were examined on the microstructural characteristics and mechanical properties. The effect HT and non-HT AA 6061 composite fabricated by reinforcing steel mesh were also investigated.

2. Experimental procedure

2.1 Fabrication process

The commercial grade AA 6061-T6 sheets of 3 mm thickness and 1 mm thick stainless-steel mesh was used for the preparation of specimens which were wire cut in dimensions of 11 mm x 5 mm x 3 mm.

The samples were fabricated by two methods. In the first method, the steel mesh, kept at 500°C in a furnace for 5hrs, was sandwiched between two AA 6061 sheets which were hot rolled to ensure fully embedment of steel mesh into AA sheets. In the second method, pouring of AA 6061 melt was done over and below the steel mesh such that the mesh gets completely incorporated at the centre of casting. Afterwards the cast sample was hot rolled to obtain MMC. One sample obtained from each route was given the T6 HT. The samples designation used in the current work is provided in Table 1.

Table 1. Sample Designation

Samples		Designation
Heat treated rolled sheet	:	HTRS
Rolled sheet with steel mesh	:	RS/SM
Heat treated rolled sheet with steel mesh	:	HTRS/SM
Heat treated cast-rolled	:	HTCR
Cast-rolled sheet with steel mesh	:	CR/SM
heat treated cast-rolled sheet with steel mesh	:	HTCR/SM

2.2 Microstructural studies

The microstructure of the samples was observed under stereo zoom microscope (Make: carl zeiss stemi 305) at different magnifications from 10x to 100x. The Field Emission Scanning Electron Microscopy (FESEM, Make: JEOL J6M-7600F) was carried to characterize the cross-sections of the samples.

2.3 Mechanical properties

I. *Micro hardness:* Omni tech micro hardness tester was used to investigate Vickers micro-hardness (HV) of the samples. The load of 100Kgf was used to make indentation on to the flat polished sample for the dwell time of 10s.

II. Tensile test: The tensile was carried out on HELCO micro tensile tester for all the samples prepared with ASTM E8 standard. In each sample loading is done at the rate of 0.01 mm/ sec to determine the UTS, yield stress, elongation and elastic modulus.

3. Result and Discussion

3.1 Microstructural studies

The Figure 1 (a-f) shows the optical microstructure of the all the composite developed. It is observed from optical microstructure images that there is coarsening of grain size of the samples which is due to the slow cooling rate rate in the second phase of T6-HT process. ture of a material strongly influences the strength, hardness, ductility, toughness, wear resistance of the material, which decides the area of application of that material.

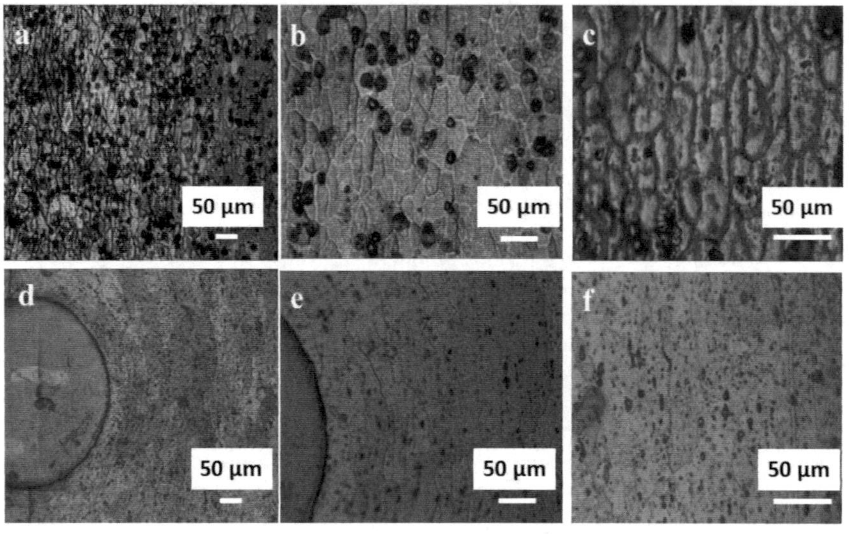

Figure 1. Optical Micrographs at (a)10x, (b) 20x and (c) 40x for rolled sheet and (d), (e) and (f) for developed composites respectively at same magnification.

Moreover as the grain size of microstructure increases, most of the mechanical properties of the specimen apart from its hardness decreases. This is because the fine grain size gives better mechanical properties [6]. The FESEM images are shown in Figure 2 (a-c). These FESEM images show that there is intact bonding between the matrix and reinforcement which indicates no sign of void formation between them.

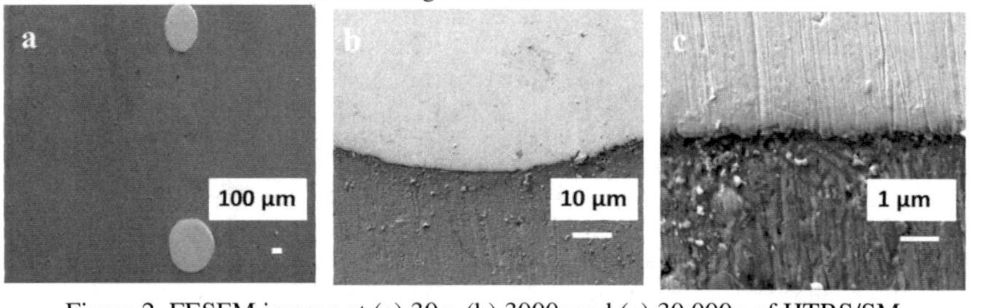

Figure 2. FESEM images at (a) 30x, (b) 3000xand (c) 30,000x of HTRS/SM.

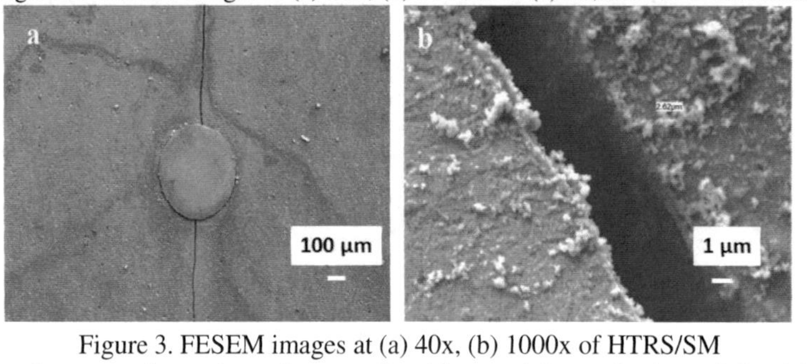

Figure 3. FESEM images at (a) 40x, (b) 1000x of HTRS/SM

Figure 3(a-b) shows there are a few points where small void around the steel wire and the AA 6061 are formed which is undesirable. The formation of these voids can weaken the composite formed [6].

3.2) Hardness

The composite formed by the rolling of the sheet was tested for Vickers hardness (HV) and a graph as shown in Figure 4(e). The graph was plotted with the HV and distance along the cross section of composite formed. Hardness is the indentation property and it is found to increase because of the rolling process. The difference in the hardness of heat treated Rolled sheet with steel mesh (HTRS/SM) sample is 94.4 HV and the Heat treated rolled sheet (HTRS) specimens is 42 HV, is because of several roll pass which the specimens has undergone while fabricating it, the T6 HT process and the aging effect. [7,8].

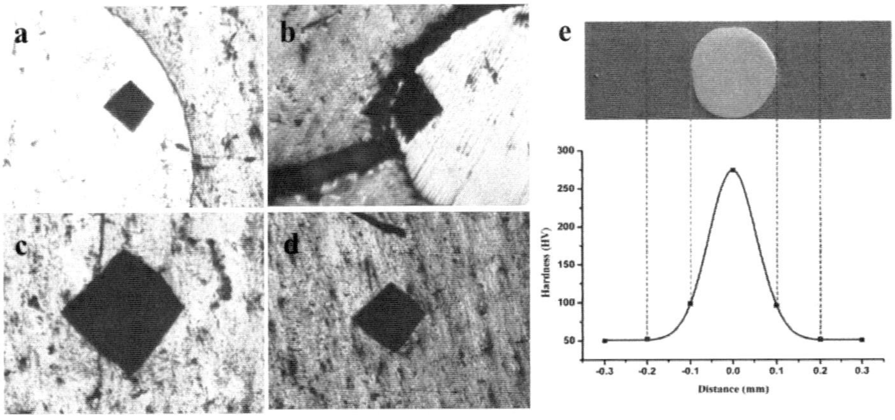

Figure 4. Vickers hardness indentation images of (HTRS/SM)

The Figure 5 shows the histogram with HV values for all the samples fabricated, it is observed that the HV increases in rolled sheet with steel mesh (RS/SM) and then (HTRS/SM) the reason for the increases T6 HT process and the aging of the specimen, which is known to increase the hardness of the sample [9].

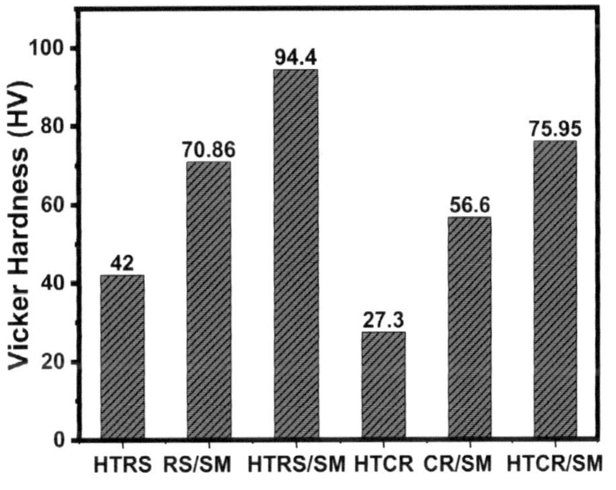

Figure 5. Vickers hardness of the samples

4) Tensile test

The tensile test was carried out to measure the tensile properties composites. Various properties such as ultimate tensile strength (UTS), elongation, elastic modulus and yield stress is shown in Table 2. The ultimate tensile stress (UTS) of HTRS is 50.9% more than the composite formed i.e. HTRS/SM [10]. As the load is applied, delamination of the composite takes place, which causes its failure. The delamination starts at the interface of steel mesh and aluminium matrix, because the point of reinforcement acts as stress concentration zone [10,11]. As per the E8 standard followed for the tensile test specimen few wire of the mesh came in the cross section of gauge length, which was not able to impart sufficient strength to the composite.

Table 2: Tensile properties of various samples

Samples	Ultimate Tensile Stress (N/mm^2)	Elongation %	Elastic Modulus (N/mm^2)	Yield stress (N/mm^2)
	Tensile Properties			
HTRS	387.33	17.2	25806	201.6
HTRS/SM	190.2	14.2	26816	83.5
RS/SM	180.6	13.6	25640	79.5
HTCR	99.46	23.25	9987	59
HTCR/SM	159.23	9.3	15145	78.6
CR/SM	144	8.6	10149	71.6

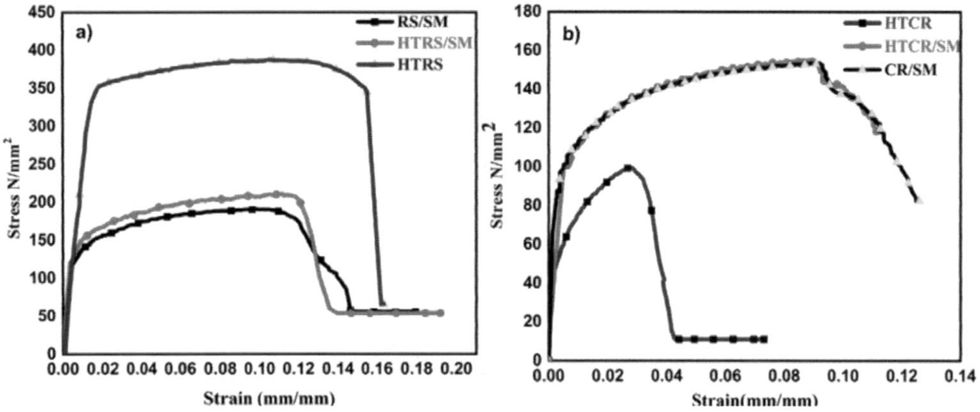

Figure 6: Stress-strain curve of (a) sheet rolled specimens and (b) cast and rolled specimens

The early stage of delamination was overcame in the specimens prepared by casting method as the results of the UTS of the heat treated Cast Rolled sheet with steel mesh (HTCR/SM) specimens are 37.7% more than that of the heat treated Cast Rolled (HTCR) sample. Therefore, it is confirmed that the bonding of the composite formed by casting is stronger as than that of rolling of sheets. The heat treated specimens have better values than non-heat treated ones due to the aging effect [12].

5) Conclusion

1. The AA 6061 composite is successfully fabricated by two routes i.e. first by casting using steel mesh and another by rolling the steel mesh between AA 6061 sheets. The T6 HT improves the mechanical properties and the FESEM images show that there is complete bonding between the reinforcement and matrix.
2. The HT process caused the coarse graining of the grains, due to which the strength is decreased and the hardness is increased.
3. The lower value of yield point and UTS in the composite made by the sheets was due to the delamination occurring at the point of reinforcement. The delamination started at reinforcement points because of the stress concentration zones formed there.
4. The UTS of HTCR/SM specimens are 37.7% more than that of the HTCR sample.

6) Acknowledgement
Authors would like to extend special thanks for the financial and technical support extended by ARMREB (DRDO), Chandigarh sponsored project.

IOP Conf. Series: Materials Science and Engineering 1219 (2022) 012027 doi:10.1088/1757-899X/1219/1/012027

Referrences

[1] Samson C, Ramesh A, Krishnamoorthy K, Selvaraj K S and Govindan P 2017 T Indian Metals 71 (5) 1041.

[2] Li Z, Jie Y, Li S, Wang Z and Wang G 2017 Trans. Nonferrous Met. Soc. 27 1498.

[3] Szczepanik, S. 2015 Mater. Today Elsevier Ltd 2 pp S9–S18.

[4] Olexandr Grydin, Gregory Gerstein, Florian Nurnberger,Mirko Schaper and Valentyn Danchenko 2013 J. Manuf. Process. 15 501–507.

[5] Zhou Li, Jingwei Zhao, Fanghui Jia, Qingfeng Zhang, Xiaojun Liang, Sihai Jiao and Zhengyi Jiang 2018 Int. J. Mech. Sci. 148 272–283.

[6] Grydina O, Gersteina G, Nürnbergera F,Schaperb M and Danchenkoc V 203 J. Manuf. Process 15 501.

[7] Zu G, Lu R, Li X, Zhong Z, Ma X, Han M and Yao G 2013 Trans. Nonferrous Met. Soc. 23 2491

[8] Barnawal V, Raghavan R, Tewari A, Narasimhan K and Mishra S 2017 Mater. Sci. Eng A 56.

[9] Khalili S, Mittal R and kalibar S 2005 Mater. Sci. Eng A 412 pp 137–140.

[10] Maurya N, Maurya M, Srivastava A, Dwivedi S, Kumar A and Chauhan S 2020 Mater. Today 25 755

[11] Kang U ,Lee H, Nam W 2012 J. Mater Sci, 7887 pp 7883.

[12] Chekuri B, S.nedkova B and Srinivas R 2005 Mater. Sci. Eng A 410–411 pp. 394–937.

Synthesis and Characterization of Graphene oxide for Gas Sensing Applications

Mathankumar.G[1], Simon Patrick D[1], Praveen. P[1], Krishna Mohan. M[1*]

[1]Functional Materials and Energy Devices Laboratory, Department of Physics and Nanotechnology, SRM Institute of Science and Technology, Kattankulathur, Chennai, 603203, Tamil Nadu, India

E-Mail ID: krishnam1@srmist.edu.in

Abstract. Organic 2D materials have gained extensive attraction in the field of the gas sensor. Because of their critical role in environmental monitoring, chemical manufacturing industries, and human health safety. In recent years most of the research groups have been interested, to fabricate the Graphene oxide-based chemi-resistive gas sensor, owing to its superior structural, mechanical, and electrical properties. Excellent mechanical strength (Young modulus ~0.05 TPA), the potentiality of ultrafast electron transport (highest mobility ~200,000 cm^2/V s) along with the best surface to volume ratio. In this paper, we have prepared Graphene oxide nanosheets by the modified hummers method. Structural properties of the as-prepared sample were confirmed by XRD and RAMAN spectroscopy. morphological analyses were done by FE-SEM analysis.

Keywords :(Graphene oxide, gas sensor)

Introduction:

Gases of Carbon (CO), Sulphur (SO_2), and oxides of nitrogen (NO, NO_2) are harmful gases that not only pollute the air but also cause global warming and it leads to several diseases headaches, lung problem, heart disease, respiratory distress, and premature deaths. In order to identify these gases, 2D materials-based gas sensors were introduced [1-3]. Sensors of these materials possess appreciable stability both chemical and physical, it is cost compatible and has good sensing properties. 2D materials such as WS_2 [4], SnS_2[5], MoS_2[6], and Graphene oxide[7], have shown excellent sensitivity towards oxidizing and reducing gases.

Among them, Graphene oxide-based materials have been attained much attention in the field of gas sensor due to the superior electrical and electronic properties of the material. Moreover, Graphene oxide is one of the multifunctional materials, which has been widely used in many applications such as solar cells[8], photocatalysis[9], thermo electrics[10], and so on. Moreover, graphene oxide is a layered material with a high surface to volume ratio. Generally, layered material with a high surface area possesses more number of surface active sites, which is helpful to enhance the gas adsorption amount and desorption properties enhancing the gas sensing effects. Herewith we made an attempt to synthesis graphene oxide using the modified hummers method. The structural and morphological properties were carried out by XRD, RAMAN, and FE-SEM with EDX spectra.

Materials used:

All the chemicals used in the present work were of analytical reagent (AR) grade.

Characterization techniques:

Powder (XRD) patterns of the as-synthesized sample were recorded by PANalytical X'Pertpro with CuKα radiation (λ = 1.5406 Å). Raman Spectroscopic analysis of the as-prepared sample was carried out by Horiba-Jobin, LabRAM HR instrument with the suitable wavelength of 533nm laser. Morphological analysis of the samples was analyzed by Field emission scanning electron microscopy (FE-SEM) (the FEG Quanta 250) using an accelerating voltage of 15 kV with an EDAX mapping system.

Synthesis procedure:

For the typical synthesis procedure, the hummers method was used; where 2g of graphite powder and 1g of sodium nitrate was dissolved into 92ml of H_2SO_4, after that above solution was stirred at room temperature for 2 h. Then, the above solution was kept in the ice bath at 5 °C for 5h, and then 6g of $KMNO_4$ was added slowly in the above solution, the solution was becoming color changed and the solution was continuously stirred at room temperature for 4h. After that, the prepared solution is diluted with 92ml of deionized water and it has been stirred for 24hours and later the solution is diluted with 250ml of distilled water. Finally, H_2SO_4 is added to the prepared solution, then washed with 10% HCL and deionized water several times until pH becomes 7. After centrifuge slurry is dried in the oven to obtained graphene oxide.

Materials characterization

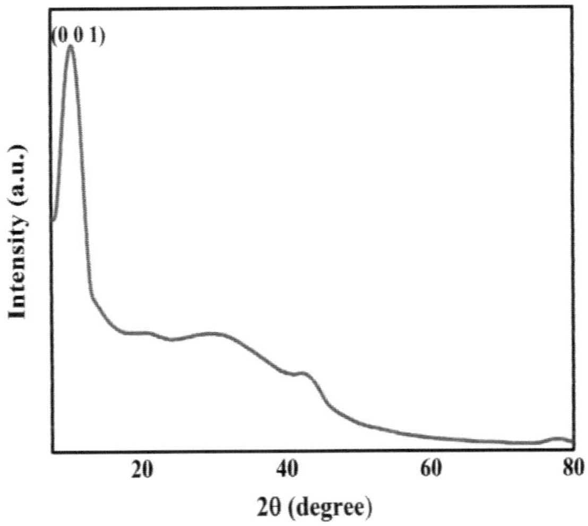

Fig.1 XRD patterns of Graphene oxide

Results and Discussion:

X ray Diffraction:

XRD is one such technique to determine the crystal structure of the prepared material. The figure shows the XRD patterns of graphene oxide, maximum intense peak placed at 2θ=11.30°, the peak corresponding to the (0 0 1) diffraction plane of Graphene oxide. This suggested a well-matched hexagonal structure.

Raman Spectroscopy

RAMAN spectroscopy is one of the potential techniques to characterize the structural and electronic properties of carbon-based materials and examine the crystal disorder, defect structures [2]. Fig.2 exhibits the Raman spectra of graphene oxide. The peak located at 1335 cm^{-1} and 1583 cm^{-1} in the inset correspond to the D band and G band of graphene oxide. The G band is associated with the symmetric stretching of sp^2 hybridization of C- C in a graphene oxide layer, the D band is related to the defects or disorder in the symmetrical hexagonal graphite layers.

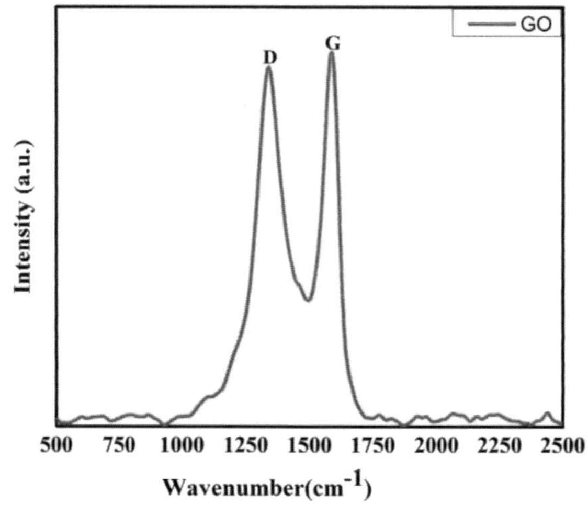

Fig.2 RAMAN spectra of Graphene oxide

FE-SEM micrograph:

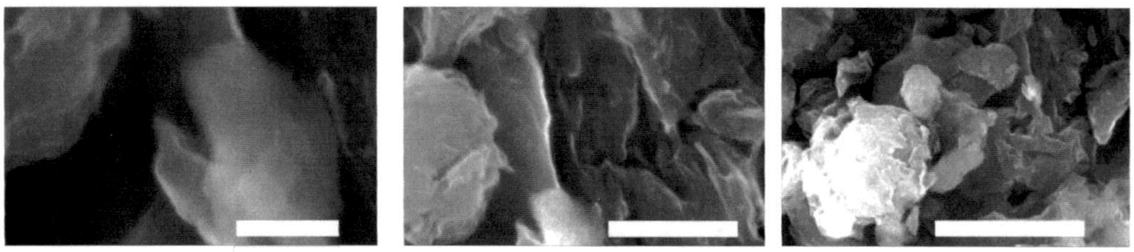

Fig.3 FE-SEM micrograph of Graphene oxide

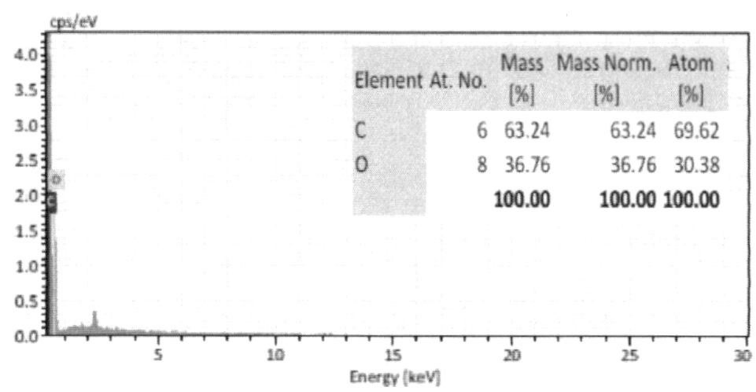

Fig.4 EDX spectra of Graphene oxide.

IOP Conf. Series: Materials Science and Engineering 1219 (2022) 012028 doi:10.1088/1757-899X/1219/1/012028

Scanning Electron Microscope (SEM) visualizes the morphology of the samples. Fig. 3 shows the different magnification of FE-SEM micrographs of as-synthesized Graphene oxide. FE-SEM micrograph exhibited that the bulk sheet-like structure. There is more number of layers joined together and showed the bulk layered structure.

Energy Dispersive Spectroscopy:

The energy-dispersive spectra (EDS) of the GO is given in Fig.4.This EDAX spectrum shows the existence of (C and O) elements present in the samples, it is confirmed that the presence of graphene oxide. This is in good agreement with our XRD results. There is no impurity peaks are observed in the EDX spectrum confirming the proper formation of graphene oxide

Conclusions:

In this present study, we have prepared Graphene oxide sheets by the hummers' method. XRD result reveals that the most intense peak located at 11.6° confirms the presence of GO. RAMAN result confirms that the presence of Graphene oxide. FE-SEM micrograph confirmed bulk layer like a morphology. EDX spectra confirmed the presence of (C&O), these results revealed the presence of graphene oxide in the given sample.

References:

1. Shukla, Vivekanand, et al. *The Journal of Physical Chemistry C* 121.48 (2017): 26869-26876.
2. Zhao, Yuxi, et al. *Nanoscale* 10.19 (2018): 9338-9345.
3. Cho, Byungjin, et al. *ACS applied materials & interfaces* 7.30 (2015): 16775-16780.
4. Late, Dattatray J., et al. *Sensor Letters* 14.12 (2016): 1249-1254.
5. Yan, Wenjie, et al. *ACS Applied Materials & Interfaces* 12.23 (2020): 26746-26754.
6. Shokri, Aliasghar, and Nadia Salami. *Sensors and Actuators B: Chemical* 236 (2016): 378-385.
7. Basu, S., and P. Bhattacharyya. *Sensors and Actuators B: Chemical* 173 (2012): 1-21.
8. Prabakaran, K., et al. *Solar Energy* 170 (2018): 442-453.
9. Hsu, Hsin-Cheng, et al. *Nanoscale* 5.1 (2013): 262-268.
10. Li, Tian, et al. *Nature Energy* 3.2 (2018): 148-156.

Structural, Luminescence and Cytotoxicity Studies of PEG Capped CaS Nanophosphors

Rekha S[1], Aravind S[1], Venkidesh T V[1], Anila E I[2,3]

[1]Department of Physics, Maharaja's College, Ernakulam -682011, Kerala, India.

[2] Optoelectronic and Nanomaterials Research Laboratory, Department of Physics, Union Christian College, Aluva-683102, Kerala, India

[3]CHRIST (Deemed to be University), Bengaluru-560029.

Abstract. Nanoparticles have remarkable applications in medical imaging and therapy owing to their enhanced optical, electronic, and structural properties. One of the challenges in nanoparticle application in the medical field is that when injected into the bloodstream, they are taken by the reticuloendothelial system (RES) of the body and removed by circulation. To avoid RES capture and protect them from the immune system, surface passivation of the nanoparticle is done with polyethylene glycol (PEG). The wet chemical synthesis of Calcium sulfide (CaS) nanophosphors capped with various concentrations of PEG is reported here. The structural and optical properties of the samples were studied by X-ray diffractogram (XRD), scanning electron microscopy (SEM), photoluminescence (PL) spectrum, and diffuse reflectance spectrum (DRS). The cell viability of the nanoparticles was estimated from the MTT assay using L929 human fibroblast cells. The cytotoxicity analysis revealed that PEG-capped CaS is biocompatible, which ensures the possibility of their prospective applications in the biomedical field.

Keywords: nanophosphors, chemical co-precipitation, photoluminescence, cytotoxicity

1.INTRODUCTION

Nanoparticles have dimensions that range from one to hundreds of nanometers, and they have significant uses in various optoelectronic devices. They play an important role in the emerging field of nanomedical imaging and therapy. Due to their small size and large surface area to volume ratio, nanoparticles exhibit novel electronic and optical properties. Calcium sulfide (CaS) is an important alkaline earth metal sulfide with a wide bandgap (4.5 eV). It serves as an excellent phosphor host material for a wide variety of dopants [1]. Many investigations have been carried out during recent years on CaS nanoparticles focusing on their luminescence applications in optoelectronics and the biomedical field [2-7].

Nanoparticles have a greater tendency to aggregate due to high surface energy. A capping agent prevents agglomeration of the nanoparticle and stabilizes them. Polyethylene glycol (PEG) is a polymer soluble in water that is widely used as a capping agent for the synthesis of nanoparticles because of its biocompatibility and biodegradability [8]. PEG decreases the surface energy of the nanoparticles and prevents aggregation, thereby reducing the nanoparticle's size [9-11]. When injected into the bloodstream, nanoparticles are taken by the reticuloendothelial system (RES) of the body and are rapidly shuttled out of circulation to the liver, spleen, or bone marrow. This results in nonspecific binding of nanoparticles to nontargeted areas of the body. To avoid RES capture and protect them from the immune system, surface passivation of nanoparticles is done with PEG. Surface passivation with PEG also decreases association with nontargeted serum and tissue [12]. In this work, the effect of PEG concentration on the structural and optical characteristics of CaS nanoparticles synthesized by wet chemical co-precipitation method is investigated using X-ray diffraction (XRD), scanning electron microscopy (SEM), room temperature photoluminescence (PL) spectroscopy, and UV-vis spectroscopy. In addition, the evaluation of cytotoxicity of the optimized sample was performed using MTT assay on L929 fibroblasts cells.

2.EXPERIMENTAL

The chemicals used for the synthesis of CaS nanoparticles are calcium chloride [$CaCl_2$. $2H_2O$, 97%Merck], sodium sulfide [Na_2S. xH_2O, Merck], PEG-200, 2-propanol and tetrahydrofuran (THF) [13,14]. The required amount of sodium sulfide is added to 50 mL of 2-propanol and stirred well using a magnetic stirrer for about one hour. To this solution, the capping agent PEG is added. It is then added dropwise to 50 mL solution of calcium chloride in 2-propanol, and the mixture is stirred well for about 2 hours. The solution is then heated at a temperature of 80^0 C so that the volume of the solution is reduced from 100 mL to about 40 mL. Then the precipitates are obtained by the addition of 20mL of THF. The solid particles are collected by filtering the precipitate through a filter paper. The precipitate is then dried overnight in an oven to obtain the nanoparticles. The experiment is repeated for various concentrations of PEG (1 mL, 2mL, 3mL and 4mL).

The XRD measurements were performed on the synthesized nanoparticles using a Rigaku Miniflex 600 X-ray diffractometer to get information about crystal structure. Meanwhile, the structure of CaS nanoparticles were examined using a scanning electron microscope (JEOL Model JSM-6390LV). Using a Flouromax4C spectrofluorometer, the PL emission spectrum was recorded. A UV-Vis-NIR spectrophotometer (Varian, Cary5000) was employed to record diffuse reflectance spectrum (DRS) of the samples. The absorbance values of the control and cells treated with samples were determined using a microplate reader (ERBA, Germany).

3. RESULTS AND DISCUSSION

3.1 XRD Studies

The XRD pattern of various concentrations of PEG capped CaS nanoparticles is depicted in fig.1. The XRD pattern is in confirmation with JCPDS file no: 77-2011 for cubic CaS. The peak broadening in the XRD pattern indicates the presence of small nanocrystals in the samples. The six peaks are indexed to (111), (200), (220), (222), (400), and (420) planes of the CaS host lattice as shown in fig. 1. The sharp diffraction peaks appearing at different 2θ values indicate good crystallinity of the prepared particles. The samples we used are CaS capped with different PEG concentrations, including 1 mL, 2 mL, 3 mL, and 4 mL, respectively. The average crystallite size (D) of all the samples estimated through Scherrer formula, was found to be in the range 30-35 nm [14,15]. The lattice parameter (a) is related to the interplanar distance d_{hkl} and the Miller indices h, k and l by the equation,

$$d_{hkl} = \frac{a}{\sqrt{h^2 + k^2 + l^2}}$$

Using this equation, the lattice constant of all the samples was determined and was found to be lower than that of the literature value 5.689 Å (JCPDS file No:77-2011). The values of average crystallite size (D) and lattice parameter (a) for the various samples are displayed in Table 1. The average crystallite size of 2 mL of PEG capped CaS nanoparticles was the least, and for concentrations of PEG lower and greater than this value, the crystallite size is more. Thus, the cubic phase and crystallinity of the samples synthesized are confirmed by XRD studies.

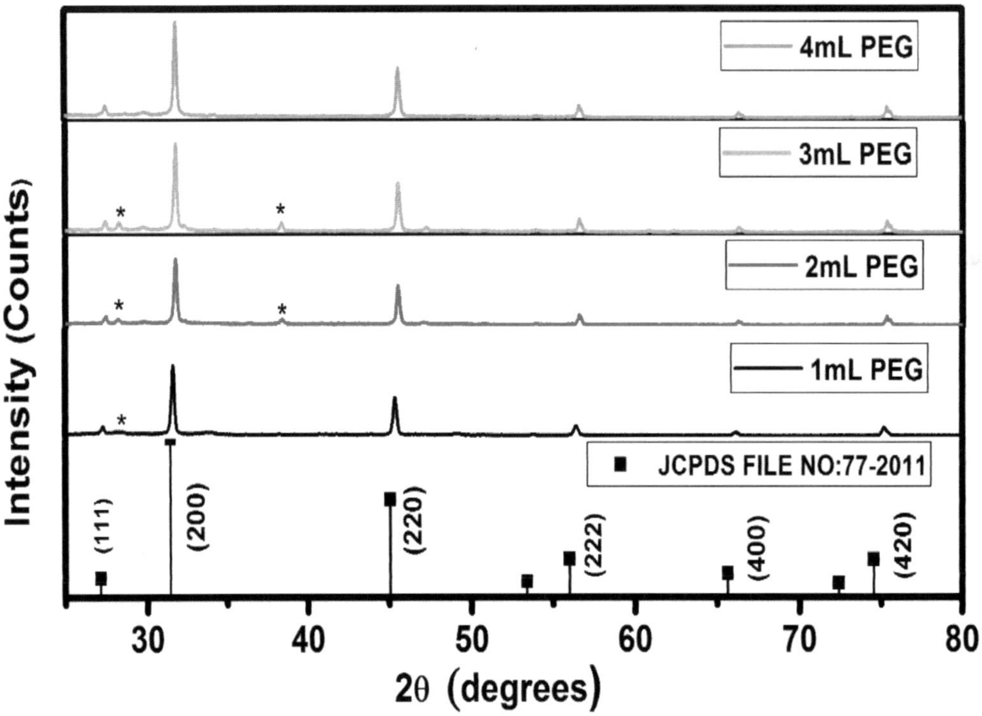

Figure 1. XRD pattern of PEG capped CaS nanoparticles (starred peaks represent impurity peaks)

Table 1. The average crystallite size, interplanar distance, and lattice constant of PEG capped CaS nanoparticles.

Sample name CaS capped with	2θ (degrees)	Average crystallite size from Scherrer formula (D) nm	Interplanar distance (d) Å	Lattice constant (a) Å
1mL PEG	31.62	31.7	2.827	5.655
2 mL PEG	31.77	30	2.814	5.628
3 mL PEG	31.78	34.7	2.813	5.626
4 mL PEG	31.81	35.1	2.810	5.621

3.2 SEM Analysis

SEM is used to analyze the surface structure and particle size of the prepared CaS nanoparticles. The SEM images of CaS nanoparticles capped with 2 mL of PEG for different resolutions are shown in fig. 2 A-B. The images show abundance of irregular spherical shape like particles with agglomeration. Therefore, each particle can be considered as an agglomerate composed of smaller crystallites having a size in the range 30 nm as obtained from XRD analysis. Although the crystallite size is in the nanometer range, it is not possible to determine the nanoparticles exact particle size due to aggregation. SEM analysis of 1, 3, and 4 mL of PEG capped CaS nanoparticles showed that morphology of these samples is almost similar to that given fig.2A-B.

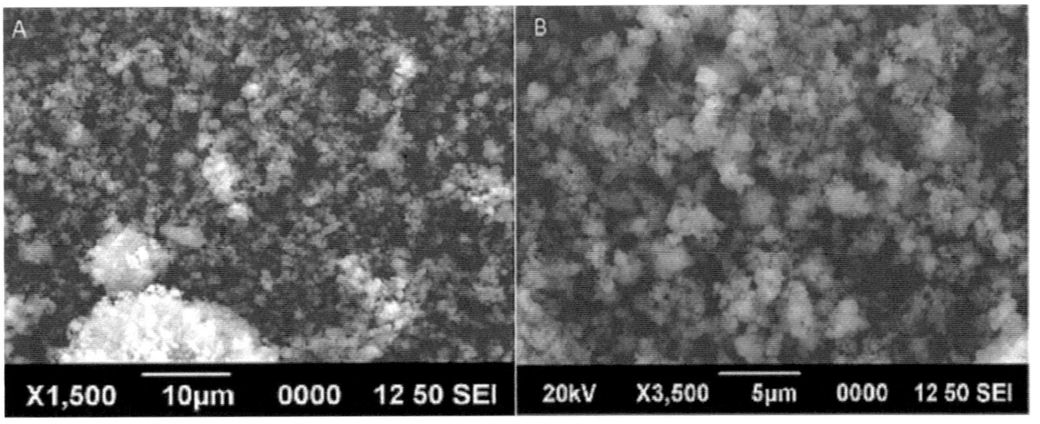

Figure 2. SEM images of PEG (2 mL) capped CaS nanoparticles.

3.3 PL Studies

Luminescence studies provide information about the defect states which take part in the radiative de-excitation of the sample. Figure 3 represents the photoluminescence excitation (PLE) spectrum of CaS nanoparticles capped with 2mL of PEG recorded for an emission wavelength of 478 nm. It consists of a maximum peak at 390 nm which is attributed to intrinsic defects. Hence, the excitation wavelength for PL emission spectral studies of PEG capped CaS nanoparticles was fixed at 390 nm.

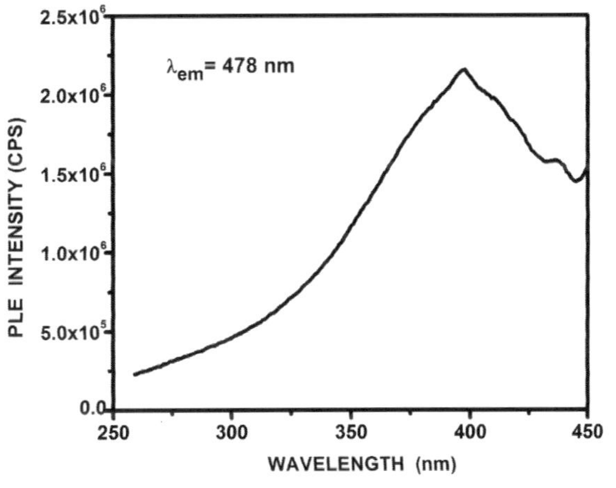

Figure 3. PLE spectrum of PEG (2 mL) capped CaS nanoparticles.

PL emission spectra of CaS nanoparticles capped with PEG at different concentrations (1,2, 3, and 4 mL) for excitation wavelength 390 nm are depicted in fig.4. It is observed that all samples exhibit a broad PL emission band with an emission maximum at 478 nm. This broad peak is due to the occurrence of intrinsic point defects like sulphur vacancies, calcium vacancies, and various other defects in the CaS host lattice [16-20]. On exciting the sample with wavelength 390 nm, excitation occurs to the various defect levels, and from these defect levels, radiative emission takes place, giving rise to the broad peak. The PL emission intensity is lowest for 1mL PEG capped CaS sample. As PEG concentration increases to 2mL and 3 mL, the PL emission intensity increases, and it is maximum for 2 mL PEG capped sample. For 4mL PEG capped sample, the PL intensity again decreases, as shown in figure 4, which may be attributed to the increase in particle size.

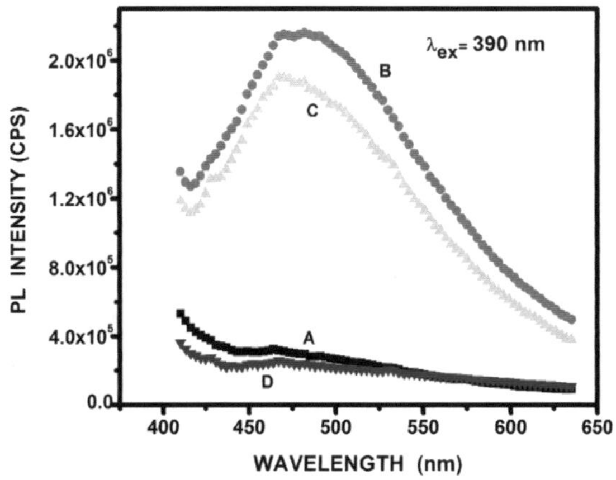

Figure 4. PL emission spectra of A) 1mL B) 2mL C) 3mL and D) 4 mL PEG capped CaS nanoparticles.

The performance of a luminescent material can be examined using the Commission Internationale de l'Eclairage (CIE) coordinates (x, y), which can be determined from the PL emission spectral data of the samples. Figure 5 illustrates the CIE chromaticity diagram of CaS nanoparticles for various concentrations of PEG for an excitation wavelength of 390 nm. Table 2 gives the values of CIE coordinates (x, y) and the emission color for various samples. It is observed that the CIE coordinates (x, y) of 1mL, 2mL, and 3mL samples are nearly equal, and it corresponds to the near green region (point a) of the chromaticity diagram. For 4mL of PEG capped sample, the emission is the near white region (point b) of the chromaticity diagram.

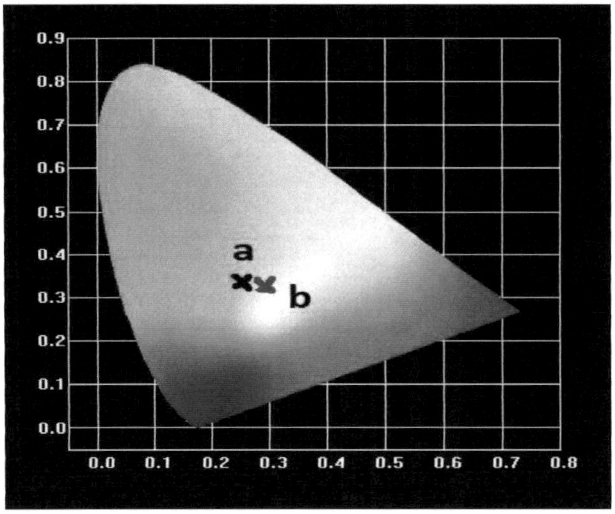

Figure 5. CIE chromaticity diagram of PEG capped CaS nanoparticles.

Table 2. The CIE chromaticity coordinates and the emission color of PEG capped CaS nanoparticles.

Sample name	CIE coordinates		Emission color
	x	y	
1 mL PEG	0.24	0.30	Near green
2 mL PEG	0.24	0.34	Near green
3 mL PEG	0.24	0.33	Near green
4 mL PEG	0.27	0.33	Near white

3.4 DRS studies

The DRS of PEG (2mL) capped CaS nanoparticles which had better structural and luminescence properties is depicted in fig.6. The reflectance of the sample was above 80% in the visible region of the spectrum. The Kubelka- Munk formula, which relates the reflectance R, the scattering coefficient s, and the absorption coefficient k was used to determine the optical band gap of CaS nanoparticles [14, 21,22]. Figure 7 shows the plot between $\{(k/s).h\nu\}^{1/2}$ and hν from which the bandgap of the sample was estimated to be 3.85 eV. The lowering of the band gap from the bulk value 4.5 eV may be attributed to the band tailing effect caused by carrier impurity interactions [23].

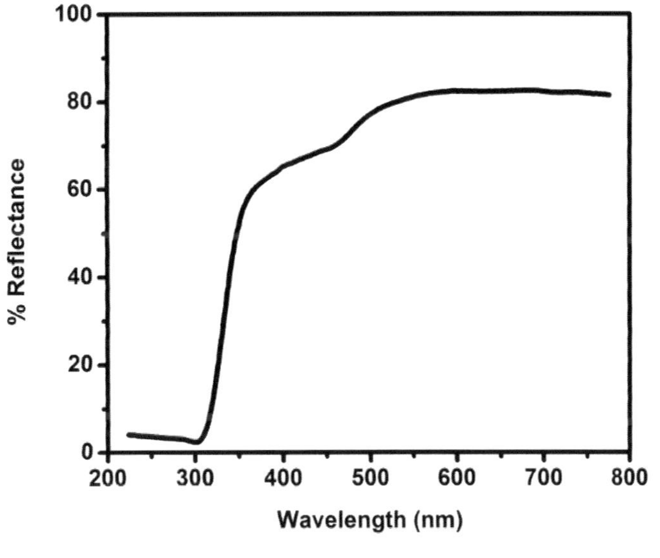

Figure 6. DRS of PEG (2 mL) capped CaS nanoparticles.

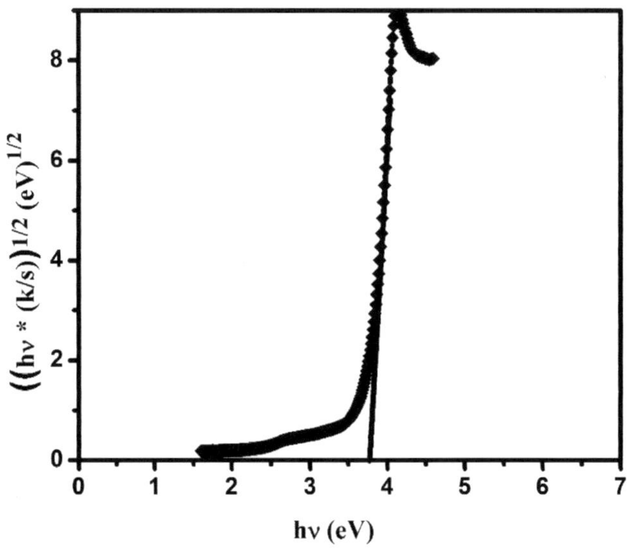

Figure 7. (k/s).hv}$^{1/2}$ versus hv diagram of PEG (2 mL) capped CaS nanoparticles.

3.5 Cell cytotoxicity evaluation by MTT assay

The cell viability of PEG (2 mL) capped CaS nanoparticles was estimated from the MTT assay analysis using L929 human fibroblasts cells [24]. The cultured cells were seeded in a 96 well tissue culture plate, and various concentrations of the sample were added to it. After 24 hours of the incubation period, the optical density of the control (untreated cells) and the treated cells were then obtained by measuring the absorbance values at 500-600 nm [25]. The following equation was employed to determine the percentage of viability :

$$\text{Percentage of viability} = \frac{\text{Mean OD of Samples x 100}}{\text{Mean OD of control group}}$$

where OD is the optical density, which can be determined from the absorbance values. Average OD and percentage of viability of the control and treated cells are given in Table 3. The viability of control cells is taken as 100 %. It is inferred from the table that for the cells treated with up to 25 (μg/mL) concentration of the sample, the viability is greater than 90%. As the concentration of the samples is increased, the cell viability decreases and reaches 79% for a concentration of 100 (μg/mL). Thus, cytotoxicity analysis reveals that PEG capped CaS nanoparticles are biocompatible at lower concentrations, which makes it a suitable material for biomedical applications at lower concentrations.

Table 3. Average OD and percentage of viability of the control and cells treated with varying concentrations of the samples.

Sample Concentration (μg/mL)	OD value I	OD value II	OD value III	Average OD	Percentage Viability
Control	0.7578	0.7596	0.7613	0.7596	100.00
6.25	0.7475	0.7432	0.7410	0.7439	97.93
12.5	0.7253	0.7214	0.7117	0.7195	94.72
25	0.6853	0.6834	0.6819	0.6835	90.00
50	0.6212	0.6251	0.6197	0.6220	81.89
100	0.6044	0.6085	0.6032	0.6054	79.70

4.Conclusions

In summary, CaS nanophosphors capped with various concentrations of PEG were successfully prepared by a low temperature wet chemical co-precipitation method. The sharp peaks in the XRD pattern confirmed the cubic phase and crystallinity of the CaS samples. The particles had almost spherical morphology, and the particle size was found to be in the nanometer scale. The PL spectra of the samples comprise of a broad emission peak, which is attributed to intrinsic point defects in the CaS host lattice. The sample capped with 2 mL of PEG concentration exhibited better structural and luminescence properties. Cytotoxicity analysis reveals that PEG capped CaS at low concentrations is biocompatible. The structural, optical, and cytotoxicity studies of PEG capped CaS nanophosphors suggest that they can be used for various applications in optoelectronics and the biomedical field.

References

[1] Lehman W 1972 J. Lumin. 5 87.

[2] Wang C, Tang K, Yang Q, An C, Hai B, Shen G and Qian Y 2002 Chem. Phys. Lett. 351 385.

[3] Wu H, Tseng C L, and Lin F H 2010 J Nanopart. Res. 12 1173.

[4] Yang P, Lu M, Song C, Gu F, Liu S, Xu D, Yuan D, and Cheng X 2002 J. Non-Cryst. Solids. 311 99.

[5]. Zhao Y, Rabouw F T, Donega C, Meijerink A, and Walree C 2012 Mat. Lett. 80 75.

[6] Roy A and Bhattacharya J 2012 Intl. J. Nanoscience. 11 250027.

[7] Bhargava R N, Gallagher D and Welker T 1994 J.Lumin. 60 275

[8] Cholan S, Shanmugam N, Kannadasan N, Sathishkumar K and Deivam K 2014 J.Mater.Res.Technol.3(3) 222.

[9] Phan C M and Nguyen H M 2017 J. Phys. Chem. A. 121(17) 3213.

[10] Ghosh G, Naskar M K, Patra A and Chatterjee M 2006 Opt.Mater.28 1047.

[11] Bowman K and Leong K W 2006 Int. J. Nanomedicine 1(2) 117.

IOP Conf. Series: Materials Science and Engineering 1219 (2022) 012029 doi:10.1088/1757-899X/1219/1/012029

[12] Jokerst J V, Lobovkina T, Zare R N and Gambhir S S 2011 Nanomedicine 6 (4) 715.
[13] Kumar V, Kumar R, Lochab S P and Singh N 2006 J. Phys: Condens. Matter. 18 5029.

[14] Rekha S, Martinez A, Safeera T A and Anila E I 2017 J.Lumin. 190 94.

[15] Cullity B D1978 Elements of X-ray diffraction,2nd edn. (Addison-Wiley Pub Co, Reading, Massachusetts) p.102.

[16] Huang B 2015 Inorg.Chem. 54 11423.

[17] Sharma G, Chawla P, Lochab S P and Singh N 2011 Physica B 406 2013.

[18] Burbano D C, Sharma S K, Dorenbos P, Viana B and Capobianca J A 2015 Ad. Opt. Mat.3 551.
[19] Burbano D C, Rodríguez E, Dorenbos P and Capobianca J A 2014 J. Mater. Chem. C.2 228.
[20]Brightwell J W, Ray B and Buckley C N 1982 J. Cryst. Growth. 59 210.
[21] Kubelka P and Munk F 1931 Zh. Tekh. Fiz. 12 593.

[22] Kubelka P 1948 Part 1, J. Opt. Soc. Am. 38 448.
[23] Awan S U, Hasnain S K and Aftab M A 2016 J.Lumin.172 231.
[24] Meerloo J V, Kaspers G and Cloos J 2011 Cell sensitivity assays: The MTT assay In: I.A. Cree (Ed), Methods in molecular biology, Cancer cell culture methods and protocols second edition, pp 237.

[25] Rekha S and Anila E I 2019 J. Fluoresc. 29 673.

IOP Conf. Series: Materials Science and Engineering 1219 (2022) 012030 doi:10.1088/1757-899X/1219/1/012030

Enzyme Free Detection of Glucose Using MgO Nanocubes Based Extended Gate N-channel MOSFET

Prasanna APS[1], Guru Prasad Kuppuswamy[1], Natarajan Pradeep[2], Velappa Jayaraman Surya[3,4], Yuvaraj Sivalingam[1*]

[1] *Laboratory of Sensors, Energy and Electronic devices (Lab SEED), Sir C. V. Raman Research Park, Department of Physics and Nanotechnology, SRM Institute of Science and Technology, Kattankulathur 603203, Tamil Nadu, India*
[2] *Department of Nanoscience and Technology, Mount Carmel College, Bengaluru 560052, Karnataka, India*
[3] *Novel, Advanced, and Applied Materials (NAAM) Laboratory, Sir C. V. Raman Research Park, Department of Physics and Nanotechnology, SRM Institute of Science and Technology, Kattankulathur 603203, Tamil Nadu, India.*
[4] *New Industry Creation Hatchery Center, Tohoku University, Aoba-Ku, Miyagi, Sendai 980-8579, Japan*

Email : yuvarajs2@srmist.edu.in

Abstract. MgO nanocubes were used as a sensing electrode material for detection of glucose in the extended gate field effect transistor (EGFET) configuration. MgO nanocubes were prepared by thermal CVD approach. Morphology and elemental composition of as-prepared MgO nanocubes were obtained using the scanning electron microscope (SEM) along with EDS. The sensing electrode was prepared by coating MgO nanocubes over the copper foil by doctor blade method and it was tested with various concentrations of glucose in Phosphate Buffer Saline (PBS) buffer solution. The MgO nanocubes coated electrode showed linear response for glucose concentrations from 1.6 mM to 25.6 mM. The sensitivity was calculated as 0.12 □A/mM.cm^2.

Keywords: MgO nanocubes, EGFET, Enzyme free detection, Glucose, Diabetes mellitus

1. Introduction

Carbohydrates are classified as monosaccharides, oligosaccharides, and polysaccharides [1]. Glucose is categorized under monosaccharides. On the other hand, biomarkers are defined as the biomolecule, cells, or biochemical alterations in the human body which are measurable in the biological media [2]. Glucose is a wellknown identified biomarker of diabetes mellitus [3]. Normal glucose concentration in the blood is 4.0 – 6.0 mMol/L. Below the 4.0 mMol/L is considered hypoglycemia. The condition above the 6.0 mMol/L is considered hyperglycemia. So, glucose detection using the portable and reliable sensor is an ongoing trend.

In recent days, various sensors have been developed, including optical sensors, gas sensors, temperature sensors, pressure sensors, biosensors, etc. Biosensors are used to sense the biomarker as analyte to identify the human body's physiochemical reaction. Metal oxide has various properties such as magnetic, dielectric, catalytic, photoluminescence, etc.[4]. With these wide ranges of properties, metal oxides are used in various applications like photocatalysis, sensors, solar cells, and other optoelectronic devices. Due to their morphological versatility, physicochemical properties, composite structure formation, chemical stability, metal oxides are used as sensing materials [5 - 13].

Magnesium oxide (MgO) has been used widely, because of its property like mechanical strength, relatively high surface area, anti-bacterial activity, high ionic character, etc. It has a wide range of applications including catalysis, petrochemical, reflecting and antireflecting coatings, detection, remediation of chemical waste and warfare agents, etc. [14,15]. Nanostructures like shells, rods, rings, beams, plates, cubes, etc., are easily synthesized with available techniques.

In this work, MgO nanocubes were synthesized and used as the sensing material. MgO nanocubes were utilized to detect glucose in phosphate buffer saline solution (pH: 7.4). Prepared material was able to measure glucose in the broad concentration range from 1.6 mM to 25.6 mM with the sensitivity of 0.12 □A/mM.cm^2.

IOP Conf. Series: Materials Science and Engineering 1219 (2022) 012030 doi:10.1088/1757-899X/1219/1/012030

2. Material and methods

2.1 MgO nanocubes preparation

Preparation technique were as per the previously reported literature [16]. MgO nanocubes were prepared by thermal CVD approach from magnesium powder under controlled conditions without a gold catalyst. The system consists of a horizontal tube furnace i.e., alumina-tube reactor. Initially, Mg powders were placed in an alumina boat, and kept into the alumina-tube reactor of a horizontal furnace. Argon and oxygen were introduced into the alumina tube through a mass-flow controller with the ratio of 3:1 respectively. MgO nanocubes were grown at the temperature of 1100°C for 60 minutes. The furnace was cooled down with argon flow. A fine white powder was collected from the boat and was given for characterization.

2.2 Fabrication of sensing electrode and characterization

For the sensing electrode preparation, 90 mg of synthesized MgO powder and 10 mg PVDF were mixed by grinding in agate mortar and pestle. Here, polyvinylidene fluoride (PVDF) was added to improve the adhesiveness of MgO on Cu foil. Also, N-methyl-2- pyrrolidone was added to the mixture to form a gelatinous solution. 0.7 mg of as-prepared MgO gelatinous solution was coated uniformly on the flexible copper electrode by the doctor blade method as shown in Fig. 1 and dried at 80°C for 12h. Finally, the MgO nanocubes coated copper electrode was characterized by SEM and EDS to confirm the morphology and elemental composition.

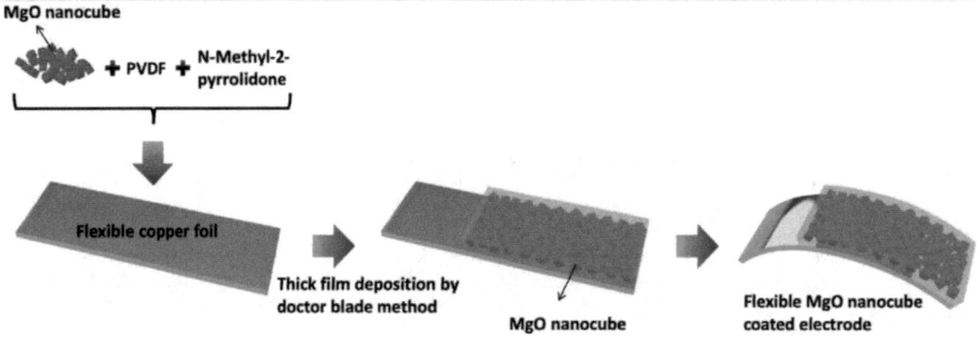

Figure 1. Schematic illustration of gelatinous solution of MgO nanocubes deposition on flexible copper electrode.

2.3 Glucose detection measurements using extended gate MOSFET based setup

The electrochemical cell containing Phosphate Buffer Saline (PBS) solution was taken for the measurements. The MgO nanocubes coated flexible copper electrode and reference electrode (Ag/AgCl) were immersed in the PBS buffer solution. The sensing electrode was connected to gate terminal of the commercial N channel MOSFET (1719 T SD21). I-V characteristics of the extended gate N- channel MOSFET was obtained using a semiconductor parameter analyzer (4200A-SCS parameter analyzer, Keithley) where voltage sweep was applied to the drain electrode, source was connected to the ground, and gate voltage was applied through Ag/AgCl reference electrode as shown in Fig. 2. For various glucose concentrations, I-V measurements was repeated and utilized for the sensitivity calculation.

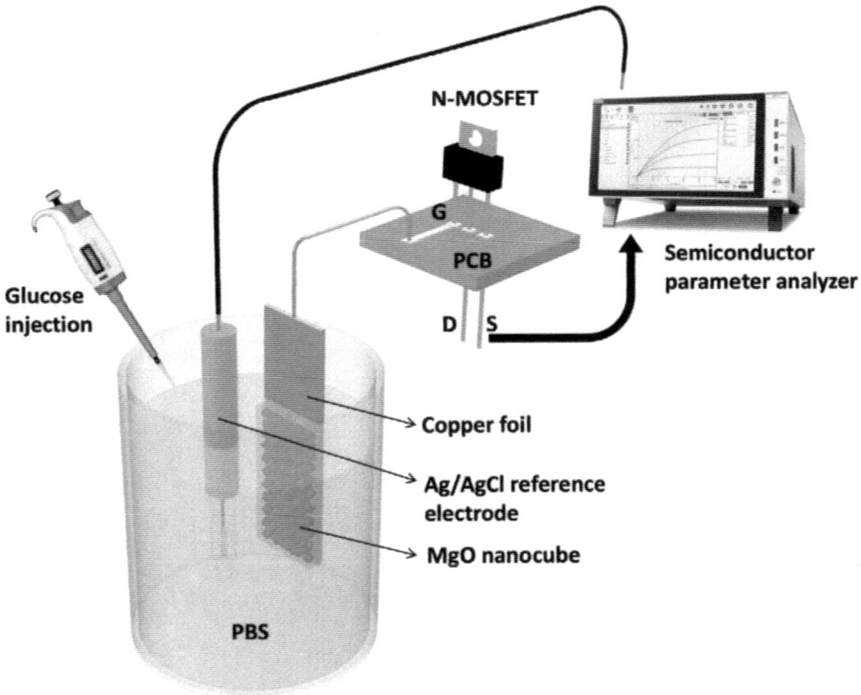

Figure 2. Pictorial representation of extended gate N-channel MOSFET based glucose sensor setup.

3. Results and Discussion

Figure 3. (a) - (c) shows the SEM images in different magnification of MgO nanocubes. The obtained MgO nanocubes size varies from 100nm to 300 nm. The SEM image clearly shows that most nanostructures are cubic and with sharp, perfect, and curved edges. It is probably due to the nucleation taking place at high temperatures and the availability of supersaturation oxygen. In the growth period, oxygen supply and diffusion in the nucleation particles support the formation of perfect shaped nanocubes.

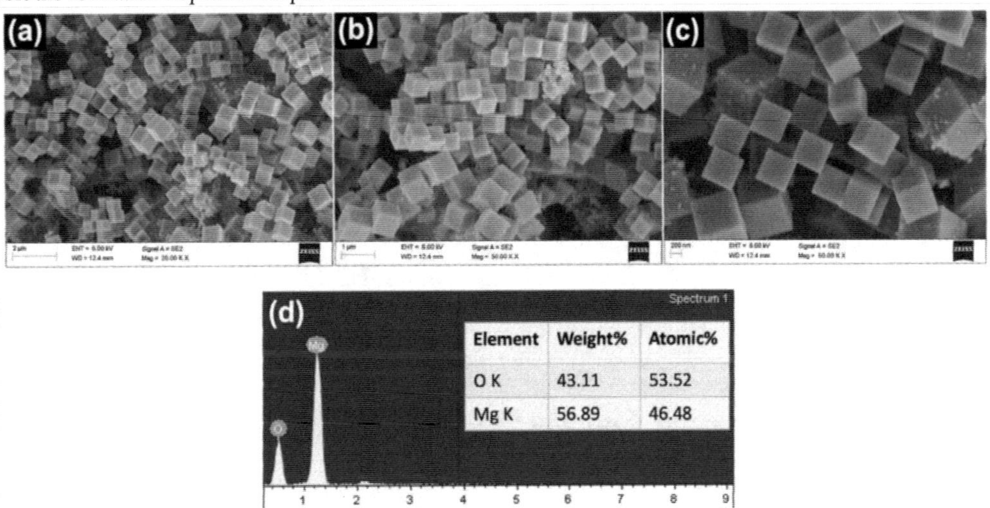

Figure 3 (a-c) SEM images of the MgO nanocube in various magnifications, (d) EDS elemental mapping and chemical composition of MgO nanocubes.

EDS analyses were done to observe the chemical compositions of the synthesized MgO nanocubes, as shown in Fig.3 (d). The result shows the atomic percentage of Mg and O as 46.48 and 53.52, respectively. The spectrum reveals the presence of Mg and O elements, and there are no other impurities peaks were observed.

Glucose response was measured using an extended gate N-channel MOSFET platform by adding different glucose concentrations in PBS buffer solution from 1.6 mM to 25.6 mM, and the respective current response was measured. Figure 4 shows the transfer characteristics for extended gate N-channel MOSFET upon addition of glucose concentrations from 1.6 mM to 25.6 mM in a PBS buffer solution at a fixed V_{DS} of 0.1 V. The sensitivity of MgO nanocubes based extended gate N-channel MOSFET glucose sensor was derived from figure 5 at reference electrode voltage of 1.45 V. The calculated sensitivity of the sensor is about 0.12 □A/mM.cm^2. The linearity of the electrode was found to be 0.9701.

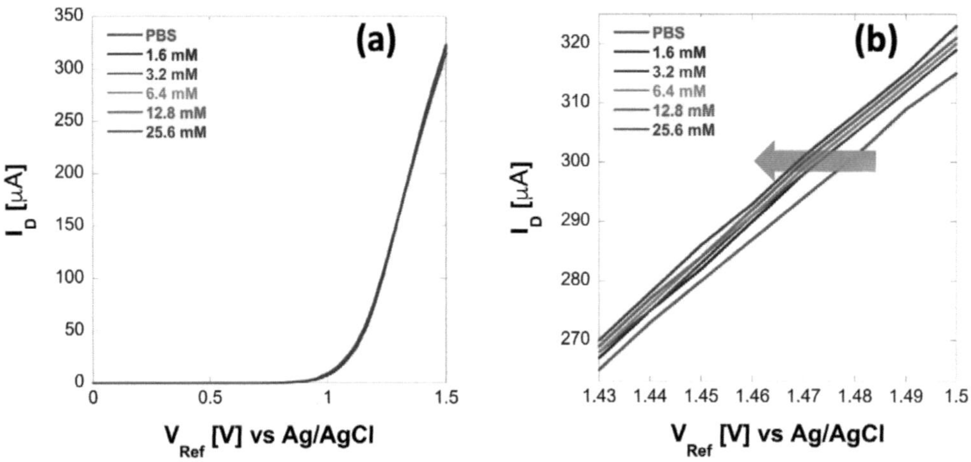

Figure 4. *Transfer characteristics for extended gate N-channel MOSFET upon addition of glucose concentrations from 1.6 mM to 25.6 mM in a PBS buffer solution.*

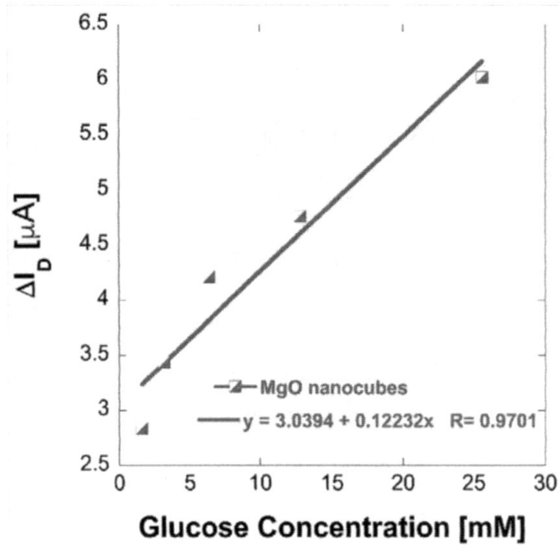

Figure 5. Changes in the I_D of extended gate N-channel MOSFET for fixed reference electrode voltage 1.45 V due to the addition of glucose concentrations from 1.6 mM to 25.6 mM in a PBS buffer solution.

The glucose detection mechanism could be expressed as two steps. As a first step oxidation reaction of MgO nanocubes occurs under a sufficient reference electrode potential.

$$MgO + OH^{-} \rightarrow MgO(OH) + e^{-} \quad (1)$$

In the second step, a nonenzymatic oxidation-reduction reaction between the formed Magnesium oxyhydroxide species and the adsorbed glucose allows for further regeneration of MgO species.

$$2MgO(OH) + glucose \rightarrow 2MgO + gluconolactone + H_2O \quad (2)$$

Due to these reactions (equations 1 and 2), a change in drain current occurs with the difference in glucose concentration, as observed in Figure 4b.

4. Conclusion

Non-enzymatic MgO nanocubes-based extended gate N-channel MOSFET based glucose sensor was studied. The SEM images were confirmed the cube shaped MgO nanostructure. EDS elemental mapping was confirmed the Mg and O atoms presence, and no impurity peaks were observed. The sensing electrode was showed significant drain current shift with the glucose concentration from 1.6 mM to 25.6 mM glucose. The sensitivity of the MgO nanocubes-based extended gate N-channel MOSFET based glucose sensor was calculated as 0.12 □A/mM.cm^2 with good linearity of 0.9701. These results suggest that the MgO nanocubes based sensing electrodes will be suitable candidate for the real time glucose detection towards diabetes mellitus application.

Acknowledgment

This work was funded by the Science and Engineering Research Board (SERB), Department of Science and Technology (DST), Government of India through the research Grant No. ECR/2017/001218. We thank SRMIST for the support of establishment of the electrical characterization facility.

References

[1] H.C. Wang, A.R. Lee, Recent developments in blood glucose sensors, J. Food Drug Anal. 23 (2015) 191–200. https://doi.org/10.1016/j.jfda.2014.12.001.

[2] R. Mayeux, Biomarkers: Potential Uses and Limitations, NeuroRx. 1 (2004) 182–188. https://doi.org/10.1602/neurorx.1.2.182.

[3] A.J. Krentz, M. Hompesch, Glucose: Archetypal biomarker in diabetes diagnosis, clinical management and research, Biomark. Med. 10 (2016) 1153–1166. https://doi.org/10.2217/bmm-2016-0170.

[4] B. Liu, J. Liu, Sensors and biosensors based on metal oxide nanomaterials, TrAC - Trends Anal. Chem. 121 (2019) 115690. https://doi.org/10.1016/j.trac.2019.115690.

[5] I. Şerban, A. Enesca, Metal Oxides-Based Semiconductors for Biosensors Applications, Front. Chem. 8 (2020). https://doi.org/10.3389/fchem.2020.00354.

[6] Sivalingam Y, Pudi R, Lvova L, Pomarico G, Basoli F, Catini A, Legin A, Paolesse R and Di Natale C, The light modulation of the interaction of l-cysteine with porphyrins coated ZnO nanorods *Sensors and Actuators B: Chemical* 209 (2015) 613–21. https://doi.org/10.1016/j.snb.2014.11.146.

[7] Kalidoss R, Umapathy S and Sivalingam, An investigation of GO-SnO2-TiO2 ternary nanocomposite for the detection of acetone in diabetes mellitus patient's breath, *Applied Surface Science* 449 (2018) 677 - 684.

[8] Kalidoss R, Umapathy S, Anandan R, Ganesh V and Sivalingam Y, Comparative Study on the Preparation and Gas Sensing Properties of Reduced Graphene Oxide/SnO2 Binary Nanocomposite for Detection of Acetone in Exhaled Breath *Anal. Chem.* 91(2019) 5116–24.

[9] Sivalingam Y, Martinelli E, Catini A, Magna G, Pomarico G, Basoli F, Paolesse R and Di Natale C, Gas-Sensitive Photoconductivity of Porphyrin-Functionalized ZnO Nanorods *J. Phys. Chem. C* 116 (2012) 9151–7.

[10] Sivalingam Y, Elumalai P, Yuvaraj S V J, Magna G, Sowmya V J, Paolesse R, Chi K-W, Kawazoe Y and Di Natale C, Interaction of VOCs with pyrene tetratopic ligands layered on ZnO nanorods under visible light *Journal of Photochemistry and Photobiology A: Chemistry* 324 (2016) 62–9.

[11] Magna G, Muduganti M, Stefanelli M, Sivalingam Y, Zurlo F, Di Bartolomeo E, Catini A, Martinelli E, Paolesse R and Di Natale C, Light-Activated Porphyrinoid-Capped Nanoparticles for Gas Sensing *ACS Appl. Nano Mater.* 4 (2021) 414–24.

[12] Magna G, Sivalingam Y, Martinelli E, Pomarico G, Basoli F, Paolesse R and Di Natale C, The influence of film morphology and illumination conditions on the sensitivity of porphyrins-coated ZnO nanorods *Analytica Chimica Acta,* 810 (2014) 86–93.

[13] Sivalingam Y, Magna G, Pomarico G, Catini A, Martinelli E, Paolesse R and Natale C D, The influence of gas adsorption on photovoltage in porphyrin coated ZnO nanorods *J. Mater. Chem.* 22 (2012) 20032–7.

[14] M.R. Bindhu, M. Umadevi, M. Kavin Micheal, M.V. Arasu, N. Abdullah Al-Dhabi, Structural, morphological and optical properties of MgO nanoparticles for antibacterial applications, Mater. Lett. 166 (2016) 19–22. https://doi.org/10.1016/j.matlet.2015.12.020.

[15] A.A. Pilarska, Ł. Klapiszewski, T. Jesionowski, Recent development in the synthesis, modification and application of Mg(OH)2 and MgO: A review, Powder Technol. 319 (2017) 373–407. https://doi.org/10.1016/j.powtec.2017.07.009.

[16] N. Pradeep, C. Venkatachalaiah, U. Venkatraman, C. Santhosh, A. Bhatnagar, S.K. Jeong, A.N. Grace, Magnesium oxide nanocubes deposited on an overhead projector sheet: synthesis and resistivity-based hydrogen sensing capability, Microchim. Acta. 184 (2017) 3349–3355. https://doi.org/10.1007/s00604-017-2348-3.

Graphene Oxide based Gas Sensor for Triethylamine Detection at Room Temperature

Haidar Abbas[1], Gobinath Marappan[2], David Chidabaram,[1] Shiva Govindasamy[2], Velappa Jayaraman Surya[2,3*], Yuvaraj Sivalingam[1*]

[1]*Laboratory of Sensors, Energy and Electronic devices (Lab SEED), Department of Physics and Nanotechnology, SRM Institute of Science and Technology, Kattankulathur 603203, Tamil Nadu, India.*

[2]*Novel, Advanced, and Applied Materials (NAAM) Laboratory, Department of Physics and Nanotechnology, SRM Institute of Science and Technology, Kattankulathur 603203, Tamil Nadu, India.*

[3]*New Industry Creation Hatchery Center, Tohoku University, Aoba-ku, Miyagi, Sendai 980-8579, Japan*

Email: yuvarajs2@srmist.edu.in; suryaj@srmist.edu.in

Abstract. We have developed a graphene oxide (GO) based sensor to detect triethylamine at room temperature. GO is synthesized by modified Hummer's method and spin-coated on fluorine-doped tin oxide (FTO) substrate. Structural and morphological characterizations of GO are done by XRD and FE-SEM. Sheet-like morphology of GO is observed in FESEM images. The existence of defects is confirmed by Raman spectra with a I_D/I_G ratio ~ 0.94. Gas adsorption studies are carried out using scanning Kelvin probe system with various volatile organic compounds (VOCs) like ethanol, acetone, n-hexane and triethylamine. Among them, GO shows more contact potential difference (CPD) response towards triethylamine. Then, a gas sensor device is fabricated using GO and its electrical characterization is done by I-V measurements. The device shows ohmic behavior with a resistance of ~ 50 MΩ. Thereafter, the sensor is exposed to triethylamine from 19 to 151 ppm and the sensitivity is 1.39×10^{-4} ppm^{-1}. Interestingly, a fast response time of 11 sec is achieved at room temperature.

Keywords: Graphene oxide; Triethylamine; Gas sensor; Room temperature; Fast response time

1. Introduction

Two-dimensional (2D) materials are considered as potential candidates for optoelectronics and gas sensor applications, because of their inherent properties [1]. Graphene oxide (GO), a 2D material with sp^2 and sp^3 bonded hybridized carbon atoms arranged in a thick honeycomb crystal structure with oxygen functional groups possess excellent properties like high stability and high surface area and hence, it offers more active sites for gas molecule interaction. Also, the presence of various functional groups such as epoxy, hydroxyl, and carboxyl groups enhance gas molecule interaction [2,3]. There are few reports on sensing of various gases using GO and its derivatives at room temperature [4–6].

Volatile organic compounds (VOCs) are carbon compounds with a high vapour pressure at room temperature (RT) that are emitted from chemical manufacturing industry, oils, food contaminants, and other sources [7]. Many researchers focus on detection of VOCs such as ethanol, acetone, nonanal, n-hexane, triethylamine, benzene, acetonitrile, and so on for environmental monitoring, chemical process control, agricultural, and medical applications [8–11]. Among them, triethylamine is a toxic organic amine released from the organic chemical processing industry, research laboratory, and sea foods, etc. The occupational safety and health administration recommends that triethylamine concentrations in the air should not exceed than 10 parts per million (ppm) [12]. If the concentration exceeds in air more than 10 ppm, it leads to health problems such as eye irritations, headaches, respiratory problem, etc. [13]. Hence, there is an urge to develop highly sensitive, selective, fast response and cost-effective sensor towards triethylamine operating at room temperature. Minitha *et al.* have used nitrogen ion implanted GO towards methanol detection at room temperature [14].

In this work, we have investigated the selective interaction of triethylamine on GO in comparison with other VOCs such as ethanol, acetone, and n-hexane using scanning Kelvin probe (SKP) system. Then, we have fabricated the GO based sensor for triethylamine detection at room temperature.

IOP Conf. Series: Materials Science and Engineering 1219 (2022) 012031 doi:10.1088/1757-899X/1219/1/012031

2. Materials and methods

2.1. Experimental section

Chemicals purchased for the synthesis of GO are graphite (98 %) from LOBA Chemie pvt. Ltd., sodium nitrate ($NaNO_3$) in sigma Aldrich (99 %), potassium permanganate ($KMnO_4$), hydrochloric acid (HCL), 30% hydrogen peroxide (H_2O_2), and sulphuric acid (H_2SO_4) from Sisco Research Laboratories Pvt. Ltd. (SRL).

GO was synthesized through modified Hummer's method [15]. Firstly, 2 g of graphite powder and 1 g of $NaNO_3$ were dissolved in 100 ml H_2SO_4 solution with constant stirring for 10 min. Then, the above solution was placed in the ice bath, followed by adding 6 g of $KMnO_4$ and kept overnight. Then, 100 ml of deionized water and 10 ml of 30% H_2O_2 were added to remove the unreacted precursor. Finally, deionized water was added to neutralize the pH to 7 through centrifugation at 10000 rpm for several times. Finally, the precipitate was dried in a hot air oven at 60 °C for overnight.

2.2. Characterization details

X-ray diffraction (XRD) pattern of GO was obtained using CuKα (λ=0.154 nm) radiation source (BRUKER USA D8 Advance, Davinci). The morphology of GO sample was examined through Field-Emission Scanning Electron Microscope (ZEISS ULTRA 55). Raman spectrum of GO was recorded with an excitation wavelength of 633 nm (HORIBA France, LABRAM HR Evolution). The gas adsorption studies were carried out by measuring contact potential difference (CPD) using SKP system (SKP5050, KP Technology Ltd, UK).

2.3. Sensor fabrication

The GO sensor device (FTO/GO/FTO) was fabricated on FTO substrate with a gap of about 200 μm to form two separate FTO electrodes [15,16]. As a first step, 2 mg of GO was dissolved in water and ethanol mixture (75 ml:25 ml). From that, 75 μl of GO solution was spin-coated on the gap and on the rims of the FTO electrodes at 1000 rpm for 30 sec about five times. Finally GO coated device was dried on the hot plate at 80 °C about 1 h. Semiconductor parameter analyzer was used to perform the I–V characterization (Keithley 4200A- SCS, Tektronix, USA).

The fabricated FTO/GO/FTO gas sensor device was placed in an 800 mL customized gas chamber. VOCs vapours are collected through a gastight syringe and then injected it into the chamber. The chamber was maintained at room temperature. The response to the VOCs were evaluated by measuring the resistance, using data logger (34972A LXI Compliant Data Acquisition along with 34901A 20 Channel Armature Multiplexer Module, Keysight, USA).

3. Results and discussion

XRD pattern of GO is shown in Fig.1a. A sharp peak is observed at 10.53°, which corresponds to (002) plane. The inter-planar distance of GO for (002) plane is 0.84 nm which confirms complete exfoliation from graphite. This observation is in line with the previous report [17]. FE-SEM images suggest the presence of homogeneous GO ultrathin sheets with wrinkle structure (Fig. 1b). Figure 1c shows the Raman spectrum of GO in the range from 1000 to 2000 cm^{-1}. D and G bands are observed at 1336 cm^{-1} and 1585 cm^{-1}, respectively. Compared with D band, the intensity of G band is higher due to the first order scattering of the E_{2g} phonon of carbon sp^2 atom at 1585 cm^{-1} [17,18]. I_D/I_G intensity ratio is approximately 0.94 which suggests the presence of defects in GO.

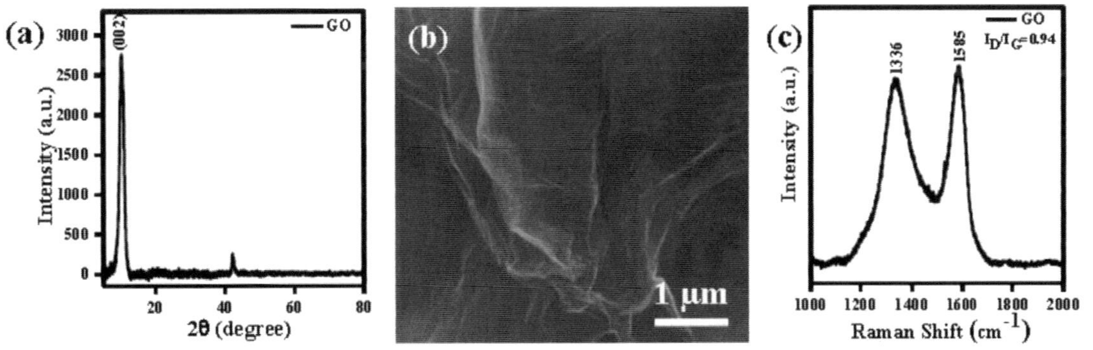

Figure. 1 shows that GO (a) XRD pattern and (b) FE-SEM image (c) Raman spectrum.

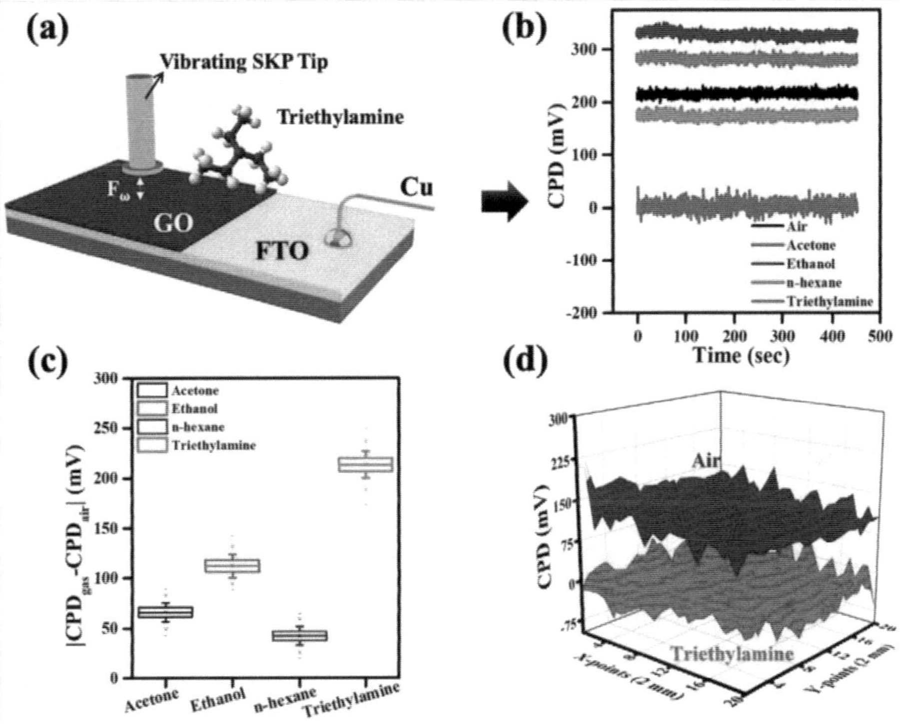

Figure 2. SKP measurement of GO on FTO substrate (a) Schematic illustration of gas adsorption measurements, (b) Time evolution of the single point CPD measurement in various VOCs atmosphere, (c) CPD difference plot of various VOCs atmosphere with respect to ambient air, (c) Raster scan measurement of GO film under ambient air and triethylamine medium.

Contact potential difference (CPD) of GO coated FTO substrate is measured using SKP system at room temperature under different VOCs (acetone, ethanol, n-hexane and triethylamine) with reference to the ambient air. The schematic illustration of SKP measurement setup is shown in figure 2a. The complete principle and working of SKP system is explained in our previous papers [18–20]. There are changes in CPD based on VOCs adsorption on the substrate (Fig. 2b). Triethylamine shows highest CPD with reference to the ambient air in comparison with other VOCs (Fig. 2c). The raster scan image confirms homogeneous deposition of GO sheets on the FTO substrate with a work function of ~ 5.46 eV

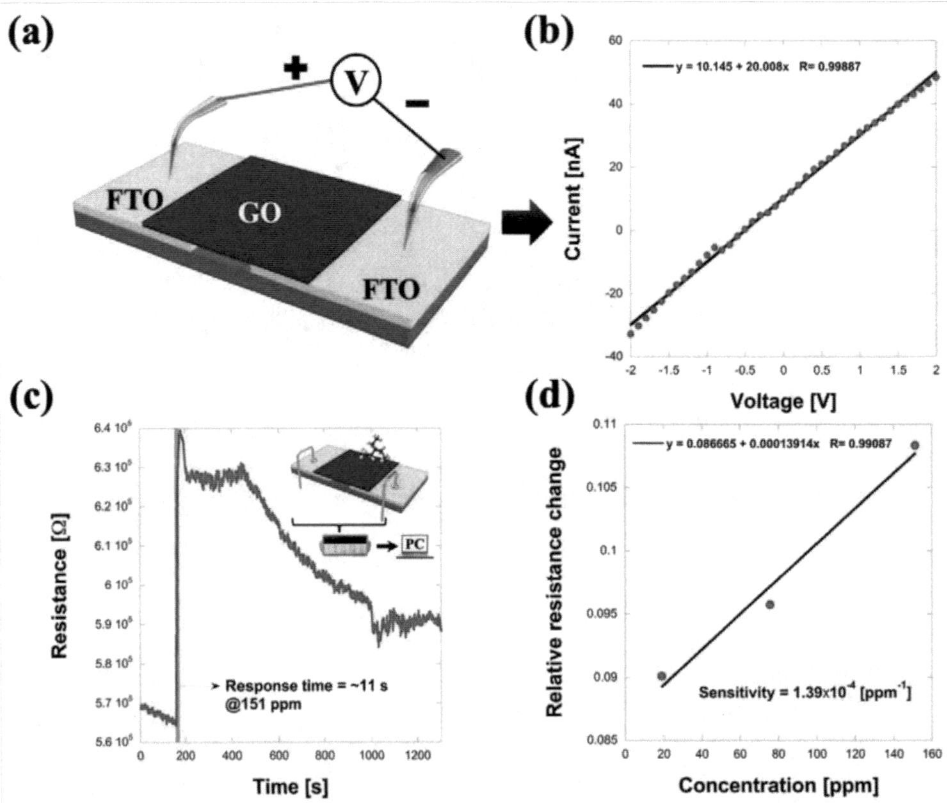

Figure 3. Conductivity and gas sensor measurements of FTO/GO/FTO device (a) Schematic illustration of FTO/GO/FTO device, (b) I-V characterization, (c) Typical sensor response of FTO/GO/FTO device with 151 ppm of triethylamine exposure, (d) The response curve plot of the device towards triethylamine exposure.

in ambient air and its response towards triethylamine with the work function of ~ 5.25 eV (Fig. 2d). This reduction in work function is due to electron transfer from triethylamine to GO.

Figure 3a shows the schematic illustration of the fabricated FTO/GO/FTO sensor device. I-V curve of the FTO/GO/FTO sensor device with the bias voltage in the interval from -2 V to 2 V is shown in Fig 3b. The results reveal the ohmic behavior with a resistance of ~ 50 MΩ. The sensor device is exposed to triethylamine from 19 to 151 ppm at room temperature. Fig. 3c shows the typical response plot of triethylamine for 151 ppm. Interestingly, we observed a fast response time of 11 sec. However, recovery is not obtained even after 800 sec. This may be due to the presence of defects in GO as confirmed by the Raman spectrum. The sensitivity is calculated from Fig. 3d and the value is 1.39 x 10^{-4} ppm^{-1}. The concentration, response time, recovery time and operating temperature are compared with the previous reports on triethylamine sensing as shown in Table 1.

Table.1 Comparison of triethylamine sensing properties of different materials

Material	Concentration (ppm)	Response time (s)	Recovery time (s)	Operating temperature °C	Reference
RGO/Au/ZnO sensor	1	30	42	50	[13]
B-RGO/SnO2@Au heterostructure	1	27	-	100	[21]
rGO/LaFeO3 nanocomposites	50	3	4	240	[22]
rGO/Fe2O3 nanocomposites	50	4	20	280	[23]
MoS_2/GO hybrid nanostructure	1	2	7	260	[25]
pine dendritic BiVO4/rGO	10	11	67	180	[25]
GO	151	11	more than 800	Room temperature	This work

4. Conclusion

We have synthesized GO by modified Hummer's method. The prepared GO has been characterized by XRD with the interplanar distance of 0.84 nm. The sheet morphology is confirmed through FE-SEM. Raman spectrum has validated the signature peaks of GO. The SKP studies on GO have projected a huge response towards triethylamine when compared to acetone, ethanol and n-hexane at room temperature. Also, the fabricated sensor device shows fast response towards triethylamine at 11 s at room temperature and the sensitivity is 1.39×10^{-4} ppm^{-1}.

Acknowledgment

This work was funded by the Science and Engineering Research Board (SERB), Department of Science and Technology (DST), Government of India through the research Grant No. ECR/2017/001218. We sincerely acknowledge SRMIST for the support of establishment of the electrical characterization facility.

References

[1] Tyagi D, Wang H, Huang W, Hu L, Tang Y, Guo Z, Ouyang Z and Zhang H 2020 Recent advances in two-dimensional-material-based sensing technology toward health and environmental monitoring applications *Nanoscale* **12** 3535–59

[2] Yao Y, Chen X, Zhu J, Zeng B, Wu Z and Li X 2012 The effect of ambient humidity on the electrical properties of graphene oxide films *Nanoscale Res. Lett.* **7** 2–19

[3] Venugopal G, Sivalingam Y, Krishnamoorthy K, Surya V J Y, Devaraju M K and Kim S J 2018 Observation of anomalous transport characteristics in graphene-oxide thinfilm *Mater. Chem. Phys.* **213** 89–94

[4] Minitha C R, Anithaa V S, Subramaniam V and Rajendra Kumar R T 2018 Impact of Oxygen

Functional Groups on Reduced Graphene Oxide-Based Sensors for Ammonia and Toluene Detection at Room Temperature *ACS Omega* **3** 4105–12

[5] Chen W, Deng F, Xu M, Wang J, Wei Z and Wang Y 2018 GO/Cu2O nanocomposite based QCM gas sensor for trimethylamine detection under low concentrations *Sensors Actuators, B Chem.* **273** 498–504

[6] Şen Z, Öztürk S, Harbeck M and Öztürk Z Z 2019 Room-temperature sensing of volatile organic compounds using graphene *Sensors Mater.* **31** 1365–74

[7] Haick H, Broza Y Y, Mochalski P, Ruzsanyi V and Amann A 2014 Assessment, origin, and implementation of breath volatile cancer markers *Chem. Soc. Rev.* **43** 1423–49

[8] Kalidoss R, Umapathy S and Sivalingam Y 2018 An investigation of GO-SnO 2 -TiO 2 ternary nanocomposite for the detection of acetone in diabetes mellitus patient's breath *Appl. Surf. Sci.* **449** 677–84

[9] Jalal A H, Alam F, Roychoudhury S, Umasankar Y, Pala N and Bhansali S 2018 Prospects and Challenges of Volatile Organic Compound Sensors in Human Healthcare *ACS Sensors* **3** 1246–63

[10] Kaushik A, Kumar R, Arya S K, Nair M, Malhotra B D and Bhansali S 2015 Organic-Inorganic Hybrid Nanocomposite-Based Gas Sensors for Environmental Monitoring *Chem. Rev.* **115** 4571–606

[11] Srinivasan P and Rayappan J B B 2020 Investigations on room temperature dual sensitization of ZnO nanostructures towards fish quality biomarkers *Sensors Actuators, B Chem.* **304** 127082

[12] Wang L, Peng R, Ci L and Li Y 2020 SnO2 microrods based triethylamine gas sensor *IOP Conf. Ser. Mater. Sci. Eng.* **772**

[13] Peng R, Li Y, Chen J, Si P, Feng J, Zhang L and Ci L 2018 Reduced graphene oxide wrapped Au@ZnO core-shell structure for highly selective triethylamine gas sensing application at a low temperature *Sensors Actuators, A Phys.* **283** 128–33

[14] Minitha C R, Nivedita L R, Asokan K and Rajendra Kumar R T 2019 Tuning the electrical properties of graphene oxide by nitrogen ion implantation: Implication for gas sensing *Nucl. Instruments Methods Phys. Res. Sect. B Beam Interact. with Mater. Atoms* **450** 257–61

[15] Hummers W S and Offeman R E 1958 Preparation of Graphitic Oxide *J. Am. Chem. Soc.* **80** 1339

[16] Choudhary M 2013 Graphene Oxide based Label Free Ultrasensitive Immunosensor for Lung Cancer Biomarker, hTERT *J. Biosens. Bioelectron.* **04**

[17] Bharath G, Madhu R, Chen S and Veeramani V 2015 *graphene oxide and Fe 3 O 4 – reduced graphene oxide nanocomposites for sensitive detection of nitrite*

[18] Teradal N L, Marx S, Morag A and Jelinek R 2017 Porous graphene oxide chemi-capacitor vapor sensor array *J. Mater. Chem. C* **5** 1128–35

[19] Elakia M, Gobinath M, Sivalingam Y, Palani E, Ghosh S, Nutalapati V and Surya V J 2020 Investigation on visible light assisted gas sensing ability of multi-walled carbon nanotubes coated with pyrene based organic molecules *Phys. E Low-Dimensional Syst. Nanostructures* **124** 114232

[20] Dutta G K, Kasthuri S, Marappan G, Jayaraman S V, Sivalingam Y, Di Natale C and Nutalapati V 2019 Aggregation behavior in naphthalene-appended diketopyrrolopyrrole derivatives and its gas adsorption impact on surface potential *J. Mater. Chem. C* **7** 9954–65

[21] Guruprasad K, Marappan G, Elangovan S, Velappa Jayaraman S, Kamala Bharathi K, Venugopal G, Di Natale C and Sivalingam Y 2020 Electrical transport properties and impedance analysis of Au/ZnO nanorods/ITO heterojunction device *Nano Express* **1** 030020

[22] Peng R, Li Y, Liu T, Si P, Feng J, Suhr J and Ci L 2020 Boron-doped graphene coated Au@SnO2 for high-performance triethylamine gas detection *Mater. Chem. Phys.* **239** 121961

[23] Hao P, Lin Z, Song P, Yang Z and Wang Q 2020 rGO-wrapped porous LaFeO3 microspheres for high-performance triethylamine gas sensors *Ceram. Int.* **46** 9363–9

[24] Sekhar C, Akimoto M, Fan X, Bivolarova M, Liao C, Lan L and Wargocki P 2020 Jo ur l P re of *Build. Environ.* **184** 107229

[25] Hou X, Wang Z, Fan G, Ji H, Yi S, Li T, Wang Y, Zhang Z, Yuan L, Zhang R, Sun J and Chen D 2020 Hierarchical three-dimensional MoS2/GO hybrid nanostructures for triethylamine-sensing applications with high sensitivity and selectivity *Sensors Actuators, B Chem.* **317** 128236

[26] Bai S, Sun L, Sun J, Han J, Zhang K, Li Q, Luo R, Li D and Chen A 2021 Pine dendritic bismuth vanadate loaded on reduced graphene oxide for detection of low concentration triethylamine *J. Colloid Interface Sci.* **587** 183–91

Disparate result of Adsorptive removal of Fluoride using Manganese oxide-coated Nano γ-alumina oxide used with a different ratio.

Diksha Choudhary[1], Deepika Tanwar[1,2], Dinesh Gupta[1,3] Pradeep Singh[4], Pankaj Raizada[4], Dr. Archana Singh[1,2] *

[1] *CSIR-Advanced Materials and Processes Research Institute, Bhopal, 462026, India*
[2] *Academy of Scientific and Innovative Research (AcSIR), Ghaziabad, 201002, India*
[3] *Rajiv Gandhi Proudyogiki Vishwavidyalaya (RGPV), Bhopal, 462033, India*
[4] *School of Chemistry, Faculty of Basic Sciences, Shoolini University, Solan (HP), India-173229*

Abstract. On exposure of the human body to drinking water containing fluoride in concentrations more than the permissible limit is very harmful to the human body. The discharge of mineral sediments and industrial effluent contaminants containing fluoride in groundwater may cause severe health problems like arthritis, skeletal fluorosis, and so on. Fluoride is beneficial in drinking water at a level up to 0.5 mg/l but is harmful above 1.5 mg/l. Herein, we report a novel nano alumina-supported manganese oxide as an efficient adsorbent that can remove fluoride up to 90% from water. The nano adsorbent was obtained by a precipitation method that was characterized using XRD, FESEM, and BET. XRD and BET results confirmed the formation of manganese oxide- Al_2O_3 composite with a surface area of 128 m^2/g. Batch adsorption studies revealed 36 mg/g fluoride adsorption capacity for the developed composite nano adsorbent.

Keywords: γ-alumina, manganese oxide, fluoride, adsorption, drinking water.

1. Introduction

A high level of fluoride is a global concern, not just harming human health, but also leading to severe environmental contamination. In-country like India, drinking water supplies especially in rural areas around 85% of the population rely on groundwater, says the Department of Drinking Water Services under the Ministry of Rural Development in India. Therefore, it is highly desirable to have fluoride concentration in groundwater well in a range of permissible limits. According to the Indian norm, the fluoride concentration has to be lowered to 1.5 mg/l (1). Since fluoride harms humans as it exceeds a suitable level, the fluoride content in the water supplies must be minimized and regulated. Among different available water treatment studies, adsorption as a treatment technique for fluoride reduction in small-scale water treatment systems is gaining popularity. This is because of its flexibility and versatility of service. In this study, γ-alumina–manganese oxide composite material was synthesized, characterized, and its defluoridation ability was evaluated in batch sorption experiments. This composite comes in light these days because of its availabilities and higher efficiency. Aluminum (Al) is the third most common metal in the Earth's crust, and Al oxides can be found in a wide variety of rocks, soils, and sediments. Manganese-containing composite metal oxides cannot only adsorb but also oxidize a variety of contaminants under certain circumstances. There has also been a lot of research done on MnO_x–Al_2O_3 nanocomposites for fluoride removal from water, and the majority of previous studies show a low adsorption potential. The estimated fluoride uptake potential of Al-Mn oxide and Al_2O_3 was determined to be 2.85mg/g and 1.08mg/g, respectively (2). Also until now, to the best of our information manganese oxide, alumina composite has been developed using large particle-sized alumina particles. To improve adsorption capacity, hereby we have developed nano alumina-supported manganese with enhanced fluoride removal capacity. To the best of our understanding, this is the first-ever report of the development of nano alumina manganese oxide composite for removal of fluoride present in contaminated water. While alumina is a well-known adsorbent, but its adsorption efficiency still needs to be enhanced significantly and therefore there is still plenty of space for development.

In the present work, we demonstrate a cost-effective, scalable process for the development of the manganese oxide-nano alumina composite as an efficient adsorbent for the removal of fluoride from contaminated water. It was demonstrated the presence of manganese oxide significantly improves the fluoride adsorption from water.

To the best of our knowledge, this is the first-ever report where nano alumina-manganese oxide has been shown as a potential adsorbent.

2. Experimental

2.1 Material

Sodium fluoride (NaF), sodium hydroxide pellets (NaOH) (Qualigens), sulfuric acid (H_2SO_4) (CDH), aluminum scrap, manganese sulphate heptahydrate ($MnSO_4.7H_2O$) (Rankem) & DI Water.

2.2 Synthesis of mesoporous γ-alumina-manganese oxide

The nanostructured γ-alumina-manganese oxide composite was synthesized in an aqueous medium under constant mixing. Nanoalumia was synthesised using precipitatipn process. In brief, for this firstly prepare γ-alumina oxide, 20 g of alumina scrap was dissolved in NaOH solution, the mixture was then filtered & neutralized by addition of H_2SO_4 was added till it reaches pH 7 (3). The mixture was homogenized for 48 h and then filtered and washed thoroughly and then dried at 80°C. This was then calcinated at 450°C for 4 hours. For the synthesis of manganese oxide-alumina composite, alumina was dispersed in the water and to this dispersion, manganese sulphate heptahydrate solution was added with continuous stirring at 300 rpm maintaining the temperature of 80°C. To this then added 2M NaOH till pH 12. The ratio of alumina oxide & manganese oxide was (1:1) was maintained.

2.3 Characterization

X-ray diffraction was recorded using Bruker AXS, D8 advance instrument with Cu Kα radiation (λ = 1.5406). Nova nano SEM-450 was used to identify the morphology of the sample. The sample was sputtered with gold under a vacuum for 10 minutes. ©Quantachrome Instruments version 11.05 model was used to determine the surface area of the developed material. Fluoride removal studies were carried out using a spectrophotometer (Hach 3900) using SPANDS solution.

2.4 Experimental procedure

Batch adsorption experiments were used to study the adsorption ability of the developed material at room temperature (30±2°C). By dissolving sodium fluoride (NaF) in distilled water, a stock solution of 100 mg/l fluoride solution was prepared. Adsorption studies were performed in 150-ml borosil-made conical flask (4). Residual fluoride was analyzed using the alizarin red visual method checked by spectrophotometer (Hach DR3900) using SPANDS solution. The adsorption efficiency of the nanoparticles was determined using equation

$$\mathbf{q_e = (c_o - c_e)\ V/m}$$

where $\mathbf{q_e}$ is the fluoride adsorbed (mg/g), $\mathbf{C_0}$ is the total fluoride concentration (mg/l) of the working solution, $\mathbf{C_e}$ is the residual fluoride concentration (mg/l) left after batch adsorption, **V** is the volume of solution in liters and **m** is the mass of the adsorbent. The removal percentage of individual adsorbate was calculated from Eq.

$$\mathbf{R\% = ((c_o - c_e)/c_o) \times 100}$$

3. Results and discussion

3.1 X-ray diffraction analysis

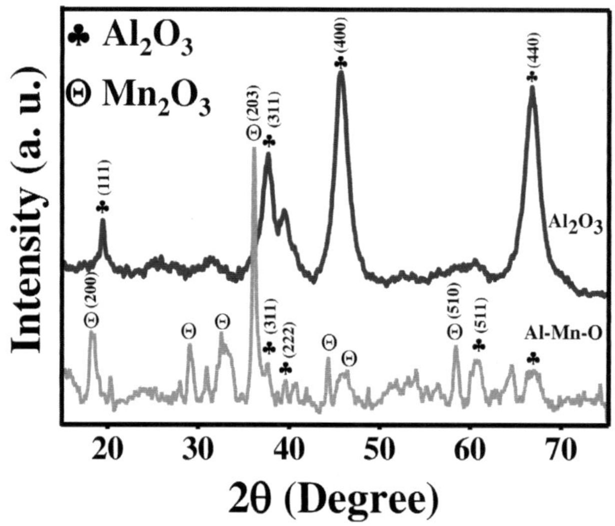

Figure 1. XRD Patterns of γ-alumina oxide & γ-alumina-manganese oxide.

Figure 1. shows the XRD diffractograms for the prepared γ-alumina-manganese oxide powders. X-ray diffraction pattern obtained shows a peak at 2θ value of 18.520°, 36.080°, 37.580°, 39.340°, 58.420°, 60.080° that matches well with JCPDS (Joint Committee on Powder Diffraction Standards) card no. (10-0425 for γ-alumina oxide) and (06-0540 for γ-manganese oxide) confirming the formation of manganese oxide-alumina composite. The presence of a major peak at 2θ=36.08° only in the composite confirms the coating of manganese oxide on alumina oxide. Also, γ-alumina oxide has a cubic structure (5) while γ-manganese oxide has a tetragonal structure respectively. Peaks obtained at 2theta degree range between 50-58 and 62–65 refers to aluminium sulfate hydrate (JCPDS No. 360569). Aluminum sulfate may have been generated by reaction of unreacted aluminium to sulfuric acid in the alumina synthesis process.

3.2 Scanning Electron Microscopy (SEM) with Energy Dispersive X-Ray Analysis (EDX)

SEM allows visual observation of an area of interest in a sample, The morphologies of γ-alumina oxide & γ-alumina-manganese oxide are shown in figure 2. γ-alumina has agglomerated tiny nanoparticles, while γ-alumina-manganese has a stone-like morphology with a variety of sizes(4) and a relatively rugged surface with many pores. (6)The signals corresponding to Al, Mn, and O were found in the EDX study of γ-alumina-manganese oxide, which is indirect evidence of the presence of manganese oxide on the surface of γ-alumina oxide.

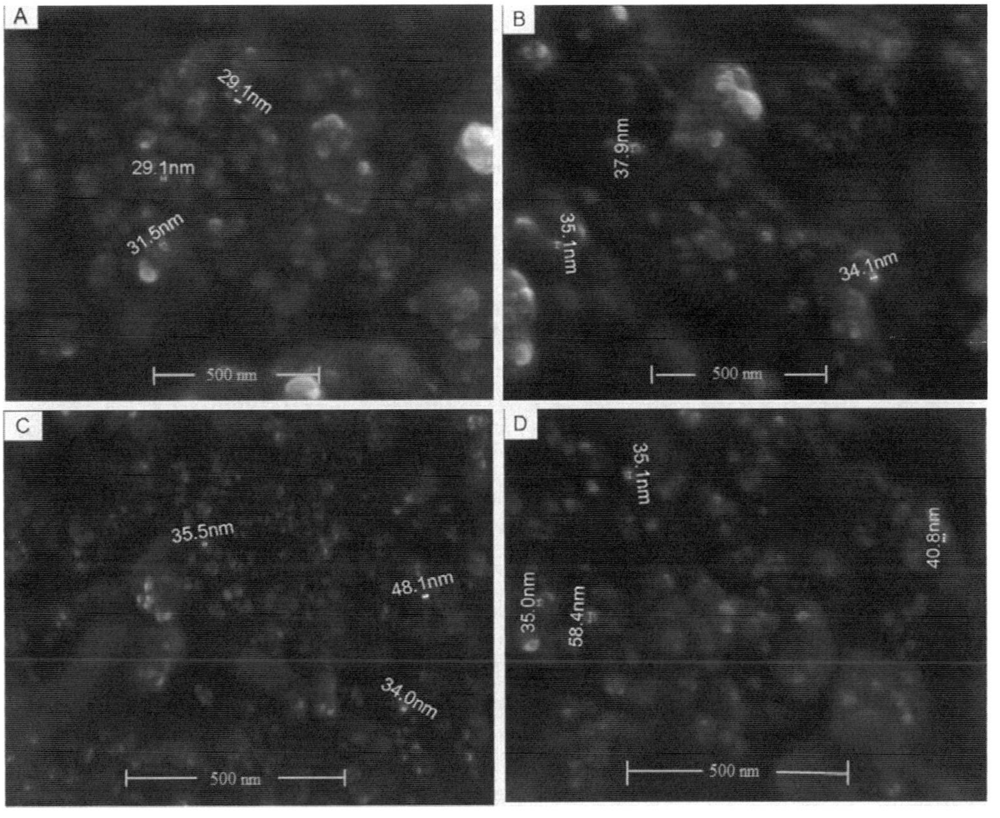

Figure 2. SEM images of γ-alumina oxide/γ-alumina-manganese oxide samples.

3.3 BET surface area analysis of developed composite

Figure 3. presents N_2 adsorption-desorption isotherms and the corresponding pore size distributions of γ-alumina-manganese oxide. The surface area of γ-alumina-manganese Oxide is 128.964 m^2/g and the pore volume was found to be 0.131 cm^3/g while the pore diameter is 3.50 nm. These results confirm the mesoporous nature of the developed composite material (7). That curve shows, The flatter region in the middle represents the formation of a monolayer. The micropores fill with nitrogen gas at very low pressure, at medium pressure monolayer formation is beginning and multilayer formation occurs, at the higher pressure capillary condensation occurs.

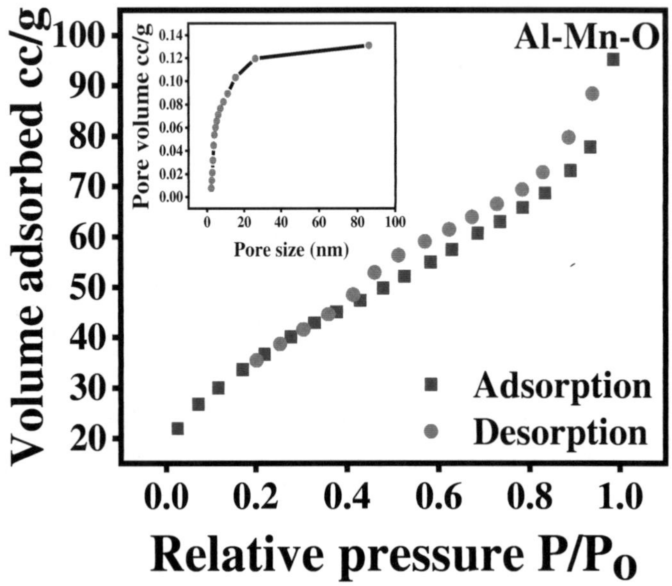

Figure 3. N_2 adsorption-desorption isotherms & PSD curve of γ-alumina-manganese oxide.

3.4 Adsorption

Adsorption increases the abundance of compounds on the surface due to a mass transfer process. The elimination of contaminants by promoting their adsorption on the adsorbent surface is the central concept of this approach to water treatment. Figure 4 shows the fluoride removal efficiency of the alumina only and the developed composite material (8). As can be seen, the fluoride removal efficiency of the composite materials increases significantly in presence of Manganese oxide. This may be due to the presence of porous manganese oxide on the surfaces of alumina oxide that improves the fluoride removal efficiency of the adsorbent. Thus, when compared to uncoated alumina oxide, manganese oxide-coated alumina oxide has been shown to be a promising adsorbent for improved defluoridation of water. (9)The γ-alumina-manganese oxide formed had wide areas of surface and well-ordered bimodal mesoporous structures and thus has greater adsorbing efficiency in water than the γ-alumina oxide for fluoride ions.

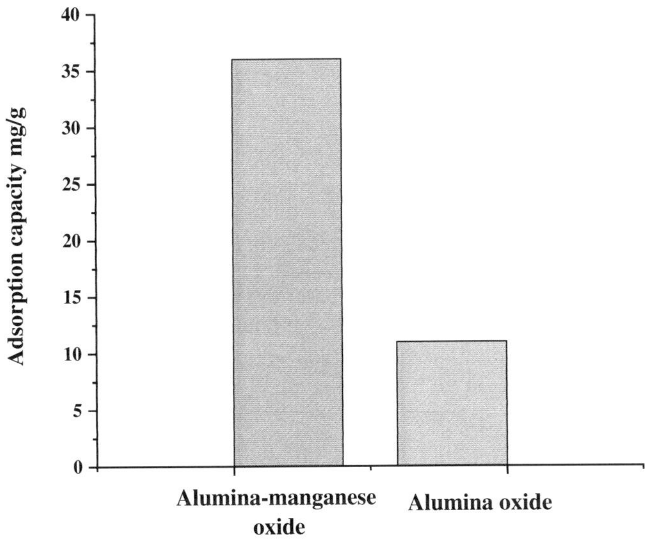

Figure 4. experimental results of fluoride adsorption capacity of alumina oxide & alumina-manganese oxide.

4. Conclusions

The present work shows the development of the nano alumina-supported manganese oxide as efficient fluoride removal media. Experiments indicate that adsorptive force like electrostatic attractive force plays an important role in the adsorption of fluoride. The material was highly mesoporous and having a high surface area. Until now, alumina-supported manganese oxide has been reported in the literature but this is the first-ever report where nano alumina was used as a support medium. This new promising adsorbent γ-alumina-manganese oxide has a good ability to remove anions from water. More importantly, the synthesis of this nanostructured bimetal composite is low cost and can easily be scalable (10). These studies will be important for the design of future nano adsorbents that can be used to remove fluoride from contaminated water.

Acknowledgments

This work was financially supported by the Council Of Scientific And Industrial Research–Advanced Materials And Process Research (CSIR–AMPR), Bhopal.

This manuscript is part of the special issue of selected papers from the 6th edition of the biennial International Conference on Nanoscience and Nanotechnology (ICONN-2021).

References

1. IS 10500 (2012): Drinking water
2. Maliyekkal, S. M., Sharma, A. K., & Philip, L. (2006). Manganese-oxide-coated alumina: A promising sorbent for defluoridation of water. Water Research, 40(19), 3497–3506. doi:10.1016/j.watres.2006.08.007
3. Alemu, S., Mulugeta, E., Zewge, F., & Chandravanshi, B. S. (2014). Water defluoridation by aluminum oxide–manganese oxide composite material. Environmental Technology, 35(15), 1893–1903. doi:10.1080/09593330.2014.885584

4. Singh, I. B., Gupta, A., Dubey, S., Shafeeq, M., Banerjee, P., & Sinha, A. S. K. (2015). Sol-gel synthesis of nanoparticles of gamma alumina and their application in defluoridation of water. Journal of Sol-Gel Science and Technology, 77(2), 416–422. doi:10.1007/s10971-015-3869-z
5. Xu, N., Liu, Z., Bian, S., Dong, Y., & Li, W. (2016). Preparation of MnO2–Al2O3 adsorbent with large specific surface area for fluoride removal. Particuology, 27, 66–71. doi:10.1016/j.partic.2016.01.005
6. TRIPATHY, S., & RAICHUR, A. (2007). Abatement of fluoride from water using manganese dioxide-coated activated alumina. Journal of Hazardous Materials. doi:10.1016/j.jhazmat.2007.09.100
7. Maliyekkal, S. M., Philip, L., & Pradeep, T. (2009). As(III) removal from drinking water using manganese oxide-coated-alumina: Performance evaluation and mechanistic details of surface binding. Chemical Engineering Journal, 153(1-3), 101–107. doi:10.1016/j.cej.2009.06.026
8. Qiu, H., Lv, L., Pan, B., Zhang, Q., Zhang, W., & Zhang, Q. (2009). Critical review in adsorption kinetic models. Journal of Zhejiang University-SCIENCE A, 10(5), 716–724. doi:10.1631/jzus.a0820524
9. Teng, S.-X., Wang, S.-G., Gong, W.-X., Liu, X.-W., & Gao, B.-Y. (2009). Removal of fluoride by hydrous manganese oxide-coated alumina: Performance and mechanism. Journal of Hazardous Materials, 168(2-3), 1004–1011. doi:10.1016/j.jhazmat.2009.02.133
10. Alhassan, S. I., Huang, L., He, Y., Yan, L., Wu, B., & Wang, H. (2020). Fluoride removal from water using alumina and aluminum-based composites: A comprehensive review of progress. Critical Reviews in Environmental Science and Technology, 1–35. doi:10.1080/10643389.2020.1769441

Enhanced Thermoelectric Power Generation Using Buck –Boost Converter

Lourdu Jame[1*], Joselin Retna Kumar[2*], Suresh Perumal[3], Jeyashree[4]

[1,4]Department of Electrical and Electronics Engineering,

[2]Department of Electronics and Instrumentation,

[3]Laboratory of Energy and Advanced Devices (LEAD), [3]Department of Physics and Nanotechnology, -

SRM Institute of Science and Technology, Kattankulathur, Chengalpattu District, Tamilnadu, India.

E-mail: lourdujame@gmail.com

Abstract: Energy generated in the vehicles depend on the engine and the fuel intake. But usually two thirds of the fuel added to the vehicle, which is, car, trucks bus etc. gets wasted as heat. The energy wasted as heat from areas like engine block, radiator, exhaust pipe engine coolant side and so on. So, to recover the waste heat as usable electrical energy back to the vehicle, Thermoelectric Generator (TEG) can be effectively utilized and its output power could be enhanced by connecting with buck boost converter. Here TEG with coupled inductor is used which produces smooth output with less ripples. TEG is placed in the engine exhaust of the vehicle and the output is connected with buck boost with coupled inductor converter so as to enhance the output voltage. Simulation using Comsol of single stage TEG without Buckboost converter produces an output of 6V.Interestingly TEG with Hotside temperature of 500K and coldside temperature of 300K with 127 legs is connected with buck boost converter with coupled inductor produces the output of 12 V and 0.65 mA is obtained to drive PIR sensor. Moreover in this paper Modified Maximum power point tracker (MPPT) algorithm is proposed to further improve the energy harvesting.

Keywords: Thermoelectric Generator, Maximum power point tracker, Buck-boost converter, PIR sensor, Coupled Inductor converter.

1. INTRODUCTION

Thermolectric generator (TEG) based waste heat recovering technology offers good fuel economy enhancements as it breaks up the primary conversion technologies. A great level of saving in fuel usage can be experienced by using it in automobile sectors. A vehicle normally utilizes only 25% of energy and wastes two third of its fuel as heat generated from radiator, engine, exhaust pipe and soon [1]. Various ways were opted for conversion of heat to electrical energy. The key function of TEG is to recover the heat produced by the engine and supply it back to the small scale electronic devices[2].TEGs are array of p- type and n-type semiconducting material which are connected thermally in parallel and electrically in series [3].They are considered to be renewable source of energy as they are environmental-friendly. But there is one major limitation of using TEG is that its low efficiency conversion of heat to electrical energy. To maximize the output signal, TEG is integrated with buck-boost converter which uses Modified Maximum power point tracker (MPPT) technique[4].As seen from Figure1 Thermoelectric Generator (TEG) consist of n-type and p-type semiconductor legs connected to each other in series through metal contact. Now the voltage is generated between the electrodes of hot side and cold side due to the moment of charge carriers. The n type legs consist of electrons whereas p type consist of holes . When the temperature difference between the hot side and cold side is sufficient for the charge carrier starts moving away from the hot side of the semiconductor and gets collected at one side which builds up voltage potential [5]. Thermoelectric material should work under specific thermodynamic conditions .The suitable thermodynamic condition can be obtained when relative current density is equal to direct material property [6].A few hundred watts of power can be generated from a light duty vehicle where exhaust pipe can carry large amount of heat [7,8].TEG is a low voltage and high current device. In order to avoid switching losses and reverse recovery problem high gain DC-DC converter should be used along with TEG and load [9].

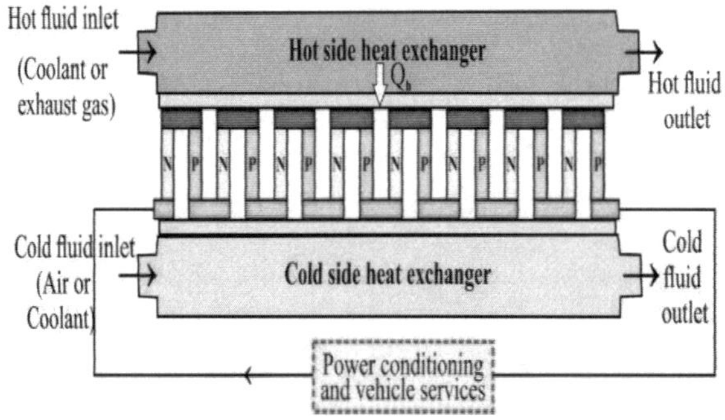

Figure1. Internal structure of TEG

As there is a continuous variation in the temperature difference the output from the TEG also keeps varying. To avoid the fluctuation of output power converter is used. CUK, Zeta, buck-boost converters if used will result in discontinuous input and output current, it will provide high stresses and large filter volume. Therefore buck-boost with coupled inductor converter is proposed. This provides a continuous input and output current due to the presence of inductor in both side which acts as filters. Many algorithm methods were implemented but due to the production of large oscillation in the output, modified perturb and observe algorithm was used. It helps in reduction of large oscillation as it uses large step perturbation and hence helps improve dynamic and steady state performance of the system. From the output of Buck boost converter 12 V, 7.8 mW power is obtained to drive Passive Infrared Sensor (PIR).

2. EXPERIMENTAL SECTION

2.1. TEG with Buck Boost converter

The hot side of the TEG module is kept near the heat source of the vehicle and the cold side of the TEG will be provided with the supply of cold water running through it shown in figure 2. The temperature difference between the plates would be enough for the working of TEG. After TEG produces electrical energy with efficiency of only 5%. It is then made to pass through buck boost converter which is responsible for providing constant supply to the load without any fluctuation in the voltage so as to power the Passive Infrared Sensor (PIR) for automatic door closing. The buck boost converter is responsible for either increasing voltage or decreasing it depending on the input supply so that the output voltage remains constant. The output of the TEG varies continuously due to the temperature variation and the amount of heat extraction from the source. MPPT algorithm is used to drive the gate signal of the MOSFET device of buck boost which enables whether to perform buck application or the boost application. This also helps in providing constant supply of current.

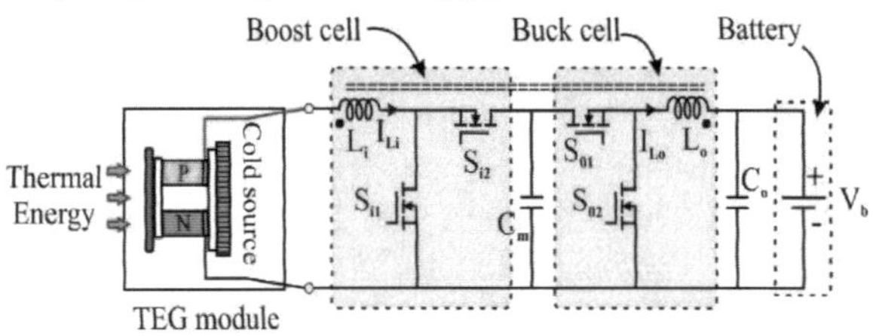

Figure 2 .TEG model of waste heat conversion

2.2. Perturb and Observation and MPPT Algorithm

Perturb and Observation MPPT algorithm is used to provide good performance of the circuit and for its simplicity. This method helps for the steady and dynamic performance of the system. TEG output is improved by applying the MPPT algorithm.

The voltage obtained from the TEG output is taken as u_{in} and the input power is calculated considering internal voltage drop of TEG to be e_{ti} and inner resistance r_{ti}.

$$P_i = \frac{u_i\,(e_{ti} - u_i)}{r_{ti}} \tag{1}$$

The maximum power depends upon the amount of internal voltage drop when $u_i = e_{ti}/2$

$$P_{inmax} = \frac{e_{ti}^2}{4r_{ti}} \tag{2}$$

Now in accordance to the MPP, when taking derivative of the p_{in}, before reaching MPP it is positive and after reaching MPP it is negative. When p_i reaches MPP point it's value becomes zero. That is,

$$\frac{dp_i}{du_i} = \frac{(e_{ti} - 2u_i)}{r_{ti}} \tag{3}$$

Hence to reach the MPP point or to make dp_i zero u_i is either increased or decreased so that the derivation of p_i reaches zero depending upon whether $\frac{dp_i}{du_i} > 0$ or $\frac{dp_i}{du_i} < 0$. When $\frac{dp_i}{du_i} > 0$, u_i is increased and when $\frac{dp_i}{du_i} < 0$, u_{in} is decreased. Now as the $\frac{dp_i}{du_i}$ is proportional to the MPP therefore input voltage is made proportional to $\frac{dp_i}{du_i}$ of perturbation step so that it can improve dynamic performance of the system.

$$\Delta u_{i\_step} = \frac{k.dp_i}{du_i} \tag{4}$$

Where, k = constant value. Now, when $\frac{dp_i}{du_i}$ is greater than zero then Δu_{i_step} will be greater than zero and hence vice versa.

$$u_{it}[n+1] = \Delta u_{i\_step}[n+1] + u_{ir}[n] \tag{5}$$

2.3 Comsol multiphysics

A single stage series TEG model is designed in a COMSOL 6.1 multi physics software. COMSOL provides a unified workflow.

2.4 Matlab Simulink

Simulink in Matlab provides a platform for modelling and simulating of physical components which provides a graphical representation. It is useful to user for better analyzing before testing it on hardware. It acts as a test bed with better visualization.

3. RESULTS AND DISCUSSIONS

3.1. Simulation of Single stage TEG in Comsol multi physics

Eight pair of unicouple TEG were designed and each leg with 2mm length, breadth and height. And the material used in the leg is Bismuth Telluride (Bi_2Te_3). The top layer is coated with copper it act as a metallic interconnect with height 0.2 mm and breadth 5mm.The same way 8 pairs of TEG were connected in series is shown in Figure 3.Hot side temperature of 500K and cold side temperature of 300 K is applied. Electric potential at corresponding leg is given. After build all the components compilation with mesh output of 8 pair series TEG is shown in Figure 4. For the corresponding input temperature and electric potential the temperature distribution and voltage output is shown in Figure 5 and 6.

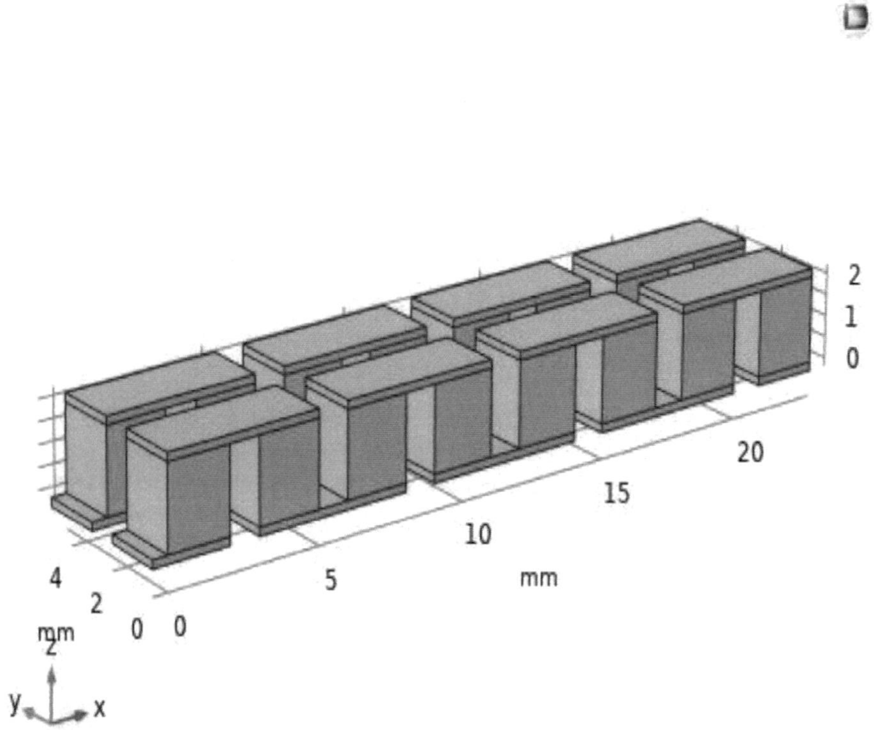

Figure 3.Simulation of 8 Unicouple series TEG

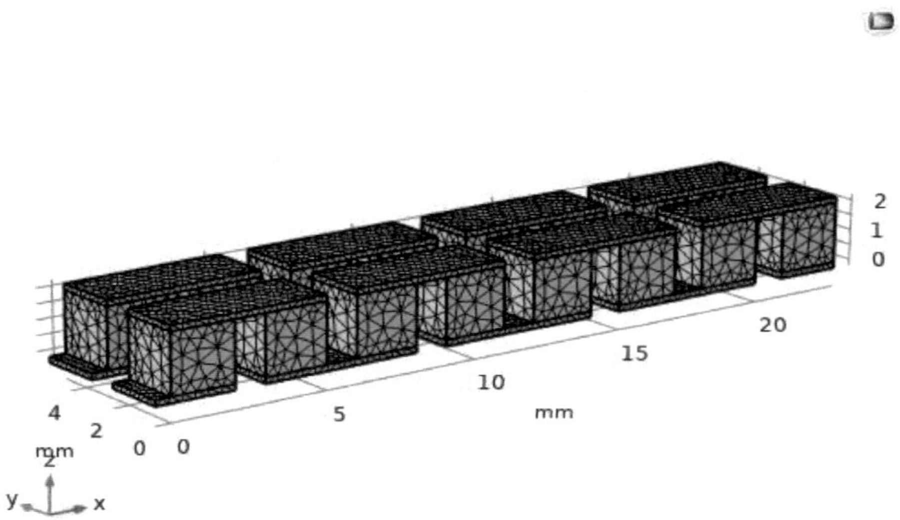

Figure 4. Mesh output of series 8 unicouple TEG

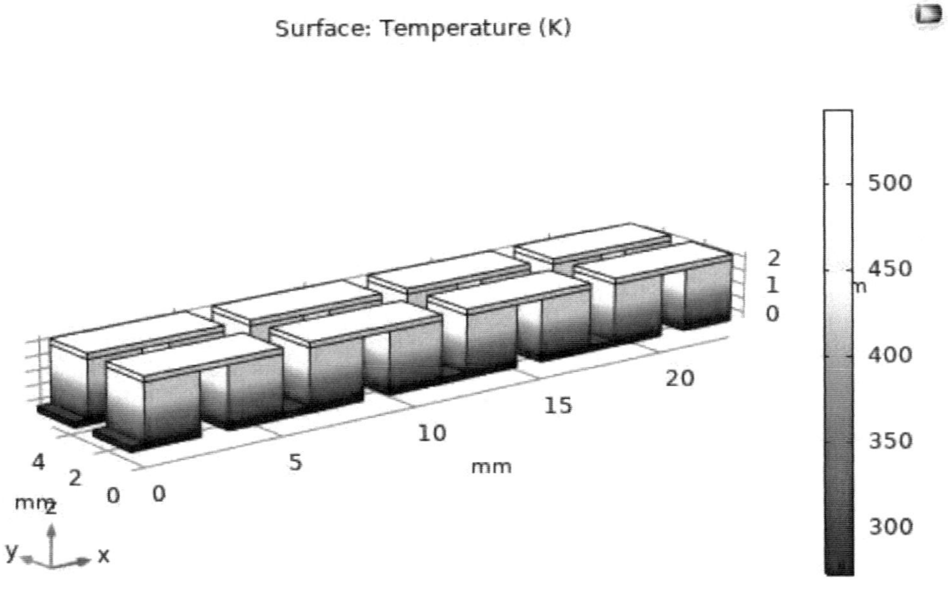

Figure 5. Temperature distribution of series 8 unicouple TEG

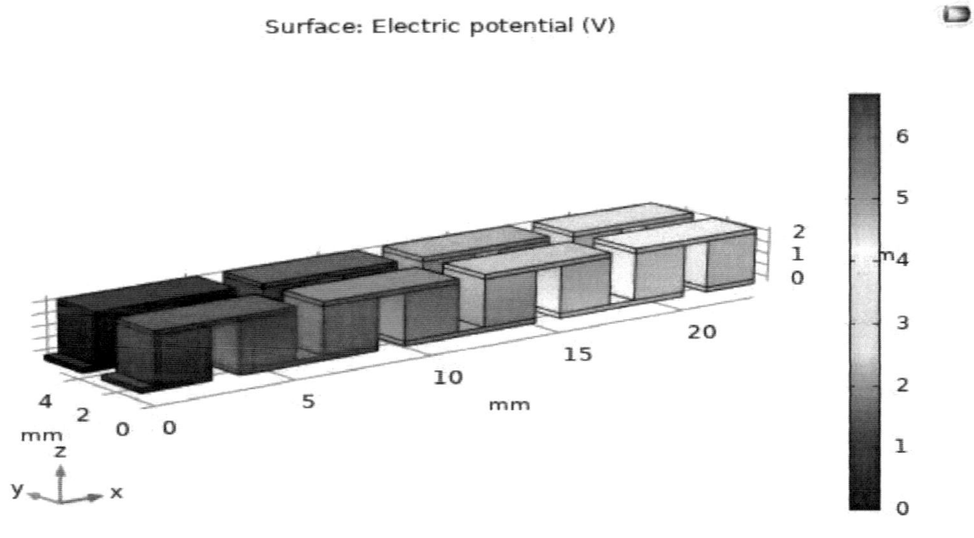

Figure 6. Output voltage of 8 series unicouple TEG

3.2 Buck-Boost Converter

Buck boost converter also called DC-DC converter, and choppers which is used to convert one voltage level to another. Buck Boost converter is used to perform either Step up conversion or Step down conversion depending upon the duty cycle (D). In this project we are implementing buck boost converter which has coupled inductor (BBC-CI). It uses boost switching cell as input and buck switching cell as output. It uses two negatively coupled inductor which have a common magnetic core. This helps to reduce direct current which in terms reduces the size of the inductor and improves power density. It also provides continuous output current. The capacitors used in the BBC-CI is used to improve the reliability of the converter. The converter operates in buck or boost mode. When S_{in1} is off and S_{in2} is in on state, the converter behaves in buck mode which has a C-L-C type filter as in input. Now, when the S_{o1} is switched on and S_{o2} is in off state, the converter operates in boost mode with C-L-C type output filter. In BBC-CI the inductors are negatively coupled with each other by 1:n ratio of turns which makes them behave as an ideal transformer. Where L_k is leakage inductor and L_m is magnetizing inductor. The inductors are coupled to each other through mutual inductance of M with coupling coefficient of k.

$$k = \frac{M}{\sqrt{L_o L_i}} \quad (6)$$

$$L_k = (1 - k^2)\, L_i \quad (7)$$

$$L_m = (1 - k^2) L_i \quad (8)$$

The negatively coupled inductor with each other produces voltage output by

$$u_{Li} = \frac{L_i di_{Li}}{dt} - \frac{M di_{Lo}}{dt} \quad (9)$$

$$u_{Lo} = \frac{L_o di_{Lo}}{dt} - \frac{M di_{Lo}}{dt} \quad (10)$$

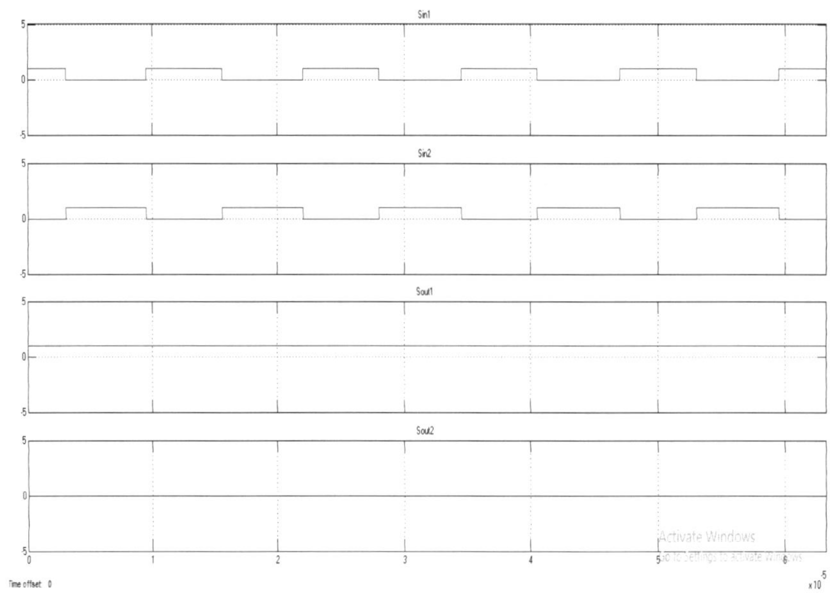

Figure 7. Boost mode output in state 1 and 2

IOP Conf. Series: Materials Science and Engineering 1219 (2022) 012033 doi:10.1088/1757-899X/1219/1/012033

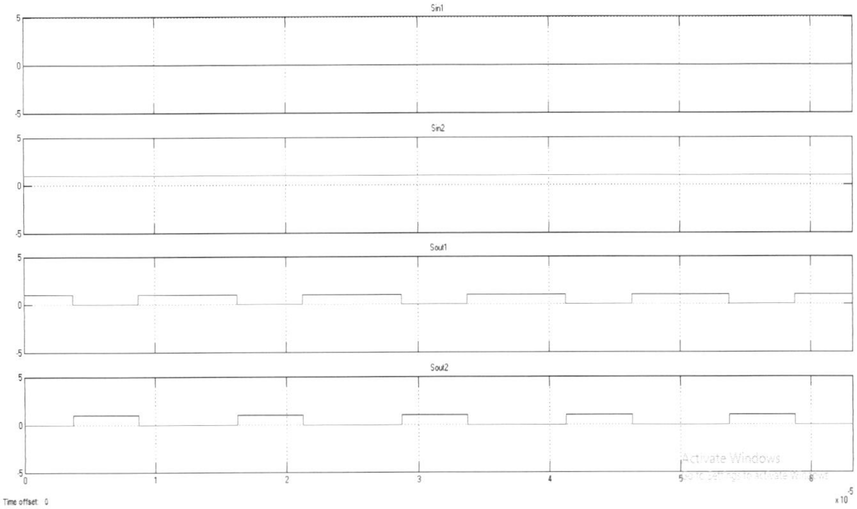

Figure 8. Buck mode output in state 1 and 2

Now, the BBCCI operates in two state. When in boost mode it consist of two switching states in a cycle. In State I S_{in1} is in on state whereas S_{in2} is off. This makes i_{Lin} and i_{Lo} which are inductor currents to increase in a linear manner with u_{in} as an input voltage. In State II Sin 1 is in off state and Sin2 is in on state at t1.This enables i_L in and i_{Lo} to decrease in a linear manner with u_b as an output voltage. The duty cycle of both the switches are Din and Do. The same analysis is used for the buck converter as well.

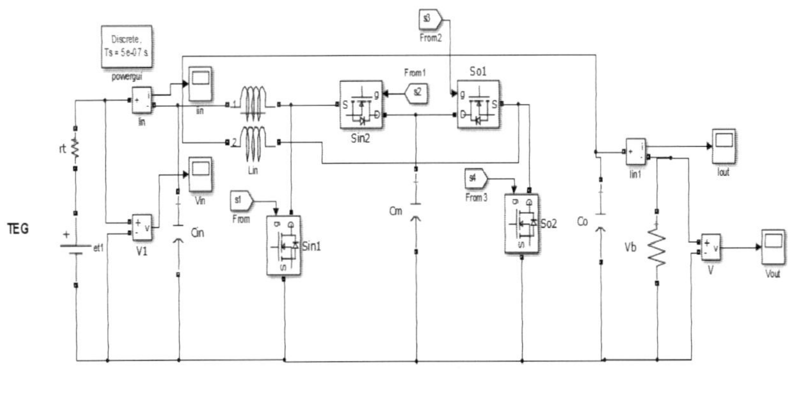

Figure 9. Simscape model of MPPT controller

IOP Conf. Series: Materials Science and Engineering 1219 (2022) 012033 doi:10.1088/1757-899X/1219/1/012033

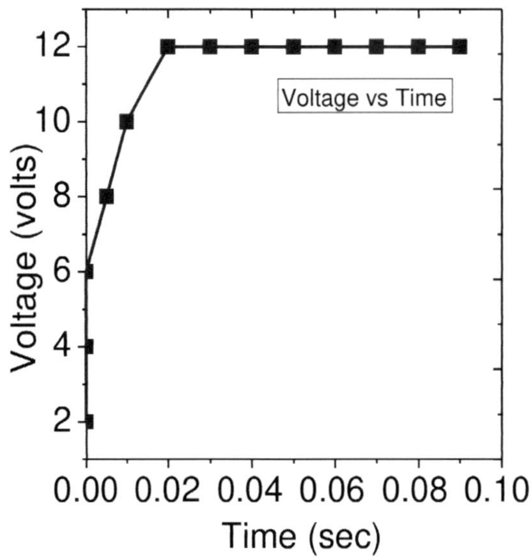

Figure 10. Graphical representation of output voltage v_s t sec of Buck-Boost converter

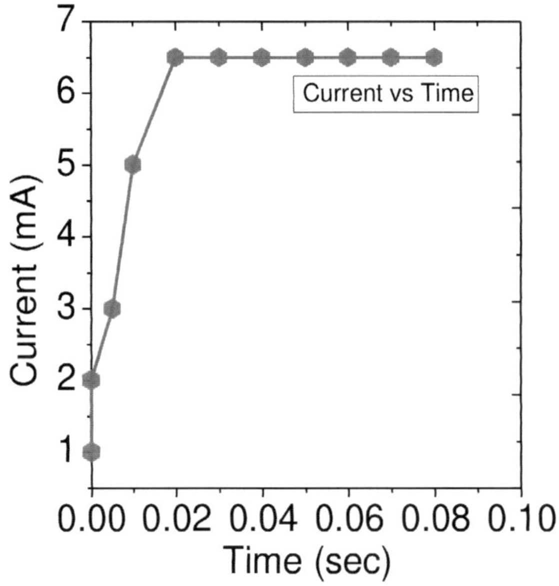

Figure 11. Graphical representation of output current v_s t sec of Buck-Boost converter

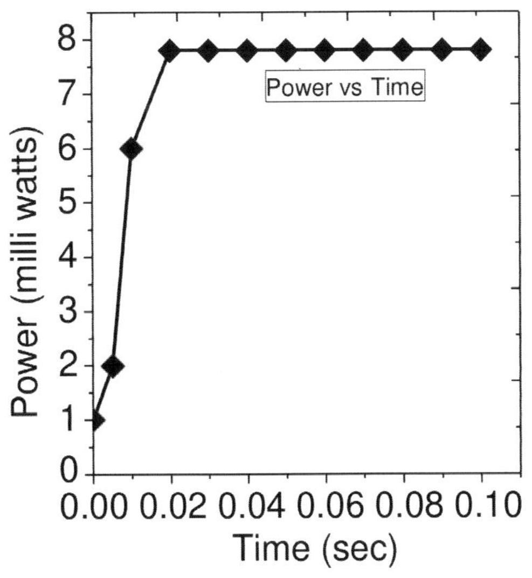

Figure 12. Graphical representation of output power v_s t sec of Buck-Boost converter.

The output current obtained from the converter increases linearly till the MPPT point is reach where it stabilizes. The stability is reached due to the use of P&O MPPT technique.

4. Conclusion

Thermoelectric generator was integrated with buck-boost converter coupled inductor so as to increase its output voltage from 6V to 12V which provides high efficiency. The converter operates over a wide range of input voltage and helps in providing maximum output power. It uses negatively coupled inductor which provides smooth power and current along with low ripples. For the maximum power extraction from the TEG we focused on P&O MPPT method which helps in improving steady and dynamic state of performance. Some strategies include, MPPT and battery voltage control. TEG waste heat recovering technology offers good fuel economy enhancements. There is a huge potential to increase the rate of conversion from heat energy to electrical energy.

References

1. R. Saidur, M. Rezaei, W.K. Muzammil, M.H. Hassan, S. Paria, M. Hasanuzzaman, Technologies to recover exhaust heat from internal combustion engines, Renew. Sustain. Energy Rev. 16 (2012) 5649–5659, .

2.M. Karvonen, R. Kapoor, A. Uusitalo, V. Ojanen, Technology competition in the internal combustion engine waste heat recovery: a patent landscape analysis, J. Clean. Prod. 112 (2016) 3735–3743

3. D.K. Aswal , , Ranita Basu , Ajay Singh ,"Key issues in development of thermoelectric power generators: High figure-of-merit materials and their highly conducting interfaces with metallic interconnects Energy Conversion and Management 114 (2016) 50–67

4. J. S. Jadhao, D. G. Thombare, Review on Exhaust Gas Heat Recovery for I.C. Engine, International Journal of Engineering and Innovative Technology (IJEIT) Volume 2, Issue 12, June 2013

5. Ms. Kiran A. Gajabe1, Ms. Kajal P. Dhanke2, Mr. S. J. Tikhe3, A Review on Utilization of Waste Heat from Automobile Based on Thermoelectric Generator, International Research Journal of Engineering and Technology (IRJET) Volume: 05 Issue: 02 Feb-2018

6. C. Goupil, W. Seifert, K. Zabrocki, E. Müller, and G. J. Snyder,Thermodynamics of thermoelectric phenomena and applications, Entropy, vol. 13, no. 8, pp. 1481–1517, 2011

7.Y. Wang, C. Dai, and S. Wang, Theoretical analysis of a thermoelectric generator using exhaust gas of vehicles as heat source,Applied Energy, vol. 112, pp. 1171–1180, 2013.

IOP Conf. Series: Materials Science and Engineering 1219 (2022) 012033 doi:10.1088/1757-899X/1219/1/012033

8. S. Kim, S. Park, S. Kim, and S.-H. Rhi,A thermoelectric generator using engine coolant for light-duty internal combustion engine-powered vehicles, Journal of electronic materials, vol. 40, no. 5, pp. 812–816, 2011.

9..Radjai, T., Gaubert, J. P., Rahmani, L., & Mekhilef, S. (2015). Experimental verifcation of P&O MPPT algorithm with direct control based on fuzzy logic control using CUK converter. International Transactions on Electrical Energy Systems, 25, 3492–3508.

10. J. Kim *et al.*, Dynamic Reconfiguration of Thermoelectric Generators for Vehicle Radiators Energy Harvesting Under Location-Dependent Temperature Variations,in *IEEETransactionson Very Large Scale Integration (VLSI) Systems*, vol. 26, no. 7, pp. 1241-1253, July2018.

11. Siyang, K. H. Lam and K. W. E. Cheng, Development of a motor waste heat power generation system based on thermoelectric generators,2016 International Symposium on Electrical Engineering (ISEE), Hong Kong, 2016, pp.1-5

Synthesis and Characterization of g-C_3N_4/ZnO Nanocomposites for Gas Sensing Application

Yethukrishnan C.S[1], Govind A.Vangari[1], Simon Patrick D[1], M. Krishna Mohan[1*]
[1] Functional Material and Energy Devices Laboratory, Department of Physics and Nanotechnology, SRM Institute of Science and Technology, Kattankulathur-603 203, India.

Abstract. Air pollution is one of the prevailing issues in our atmosphere, which greatly affects humans and causes dangerous diseases. Toxic and green gases (NO_2, CO_2, NO_x, H_2S, CO, Sulphur dioxide, and CH_4) are present in the ambient atmosphere arising from different industries, vehicles, and fossil fuel combustion. g-C_3N_4/ZnO nanocomposites were prepared via a facile hydrothermal method. The synthesized g-C_3N_4/ZnO samples were analyzed using FE-SEM for morphological analysis and structural parameters are confirmed by XRD. Further investigation is done on sensing films for toxic gases by varying temperature, resistance change, and gas type.

Keywords: g-C_3N_4/ZnO nanocomposites, gas sensor

1. Introduction

The swift growth of human living standards has revealed significant environmental pollution issues in modern ages with the growth of popularization and industrial development, showing a major challenge to nurturing and hampering the sustainable growth of humanity (1). Several researchers have made prominent efforts to discover new materials to improve the properties of gas sensing. Research has revealed that the use of nano regime metal oxide semiconductors (MOS) can increase the gas sensor's efficiency and resolve or minimize several of its difficulties in part (2). Experiments and subsequent theoretical simulations have displayed that, nanomaterials are generally especially encouraging for the production of gas sensors because, when the surface to volume (S2V) ratio is very high, the nanoscale amplifies surface-related properties that remain critical for gas sensing. (3). Researchers have developed several improvements, such as building a heterostructure, doping metal ions **(5),** controlling the size and morphology (4), to enhance the sensing efficiency of the metal oxide-based gas sensors. As a result, the properties of gas sensing increased significantly. The metal oxide semiconductor has become a hotspot in gas sensors application.

To detect harmful gases such as NO_2, H_2S, NH_3, CO_2, CO, ethanol, etc., various metal oxides like CuO, ZnO, Fe_2O_3, In_2O_3, WO_3, and SnO_2 are carried out for exploring gas sensing (6), (7), (8), (9), (10), (11). ZnO is the most fascinating metal oxide material among these various materials because of its gorgeous and special characteristics, such as low-cost preparation, eco-friendly, tuning the bandgap **(12)**, etc., Moreover, Zinc Oxide is also a well-known material in the application of gas sensors possessing high sensitivity and selectivity towards various toxic and polluted gases (13). Also, 1D zinc oxide nanostructures like nanowires or nanorods have enhanced properties of gas sensing than nanoparticles, as 1D zinc oxide nanostructures can improve gas sensing stability.

A conjugated carbon-based semiconductor composed of sp^2 hybridized nitrogen and carbon atoms is graphitic natured carbon nitrides (g-C_3N_4) (14). Researchers are increasingly paying consideration to its application in numerous fields such as water splitting (16), dye degradation (15), and gas sensors (17), because of its low-cost preparation method, easy synthesis technique, and special layer structure. Owing to its narrow bandgap, it shows the outstanding catalytic activity in the visible light area is the most significant characteristic for g-C_3N_4 (18).

Wang et. al, 2D/2D ZnO/g-C3N4 heterojunction was successfully synthesized via ultrasound mixing subsequently calcination process. At room temperature, under the 460nm light excitation, the sensing properties in NO2 were systematically examined (19). Cao et. al., constructed Cocoon-like ZnO ornamented graphitic carbon nitride nanocomposites through the hydrothermal process, for ethanol gas sensing at lower optimum operating temperature (20).

In this paper, we have successfully prepared nanocomposites of the above-discussed materials via the hydrothermal method for gas sensing applications and their characteristics were studied using techniques such as XRD, FE-SEM.

2. Experimental Section

Preparation of g-C_3N_4

The pristine g-C_3N_4 prepared by calcination of melamine at atmospheric pressure. In general, 10 g of melamine was placed in an enclosed alumina crucible and calcined in a muffle furnace at 550 °C for 4 h in inert air at a rising rate of 5 °C min^{-1} and cooled to room temperature naturally.

Preparation of ZnO/ g-C_3N_4 nanocomposites

In 20 mL of deionized water, 10 mL of 4 M NaOH solution was gradually added. After that, in the NaOH solution with magnetic stirring, 10 mL of 0.2 M $Zn(NO_3)_2$ $6H_2O$ was added, followed by the adding of a certain quantity of 0.03 mol/L CTAB and g-C3N4. The solution was stirred uniformly for 10 minutes and the mixture was moved to the autoclave, and placed in a muffle furnace at 160 °C for 6h. The suspension was centrifuged at room temperature and washed several times with DI water and ethanol as well. Finally, dried for 12 h at 80 ° C.

3. Result and Discussion

Fig. 1 presents the ZnO, g-C_3N_4, and ZnO/ g-C_3N_4 nanocomposite XRD patterns. The result shows that the ZnO/ g-C_3N_4 nanocomposites are indexed with (JCPDS No. 079-0206) showing typical three high diffraction peaks (100), (002), and (101) at 31.62°, 34.15° and 36.05 respectively. The distinctive diffraction peak at 27.5 ° is due to an inter-planar stacking peak representing the aromatic structure (002). The poly-aromatic rings present in graphitic materials with interlayer stacking and the g-C_3N_4 interlayer distance are correlated with two characteristic peaks at 12.50 ° and 27.5 ° in the g-C_3N_4 sample.

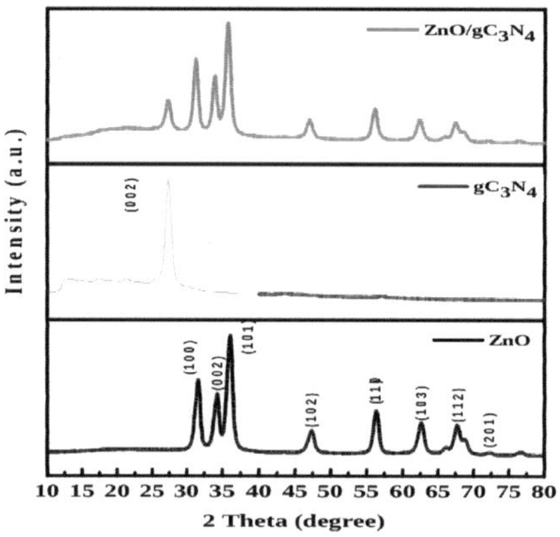

Fig. 1: XRD Spectra as prepared samples

The morphologies analysis of the as-prepared nanocomposites material was observed in FE-SEM and corresponding images are displayed in (Fig. 2). The FE-SEM images of nanocomposites g-C_3N_4 and ZnO/ g-C_3N_4 demonstrate the creation of noticeable 2D layered bulk sheets that remain gathered together, as shown in (Fig 2a,b). The intermediate junctions formed between the two materials develop active sites for gas interactions. Hence, this composite material might act as an effective sensing material.

IOP Conf. Series: Materials Science and Engineering 1219 (2022) 012034 doi:10.1088/1757-899X/1219/1/012034

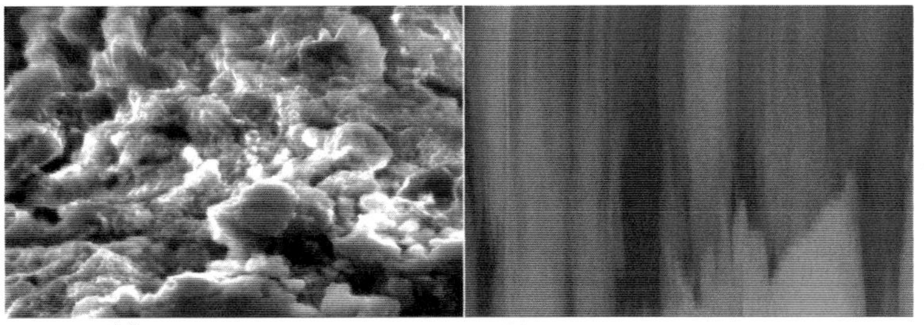

Fig. 2: FE-SEM images of (a) g-C_3N_4, (b) ZnO/g-C_3N_4 nanocomposites.

Conclusion

In summary, ZnO/g-C3N4 composites were synthesized via calcination and hydrothermal method. The structural characteristics and morphology properties were distinguished respectively by XRD and SEM. The SEM results indicate g-C3N4 and ZnO form the heterojunction successfully. The ZnO/g-C3N4 nanocomposites would be the ideal peak for the application of gas sensors.

References

1. Bhati, Vijendra Singh, Mirabbos Hojamberdiev, and Mahesh Kumar. "Enhanced sensing performance of ZnO nanostructures-based gas sensors: A review." *Energy Reports* 6 (2020): 46-62.
2. Guan, Wei, Na Tang, Kuang He, Xiaoying Hu, Mingshan Li, and Kaiming Li. "Gas Sensing Performances of Metal Oxides Nanostructures for Detecting Dissolved Gases: A Mini Review." *Frontiers in chemistry* 8 (2020): 76.
3. Ali, Ali Salman. "Application of nanomaterials in environmental improvement." In *Nanotechnology and the Environment*. IntechOpen, 2020.
4. Zhang, Jian, Ziyu Qin, Dawen Zeng, and Changsheng Xie. "Metal-oxide-semiconductor based gas sensors: screening, preparation, and integration." *Physical Chemistry Chemical Physics* 19, no. 9 (2017): 6313-6329.
5. Ganesh, R. Sankar, E. Durgadevi, M. Navaneethan, V. L. Patil, S. Ponnusamy, C. Muthamizhchelvan, S. Kawasaki, P. S. Patil, and Y. Hayakawa. "Tuning the selectivity of NH3 gas sensing response using Cu-doped ZnO nanostructures." *Sensors and Actuators A: Physical* 269 (2018): 331-341.
6. Shahabuddin, Md, Anjali Sharma, Jitendra Kumar, Monika Tomar, Ahmad Umar, and Vinay Gupta. "Metal clusters activated SnO2 thin film for low level detection of NH3 gas." *Sensors and Actuators B: Chemical* 194 (2014): 410-418.
7. Zhang, Baohui, Min Li, Zhilong Song, Hao Kan, Haoxiong Yu, Qian Liu, Guangzu Zhang, and Huan Liu. "Sensitive H2S gas sensors employing colloidal zinc oxide quantum dots." *Sensors and Actuators B: Chemical* 249 (2017): 558-563.
8. Mirzaei, Ali, Kamal Janghorban, Babak Hashemi, Anna Bonavita, Maryam Bonyani, Salvatore Gianluca Leonardi, and Giovanni Neri. "Synthesis, characterization and gas sensing properties of Ag@ α-Fe2O3 core–shell nanocomposites." *Nanomaterials* 5, no. 2 (2015): 737-749.
9. Prajapati, Chandra Shekhar, and Navakanta Bhat. "ppb level detection of NO 2 using a WO 3 thin film-based sensor: material optimization, device fabrication and packaging." *RSC advances* 8, no. 12 (2018): 6590-6599.
10. Lu, Rui, Xiaoling Zhong, Shiguang Shang, Shan Wang, and Manling Tang. "Effects of sintering temperature on sensing properties of WO3 and Ag-WO3 electrode for NO2 sensor." *Royal Society open science* 5, no. 10 (2018): 171691.
11. Xing, Ruiqing, Lin Xu, Jian Song, Chunyang Zhou, Qingling Li, Dali Liu, and Hong Wei Song. "Preparation and gas sensing properties of In 2 O 3/Au nanorods for detection of volatile organic compounds in exhaled breath." *Scientific reports* 5, no. 1 (2015): 1-14.
12. Klingshirn, Chem. "ZnO: material, physics and applications." *ChemPhysChem* 8, no. 6 (2007): 782-803.
13. Bhati, Vijendra Singh, Mirabbos Hojamberdiev, and Mahesh Kumar. "Enhanced sensing performance of ZnO nanostructures-based gas sensors: A review." *Energy Reports* 6 (2020): 46-62.

14. Dong, Guoping, Yuanhao Zhang, Qiwen Pan, and Jianrong Qiu. "A fantastic graphitic carbon nitride (g-C3N4) material: electronic structure, photocatalytic and photoelectronic properties." *Journal of Photochemistry and Photobiology C: Photochemistry Reviews* 20 (2014): 33-50.s

15. Ngullie, Renathung C., Saleh O. Alaswad, Kandasamy Bhuvaneswari, Paramasivam Shanmugam, Thangavelu Pazhanivel, and Prabhakarn Arunachalam. "Synthesis and Characterization of Efficient ZnO/g-C3N4 Nanocomposites Photocatalyst for Photocatalytic Degradation of Methylene Blue." *Coatings* 10, no. 5 (2020): 500.

16. Yu, Fan, Laichun Wang, Qiuju Xing, Dengke Wang, Xunheng Jiang, Guangchao Li, Anmin Zheng, Fanrong Ai, and Jian-Ping Zou. "Functional groups to modify g-C3N4 for improved photocatalytic activity of hydrogen evolution from water splitting." *Chinese Chemical Letters* 31, no. 6 (2020): 1648-1653.

17. Wang, Hongtao, Jihao Bai, Meng Dai, Kunpeng Liu, Yueying Liu, Linsheng Zhou, Fengmin Liu et al. "Visible light activated excellent NO2 sensing based on 2D/2D ZnO/g-C3N4 heterojunction composites." *Sensors and Actuators B: Chemical* 304 (2020): 127287.

18. Dong, Guoping, Yuanhao Zhang, Qiwen Pan, and Jianrong Qiu. "A fantastic graphitic carbon nitride (g-C3N4) material: electronic structure, photocatalytic and photoelectronic properties." *Journal of Photochemistry and Photobiology C: Photochemistry Reviews* 20 (2014): 33-50.

19. Wang, H., Bai, J., Dai, M., Liu, K., Liu, Y., Zhou, L., ... & Geyu, L. (2020). Visible light activated excellent NO2 sensing based on 2D/2D ZnO/g-C3N4 heterojunction composites. *Sensors and Actuators B: Chemical*, *304*, 127287.

20. Cao, J., Gong, Y., Wang, Y., Zhang, B., Zhang, H., Sun, G., ... & Zhang, Z. (2017). Cocoon-like ZnO decorated graphitic carbon nitride nanocomposite: Hydrothermal synthesis and ethanol gas sensing application. *Materials Letters*, *198*, 76-80.

IOP Conf. Series: Materials Science and Engineering 1219 (2022) 012035 doi:10.1088/1757-899X/1219/1/012035

First Order Hyper polarizability and Intramolecular Charge Transfer of N-Ethyl-N-(2-Hydroxyethyl)-4-(4-Nitrophenylazo) Aniline for Photonic Applications

B.Raaghul, M.R.Kannan, T.Vijayakumar*

Futuristic Materials Research Centre for Planetary Exploration, Department of Physics and Nanotechnology, College of Engineering and Technology, Faculty of Engineering and Technology, SRM Institute of Science and Technology, Kattankulathur - 603 203, Kancheepuram, Tamil Nadu, India.

E-mail: *vijayakt1@srmist.edu.in*

Abstract. We report a computational methodology for estimating the first order hyperpolarizability (β) of N-ethyl-N-(2-hydroxyethyl)-4-(4-nitrophenylazo) aniline, a nonlinear optical (NLO) chromophore being used as anazobenzene dye. The planar and non-planar structures of N-ethyl-N-(2-hydroxyethyl)-4-(4-nitrophenylazo) aniline were optimized by B3LYP/6-311++G (D,P) basis set using GAUSSIAN'09W program package. The dipole moment, static polarizability, first order hyperpolarizability were studied using HF/6-31G (D) basis set. The β value of the planar and non-planar structure of N-ethyl-N-(2-hydroxyethyl)-4-(4-nitrophenylazo) aniline is 196 and 124 times higher than that of urea standard, respectively, which qualify both of the novel molecular systems as highly efficient for second order NLO applications and adds evidence to the importance of delocalization of π-electrons in a system for effective second order NLO processes. Experiments were also performed using an Nd-YAG laser of pulses in the range of 5-10 nanoseconds with a tunable peak power of 4.7 W.

Keywords: DFT;Nd-YAG laser; hyper polarizability;

1. Introduction

Organic chromophores have turned out to be an interesting component of research in the field of non-linear optics.It is established that for a material to exhibit high efficiency of second order non-linear optic processes, it must carry a noncentrosymmetric bulk structure as well as a large intrinsic non-linearity that results from ionic and vibronic contributions. Polymers, in particular, can be geared on a molecular scope to satisfy the discrete requirements of the desired exertion due to its readily accessible and manipulative nature. Mostly, such modifications are done in order to influence the refractive index, dielectrics and the optical non-linearity of the material by introducing the non-linear optic moieties as side groups that are functionalized covalently and this has proved to be a current research trend in an effort to substantially improve the sum non-linearity of the material.It has been well demonstrated that organic chromophores with conjugated donor- and acceptor-substituted moieties have large molecular hyperpolarizabilities [1-7].

Even though a large number of non-linear optic processes are available to determine and study the efficiency of any material, it is primarily important to determine the hyperpolarizabilities of the order corresponding to the order of non-linear optic process.In this paper, we analyze thefirst order hyperpolarizability (β_{ijk}), a tensor quantity of rank three,which is a consequence of the transfer of energy in second order non-linear optic processes as electronic wave functions of the molecules in the medium are perturbed using photons. For computational calculations, the codes used in Density Functional theory are used as a means to study the specifics of materials such as its structural or electronic and magnetic components and characteristics.

2. Theoretical and Experimental work

Material

N-Ethyl-N-(2-Hydroxyethyl)-4-(4-Nitrophenylazo) Aniline, Disperse Red 1 (DR1), is a nonlinear optical (NLO) chromophore (Fig.1) being used as an azobenzene dye. It can form a dipole network that enhances the electro-optic effect of the non-linear components and also improves the photorefractive effect. Two structures of DR1 were used in the studies, *viz*, the Planar structure and the Non-planar structure. In the planar structure, all of the functional groups are in the same plane whereas in the non-planar structure, benzene and other functional groups are slightly deviated and are not on the same plane.

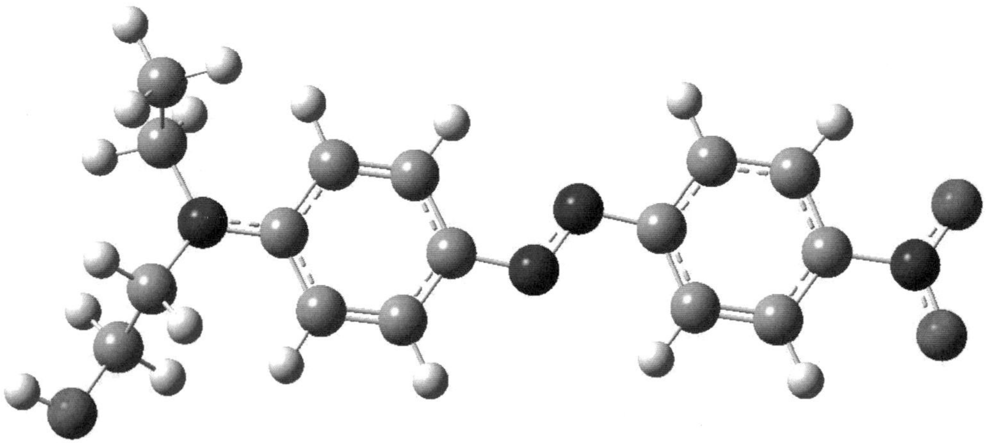

Fig.1 Optimized Molecular Structure of Disperse Red 1

Computational method to determine the first order hyperpolarizability

Density Functional Theoretical (DFT) computations were executedimplementing the closed-shell Becke-Lee-Yang-Parr hybrid exchange-correlation three-parameter functional (B3LYP) conjointly with 6-31G (d) basis set to procure the full bore geometry optimizations and normal-mode scrutiny on isolated entities. MP2 and *ab initio* computations have also been accomplishedin order to contrast the reliability of non-identical kinds of computational mechanisms. Fore by, DFT force fields,calibrated using the B3LYP functional, flex infrared spectra that are in sublimeaccordance with experiments. Molecular geometries were extensively optimized by Berny's optimization algorithm employing redundant internal co-ordinates. All of the optimized structural entities were affirmed to be minimum energy conformations.

For the frequency reckonings, the B3LYP/6-311++G(d,p) theory was utilized, while the Harmonic vibrational wave numbers were reckoned using analytic second derivatives to affirm the convergence to minima on the potential surface and to enumerate the zero-point vibrational energies (ZPVE). The encompassing of 'd' polarization in the split valance basis set wasforeseen to produce a marked refinement in the reckoned geometry. The reckoned harmonic vibrational frequencies were impartially scaled down, to elucidate for systematic errors spawned by basis set inadequacy, oversight of electron correlation and vibrational anharmonicity. All the theoretical reckonings have been carried out utilizing the GAUSSIAN'09 program package. The correspondences on the exchange correlation energy are perpetrated by a class of approximations called Hybrid Functionals. We, in this paper, use the B3LYP [8-10]functional:

$$\mathbf{E_{XC}^{B3LYP}=E_X^{LDA}+a_0\,(E_X^{HF}-E_X^{LDA})+a_X\,(E_X^{GGA}-E_X^{LDA})+E_C^{LDA}+a_C\,(E_C^{GGA}-E_C^{LDA})}$$

The first hyperpolarizability (β) and related properties (α and μ) have been studied and estimated using the basis set HF/6-31G (D) based on finite field approach. In our case, the number of components of the 3-dimensional matrix of the first order hyper polarizability gets cut short from 27 to just 10 as a result of Kleimman symmetry. The energy in the external electric field when expressed in the form of Taylor series, the coefficients that are a part of the same are basically the components of β. When the homogeneous external electric field is weak, the Taylor series expansion transforms to:

$$\mathbf{E=E^0-\mu_iF_i-(1/2)\,\alpha_{ij}F_iF_j-(1/6)\,\beta_{ijk}F_i\,F_j\,F_k-\ldots}$$

Where E° is the energy of the unperturbed molecules, F_i is the field at the origin;μ_i, α_{ij}, β_{ijk} correspond to the dipole moment, polarizability and the first hyper polarizabilities of the molecule taken into consideration respectively. The total static dipole moment μ, the mean polarizability $\alpha_0$and the first hyper polarizability β_{tot}, using the x-, y- and z- components are defined as:

$$\mathbf{\mu=(\mu_x^2+\mu_y^2+\mu_z^2)^{1/2}}$$

$$\mathbf{\alpha_0=(\alpha_{xx}+\alpha_{yy}+\alpha_{zz})/3}$$

$$\beta_{tot} = [(\beta_{xyy}+\beta_{xzz}+\beta_{xxx})^2 + (\beta_{yzz}+\beta_{xxy}+\beta_{yyy})^2 + (\beta_{xxz}+\beta_{yyz}+\beta_{zzz})^2]^{1/2}$$

3. Results and Discussion

The dipole moment (μ) and the non-linear optical properties- Static polarizability (α) and the first order hyper polarizability (β) of Disperse Red-1 (both planar and non-planar structures) are provided as follows. All the theoretical calculations have been carried oututilizingthe GAUSSIAN09W program package.

Dipole moment (μ)

Among the above two systems, the Non-planar system has a lesser dipole moment compared to the planar system. The probability of redistribution of charges symmetrically around the geometric center is higher in case of the non-planar system compared to the planar system which may be one of the reasons for the comparatively lesser dipole moment.Table 1 shows the dipole moment components of DR1 for theplanar structure and non-planar structure and the effective dipole moment is calculated to be 4.29172019 Debye and 3.78940448 Debye,respectively.

Table 1 Dipole Moment Components of Disperse Red 1

Dipole Moment Components		
	Planar	Non-planar
μ_x	-3.9142882	-3.5026339
μ_y	+0.1435188	+0.0036740
μ_z	+1.7540275	+1.4460735

Polarizability (α)

As polarizability is a phenomenon that is majorly dependent on the delocalization of valence electrons in the system, it is observed that the polarizability is larger in case of the planar structure though the polarizability of the non-planar structure is not too far off from that of the planar structure. This proves the comparatively higher sensitivity of the planar system to an external perturbation and explains the extent of delocalization of the electrons in the system. Table 2 shows the polarizability components of DR1 for the planar structure and non-planar structure and the static polarizability is found to be $3.61487937\text{x}10^{-23}$e.s.u. and $3.2991076\text{x}10^{-23}$e.s.u.

Table 2 Polarizability Components of Disperse Red 1

Polarizability Components		
	Planar	Non-Planar
α_{xx}	+346.4684708	+309.7141333
α_{xy}	-002.1204062	-001.5789133
α_{yy}	+091.9730263	+101.4879496
α_{xz}	-120.9118342	-097.2050684
α_{yz}	-003.1724061	-028.2733634
α_{zz}	+293.3154608	+256.6334627

First order Hyperpolarizability (β)

Both the planar and non-planar systems of Disperse Red-1 exhibit noticeable values for the first order hyper polarizability (β). The recommended organic molecule to be used as a reference for effective comparison of Non-linear optic properties is Urea. The β value for Urea is 3.728×10^{-31}e.s.u. or 43.1516442 a.u. The β value of the planar structure of Disperse Red-1 is 196 times higher than that of Urea and the β value of the non-planar structure of Disperse Red-1 is 124 times higher than that of Urea which qualifies both of the novel molecular systems as highly efficient for second order NLO applications like Second Harmonic generation (SHG) and also adds evidence to the importance of delocalization of π-electrons in the system for effective second order non-linear optic processes.

It is also noticeable from the results of both the systems that contribution to β is mainly from the β_{xxx} tensor component with additional contributions from the β_{xxz}, β_{xzz} and β_{zzz} tensor components. This also hints the dominant charge transfer along these directions or planes. The large values of Non-linear optic responses of the Disperse Red-1 systems make them highly suitable for designing Non-linear optic systems or other Non-linear optic applications. Table 3gives the hyperpolarizability components of DR1 for the planar structure and non-planar structure and the total hyperpolarizability is computed to be $7.31196134 \times 10^{-29}$e.s.u. and $4.6313023 \times 10^{-29}$e.s.u.

Table 3 Hyperpolarizability Components of Disperse Red 1

Hyperpolarizability Components		
	Planar	Non-planar
β_{xxx}	+4501.3902941	+2904.8040706
β_{xxy}	+0016.7662298	+0215.6936782
β_{xyy}	-0028.0825911	-0027.1108685
β_{yyy}	-0000.4843185	-0003.8512413
β_{xxz}	-3685.1369542	-2361.7967888
β_{xyz}	-0014.6613299	-0124.6016940
β_{yyz}	+0011.7283414	+0012.9168269
β_{xzz}	+2414.6929921	+1491.2978233
β_{yzz}	+0014.0470627	+0054.9218677
β_{zzz}	-1244.6314480	-0745.9972697

4. Conclusions

Detailed computational studies of Disperse Red-1 have been performed to work towards the goal of finding materials with high nonlinear optic responses using Density functional theory (DFT) and experiments on the same were attempted using a Nd-YAG laser of pulses in the range of 5-10 nanoseconds with a tunable peak power of 4.7W.The nonlinear optic responses of Disperse Red-1 in its planar and non-planar forms were studied and it was found that both the systems showed very high non-linear optic responses and efficiency than the reference molecular system-Urea which qualify both of the systems to be very useful for non-linear optic applications such as non-linear optoelectronics, etc. The linear and non-linear optic responses, *viz*, Dipole moment (μ), Polarizability (α) and the first order hyperpolarizability (β) were estimated to be larger in case of the planar system of Disperse Red-1 compared to the non-planar system of Disperse Red-1 by a considerable margin making it much more efficient for Second Harmonic generation applications.

Acknowledgements:

We would like to express our sincere thanks to Dr. Junaid MasudLaskar, Assistant Professor, Department of Physics and Nanotechnology, SRM-IST for his assistance in the experimental work involving lasers. Authors thank High Performance Computing Center, SRM Institute of Science and Technology for providing the

IOP Conf. Series: Materials Science and Engineering 1219 (2022) 012035 doi:10.1088/1757-899X/1219/1/012035

computational facility. The author T.Vijayakumar acknowledges the financial support obtained from the Department of Space, Government of India [Grant no.B.19012,57,2016-II] through RESPOND project and the selective excellence initiative award received from SRM Institute of Science and Technology.

References

1. D. J. Williams, "Organic polymeric and non-polymeric materials with large optical nonlinearities," Angew. Chem. Int.Ed. Engl. 23, 690-703 (1984).
2. P. Pantelis, J. R. Hill, S. N. Oliver, and G. J. Davies, "Organicpolymer films for nonlinear optics," Br. Telecommun. Technol. J. 6, 5-17 (1988).
3. D. S. Chemla and J. Zyss, eds., Nonlinear Properties of Organic Molecules and Crystals (Academic, Orlando, Fla., 1987), Vols. 1 and 2.
4. R. A. Hann and D. Bloor, eds., Organic Materials for Nonlinear Optics, Spec. Pub. 69 (Royal Society of Chemistry, Oxford, 1989).
5. P. V. Kolinsky and R. J. Jones, 'Advances in materials for nonlinear optics," GEC J. Res. Inc. Marconi Rev. 7, 46-58 (1989).
6. K. D. Singer, J. E. Sohn, L. A. King, H. M. Gordon, H. E.Katz, and C. W Dirk, "Second-order nonlinear-optical properties of donor- and acceptor-substituted aromatic compounds"J. Opt. Soc. Am. B 6, 1339-1350 (1989).
7. Ajay Nahata, Jianhui Shan, James T. Yardley, and Chengjiu Wu, "Electro-optic determination of the nonlinear-optical properties of a covalently functionalized Disperse Red 1 copolymer," J. Opt. Soc. Am. B 10, 1553-1564 (1993).
8. Becke, Axel D., "A new mixing of Hartree–Fock and local density-functional theories", J. Chem. Phys., 98(2) 1372-1377.
9. Lee, Chengteh and Yang, Weitao and Parr, Robert G., "Development of the Colle-Salvetti correlation-energy formula into a functional of the electron density", Phys. Rev. B, 37(2),785-789.
10. P. J. Stephens, F. J. Devlin, C. F. Chabalowski, and M. J. Frisch, "Ab Initio Calculation of Vibrational Absorption and Circular Dichroism Spectra Using Density Functional Force Fields", J. Phys. Chem., 98(45),11623-11627.

3.2% efficient cadmium free Cu_2ZnSnS_4/ZnO solar cells fabricated using solvothermally synthesized nanoparticles

A. G. Kannan[1], T. E. Manjulavalli[1,*], M. Thambidurai[2,3], K. Habeeba[1] and D. V. Ezhilarasi GnanaKumari[1]

[1]Department of Physics, Nallamuthu Gounder Mahalingam College, Pollachi 642001, Tamil Nadu, India
[2]Energy Research Institute @NTU (ERI@N), Research Techno Plaza, X-Frontier Block, Level 5, 50 Nanyang Drive, 637553, Singapore
[3]LUMINOUS! Centre of Excellence for Semiconductor Lighting and Displays, School of Electrical and Electronic Engineering, The Photonics Institute (TPI), Nanyang Technological University, 50 Nanyang Avenue, 639798, Singapore

E-mail: emanjhu@gmail.com

Abstract: In this paper, we report the fabrication of a hetero-junction solar cell using nanocrystalline p-type Cu_2ZnSnS_4 (CZTS) as absorber layer and ZnO as buffer layer. The effect of copper ratio on the properties of copper zinc tin sulfide nanoparticles synthesized using solvothermal method and the influence of Cu ratio on the cell efficiencies are also systematically investigated. The structural, optical and electrical properties of prepared nanoparticles were studied using X-ray powder diffraction, Raman analysis, scanning electron microscopy, UV-vis absorption and J-V characteristic studies. The device fabrication and conversion efficiency of cadmium free CZTS/ZnO solar cells are also discussed.

Keywords: Cu_2ZnSnS_4, Nanomaterials, Energy harvesting, Low-cost

1. Introduction

The non-toxic elements of Cu_2ZnSnS_4 has captivated an enormous amount of effort in the study of the semiconducting material in recent years. Direct bandgap value and high absorption coefficient in the visible range made a p-type material, kesterite CZTS, the pertinent choice of absorbers for solar cells [1, 2]. A pure CZTS solar cell was fabricated by Chang Yan et al [3] with highest power conversion efficiency (PCE) of more than 11%. Nevertheless, desirable potential is essential to enhance the efficiency of the CZTS solar cells collated with an efficiency of 32.4% theoretically and the associated solar cells [4-6]. Notwithstanding, the ideal stoichiometry of CZTS goes in agreement with the ratios of Cu/(Zn+Sn) and Zn/Sn equal to 1 [7] but only Cu-poor and Zn-rich CZTS exhibits best device efficiency [3]. Consequently, enhanced optoelectronic properties appropriate for solar cell applications can be pulled off by regulating the chemical composition of CZTS nanocrystals, as properties of CZTS chalcopyrite semiconductors strongly depend on the compositional ratios of Cu/(Zn+Sn) [4, 5]. Optimization of copper (Cu) concentration is also one of the essential parameter to get highly efficient photovoltaics. A majority of the research groups used non-stoichiometric Cu-rich and Zn-poor starting molar ratios for reporting the high photovoltaic effect of CZTS [6, 8, 9]. The data collected from the literature connoted that the predominant formation of kesterite phase occurs only under Cu-poor and Zn-rich conditions [4]. Under these circumstances, the predominant Cu_{Zn} antisite is expected to exist in excess and has the minimum energy of formation. Besides, ternary phases of Cu_2SnS_3 are inadequate or missing in Cu-poor/Zn-rich samples which are accountable for the low open circuit voltage (V_{oc}) seen in Cu-rich devices [10]. All these parameters, clearly suggest that the copper content extensively influences the p-type semiconductor and its electronic properties. On that account, the improved efficacy of CZTS solar cells demands the composition dependent studies on the properties. In our past report, we have engrossed on the influence of solvent on the properties of CZTS nanoparticles synthesized by solvothermal method. It has been demonstrated that the CZTS nanocrystals synthesized in ethylene glycol

(EG) were utilized significantly in solar cells as the absorber layer [11]. This paper presents a detailed investigation on the properties of CZTS nanoparticles by varying the ratio of Cu/(Zn+Sn) using solvothermal method with EG as solvent.

2. Experimental procedure

This method involves dissolving of 0.6 mmol of Cu $(CH_3COO)_2$, 0.5 mmol of $Zn(CH_3COO)_2.2H2O$, 0.5 mmol of $SnCl_4.5H_2O$ and 2 mmol of CH_4N_2S solutions in EG. A clear solution was found after constant stirring of the reaction mixture. The mixture is then poured in to a autoclave and the temperature was sustained at 180°C for 6 hours. The procured precipitate was filtered and rinsed with deionized water followed by further rinsing with ethanol. The final products were heated at 60°C for 3 hours in vacuum. To study the influence of ratio of copper on the properties of the CZTS nanoparticles, an analogous experiment was carried out by changing the concentration of copper (0.8, 0.9, 1.0, 1.1and 1.2 mmol).

3. Results and discussion

3.1. Structural properties of CZTS nanoparticles

XRD pattern of Cu_2ZnSnS_4 nanoparticles prepared with copper ratios of 0.7 (Cu-very poor), 0.8 (Cu-poor), 0.9 (Cu-slightly poor), 1.0 (Cu-normal), 1.1 (Cu-rich) and 1.2 (Cu-very rich) are shown in Figure 1. Altogether the samples reveal peaks corresponding to (112), (200), (220), (312) and (332) planes of kesterite structure (Card No. 26-0575).

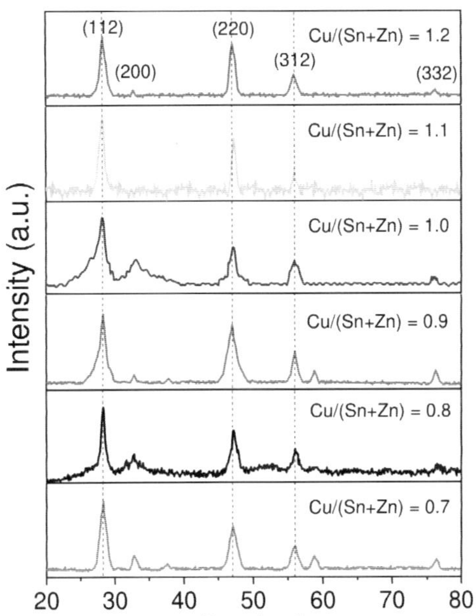

Figure 1. XRD spectra of CZTS nanoparticles with different copper ratios

It has been noted that all the nanoparticles exhibit basic chalcopyrite crystal structure, and the intensity of the predominant (1 1 2) orientation increases as the copper ratio increases except for 1.2 (Cu-very rich). Debye-Scherrer equation D=0.9□□□cos□, is used to calculate the average crystallite size [12]. The mean crystallite sizes are calculated as 9 nm, 11 nm, 8 nm, 7 nm, 10 nm and 9 nm for the prepared samples with copper ratios of 0.7, 0.8, 0.9, 1.0, 1.1 and 1.2, respectively. These results show that, a pivotal role is played by the different copper ratio in the precursor solution, in the formation of CZTS. It is noteworthy that the calculated lattice constants (a=5.432 Å and c=10.844 Å) were in agreement with the value from the standard card (a = 5.427 Å and c = 10.848 Å) substantiating the kesterite structure and complementing the earlier reports [13-15].

3.2. Raman analysis

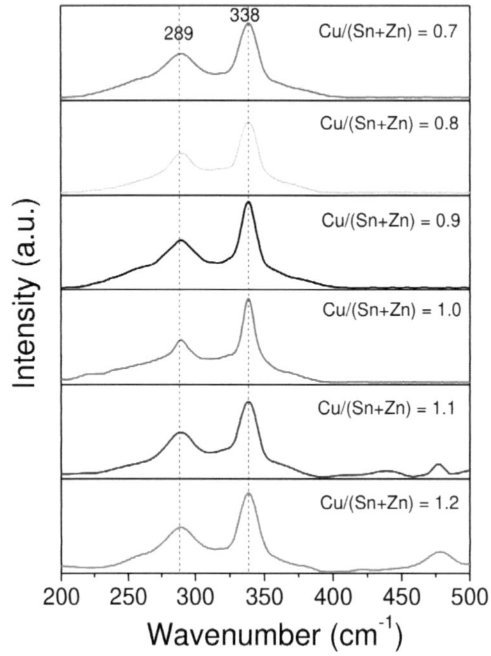

Figure 2. Micro-Raman spectra of CZTS nanoparticles with different copper ratios

Raman spectroscopy (Figure 2) was used to ascertain the existence of the secondary phase. Existence of two main peaks at 338 cm^{-1} and 289 cm^{-1} are observed for the samples altogether and it complements the reports by earlier researchers [16, 17]. Two of the peaks are in accordance with the CZTS phase. A_1 symmetry is accountable for the stronger peaks at 338 cm^{-1} and the vibration of the S atoms in CZTS is related to it [18, 19]. In CZTS lattice, the peaks at 289 cm^{-1} features the vibration of the Zn atoms and S atoms along with the contribution of Cu atoms. It is also obvious that an additional peak at 475 cm^{-1} is observed for copper ratio 1.1 (Cu-rich) and 1.2 (Cu-very rich) assigned to Cu_xS.

3.3. Morphological studies

Figure 3 shows the scanning electron microscope images of CZTS nanoparticles with copper ratios of 0.7, 0.8, 0.9, 1.0, 1.1and 1.2. From the micrographs the average nanoparticle sizes are approximately 50-100 nm. It is observed that the CZTS particles with different copper ratios have different surface morphologies. For compositional ratio of 0.7 (Cu-very poor), 0.8 (Cu-poor) and 0.9 (Cu-slightly poor), the nanoparticles exhibit non-uniform distribution and are closely packed. On the other hand, for stoichiometric ratio (Cu-normal), the nanoparticles exhibit homogenous nature with spherical shape and are found to be well separated from each other. However, when increasing the copper ratio to 1.1 (Cu-rich) and 1.2 (Cu-very rich), the nanoparticles contain agglomerated particles of larger size.

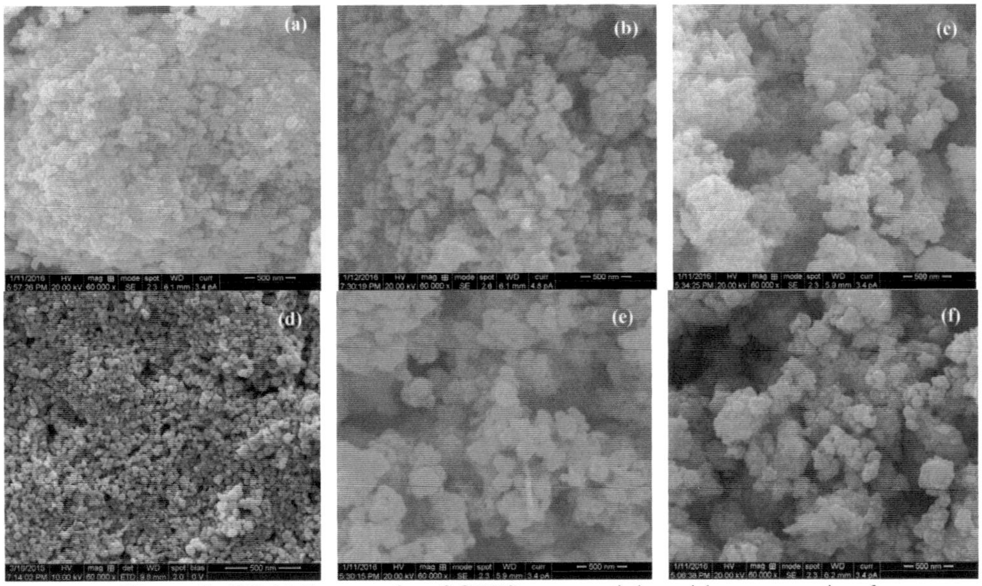

Figure 3. SEM Micrographs of CZTS nanoparticles with copper ratio of a) 0.7, b) 0.8, c) 0.9, d) 1.0, e) 1.1 and f) 1.2

3.4. EDX analysis

EDX spectra (Figure 4) confirm the existence of Cu, Zn, Sn and S for all the prepared samples. For CZTS nanocrystals decreasing the copper content from 0.7 (Cu-very poor) to 0.9 (Cu-slightly poor) shows some deviation from the expected elemental target ratio. At the compositional ratio of 1.0 (Cu-normal) the values are close to the stoichiometry, on further increase of the compositional ratio 1.1 (Cu-rich) to 1.2 (Cu-very rich) the substantial decrease in Zn and an increase in Cu has been observed. It can be inferred from the table that the optimal composition for high efficiency photovoltaic device has been achieved for the compositional ratio of 0.9 (Cu-slightly poor).

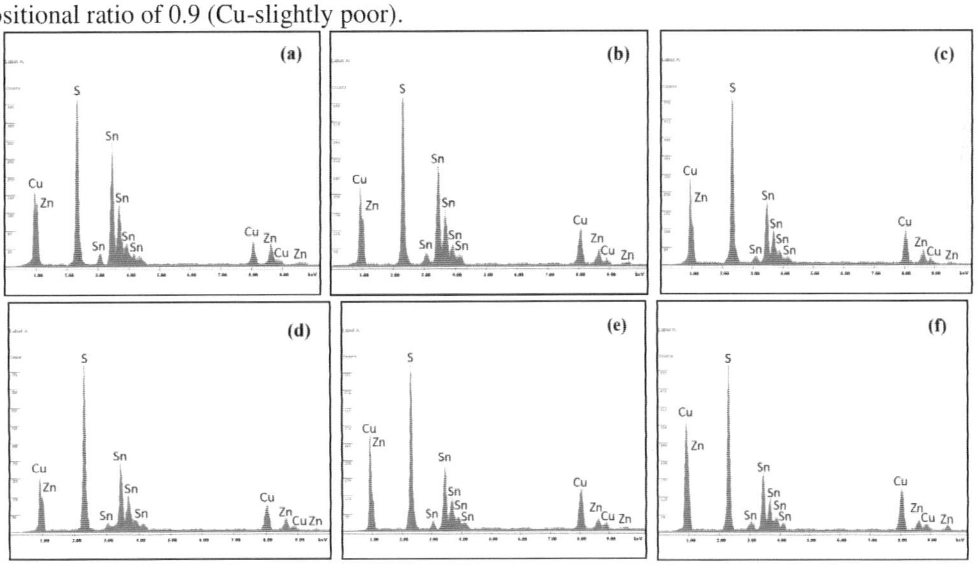

Figure 4. EDX spectra of CZTS nanoparticles with copper ratio of a) 0.7, b) 0.8, c) 0.9, d) 1.0, e) 1.1 and f) 1.2

IOP Conf. Series: Materials Science and Engineering 1219 (2022) 012036 doi:10.1088/1757-899X/1219/1/012036

Table 1. Atomic percentage of Cu, Zn, Sn and S in CZTS nanoparticles with different copper ratios

Cu/(Zn+Sn) ratio	Atomic percentage				Cu/(Zn+Sn)	Zn/Sn	S/Metal
	Cu	Zn	Sn	S			
0.7	17.99	13.12	13.86	55.03	0.67	0.95	1.22
0.8	20.11	13.48	13.82	52.59	0.74	0.97	1.11
0.9	23.16	15.23	11.78	49.83	0.82	1.19	0.99
1.0	24.85	12.37	12.72	50.06	0.99	0.97	1.00
1.1	27.01	13.65	11.80	47.54	1.06	1.16	0.91
1.2	28.19	11.35	12.77	47.69	1.17	0.89	0.91

3.5. Optical absorption and band gap

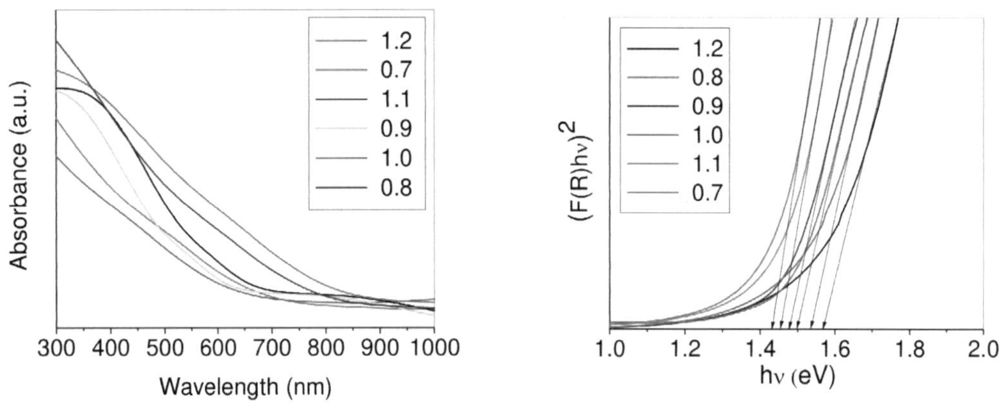

Figure 5 (a) Absorption spectra and **(b)** Tauc plot of CZTS nanoparticles with different copper ratios

The absorption spectra are recorded for the prepared samples in diffuse reflectance mode, where $BaSO_4$ is used as reference. Kubelka-Munk function is used to convert the reflectance values to absorbance [20]. The absorption spectra of prepared samples with copper ratios of 0.7, 0.8, 0.9, 1.0, 1.1and 1.2 is shown in Figure 5. A strong absorption ranging from UV to visible spectral region which is further extended to the NIR region is exhibited by the nanoparticles. Figure 5 (b) shows the band gap values of the prepared samples. the band gap values of the prepared samples are found to be 1.50, 1.43, 1.48, 1.53, 1.45 and 1.57 eV for the copper ratios of 0.7, 0.8, 0.9, 1.0, 1.1and 1.2 correspondingly, and is in accordance with the past reports [21-23]. For the compositional ratio of 0.7 to 0.8 decreases in band gap value and may be due to increase in the crystallinity or may be the existence of narrow band gap secondary phases. The raise in band gap value, when the copper ratio is increased from 0.8 to 1.2, might be due to the secondary phases. In the present work, from the Raman analysis (Section. 3.2) presence of secondary phases were observed for the Cu-rich (1.1) and Cu-very rich (1.2) sample.

3.6. Photovoltaic performances of CZTS/ZnO solar cells

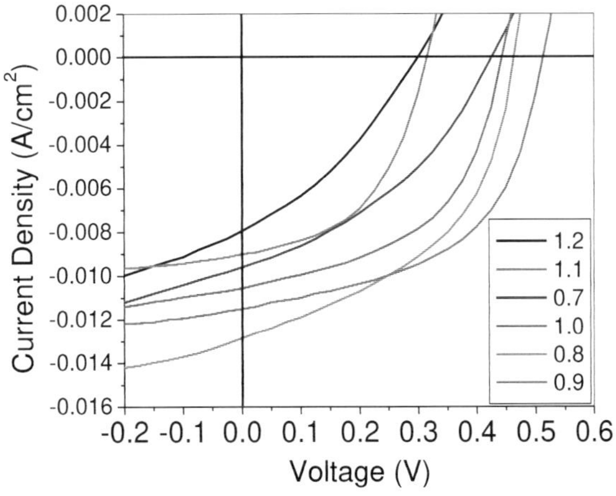

Figure 6. J-V curves of CZTS solar cells with different copper ratios

The typical current-voltage (J-V) characteristic curves for devices fabricated using CZTS absorbers with different copper compositional ratio from 0.7 (Cu-Very poor) to 1.2 (Cu-very rich) measured under AM1.5G solar illumination at 100 mW.cm^2 are shown in Figure 6. Table 2 summarizes the device parameters of the solar cell calculated [24] from J-V curves. It is observed from the table that as the compositional ratio increases from 0.7 (Cu-very poor) to 0.9 (Cu-slightly poor), there is an raise in V_{oc} (from 420 to 510 mV), J_{sc} and FF (from 38.04 to 52.97) and as a result of this, the efficiency increases from 1.55% to 3.2%. The reduced series resistance from 20.17 to 8.13 Ω.cm^2 and increased shunt resistance from 101.60 to 175.01 Ω.cm^2 resulted in the improvement in FF from 38.04 to 52.97. When the compositional ratio of copper increased from 0.9 (Cu-Slightly poor) to 1.1 (Cu-rich), FF, J_{sc} and power conversion efficiency decreases and corresponding increase in the series resistance. On further increase of the compositional ratio to 1.2 (Cu-very rich), power conversion efficiency drastically reduces and corresponding J_{sc} and FF also decreases. An enormous increase in series resistance and a decrease in shunt resistance are also observed. In the present work, presence of secondary phases for very rich compositional ratio (1.2; Cu-very rich) and a high value of band gap have been observed from the Raman studies and optical measurement.

Table 2. Device parameters of CZTS solar cells with different copper ratios

Cu/(Zn+Zn) Ratio	J_{sc} (mA cm^{-2})	V_{oc} (V)	FF (%)	PCE (%)	R_s (Ω cm^2)	R_{sh} (Ω cm^2)
0.7	09.59	0.42	38.04	1.55	20.17	101.60
0.8	12.80	0.46	47.44	2.81	6.89	104.42
0.9	11.49	0.51	52.97	3.2	8.13	175.01
1.0	10.54	0.44	50.84	2.37	9.19	154.14
1.1	9.01	0.31	48.63	1.38	10.76	142.05
1.2	7.93	0.30	32.96	0.78	26.24	60.83

4. Conclusion

CZTS nanoparticles were synthesized using solvothermal method by varying the compositional ratio of copper. From different copper ratios, the highest PCE of the device was found to be 3.2%, and the V_{oc}, J_{sc}, and FF were measured to be 510 mV, 11.49 mA/cm^2, and 52.97%, respectively for compositional ratio of 0.9 (Cu-slightly poor). 3.2% is the maximum PCE reported so far for Cu_2ZnSnS_4 solar cells fabricated by doctor bladed absorber and drop casted ZnO buffer layer prepared using solvothermal method nanoparticles. This study demonstrates a simple and environmentally benign approach towards the production of CZTS/ZnO solar cell device with efficacy of 3.2 %, which will open a new avenue for establishing low-cost high performance solar cells in near future.

References

[1] Tong Z, Liu F, Jiang L and Lai Y 2019 Mater. Lett. 254 50
[2] Gezgin S Y and Kilic H S 2020 Opt. Mater. 101 109760-1
[3] Yan C, Huang J, Sun K, Johnston S, Zhang Y, Sun H, Pu A, He M, Liu F, Eder K, Yang L, Cairney J M, Ekins-Daukes N J, Hameiri Z, Stride J A, Chen S, Green M A, and Hao X 2018 Nat. Energy 3 764
[4] Mukesh Kumar, Ashish Dubey, Nirmal Adhikari, Swaminathan Venkatesan and Qiquan Qiao 2015 Energy Environ. Sci. 8 3134
[5] Nessrin Kattan, Bo Hou, David J. Fermín and David Cherns 2015 Appl. Mater. Today 1(1) 52
[6] Daniel Vaccarello, Amy Tapley and Zhifeng Ding 2013 RSC Adv. 3 3512
[7] Mou Pal, Mathews N R, Silva Gonzalez R and Mathew X 2013 Thin Solid Films 535 78
[8] Jie Zhong, Zhe Xia, Miao Luo, Juan Zhao, Jie Chen, Liang Wang, Xinsheng Liu, Ding-Jiang Xue, Yi-Bing Cheng, Haisheng Song and Jiang Tang 2014 Sci. Rep. 4 6288
[9] Daniel Vaccarello, Lijia Liu, Jigang Zhou, Tsun-Kong Sham and Zhifeng Ding 2015 J. Phys. Chem. C 119(21) 11922
[10] Giovanni Altamura and Julien Vidal 2016 Chem. Mater. 28(11) 3540
[11] Kannan A G, Manjulavalli T E and Chandrasekaran J 2016 Procedia Eng. 141 15
[12] Holzwarth U and Gibson N 2011 Nature Nanotech. 6 534
[13] Wen-Hui Zhou, Yan-Li Zhou, Jun Feng, Ji-Wei Zhang, Si-Xin Wu, Xiu-Chun Guo and Xuan Cao 2012 Chem. Phys. Lett. 546 115
[14] Prabhakar T and Jampana N 2011 Sol. Energy Mater. Sol. Cells 95 1001
[15] Mali S S, Shinde P S, Betty C A, Bhosale P N, Woo Oh Y and Patil P S 2012 J. Phys. Chem. Solids 73 735
[16] Sousa M G, da Cunha A F, Fernandes P A, Teixeira J P, Sousa R A and Leitão J P 2014 Sol. Energy Mater. Sol. Cells 126 101
[17] Tiwari D, Chaudhuri T K, Rayb A and Tiwaric K D 2014 Thin Solid Films 551 42
[18] Kheraj V, Patel K K, Patel S J and Shah D V 2013 J. Cryst. Growth 362 174
[19] Dalapati G K, Batabyal S K, Masudy-Panah S, Su Z, Kushwaha A, Wong T I, Liu H F, Bhat T, Iskander A, Lim Y, Wong L H, Tripathy S and Chi D 2015 Mater. Lett. 160 45
[20] Michael L Myrick, Michael N Simcock, Megan Baranowski, Heather Brooke, Stephen L Morgan and Jessica N McCutcheon 2011 Appl. Spectrosc. Rev. 46:2 140
[21] Sarkar S, Bhattacharjee K, Dasb G C and Chattopadhyay K K 2014 Cryst. Eng. Comm. 16 2634
[22] Yang H R, Jauregui L A, Zhang G, Chen Y P and Wu Y 2012 Nano Lett. 12 540
[23] Lin X, Kavalakkatt J, Kornhuber K, Levcenko S, Lux-Steiner M C and Ennaoui A 2013 Thin Solid Films 535 10
[24] Thambidurai M, Muthukumarasamy N, Velauthapillai D and Lee C 2013 J. Mater. Sci: Mater. Electron. 24 1921

Growth and Gd doping of ZnO nanostructures with enhanced structural, optical properties and photocatalytic applications

Debasrita Dash and Dojalisa Sahu*

School of Applied Sciences, Centurion University of Technology and Management, Odisha, India

E-mail: dojalisa.sahu@cutm.ac.in

Abstract
In this study, pure and Gadolinium doped ZnO (Gd-ZnO) nanoparticles have been synthesized by wet chemical method and the effect of Gadolinium incorporation on the structure and optical activity of ZnO was examined. The average size of the undoped ZnO and Gd-ZnO (0.002 M and 0.003 M) nanoparticles was found to be 48 nm, 44 nm and 42 nm respectively which shows the lattice compression of ZnO after Gd doping. TEM study shows the hexagonal nanocrystallites of Gd-ZnO with size around 45 nm is formed. FTIR study shows the metal-oxygen band below 600 cm^{-1} and absorbance study gives the information about the narrowing of band gap of ZnO after Gd doping. The photocatalytic degradation efficiency of Gd-ZnO (92 %) has been found to be better as compared to undoped ZnO (89 %) in degrading methyl blue (MB). The photodegradation of MB is found to follow first order kinetics.

Keywords: Gd-ZnO, XRD, FESEM, Photocatalysis, Methylene blue

1. Introduction
Semiconductor nanomaterials are considered to be novel functional material and possess interesting optical, electrical and magnetic properties which find application in next generation optoelectronic and spintronic devices. Wide band gap semiconductors like ZnO, SnO_2 and GaN are extensively studied and drawn considerable attention of research community owing to their technological importance and versatility [1-2]. Zinc oxide has the richest morphology and enthralling properties like large exciton binding energy (60 meV), wide band gap (3.37 eV) outstanding chemical and thermal stabilities. These are some of the captivating attributes of ZnO that largely influences the market value for the fabrication of ZnO based devices for different optoelectronic applications [3]. The incorporation of suitable and selective dopant sources substantially alters the above noted properties of ZnO with its size changing to nanometer regime [4]. Presently, intense research has been carried out in the above area and doping of elements especially from the rare earth family into ZnO has resulted multidimensional properties like photoluminescence, antibacterial activity and photocatalysis. A number of research studies reveal that the rare earth elements with 4f–5d and 4f–4f electronic transitions can be excellent additives to metal oxide semiconductors like ZnO as compared to other elements [5–7]. In addition, the above elements are well known anti-tumor, anti-inflammatory, anticoagulant, antibacterial and anti-cancer agents and a modulated response from these elements can be obtained if incorporated into ZnO host [8]. The lanthanide ions such as; Pr^{3+}, Er^{3+}, Sm^{3+}, Eu^{3+}, Ce^{3+}, Tb^{3+}, Dy^{3+}, Nd^{3+}, Yb^{3+} and Tm^{3+} are utilized mostly to obtain highly attractive properties of ZnO [9]. Advanced and improved optical, magnetic and catalytic properties may be obtained with rare-earth (RE) doping into metal oxide semiconductors. Structural changes in ZnO nanocrystal is expected as the distribution of rare-earth dopants takes place on ZnO surface or in core lattice. Gadolinium (Gd^{3+}) is treated as a key RE element having very good scintillation properties which is helpful in making phosphor films and luminescent materials [10-14]. The energy levels of Gadolinium are positioned closed to the conduction band of ZnO leading to unique change in energy distribution with the host like semiconducting ZnO [11]. The structural and optical properties of Gd doped ZnO has been

studied by Mithal *et al.* for optoelectronic application [15]. The luminescence and spectroscopic properties of ZnO doped with Gd ions has been investigated by Reddy *et al.* demonstrating phosphor application [16]. Dilute magnetic semiconductor (DMS) properties of ZnO doped Gd nanorods have been studied by Obeid et al. and Isik et al. studied the thermoluminescence properties of Gd-ZnO [17, 18]. But, the photocatalytic activity of Gd doped with ZnO is not studied much to unveil this property. So, it is very interesting to study the effect of Gd doping in altering the structural properties and photocatalytic activity of ZnO which has been the prime focus of this study.

2. Materials and method

In present work, pure and Gd doped ZnO were prepared using co-precipitation method. Zinc acetate dihydrate [Zn (CH3COO)2·2H2O], Gadolinium nitrate, ammonium oxalate and ethanol were used without further purification. 2.5 gm of zinc acetate dihydrate was dissolved in 10 ml of DDW with 10 ml of ammonium oxalate and a magnetic stirrer was deployed to form a clear solution. The mixture was stirred continuously while heating at 100 °C for 4 hours. Similarly, zinc acetate (0.2 M) and gadolinium nitrate (0.002 M) was dissolved in 20 ml of DDW and ammonium oxalate with a pH of about 8.5 to 11. The above process was followed taking 0.003 M gadolinium nitrate. The prepared particles of pure and doped ZnO were further kept in an oven at 400 ^{0}C for 4 h and collected for characterization.

We used a X-Pert Pro diffractometer (M/s Panalytical) for XRD measurement, SMT AG (Carl Zeiss) and TechnaiG^2 (FEI) electron microscopes to study the surface morphology and microstructure respectively. The absorption spectra were recorded by a Shimadzu UV-1700 spectrophotometer and a Perkin-Elmer spectrometer (Spectrum GX) helped to record the FTIR spectra. The photocatalytic experiment was performed with the prepared catalysts (ZnO and G-ZnO) by the same procedure as discussed in our earlier report [19].

3. Results and Discussion

X-ray diffraction (XRD) study gives an idea to analyze the structural parameters by identifying different crystallographic planes in the spectra. Figure-1 shows the XRD spectra of pristine ZnO and Gd doped ZnO (G-ZnO) samples. Clear and sharp peaks corresponding to hexagonal wurtzite structure of ZnO have been observed in the above XRD pattern. Highly crystalline nature of the samples is also concluded from the nature of the spectra. Standard JCPDS pattern (card no-00-005-0664) was used to compare the XRD patterns and accordingly the peaks are indexed. In the XRD spectra of doped samples, no signature of impurity peaks or peaks related to Gd clusters are observed which hits that the quantity of dopants is less than the solubility limit [20]. But, the possibility of formation such phases cannot be ruled out which may be very small enough to be detected by XRD measurement [17].

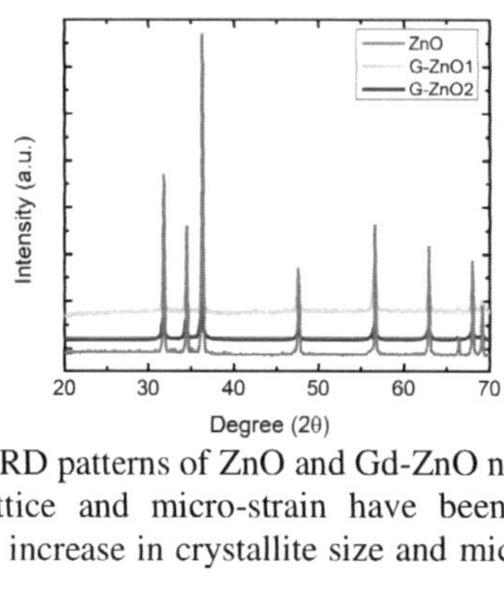

Figure-1: XRD patterns of ZnO and Gd-ZnO nanoparticles.

The average crystallite size lattice and micro-strain have been calculated by using Scherrer's equations. Gd doping leads to an increase in crystallite size and microstrain as per the result obtained from XRD data. The average crystallite size for pristine ZnO is 48 nm but with Gd doping it increases to 44 nm for G-ZnO1 and 42 nm for G-ZnO2 respectively. This indicates that insertion of Gd ions leads to the compression of ZnO lattice with more lattice strain. Distortion of ZnO lattice structure by Gd^{3+} with higher ionic radius (0.093 nm) compared to Zn^{2+} (0.074 nm) leads to check the nucleation process and subsequent growth of ZnO. This results to such reduction in crystallite size of ZnO nanocrystallites with Gd doping. Similar observation has been reported by other groups studying the effect of Gd doping on the structural modification incurred in ZnO [17, 21]. The morphology of the doped sample (G-ZnO2) has been studied by FESEM and illustrated in Figure-2a. Uniformity in particle distribution has been observed as well as the hexagonal morphology. Further study on the microstructure assessment of the same sample by TEM gives a clearer picture on the size and distribution of the nanoparticles. The TEM image (Figure-2b) depicts the formation hexagonal nanoparticles of size ranging from 20 – 70 nm with an average size of 45 nm which closely matches with XRD data.

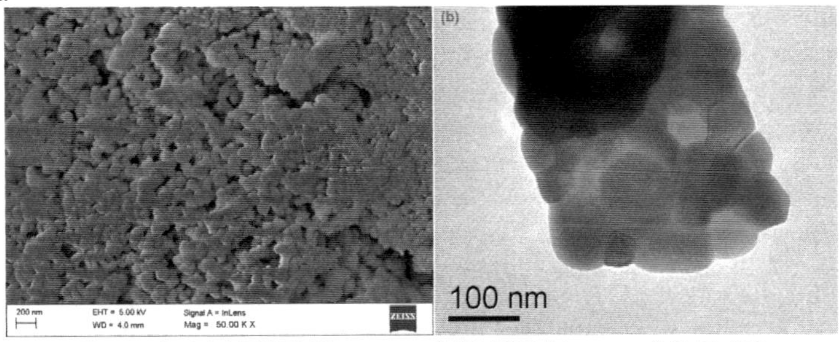

Figure-2: (a) FESEM image and (b) TEM image of G-ZnO2.

The FTIR spectra represent the presence of number of functional groups along with the metal –oxide stretching and bending modes. The FTIR spectra of pristine ZnO, Gd_2O_3 along with the doped samples have been illustrated in Figure-3a. The FTIR spectra in the region of significance i.e. 2000 – 400 cm^{-1} are being highlighted. The FTIR band around 1634 cm^{-1} is linked with the molecular adsorption of H_2O on the ZnO surface [17, 18]. The peaks around 2929 cm^{-1} and 2850 cm^{-1} are ascribed to the symmetric and asymmetric stretching vibrations of CH_2 [22, 23]. The carboxylic groups of acetate attached to ZnO with their symmetrical and asymmetrical stretching modes have appeared in the form of a doublet around 1500 cm^{-1} and 1400 cm^{-1} [23]. The region below 700 cm^{-1} generally covers the metal-oxygen (Zn–O and Gd–O) bonds [24]. For the doped samples, new bands corresponding to Gd-O modes have observed at 464 cm^{-1}.

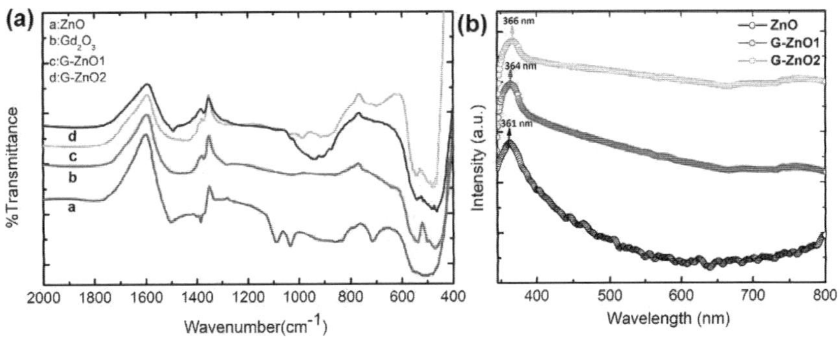

Figure-3: (a) FTIR spectra, (b) UV-VIS spectra of ZnO and Gd-ZnO nanoparticles.

The optical absorption properties of pristine and doped ZnO nanoparticles were examined by UV–Vis absorption spectroscopy and shown in Figure-3b. The synthesized nanoparticles were dispersed in acetone prior to the above measurement. All the three samples depict a robust UV absorption band in the region of 360 – 370 nm and the absorption band gets red shifted with Gd doping. The band for ZnO is observed around 361 nm which switches to 364 nm and 366 nm for G-

IOP Conf. Series: Materials Science and Engineering 1219 (2022) 012037 doi:10.1088/1757-899X/1219/1/012037

ZnO1 and G-ZnO2 respectively. This intends a compression of ZnO band gap with the incorporation of gadolinium ions. The modification in the electronic structure of ZnO with impurity related defect centres upon Gd doping may be the origin of band gap modification in ZnO [25].

The photocatalytic (PC) measurement was carried out with a purpose to degrade the neurotoxic hazardous dye methylene blue. ZnO and Gd doped photocatalysts were used for the above experiment. The results have been depicted in Figure-4. It is clearly observed that the rate of photocatalytic degradation of Gd doped ZnO catalyst is much higher when compared with that of undoped ZnO. It is interesting to note that the degradation rate of the catalysts increases with increase concentration of photocatalyst and then decreases. So, 10 mg of photocatalyst may be more efficient for separating photo induced electron-hole pairs and show enhanced photocatalytic activity because of small particle size. The improvement in the photocatalytic activity of Gd doped ZnO is due to the formation of Zn–O–Gd bonds formed on the surface of ZnO [26]. The reason for the high activity of the doped samples and the result of Gd loading on the PC property can be explained by the following mechanism. Upon UV irradiation on the photocatalyst, Gd^{+3} behaves as an electron scavenger and prevents the electron-hole recombination by reacting with the superoxide species and, thus increases photodegradation efficiency. With the increase in Gd^{+3} concentrations, the surface barrier becomes higher by narrowing the space charge region, and therefore separating the electron-hole pairs efficiently by the large electric field [27]. Hence, 0.002 M Gd doping concentration may be considered as ideal amount for good photocatalytic activity as depicted in Figure-4. Figure-4a shows how the absorbance decreases significantly with irradiation time. The degradation rate was calculated and found that almost 92 % degradation was recorded for Gd doped sample which is 89 % for pristine ZnO as depicted in Figure-4b. The reaction kinetics for the photocatalytic degradation of methylene blue follows first order with rate constant (k) obtained as 0.1001 and correlation constant (R^2) as 0.9668. The rate constant was found to be 0.021 min^{-1} for doped sample which is higher than the undoped ZnO (0.00942 min^{-1}).

During the photocatalytic process, the valence band electrons of ZnO crystal lattice are excited to reach the conduction band upon the exposure of light. The generation of hydroxyl radicals by the holes in the valence band leads to the degradation of MB as these radicals react with water [26]. Gd^{2+} ions are created as the reaction of Gd^{3+} ions take place with the electrons in the conduction band of ZnO. Super oxide radicals are created when the Gd^{2+} ions are diffused into the atmospheric oxygen and react with the water molecules to produce H_2O_2 which further leads to the formation of hydroxyl radicals. The dye molecules are degraded upon interacting with these radicals [19, 25, 26]. The overall process leads to the trapping of electrons and the recombination of photo-induced charge carriers is vanquished. To estimate the stability of the photocatalyst, we performed the recycle test in four successive sessions under identical environmental condition. The Gd doped ZnO photocatalysts show ≈ 80 % efficiency after four cycles which enhances its practical application.

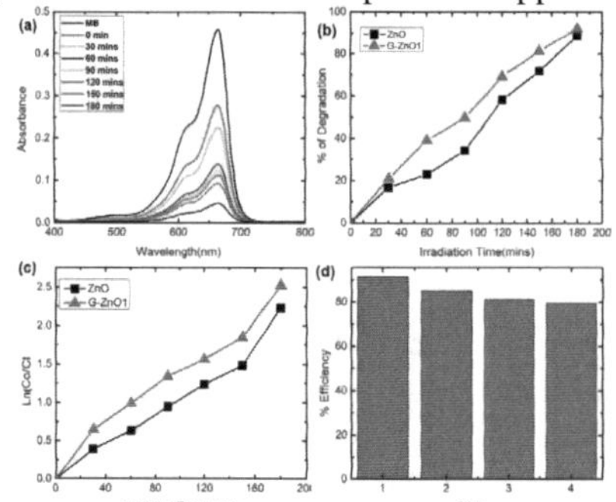

Figure-4: (a) Absorption spectra of MB dye with different exposure time, (b) variation of degradation (%) with irradiation time, (c) the plot of reaction kinetics for MB degradation and (d) efficiency of Gd doped ZnO photocatalysts in different cycles.

Conclusion

Pure ZnO and Gd doped ZnO nanoparticles were prepared by wet chemical method. The crystallite size of ZnO decreases with Gd doping as observed from XRD results. Uniform distribution of hexagonal nanocrystallites of Gd doped ZnO is observed from FESEM and TEM study. FTIR spectra indicate the formation new bands linked with Gd-O in the doped samples. UV–Vis spectroscopy exhibits a considerable decrease in band gap after doping of Gd ions into ZnO nanoparticles. PC measurement shows that the Gd doped ZnO sample demonstrates better degradation efficiency of MB as compared to pure ZnO. Doping of Gd ions leading to the suppression of recombination of photogenerated carriers leads to such enhanced photodegradation efficiency. The prepared photocatalysts are also stable after fourth cycle of degradation hence useful in their practical application.

References:

[1] Kaur G, Mitra A, Yadav K L 2015 Adv. Mater. Lett. 6 73–79.
[2] Rekha K, Nirmala M, Nair M G, Anukaliani A 2010 Phys. B 405 3180–3185.
[3] Wagh M S, Jain G H, Patil D R, Patil S A, Patil L A 2006 Sens. Actuat. B Chem. 115 128–133.
[4] Zargar R A, Chackrabarti S, Shahabuddin M, Kumar J, Hafiz A K 2015 J. Mater. Sci. Mater. Electron. 26 10027–10033.
[5] Lucovsky G, Phillips J C 2005 Thin Solid Films 486 200–204.
[6] Frindell K L, Bartl M H, Matthew R, Bazan G C, Popitsch A 2003 J. Solid State Chem. 172 81–88.
[7] Carreno N V, Fajardo H V, Maciel A P, Valentini A, Pontes F M, Probst L F D, Leite E R, Longo E 2004 J. Mol. Catal. A: Chem. 207 89–94.
[8] Subash B, Senthilraja A, Dhatshanamurthi P, Swaminathan M and Shanthi M 2015 Spectrochimica Acta Part A: Molecular and Biomolecular Spectroscopy 115175-182.
[9] Reisfeld R, Gaft M, Boulon G, Panczer C and Jorgensen C K 1996 J. Lumin. 69 343–353.
[10] Ney V, Ye S, Kammermeier T, Ollefs K, Wilhelm F, Rogalev A, Lebegue S, Rosa A L D and Ney A, 2012 Phys. Rev. B 85 235203.
[11] Mithal D and Kundu T, 2013 Asian J. Chem. 25 S12–16.
[12] Murmu P P, Mendelsberg R J, Kennedy J, Carder D A, Ruck B J, Markwitz A, Reeves R J, Malar P and Osipowicz T, 2011 J. Appl. Phys. 110 033534.
[13] Maa X and Wang Z 2012 Mater. Sci. Semicond. Process. 15 227–231.
[14] Aggarwal N, Kaur K, Vasishth A and Verma N K, 2016 J. Mater. Sci., Mater. Electron. 27 13006–11.
[15] Mithal D and Kundu T 2017 Solid State Sci. 68 47–54.
[16] Reddy G K, Reddy A J, Krishna R H, Nagabhushana B M and Gopal G R, 2017 J. Asian Ceram. Soc. 5 350–356.
[17] Obeid M M, Jappor H R, Al-Marzoki K, Al-Hydary I A, Edrees S J and Shukur M M 2019 RSC Adv. 9 33207-33221.
[18] *Isik M* and *Gasanly N M* 2019 ***J. Lumin*. *207*** 220-225.
[19] Dash D, Panda N R and Sahu D 2021 Nano Express 2 010007.
[20] Dakhel A and El-Hilo M 2010 J. Appl. Phys. 107 123905.
[21] *Vijayaprasath G, Murugan R, Hayakawa Y, Ravi G*, 2016 ***J. Lumin*. *178*** 375-383
[22] Coates J 2000 Interpretation of infrared spectra, a practical approach Encyclopaedia of Analytical Chemistry ed. Meyers R A (New York: Wiley) pp 10815–37
[23] Jabri S, Souissi H, Sallet V, Lusson A, Meftah A, Galtier P and Oueslati M 2017 J. Raman Spectrosc. 46 4690–4694.
[24] Sahu D, Acharya B S, Bag B P, Basanta T and Gartia R K 2010 J. Lumin. 130 1371.
[25] Dash D, Pand N R, Sahu D 2019 Appl. Surf. Sci. 494 666–674.
[26] *Selvaraj S, Krishna Mohan M, Navaneethan M, Ponnusamy S, Muthamizhchelvan* C 2019

Mat. Sci. Semicon. Proc. 103 104622.
[27] Liang C H, Li F B, Liu C S, Lü J L, Wang X G 2008 ***Dyes and Pigments*** 76 477-484.

IOP Conf. Series: Materials Science and Engineering 1219 (2022) 012038 doi:10.1088/1757-899X/1219/1/012038

Synthesis and characterization of mesoporous SiO_2 nanoparticles for bio medical applications

VigneshKrishnan[1], M Suresh, T Kalaivani

[1, 2]Departmsent of Physics and Nanotechnology, SRM Institute of Science & Technology, Katankulathur, Tamil Nadu, India

E.Mail: kalaivat@srmist.edu.in

Abstarct. In the present work, synthesis and characterization of mesoporous silica nanoparticles (MSNs)were examined. In this work, the synthesis of silica by co-precipitate process were used for synthesis fromPrecursor of silica tetraethyl orthosilicate (TEOS),NH_4OH, surfactant (CTAB). Silica dioxide were confirmed byX-Ray Diffraction (XRD) analysis and Fourier Transform Infra-Red (FTIR) analysis. The MesoporesSizes and N2 adsorption/desorption are carried out by BET analysis.

Key words: silica dioxide,mesoporous, BET

1. Introduction

In recent centuries bio medical applications of nanomaterialshas inspiredin the Targeted drug delivery (TDD). MSNs is the greatest drug carrier in TDD system. Drug loading is the main challenge in TDD but when mesoporous nanoparticle used as a carrier, we can load extra drugs due to its porosity.MSNs hasmany noble properties. The mesoporous silicon dioxideis novel at TDD. MNS's can be allow in biomedical applications because of mesoporessize and huge porevolume in pore.GenerallyMSN'swill have controlled pore size and volume, this leads to allowmore drugs to load [1]. There are variousoxides are using in bio medical application like CaO, TiO_2,and SiO_2etc. Silica Np'sare used in several fields like catalysis, andTDD. MSNs are biocompatibility and nontoxic. SiO_2isnoble drug carrier due to its unique mesoporous structure.[2]. in our environment SiO_2 spreads universally in the sand[3]. Thegenerally used process to prepare silica nanoparticles are co-precipitationand wet chemical process [4]. The nanostructures of silica's preparation isthe cost effect is low and easy to prepare on experimental concepts in a co-precipitation process [5]. The co-precipitation process is commonly used to prepare pure silica Particles because of is to its capacity to control particle size, Metaphor by regulatory surveillance Reaction parameters. [4]In this work we have prepared mesoporous SiO_2in two different annealing temperature.

2. MATERIALS:

Tetraethyl orthosilicate(TEOS), EtOH,cetyltrimethylammonium bromide (CTAB), Ethanol,NH_4OH Were bought from SRL chemicals, India with AR grade (>99%) and used without additional purification

SYNTHEIS PROCEDURE OF MESOPHOROUS SiO_2:

4 ml Ethanol and 2ml ammonia solution mixed in 100 ml DI water. Then 0.5 CTAB were mixed in to the solution. The solution stirred at 400 rpm in room temperature for 15 minutes to dissolve the CTAB. Aster dissolved the CTAB 0.5 ml TEOS added drop by drop. The solution slowly changed to milky color within 10 minutes. Then the solution continuously stirred in 4 hours in room temperature. Finally the solution centrifuged with water and ethanol several times and let the nanoparticles dry at room temperature overnight. The obtained nanoparticle calcined at 500° and 1000^oC in the muffle furnace.

3. result and discusion:

XRD(X-Ray Diffraction) analysis:
XRD image of SiO_2 are shown in fig (1). The amorphous peak shows the in the SiO_2nanoparticles. This shows that the annealingtemperatures up to 1000^0 C have a nonoteworthyinfluenceon a SiO_2nanoparticles particles [5]. Broad Peak 23° is typical peak of SiO_2.

FTIRanalysis:
Fig (2) shows FT-IR spectrum of amorphous SiO_2.The strong and broad IR band in 1078 cm^{-1} is generally reserved for TO andLO systems of Si-O-Si asymmetric stretch vibrations. The IR band at 962 cm^{-1} can be designatedto silanol groups. The IR band at 793 cm^{-1} is assigned to the Si-O-Si Symmetric Extension vibrations [6]. The strongest peaks in 548 cm-1 this is different frequency. These frequency are allocated to the mixed Si-O-Si and O-Si-O bending system [7].

BET (Brunauer-Emmett-Teller) analysis:
Fig (3) summarizes the adsorption and desorption isothermusing N2 adsorption/desorption measurements. Pore diameter, volume and surface area of the mesoporous SiO_2 nanoparticles are illustrated in table 1. The pore size has been measured by BJH pore distribution. The isotherm exhibits Type IV isotherm and the average pore size is2.4 nm for S1 and 2.1 for S2and it has a high surface area about56.0 m²/gfor S1, 72.3 m²/gfor S2 andpore volume to be 0.1 cc/gfor S1, 1.0 for S2. The pore size indicates that it is in the range of mesoporous in between 2nm and 50nm. From BJH) distribution, it shows that the CTAB enhances the pore volume, surface area and pore size, in whirl which it is worthy for loading the drug for TDD systems.

Conclusion
Mesoporous SiO_2 nanoparticles were prepared by co-precipitate process, which have high pore volume. From this it has been summarized, MSNs is good for drug loading process. In future this nanoparticles leads to load drug.

References:
[1] Si-Han Wu, Chung-Yuan Moua and Hong-Ping Lin Synthesis of mesoporous silica nanoparticles. Chemical Society Reviews, RSC publication, 2012

[2] A. Lodha, M. Lodha , A. Patel1, J. Chausdhuri, J. Dalal1, M. Edwards, D. Douroumis , Synthesis of mesoporous silica nanoparticles and drug loading of poorly water soluble drug cyclosporinAJournal of Pharmacy and Bioallied Sciences , 2012.

[3] Dharani Das, Yong Yang, Julie S. O'Brien, DaliborBreznan, SurendraNimesh, StéphaneBernatchez, Myriam Hill, AbdelhamidSayari, Renaud Vincent, and PremKumarathasan , Synthesis and Physicochemical Characterization of Mesoporous SiO2 Nanoparticles , Journal of Nanomaterials , 2014 .

[4] Ismail AbRahman and VejayakumaranPadavettan ,Synthesis of Silica Nanoparticles by Sol-Gel: Size-Dependent Properties, SurfaceModification, and Applications in Silica-Polymer Nanocomposites ,Journal of Nanomaterials , 2012 .

[5] Olfat M. Sadek, Safenaz M. Reda, Reem K. Al-Bilali ,Preparation and Characterization of Silica and Clay-Silica Core-Shell Nanoparticles Using Sol-Gel Method , Advances in Nanoparticles, 2013.

[6] S. Musić, N. Filipović-Vinceković L. Sekovanić , Precipitation of amorphous SiO_2

particles and their properties , Brazilian Journal of Chemical Engineering , 2011.

[7] Ellis R. Lippincott Alvin Van Valkenburg, Charles E. Weir, and Elmer N. Bunting , Infrared Studies on Polymorphs of Silicon Dioxide and Germanium Dioxide , Journal of Research of the National Bureau of Standards , 1958.

Figure Legends:

Figure 1.XRD of mesoporousSiO_2 (S1-500° C, S2-1000°C)
Figure 2.FTIR of mesoporous SiO_2
Figure 3. Adsorption isotherms for SiO_2

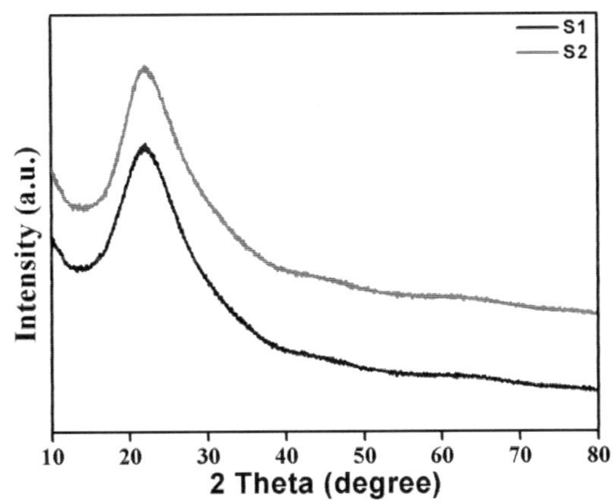

Figure 1.X-ray diffraction patterns for SiO_2

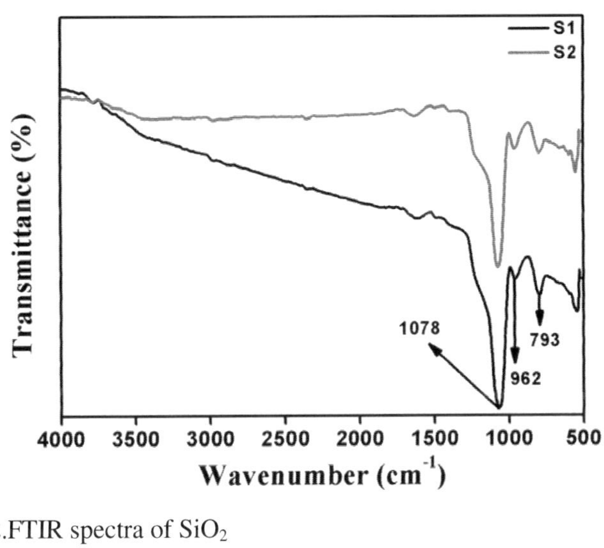

Figure 2.FTIR spectra of SiO_2

IOP Conf. Series: Materials Science and Engineering 1219 (2022) 012038 doi:10.1088/1757-899X/1219/1/012038

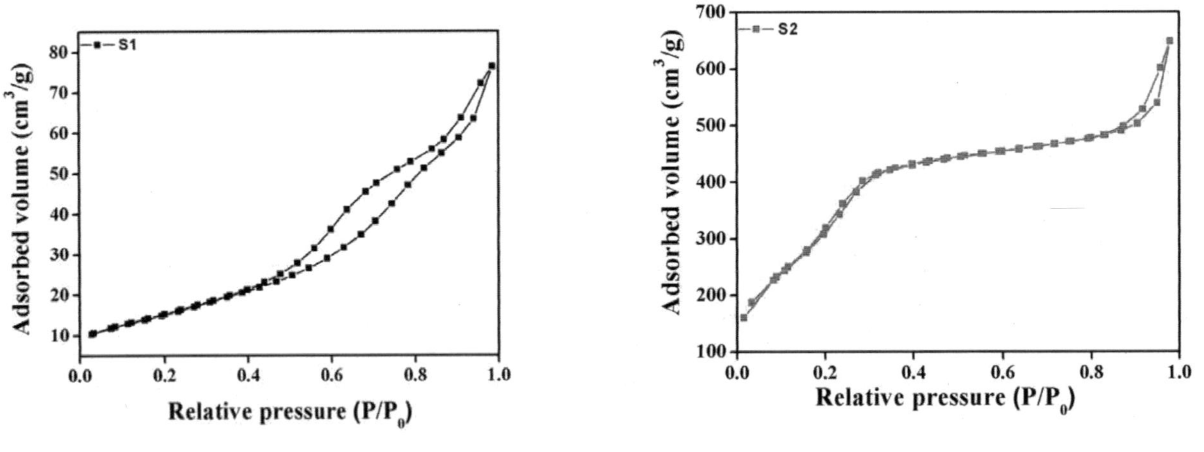

Fig. 3Adsorption isotherms for SiO_2

TABLE 1

SAMPLE	Pore Volume cc/g	pore diameter nm	Surface Area m²/g
S1	0.1	2.4	56.0
S2	1.0	2.1	72.3

Process Variation analysis of 1-bit ALU based on GAA-CNTFET

Pragya Sharma[1*], Singh Rohitkumar Shailendra[2, 3], Dr.M. Aarthy[1], Dr. Hidenori Mimura[3]

[1]Department of Nano & Micro Electronics, SENSE, VIT, Vellore 632014, TN, India
[2]Graduate School of Science and Technology, Shizuoka University, Hamamatsu, Japan- 432-8018
[3]Research Institute of Electronics, Shizuoka University, Hamamatsu, Japan-432- 8018

[1]Email: pragyasharma1696@gmail.com

Abstract. This paper is based on a technique of Carbon Nanotube Field-Effect Transistors (CNTFET) for Arithmetic Logical Unit (ALU) using 10nm technology. For the past few decades, global demand for smaller dimension transistors with ultra-speed and low power consumption processor is a challenge for device designers. According to Moore's law, the number of transistors on an integrated circuit (IC) doubles every two years. Traditional Si transistors, scaled down to its limit sub 10nm, suffer from short channel effects, direct tunneling from source to drain, drain induced barrier lowering, and hot carrier degradation. Hence device designer is looking for nano devices which can replace traditional Si. Carbon Nanotube Field Effect Transistors (CNTFET) are promising Nano device for the next generation. A parametric analysis has been performed for various measurements such as delay, power, and PDP by varying chirality vector. Here we discuss four Different configurations such as Single channel Single chirality (SCSC), Single Channel Dual chirality (SCDC), Double Channel Single Chirality (DCSC), Double Channel Dual chirality (DCDC). Based on the results it was observed that as the chirality vector increases delay decreases. It is also observed that the delay in Double channel configuration is less than in Single-channel configuration.
Keywords CNTFET, Chirality, CNT, ALU, GAA-CNTFET

1. Introduction

With the advancement of technology accompanied by rapid innovation in the semiconductor industry, the economics of device scaling remain lucrative due to exhaustive demand in the consumer market across the sectors, be it automotive, avionics, PCs or mobile devices. Device miniaturization has become a custom in today's IC Industry due to complex and extensive functionality integration on a chip at higher performance, with lower power per function, and at lower cost per circuit with an improved reliability. To enable efficient 10nm and beyond process node development and to overcome architectural limitations due to increase in 3D complexity, CNTFET could be an alternative with excellent performance across all the metrics of measurement, be it power, speed or area [1]. Over the last 30 years (1970-2000), designers tried to reduce the oxide width and channel length of conventional silicon transistors without changing the material used in fabrication. In 2001, it was found that traditional silicon transistors scaled down to its limit sub10nm suffered from high leakage current, mobility degradation, and threshold voltage instability [2][3]. The following strategies can be accepted to address these issues, first change in material which replace SiO_2 having (K =3.9) with HfO_2 (K =16) that will help to reduce the area without changing the thickness of oxide and second change in geometry whereas the second is the change in geometry through one can shift to a nanoscale regime such as FINFET, CNTFET, SET, FD-SOI, PD-SOI.
All transistors have different geometrical properties. Transistor such as FINFET is being used at INTEL, SAMSUNG. IBM is shifting to CNTFET and FD-SOI is used by ST-Microelectronics. Among all the alternatives of CMOS, CNTFET is the most promising nanoscale regime (beyond 32nm) due to its unique electrical properties such as high thermal conductivity, high tensile strength, high mobility, flexible threshold voltage, superior threshold slope, high current density, strong covalent bond, high I_{on}/I_{off} ratio [4][5].
ALU is considered to be the core of any a computer. The efficiency and performance of computer can be enhanced by scaling down the process node of ALU. CNTFET provides user defined chirality which allows circuit to scale down to 32nm and provide a wide range of diameter from 0.2nm to 2nm and because of this unique property the article proposes compact and compatible model, 1- Bit ALU based on CNTFET with intrinsic channel region only

IOP Conf. Series: Materials Science and Engineering 1219 (2022) 012039 doi:10.1088/1757-899X/1219/1/012039

applicable in 1-D devices to make this model dynamically applicable for analog and digital circuit, dynamic gate capacitor has been described.
The arithmetic logic unit is exposed to arithmetic operation like addition, subtraction. Logical operation includes AND and OR operation. Optimization of CNTFET based 1-Bit ALU simulated on virtuoso cadence by importing Hspice compatible Verilog-A model developed by Stanford University. The model includes multiple CNTs per device and Schottky barrier resistance, traditional MOSFET enhancement mode device with carbon nanotube channel which supports unidirectional current flow. Every P or N type GAA-CNFET contains single or multiwall carbon nanotube with user define chirality. This indicates that scope of carbon nanotube as a promising alternative to replace silicon transistor as a fundamental building block of integrated circuits in future.
Section II deals with CNT and its properties whereas Section III describes GAA-CNTFET model used in simulation based on SPICE-compatible Verilog-A model which has been developed by Stanford University. Section IV is comprehensive explanation of proposed design of Arithmetic Logic Unit, Section V signifies simulation result of 1-Bit ALU and section VI involves conclusion and future work.

2. Carbon Nano Tube Field Effect Transistor

CNTFET is nothing but CNT as a channel in traditional MOSFET. CNT stands for Carbon nanotube, which graphene sheet rolled on conventional MOSFET [5] as shown in Figure 1, 2, the graphene hollow sheet rolled in such a way that inherited unique electrical property. Characteristic and behavior of nanotube depends upon chirality vector (n, m), which is nothing but arrangement of atoms in hallow sheet.

2.1 Properties of CNT

2.1.1 Behavior of CNT

Most interesting property of CNTFET is either work as semiconductor or metallic. If both chirality vectors are equal then it acts as metal with the small bandgap and if Difference between vectors are equal to odd multiple of K ($K \in Z$) it will behave as semiconductor whose bandgap is greater than metal refer Table1. Arrangement of atom not only affects chirality but also band gap [6].

2.1.2 Ballistic transport

CNTFET is known as quasi-1-D material which means CNT is hollow cylinder having no boundary so no random scattering only forward and backward scattering will occur. Because of this materialistic property CNT shows quasi ballistic transport which provides high current [7]. Most interesting thing about CNTFET is it works like MOSFET means it will give rail to rail swing, low static power consumption [8].

2.1.3 Low V_{th}

For making circuit faster V_{th} should be as low as possible, CNTFET gives us ability to do that by varying chiral vectors. Increases the chirality vectors, V_{th} decreases [9].

2.1.4 Types of wall

CNT refers as tubes having undetermined carbon structure means CNT can be refer as single and dual wall carbon nanotubes with large range of diameter. In Single wall carbon nanotube (SWCNT), carbon forms 2-D hexagonal cylindrical shape having no boundaries and Dual channel CNT or multi-channel CNT (DCCNT or MWCNT) is nothing but nested single wall CNTs [10][11].
According to difference of geometrical formation of channel of CNTFET can be many types like Back-gated CNTFET, Top-gated CNTFET, Gate all-around CNTFET, Suspended CNTFET. Among them GAA-CNTFET gives the improved result.

Table 1. behavior of CNTFET

Chirality vector configuration	Behavior of CNTFET
n = m	Metallic
n ≠ m or n-m = 3 K	Semiconductor

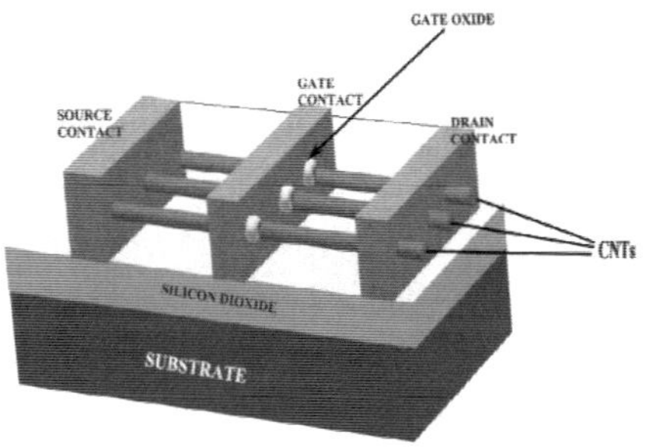

Figure 1. Diagram of CNTFET

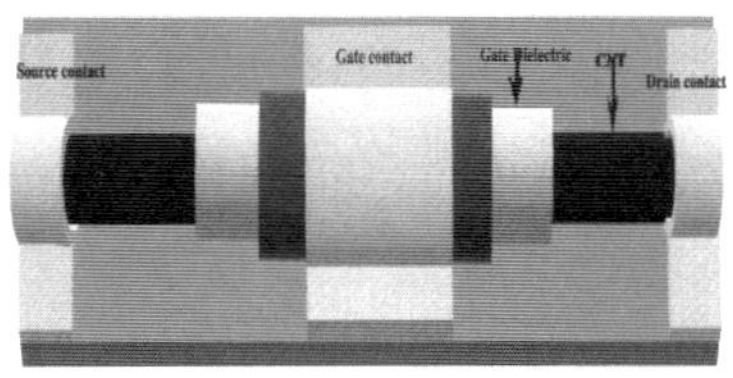

Figure 2. Gate All Around CNTFET

3. GAA-CNTFET

Gate all around CNTFET is also called as wrap around CNTFET. Main feature of GAA-CNTFET is good ability of controlling of gate, which improves the nanodevice below 10nm process node [12]. Invention of this type of CNTFET is held in 2008 by improving top gated CNTFET. In GAA-CNTFET CNT's surrounded by thick gate whose width is Wg and to cancel the body effect insulator applied between them, this will make transport easy it improves ON current (I_{ON}) and reduce leakage current (I_{OFF}) so ratio of I_{ON}/I_{OFF} also improves [13]. When chiral vector m=0 CNTFET will behave as semiconductor and chiral vector representation will be (n, 0) [14]. Dividing n vector in such a way it will represent the chirality vector of P-N type CNTFET used in any circuit.so n will divide as n_p, n_n. (n_p,0) is representation of chirality vector for P-GAACNTFET and (n_n,0) describe the chirality vector for N-GAA CNTFET so (n_p, n_n,0) will represent the chirality vector used for the proposed design contains both P and N type CNTFET.
(n_p, n_n, 0) will decide diameter of CNT in and CNTFET switching time will change according to diameter. when n_n is equal to n_p then CNTFET act as symmetrical device and when n_n is not equal to n_p then CNTFET act as asymmetrical device. D_{CNT} is diameter of Carbon nanotube expressed in equation and C_{ox} is oxide thickness which can be calculated as shown in Equation 2, For calculation of Vth we use the Equation 3 where Vπ is carbon bond energy.

$$D_{CNT} = \frac{\sqrt{3}.a\sqrt{n^2+m^2+nm}}{\pi} \quad [13] \qquad (1)$$

$$C_{ox} = \frac{2.\epsilon ox.\pi}{2a\ln\{(tox/r)+1\}} \qquad (2)$$

$$Vth = \frac{a.Vth}{\sqrt{3}.q.Dcnt} \qquad (3)$$

Where a = 0.142nm, V_π = 3.033eV

IOP Conf. Series: Materials Science and Engineering 1219 (2022) 012039 doi:10.1088/1757-899X/1219/1/012039

4. 1-Bit Arithmetic Logic Unit

ALU is the most versatile and important combinational circuit stands for arithmetic logic design unit integrated into control processing unit and GPU, ALU uses arithmetic instruction to perform Different type of operation like addition, subtraction, logical operation (OR, AND), shifting and so on. It works asynchronously means output changes asynchronously with the change in input [15].

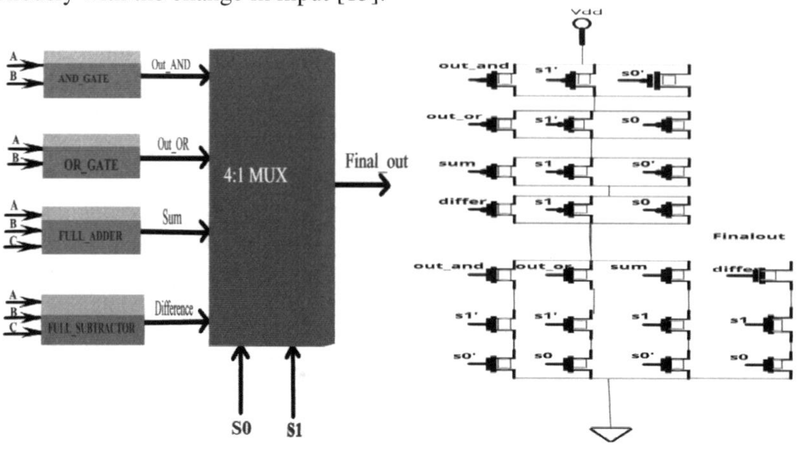

Figure 3. Design of arithmetic logic unit

Proposed 1-bit ALU contains AND, OR, Full Adder, Full subtractor, to decide which operation should perform we use 4:1 MUX as shown in Fig. 3. 4:1 MUX will work according to selection line S_0, S_1.it has 4 inputs and give one output at a time. Every operation has P and N-type GAA-CNTFET. AND gate designed by using 2 pull-ups CNTFET and 2 pull-downs CNTFET. The function of PUN is connected output to V_{dd} gives the value of 1. PDN use to connect the output to V_{ss} gives the value of 0 [16]. PDN and PUN are placed in a circuit in such a manner that one network will work at a time. 1-bit ALU consists of 150 CNTFET [17], whose primary inputs are A, B, C, and primary output will be either Sum, Difference, Out_AND, Out_OR. AND and OR take two inputs A, B and gives the output as Out_And shown in Figure 4(a) and Out_OR Figure 4(c). Full adder and Full subtractor will take three inputs A, B, C, and Full adder gives the output as the sum and carries shown in Figure 4(b) and Full subtractor will give output as Difference and borrow shown in Figure 4(d). Working of 4:1 MUX is explained in Figure 3 and operation explained in Table 2.

Table 2. Operation performed by 4:1 mux in 1-Bit ALU

S0	S1	Operation	OUTPUT
0	0	AND	Out_AND
0	1	OR	Out_OR
1	0	Full adder	Sum
1	1	Full subtractor	Difference

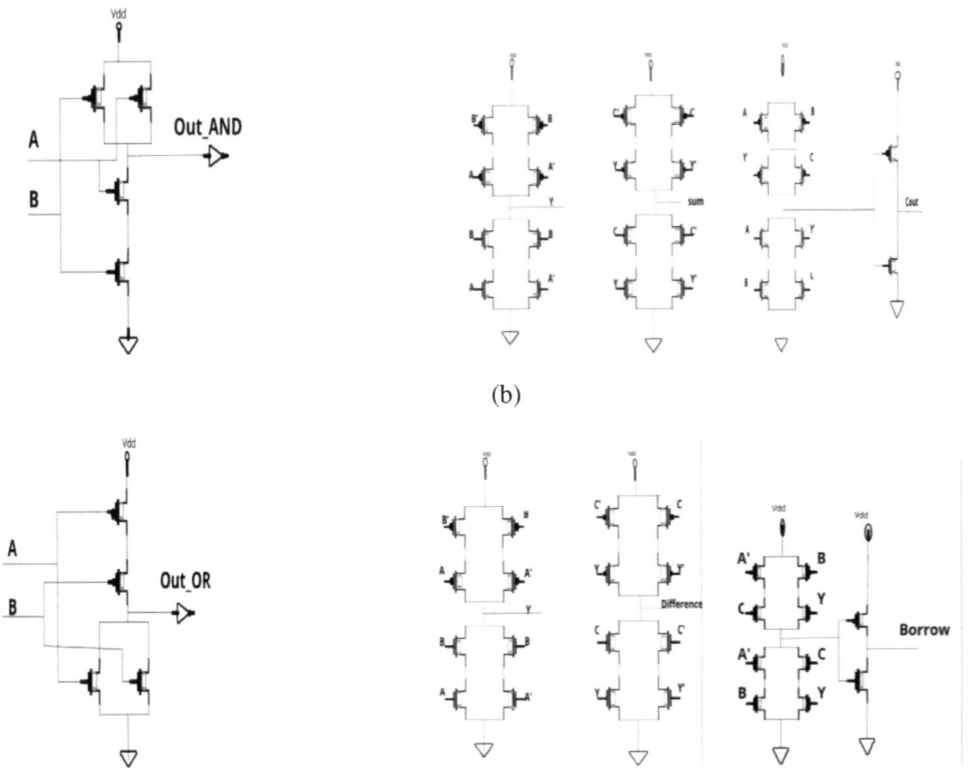

(a) (b)

(c) (d)

Figure 4. Circuit representation of (a) AND gate ;(b) Full Adder ;(c) OR gate and (d) Full Subtractor.

5. Simulation Results

In this section GAA-CNTFET model used for functional simulation is based on Verilog-A model compatible with SPICE developed by Stanford University. User-defined chirality vector and different type CNT channels used for the 1-bit ALU constructed by use of P and N-type GAA-CNTFET, this allows the designer to process node as short channel up to 11nm technology having power supply 0.8 V [18][19]. Simulation of 1-bit ALU shows different types of CNT walls such as single wall and double wall. GAA-CNTFET 1- Bit ALU describes comparative analysis evaluated in terms of delay, power, and power delay product using transient response graph. Delay is calculated as taking the time difference of transient response of signals between 50% of a rising edge of an input signal to 50% of the falling edge of the output signal. Power is calculated by using a PSF (power storage format) file generated by the virtuoso tool. Power delay product (PDP) is a very important parameter for any circuit to knowing its performance and can be calculated by multiplying power and delay it explains low PDP means better performance to a circuit. Delay of circuit generally depends upon the capacitor of design and by increasing diameter of CNT using chirality vector, drain current and capacitance can decrease. Figure 5 will represent two types of chirality which are single chirality ($n_n = n_p$) and dual chirality ($n_n \neq n_p$). Figure 5(a)is a comprehensive analysis of the delay of the single wall CNT with single chirality and dual chirality. Figure 5(b) describes the comparative analysis of PDP of single and double-channel CNT with single chirality and dual chirality. The comparative analysis shown in Figure 5 makes clear that (20,20,0), (20,16,0) offers less delay but not power and (8_8_0), (10_14_0) offers the best power but now best delay so tries to find a configuration that gives moderate power with moderate delay. (20,20,0) gives the better result for single chirality in terms of delay and power. (20,16,0) is dual chirality configuration which gives a better result. k_{ox} is the relative dielectric constant, by increasing k_{ox} in 1-bit ALU design like 3.9, 7.8, 11, 27 delay

IOP Conf. Series: Materials Science and Engineering 1219 (2022) 012039 doi:10.1088/1757-899X/1219/1/012039

decreases but power increases. Variation of delay and PDP with k_{ox} presented in Table 3, 4. Table 5 presented variation in delay and PDP by changing t_{ox} value 1, 2.

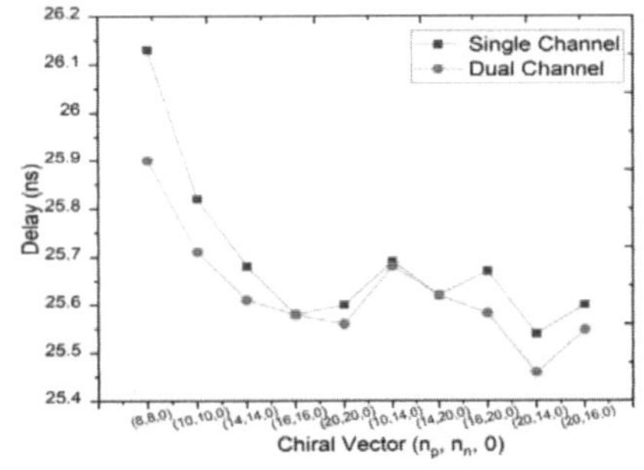

(a)

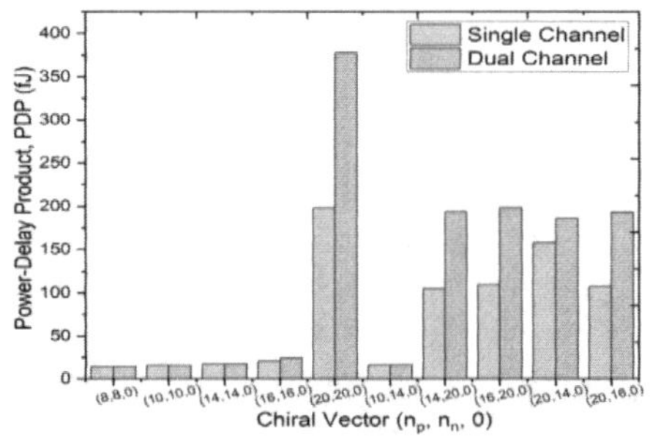

(b)

Figure 5. Comparative analysis of (a) delay of single and dual chirality with single and double channel; (b) PDP of single and dual chirality with single and double channel.

Table 3. Delay and PDP calculation of (20,20,0) with single and double channel by varying k_{ox}

Value of k_{ox}	Chirality with channel	Delay (ns)	Power (μW)	PDP (fJ)
7.8	20_20_0 Single channel	10.54	9.192	96.883
11	20_20_0 Single channel	10.52	9.408	98.972
27	20_20_0 Single channel	10.5033	10.040	105.45
3.9	20_20_0 Double channel	10.52	15.816	166.384
7.8	20_20_0 Double channel	10.50	17.136	179.928
11	20_20_0 Double channel	10.49	17.856	187.309
27	20_20_0 Double channel	10.472	19.400	203.156

IOP Conf. Series: Materials Science and Engineering 1219 (2022) 012039 doi:10.1088/1757-899X/1219/1/012039

Table 4. Delay and PDP calculation of (20,16,0) with single and double channel by varying k_{ox}

Value of k_{ox}	Chirality with different channel	Out_and _delay (ns)	Power (uW)	PDP (fJ)
3.9	20_16_0 Single channel	10.58	5.1472	54.457
7.8	20_16_0 Single channel	10.54	5.52	58.180
11	20_16_0 Single channel	10.52	5.8024	61.157
27	20_16_0 Single channel	10.5033	9.2	96.630
3.9	20_16_0 Dual channel	10.51	8.856	93.076
7.8	20_16_0 Dual channel	10.49	9.672	101.45
11	20_16_0 Dual channel	10.48	10.136	106.22
27	20_16_0 Dual channel	10.46	10.631	113.80

Table 5.PDP and delay calculation by varying t_{ox} of (20,20,0), (20,16,0) with single and double channel

Value of tox	Chirality with channel	Delay(ns)	Power(μW)	PDP (fJ)
1	20_20_0 Single channel	20.509	10.888	223.30
1	20_20_0 Dual channel	20.478	20.824	426.433
1	20_16_0 Single channel	20.506	6.5328	133.96
1	20_16_0 Dual channel	20.4727	12.064	246.97
2	20_20_0 Single channel	20.511	8.6	176.39
2	20_20_0 dual channel	20.4801	16.256	332.924
2	20_16_0 Single channel	20.509	6.3296	129.813
2	20_16_0 Double channel	20.473	5.49	112.396

6. Conclusion

CNTFET is nanodevices which have distinctive properties such as small dimension, provides less delay and power by operating in low power. By the reason of this unique property implemented 1-bit ALU to perform arithmetic and logical operation. The different diameter has been analyzed by using the virtuoso tool at 11nm technology node with 0.8 dc voltage. 1-bit ALU based on CNTFET designed for single chirality and double chirality with single wall and double wall simulation results concluded that (20,20,0) gives optimized PDP 197.708fJ, Power 7.723μW, delay 25.60 ns in single chirality with a single channel, dual chirality (20,16,0) with single-channel gives PDP 107.084fJ Power 4.183μW, Delay 25.60ns. Dual-channel with single chirality (20,20,0) provides good stability with PDP 377.87fJ, Power 14.784μW, delay 25.56ns for dual chirality (20,16,0) gives PDP 193.24fJ, Power 7.5641μW, Delay 25. 548ns.based on this simulation results conclude that by increasing the value of k_{ox} delay decreases but power increases and by increasing the value of t_{ox} delay and power decreases. The proposed CNTFET based 1-Bit ALU with (20,16,0) chirality can be further used for implementing ALU in Computers and Microprocessors.

IOP Conf. Series: Materials Science and Engineering 1219 (2022) 012039 doi:10.1088/1757-899X/1219/1/012039

References

[1] S Iijima, 1991 Nature **354** pp. 56-58.
[2] S M Kang, Y Leblebici 2002 *Tata McGraw Hill Education Private Limited.*
[3] K Nehru, T Nagarjuna, 2017 *Journal of Nano and Electronic Physics* **09** (04).
[4] Sinha Sanjeet Kumar, Chaudhury Saurabh, 2014 *Journal of Semiconductors* **35** (11).
[5] S Farhana, A Z Alam, S Motakabber, S Khan,2013, *International Nanoelectronics Conference* pp.329-332.
[6] Soheli Farhana, AHM Zahirul Alam,2013 *IEEE* pp. 589-592.
[7] A Raychowdhury 2004 *IEEE Trans.on design of integrated circuits and system* **23**(10).
[8] Gordon E Moore, 1975 *IEEE*, **11** pp.11-13.
[9] A. Giorgio 2014 *International Journal of Engineering and innovative Technology* **4**(5) pp. 165-168.
[10] Shima Ibrahim Aayed, M M Abutaleb,2016 *Advance in Material Science and Engineering.*
[11] Greeni Navin, K Subhulakshmi 2016 *IEEE* pp. 1-3.
[12] Singh Rohit Kumar Shailendra, Dr V N Ramakrishnan 2017 *IEEE* pp 213-217.
[13] Hoeffinger B 2012 *Springer*r pp. 477
[14] G R Ahmed Jamal, S M Mominuzzaman 2015 *Journal of nanoscience and nanoengineering* **1**(2).
[15] Sharvani Gadgil and Chetan Vudadha 2020 *IEEE* **19**.
[16] Ashish Sharma, Umesh Dutta,2016 *IJRCCE* **5**(5).
[17] A Nagendrababu, Ch P Amuleti,2017 *IJRTI* **2**(1).
[18] Rashimika Sahoo, 2009 *International Journal of Electronic Engineering Research* **1**(2).
[19] Dang T, Anghel L, Leveugle R, 2006 *IEEE conference on design and test of integrated systems in nanoscale* pp. 28-33.

IOP Conf. Series: Materials Science and Engineering 1219 (2022) 012040 doi:10.1088/1757-899X/1219/1/012040

Synthesis and characterization studies of polypyrrole/iron tungstate nanocomposites for interfacial water evaporation

V Robert Bellarmin[1*], Arjun Moorthy[2], M.E. Raja Saravanan[3]*

[1] Department of Physics, Periyar University, Salem, Tamil Nadu, India

[2] Independent Researcher, Arjun's nanomaterials research lab, Salem, Tamil Nadu, India
[3] Government Arts College (Autonomous), Salem, Tamil Nadu, India

*Email: physicsrobi@gmail.com, raja_megs1@yahoo.com

Abstract. PPy-$FeWO_4$ composites were synthesized by polymerization of pyrrole in presence of varying weight percentage of $FeWO_4$ nanoparticles using $FeCl_3$ as both catalyst and dopant. The synthesized polymer composites were characterized by FTIR, XRD, FESEM-EDAX, TEM and TGA analysis. The XRD and FTIR analysis confirmed showing near pure samples were achieved. The FESEM-EDAX and TEM studies showed the presence of $FeWO_4$ nanoparticles which were well dispersed in the polymer matrix. The thermal degradation studies showed similar stability of polymer matrix in the presence of nanoparticles. Interfacial solar water evaporation devices were fabricated by depositing the polymer composite on floatable PU sponges. The studies revealed that 50-50 wt% PPy-$FeWO_4$ showed maximum solar to vapour conversion up to 86% with an evaporation rate up to 1.2 kg m^{-2} hr^{-1}.

1. Introduction

In recent decades, with a rising population, the scarcity of freshwater to meet the human and environmental demand is a priority for global development[1]. Even though 70% of our planet is covered with water, only 3% is fresh water and two-thirds of that are unavailable, trapped in glaciers and the remaining considerable percentage is polluted. As a result, almost 1.1 billion people lack access to water at this current consumption rate. The shortage of clean freshwater has become a severe problem that requires a conventional process with economic viability and reliability to be urgently solved. Desalination to provide fresh water from the seawater and reusage of contaminated freshwater are the promising approaches through electrical, membrane filtration or thermal process to meet this devasting challenge[2-4]. Since most of these process involves massive energy consumption, environmental pollution and lower energy conversion efficiency due to moving mechanical parts, a prominent and rational procedure is essential for the production of freshwater via green and zero-emission manner[5,6]. Although solar energy is considered a green and sustainable energy source, in natural water evaporation, an endothermic process, there is an inevitable loss of energy by transferring it to the surrounding that is the entirety of bulk water. To achieve higher solar-to-vapour conversion efficiency, many photothermal materials such as graphene-oxide[7,8], semiconductors[9-15], metallic nano particles[16,17], carbon-based materials[1824],ceramic-based materials[25,26], plasmonic metal nanomaterials[27,28] and polymers have been developed recently, they possess broadband and high light absorption ability, thermal insulation, floatable with low density, hydrophilicity and porous structure for water transportation and steam escape to maximize solar harvesting and minimize thermal loss by localizing the solar heating at the interface between the air and water. These materials pay the way to the researchers for the fabrication of high-efficiency solar steam generation devices. Solar steam generation technology is more promising and one of the effective methods for its potential application in seawater desalination, power generation, atmospheric water harvesting, actuators and heavy oil spill clean-up work to avert famine. Productive collection of solar energy is an indispensable part of these solar steam generation devices. The elemental part of these devices is the substrate whose top surface provides a platform for the efficient collection of solar energy and maintain hydrophobicity to facilitate the vapour escape whereas the bottom surface retains hydrophilicity to enhance the upward water transport pathway. Besides this, a firmly coated thin optical absorption layer at the top provides a complete distillation system. Filter paper, wood, polystyrene foam, melamine foam, cotton cloth, stainless steel mesh, aerogel, hydrogel, cellulose membrane, polyimide membrane, polyethersulfone membrane, cellulose acetate membrane has been already reported as a substrate.

Herein, for the first time, we fabricated a solar-to-steam generator device using polyurethane (PU) sponge as a substrate material and polypyrrole-Iron tungstate (PPy/ $FeWO_4$) composite as a light absorption layer. PU sponge which has low heat conduction property, high hydrophilic property, low density, low cost with excellent durability and reversibility was adopted as floating substrate with a thin layer of rationally developed photothermal conversion material PPy/ $FeWO_4$ composite as a strong light absorber, prepared by varying the concentration of iron tungstate with the standard weight of polypyrrole. The composite showed solar -to-heat conversion efficiency[29,30] increases with the percentage of $FeWO_4$ in them up to 86.9% under irradiation of 1 sun. It is also noted that the evaporation rate exhibited by the composite is remarkable compared with their counterpart materials and the performance of isolated $FeWO_4$ has an edge over polypyrrole in the steam generation process.

PPy as a conjugated conductive dark black polymer is known to possess excellent broadband light absorption and high energy conversion performance explored in the field of light-emitting diodes[31,32], catalysts[35], supercapacitors[33,34], corrosion protection[36,37] and photothermal agent for cancer removal[38], whereas Iron tungstate nanoparticles belong to the family of wolframite type material, for their excellent optical and electronic properties have obtained wide applications such as photocatalyst for dye degradation[39], energy storage electrodes[41], a catalyst for peroxide oxidation[40], and bipolar switching devices[42]. Considerable efforts have been devoted to engineering malleable nanoplatforms by combining two or more materials into one single platform which is suggested to inherit the talents of each source

PPy based Solar-to-heat conversion evaporators have been reported recently with considerable efficiency. Assessing the appropriateness of iron tungstate as a photoelectrode material for water oxidation, the PPy /$FeWO_4$ composite can become a promising candidate for seawater desalination and sewage treatment to produce clean water with high efficiency. It provides a new approach for interfacial solar vapour generation, as they required an easy fabrication process without any complicated pieces of equipment. Furthermore, the PU sponge, also a linear polymer, is a lightweight porous material whose wide range of properties such as low density, low thermal conductivity, excellent chemical and mechanical stability, sustainable even after many cycles of performance, age-resistant, cost-effective, making our design a promising and more practical for interfacial solar vapour generation.

2. Materials and methods

Pyrrole (C_4H_5N), ferric chloride anhydrate ($FeCl_3$), ferrous sulfate heptahydrate ($FeSO_4.7H_2O$), sodium tungstate (Na_2WO_4), sodium hydroxide (NaOH), cetyltrimethyl ammonium bromide ($[(C_{16}H_{33})N(CH_3)_3]Br$) (CTAB) and ethanol ($C_2H_6O$) were purchased from Avra and Nice chemicals of minimum 95% purity. All the chemicals purchased were used directly without any further purification. Commercial grade polyurethane sponge, Double distilled water was used in the experiment.

2.1 Iron tungstate and composite synthesis

Iron tungstate nanoparticles were prepared by simple sol-gel method following our previously reported work[43], where 0.003 moles of $NaWO_4$ and 0.0012 moles of CTAB were dissolved 120 ml distilled water, after 15 minutes freshly prepared 5 ml, 0.6 M ferrous sulphate solution is added to the reaction mixture. The reactant mixture was heated around $80^{O}C$ using a hot plate and freshly prepared sodium hydroxide solution (0.82 g in 2.5 ml DDW) was swiftly added resulted in dark-brown dispersion showing the formation of iron tungstate, were cleaned, dried and labelled as Fwo.

PPy/$FeWO_4$ nanocomposites were synthesized using the emulsion polymerization technique, where $FeCl_3$ was used as both initiator and dopant in the formation of polypyrrole. $FeCl_3$ solution is freshly prepared by mixing 9.6 g $FeCl_3$ in 100 ml double distilled water. The prepared solution is filtered using filter paper. the solution containing beaker is placed in a bath ultrasonicator. The previously prepared Iron tungstate nanopowder (Fwo) weighed 0.22 g and added to the above solution and ultrasonicated for 30 minutes. 2 g of pyrrole monomer is dropped slowly to the sonicating reaction mixture. The slowly black precipitate formation is inferred due to the formation of polypyrrole. The prepared composite was labelled PPy10Fwo and all materials characterization except XRD (PPy-50wt%Fwo was used to perform XRD analysis) was performed for this sample. For PPy30wt%$FeWO_4$, PPy-50wt%$FeWO_4$ composites, keeping pyrrole concentration constant, Fwo was varied as 0.857 g and 2 g respectively. For the case of the pure PPy sample, no Fwo was added in the polymerization process. The above-prepared samples were washed and centrifuged several times using DDW and the final wash was done using ethanol. The obtained precipitates were dried in a hot air oven for 12 hours overnight and labelled respectively as PPy, PPy10Fwo, PPy30Fwo and PPy50Fwo.

2.2 Device fabrication

Polyurethane sponge (PU sponge) with the size of 1 inch (length) x 1 inch (breath) x 1 inch (width) is a self floating solar steam generating device on which prepared polypyrrole composite with different concentration of $FeWO_4$ viz. 10 wt%, 30 wt% and 50 wt% coated (The PU sponge was stained by drop-casting 10 mg of suitable composite dispered in water) whereas the bottom surface being highly porous keep contact with the water surface in the beaker and the experiment is done in outdoor under the illumination of 1 sun (1 kW m^{-2}), checked by an solar power meter. Once the irradiation started, water absorbed through the capillaries would evaporate into steam immediately and the real-time mass changes of water are determined using an electronic analytical balance over 60 minutes.

The phase composition and material purity for nanocomposites were analyzed by powder X-ray diffraction (XRD) technique (Bruker D8 using Cu-Kα targets with 2°/min scan rate). The polymer formation was studied using Fourier transform infrared spectrometer (FTIR) (Thermo Fisher Nicolet iS10). surface morphology, particle size, particle distribution in polymer matrix were analysed by field emission scanning electron microscope (FESEM) using Carl Zeiss SIGMA coupled with Bruker Quantax 200—Z10 EDS detector and highresolution transmission electron microscope (HRTEM) (JEOL JEM 2100). Thermal studies are carried out using thermogravimetry-differential thermal analysis (TG-DTA) (TA Instruments SDT Q600). The solar power was monitored using Kusam Meco KM-SPM-11BTU Solar Power Meter. water evaporation rate was evaluated by an analytical balance. The temperature was recorded using a thermocouple (Tanotis TM902C LCD thermometer), ion concentration was recorded using a TDS meter (generic TDS-3).

3. Results and discussion

3.1 materials characterization

Fig. 1 (a) shows the X-ray diffraction pattern of PPy, Fwo, PPy-Fwo composite. The PPy sample showed a broad absorption peak from 20° to 30°, which strongly shows the amorphous nature of prepared polypyrrole using $FeCl_3$ as dopant and initiator, no other characteristic peaks were absorbed indicating the purity of the as synthesized sample. The XRD pattern of prepared Fwo shows diffraction peaks at 35.47°, 43.11°, 62.52° well matched with the standard JCPDS data (JCPDS data number 74-1130) showing the corresponding planes of (021), (112) and (113) respectively confirming the monoclinic structure. the Scherrer equations were applied to measure the crystalline size[44], which revealed the particles ranges in 17 nm. The peak obtained from the PPyFwo composite revealed the individual characteristics of PPy and iron tungstate nanoparticles with no other peaks confirming the purity of the composite prepared.

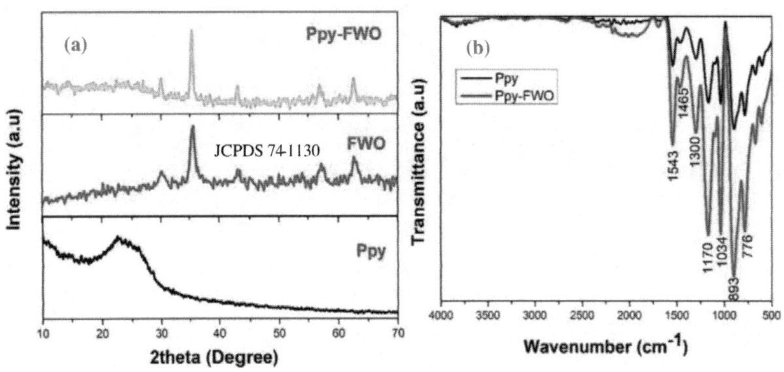

Figure 1(a) & (b) shows the XRD analysis of PPy, Fwo and PPy-Fwo composite and FTIR analysis of PPy and PPy-Fwo composite prepared.

Fig 1. (b) shows the FTIR spectrum of PPy and PPy-Fwo composite obtained from 4000-500 cm^{-1}. Observed peaks in the FTIR spectrum of the samples are 1543, 1465, 1300, 1170, 1034, 893, and 776 cm^{-1}. The band at 1543 cm^{-1} corresponds to the pyrrole ring C-C bonds, the band 1465 cm^{-1} corresponds to the C-N bond of the pyrrole ring. The 1300 cm^{-1} and 1034 cm^{-1} bands correspond to the =C-N plane vibration mode. The band at 1170 cm^{-1}corresponds to the C-H out of plane deformation vibration, the bands at 776 cm^{-1} and 893 cm^{-1} attributed to the C-H stretching. Both the composite and pure PPy samples showed a similar band structure confirming the purity and presence of polypyrrole in both the samples.

Fig. 2 (a) shows the Thermogravimetric analysis (TGA) of Fwo, PPy and PPy-Fwo composite samples. All the three samples showed a minor weight loss from 30-100°C which is due to the evaporation of water molecules present in the samples. Fwo, PPy and PPy-Fwo showed 6.7%, 15% and 11% respectively. They indicate that the prepared composite is less hygroscopic when compared to the PPy sample. Fwo showed a weight loss of 9% from 200-300^{0}C which may be due to the decomposition of CTAB used in the synthesis and no significant weight loss was observed beyond this up to 900^{0}C. PPy and PPy-Fwo samples showed similar decomposition nature of around 8% and 6% from 110-250°C which shows the structural decomposition of the polymer, while a major weight loss is observed from 250^{0}C which attributes to the skeletal decomposition of pyrrole where PPy showed 53.6% and PPy-$FeWO_4$ showed 35.3% from 250-900^{0}C this shows that the presence of iron tungstate nanoparticles enhances the stability of polypyrrole.

TEM images of prepared PPy, Fwo and PPy-Fwo composite are shown in figure 2 (b). The PPy sample showed a large agglomerated of polypyrrole with amorphous nature is observed, Fwo sample showed polydispersed and polycrystalline natured $FeWO_4$ nanoparticles ranging from 2-25 nm. the PPy-Fwo composite shows the presence of $FeWO_4$ nanoparticles within the PPy polymer matrix. Good dispersion of nanoparticles is observed throughout the polymer matrix.

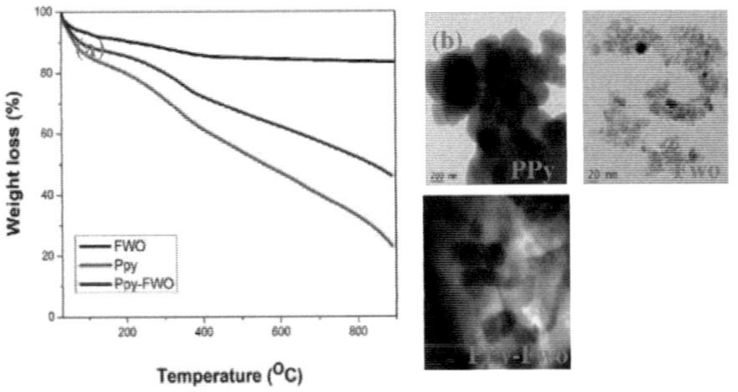

Figure 2 (a) shows the TGA curves and (b) shows the TEM images of PPy, Fwo and PPy-Fwo samples.

FESEM and elemental mapping images PPY and PPy-Fwo composite is shown, the sample showed the presence of C, O, Cl, Fe, W. the carbon and chlorine correspond to the polypyrrole, where chlorine was used as a dopant, Fe, W and O shows the presence of iron tungstate in the sample, the mapping reveals that the atoms are well dispersed in the sample which infers that the nanoparticles are well dispersed in the composite prepared.

Figure 3 (a), (b) shows the EDAX mapping of PPy and PPy-Fwo composite prepared.

3.2 Interfacial solar evaporators

The mass change caused by water evaporation were recorded under varied experimental conditions through six vapor evaporating systems, they are bare PU sponge, PPy coated PU sponge, Fwo coated PU sponge, PPy10Fwo composite coated PU sponge, PPy30Fwo composite coated PU sponge, PPy50Fwo composite coated PU sponge, and light-to-vapor efficiency is calculated for every reading, at average temperature ~35°C, using the formula[45]

$$\eta = mH_{LV}/I$$

Where m is the mass flux, I is the power density of solar energy(1000 W m^{-2}) and H_{LV} is the total enthalpy of liquid-vapor complete phase change (2455.6 kJ kg^{-1})
The below tabular Colum shows the optioned mass change of samples tested and their calculated efficiency.

substrate	Mass change ($g/inch^2$)	Mass change (Kg/m^2)	Efficiency (%)
PU	0.608	0.942	64.25
PPy	0.728	1.128	76.94
Fwo	0.732	1.135	77.41
PPy10Fwo	0.738	1.144	78.03
PPy30Fwo	0.786	1.218	83.08
PPy50Fwo	0.822	1.274	86.90

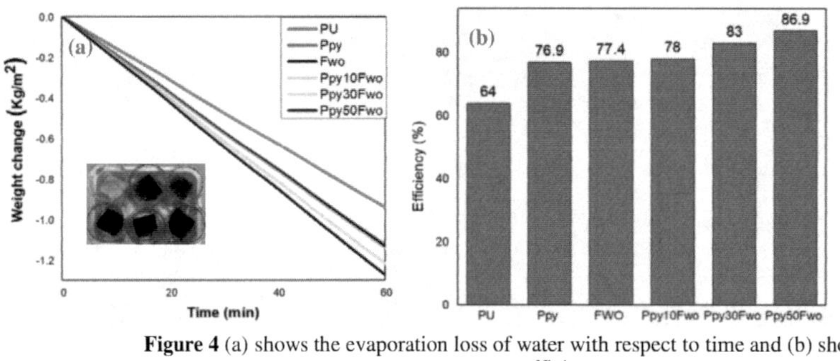

Figure 4 (a) shows the evaporation loss of water with respect to time and (b) shows the calculated efficiency.

The steam generation performance is studied by marking the total mass change due to water evaporation as a function of the solar illumination time. Upon the light illumination, the bare sponge showed a relatively less amount of weight loss, sponges coated with only PPy and only Fwo exhibits almost similar weight loss, with increasing the concentration of $FeWO_4$ nanoparticles in PPy, the solar vapour generation performances improve sharply and reach 86.90% for PPy50Fwo composite which is very promising. This reveals the fact that the optical properties of composites are enhanced by the individual counterparts. The evaporation rates and efficiencies of PU sponge remain nearly unchanged after few cycles (4 times) of usage and long-term storage (45 days), indicating the long-term stability and recyclability of the device.

The salt water distillation experiment and sewage treatment experiments were performed using the PU sponge with PPy50Fwo composite. Under one sun illumination the samples exhibited ~1.101 kg m^{-2} and 1.26 kg m^{-2} mass change for the periods of 60 minutes. The freshwater collected could meet the drinking water standard confirmed using a TDS meter, from both evaporators with efficiencies of 75.1% and 85.94% respectively. The saltwater taken has the NaCl concentration of 3.5 wt% and the sewage water was contaminated with mud, dust, etc. During the evaporation, the blockage of pores of the sponge by the sodium chloride salt is the reason behind the considerable drop in absorbed efficiency of saltwater.

4. Conclusion

In this study, we synthesized PPy/$FeWO_4$ composite by polymerization of pyrrole with a varying weight percentage of $FeWO_4$ and demonstrated better performance photothermal property using the polyurethane sponge as a substrate on which PPy-50 wt% $FeWO_4$ composite is deposited as a light sensitive layer. The asprepared solar evaporator possesses an evaporation rate of 1.35 kg m^{-2} h^{-1} under 1 sun illumination with a better solar thermal conversion efficiency of 86.9% and provided a new approach for the fabrication of a photothermal layer to design a solar steam generator which has the advantages of improved efficiency, low price, good durability which may be suitable for clean water production to address the water scarcity issues.

5. Acknowledgement

We gratefully thank MHRD, COE-AMGT, Coimbatore and DST-SAIF, Kochi for XRD, FTIR, TGA, FESEMEDAX and TEM supports.

References

[1] Bartram J, Brocklehurst C, FisherM B, Luyendijk R, Hossain R, Wardlaw T and Gordon B 2014 International journal of environmental research and public health. 11 8137.
[2] Wang L,Boutilier M S, H Kidambi P R, Jang D,Hadjiconstantinou N G and Karnik R 2017 Nature nanotechnology. 12 509.
[3] Yin J and Deng B 2015 Journel of Membrane Science. 474 39.
[4] Cohen-Tanugi D and Grossman J C 2015 Desalination. 366 59.
[5] Zong L,Li M and Li C 2018 Nano Energy. 50 308.
[6] Santosh R, Arunkumar T, Velraj R and Kumaresan G 2019 Journel of Cleaner Production. 207 826.
[7] Guo A, Ming X, Fu Y, Wang G, Wang X 2017 ACS Appl. Mater Interfaces. 29958.
[8] Zhu X Y, Yang K J, Chen B L 2017 Environ-Sci Nano. 4(12) 2267.
[9] Fu Y, Mei T, Wang G ,Guo A, Dai G, Wang S, Wang J, Li J, Wang X 2017 Appi.Therm. Eng. 114 961.
[10] Guo Z, Ming X, Wang G, Hou B, Liu X, Mei T, Li J,Wang J, Wang W 2018Semicond.Sci.Technol. 33(2) No.025008.
[11] Guo.Z, Wang.G, Ming.X, Mei.T, Wang.J, Li.J,Qian.J, Wang.W 2018 ACS Appl.Mater.Interfaces 10(29) 24583.
[12] Liu X, Hou B ,Wang G, Cui Z, Zhu X, Wang X 2018 J.Matter.Res. 33(06) 674.
[13] Ma Q,Yin P,Zhao M,Luo Z,Huang Y,He Q,Yu Y,Liu Z,Hu Z,Chen B,Zhang H 2019 Adv.Mater. 31(17) 1808249.
[14] Liu X,Cheng H,Gue Z,Zhan Q,Qian J,Wang X 2018 ACS APPL.Mater. Interfaces. 10(46) 39661.
[15] Ming X,Guo A,Wang G,Wang X 2018 Sol.Energy Mater.Sol.Cells. 185 333.
[16] Guo A, Fu Y, Wang G, Wang W 2017 RSC Adv. 7(8) 4815.
[17] Xu Y,Ma J,Han Y,Xu H,Wang Y,Qi D,Wang W 2020 Chem.Eng.J. 384 123379.
[18] Wang X Z, He Y R, Cheng G, Shi L, Liu X, Zhu J Q 2016 Energy Conv. Manag. 130 176.
[19] Zhan Q,Xu W,Wang X 2018 Sci.China Mater.61(7) 905.
[20] Wang G,Fu F, Guo A,Mei T ,Wang J,Li J,Wang W 2017 Chem.Matter. 29 5629.
[21] Fu Y,Wang G,Ming X,Liu X,Hou B,Mei T,Li J,Wang J,Wang W 2018 Carbon. 130 250.
[22] Ding T, Zhu L, Wang X, Chan K, Lu X, Cheng Y, Ho G 2018 Adv.Energy Mater. 8 1802397.
[23] Zhang P, Li P, Lv L, Zhao Y, Qu L 2017 ACS Nano. 11(5) 5087 .
[24] Xu N, Hu X, Xu W,Li X ,Zhou L,Zhu S,Zhu J 2017 Adv.Matter. 29 160676.
[25] Zhao J Q, Yang Y W, Yang C H 2018 J Mater Chen A. 6(33) 196.
[26] Kaur M, Ishii S, Shinde S L, Nagao T 2017 ACS Sustain Chem Eng. 5(10) 8523.
[27] Zhou L, Zhaung S D, He C Y, Tan Y L, Wang Z, Zhu J 2017 Nano Energy. 2 195.
[28] Wang H, Lei M, Tanemura S 2017 Solar RRL. 1(3-4) 1600023.
[29] Qin G , Qiu J, 2019 Chemical Engineering Journal. 359 652.
[30] Zhou Y, Hu Y, Sun W, Lu S, Cai C, Peng C, Yu J, Popovtzer R, Shen M , Shi X, 2018 Biomacromolecules. 19 2034.
[31] Greenham N C, Moratti S C, Bradley D DC, Friend R H, Holmes A B. 1993 Nature. 365 628.
[32] Rasouli H, Naji L, Hosseini M G. 2019 J Electroanal Chem. 836 165.
[33] Chaudhary S, Kiran Kumar A BV, Sharma N D, Gupta M. 2019 Int. J Energy Res. 43(8) 3446.
[34] Bae K L, Kim K. 2017 Int .J Energy Res. 41(9) 1335.
[35] Radhakrishnan S, Adhikari A. 2006 J Power Sources. 155(2) 157.
[36] Lei Y H, Sheng N, Hyono A, Ueda M, Ohtsuka T 2013 Corros Sci. 76 302. [37] Liu J R, Lei Y H, Qiu Z Z 2018 J Electronchem Soc. 165(13) C878.
[38] Ramanaviciene A ,Kausaite A ,Taurus S, Ramanavicius A 2007 J Pharm Pharmacol. 59(2) 311.
[39] Zhou Y X, Bin Yao H, Zhang Q, Gong J Y 2009 Chem. 48 1082.
[40] Tian T, Ai L, Liu X, Li L, Li J, Jiang J 2015 Ind.Eng.Chem.Res. 54 1171.
[41] Wang W, Hu L, Ge J, Hu Z, Sun H, Zhang H, Zhu H, Jiao S 2014 Chem.Matter. 26 3721.
[42] Sun B, Li C M 2014 Chem. Phys.Lett. 604 127.
[43] Arjun M, Robert Bellarmin V, Balaji Kumaran, Raja Saravanan M E 2020 Inorg and Nano-Metal Chem. 50:10 1012
[44] Bujanos Buenrostro I, Rivera Morales I M, Ramos Jimenez J, Erhard Ramırez A 2014 Microbiology. 34 54.
[45] Eliodoro C, Matteo M, Francesca V, Matteo F, Pietro A 2018 nature sustainability. 1 763.

Bandgap Engineering of TiO_2 as Visible light Photocatalyst with Pd Doping Using First Principles

A.Angeline Dorothy[1], Puspamitra Panigrahi[2]**

[1,2]Hindustan Institute of Technology and Science, Chennai, 603103, India

Email: rs.ad0914@hindustanuniv.ac.in

Abstract. Doping of Anatase TiO2 is a successful method to facilitate its photocatalytic activity to shift to the visible region. While noble metals in general can effectively slow down carrier recombination, the study of Pd doping in anatase TiO2 can unravel its potential as a visible light photocatalyst. Our first principles study indicate that the electronic structure is altered by the Pd 3d orbitals and bandgap narrowing is successively increased as dopant concentration of Pd is increased from 4.17% to 8.33% to 12.5%.Charge density plots of the doped systems indicate an electron cloud over the Pd- O bond which is now more covalent. In the visible region, absorption coefficient of Pd doped TiO2 supercell is around 1.2 x $10^{-5}cm^{-1}$ which is comparatively greater than that of pure TiO2 confirming its utility as a versatile and viable visible light photocatalyst.
Keywords: electronic structure, photocatalyst, Pd doped TiO2, First Principles Study

1 Introduction

Numerous methods have been tried to modify TiO2 as a visible light photocatalyst such as cation doping, anion doping[1], transition metals[2], lanthanides[3] etc. Noble metals like Pt, Au, Pd, and Ag have been routinely used in TiO2 doping to enhance charge carrier separation. The noble metals have higher work functions but lower Fermi energy level compared to TiO2. Thus they enable electrons to be transferred from the CB of TiO2 by means of the Schottky barriers to get trapped by these noble metals[4]. Hence, they promote ample separation of charges while minimizing charge carrier recombination, thereby improving photocatalytic activity of TiO2.

In our earlier paper [5], we had envisaged using first principles DFT calculations how the electronic and optical properties of anatase TiO2 are effectively tuned by controlled concentrations of Pt and Ag as dopants. The modified electronic structures end up in successful operation of the optical and electronic properties in the favourable visible region.

Bai et al[6] have studied the Pd doped flower like structures anatase TiO2 with exposed (001) facets. They synthesized Pd/TiO2 nanocomposite catalysts which exhibited strong absorption in the desirable visible light and remarkable photocatalytic activity for BPA degradation under UV and visible light compared to bare TiO2. Dong Li et al [7] have successfully prepared a visible light active Pd/C-N-S-TiO2 NS photoelectrode which can function as a viable visible light based photocatalyst. Studies also show Pd-doped TiO2 nanofibers can be utilized as highly sensitive gas sensors[8][9].

In this paper, we have doped TiO2 with increasing concentrations of 4.17%, 8.33% and 12.5% of Pd atoms respectively and using first principles calculations, studied the corresponding energy band gap variation. We have also compared the modified optical properties for bulk TiO2 system to the 4.17% Pd doped TiO2. Our electronic structures prove that increasing Pd dopants result in successive narrowing of band gap in TiO2. The encouraging electronic and optical properties hold promise tothe efficient photocatalytic applications of Pd doping in TiO2.

2. Computational Method

To examine the impacts of doping with Pd on the electronic structure of TiO2, the calculations were performed using the VASP[10] code based on first-principles density functional theory. The ground state structure optimization and the corresponding electronic properties calculations were performed based on the Perdew-Burke-Ernzhof (PAW-PBE) exchange functional [11] under the Generalized Gradient Approximation (GGA) [12] . For calculations, cutoff energy for the plane-wave basis set was set to be 550 eV after convergence analysis.
The ground state structure optimization for the Brillouin zone of the supercell was performed by a (3 × 3 × 3) Monkhorst-Pack K-point mesh whereas for electronic properties calculation a higher mesh of (5 × 5 × 5) and were used. The optical programs were carried out separately with (6 x 6 x 6) K-Points. As shown in Figure 1, we have used a (2 x 1x 1) anatase supercell of TiO2 consisting of 8 Platinum and 16 Oxygen atoms to construct the doped structures, where one Ti atom is replaced by one, two and three Pd atoms, corresponding to a doping concentration of about 4.17%, 8.3% and 12.5% respectively. To explicitly treat the valence electrons of Ti (3d2 4s2), O (2s2 2p4), Pd (5s04d10), the projected augmented wave (PAW) method is used. Both pristine Ti8O16 and substitutional doped Ti8−xO16Pdx (x = 1, 2, 3) structures along with lattice parameters and ionic positions were fully relaxed so that energies and forces reached convergence values. The criterion of electronic convergence in the self-consistent field was 10^{-7} eV and the force convergence was set to be 0.001 eV/Å.
Optical properties of solids are given by the response of the electron system to a time-dependent electromagnetic perturbation occuring by the incoming light. To calculate optical properties we need to calculate this response function which is the complex dielectric tensor. The VASP 5.4 optical programs were run to obtain the frequency dependent dielectric matrix.

3 Results and Discussion

3.1 Electronic Structure

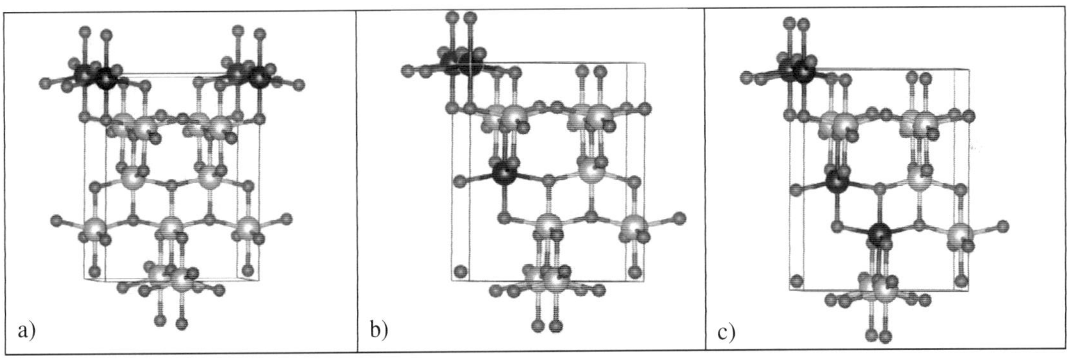

Fig 1 Optimized structures of (2x1x1) Supercell of TiO2 doped with a) 1 Pd atom b) 2 Pd atoms and c) 3 Pd atoms. Colour: Blue: Ti, Red: Oxygen, Violet: Pd.

The tetragonal shaped anatase TiO2 belongs to the space group I41/amd (141). In the anatase structure, every Ti atom is octahedrally bonded to six O atoms where four Ti-O bonds are shorter at
1.93 Å while the other two Ti-O bonds are longer at 1.98 Å[13]. The Total density of states of the Pd doped TiO2 supercells are shown in Figure 2. It is clearly seen that as doping Pd atoms increase in the TiO2 supercell, the band gap narrowing occurs. To understand this phenomenon, let us investigate the partial density of states as given in Figures 3-5.From Figure 3, we see the O 2p orbitals are present significantly in valence band (VB) region of -0.2 to -1.8 eV. The Ti 3d is prominent in the conduction band (CB) region. The Pd 3d states are intense but exist as distinct peaks in CB region of 1.2 -1.8 eV where there is strong hybridization between Ti 3d and Pd 3d orbitals along with O 2p. Hence the bonding is expected to be quite strong. In Figure 4, when we consider the 2 Pd doped case, the electronic structure shows again the O 2p orbitals are intense in VB region compared to CB region. The hybridized orbitals which are extended in the CB

1.2-1.8 eV region occur with higher intensity compared to the 1 Pd case. Distinct peaks are not only absent but the states in the 0.6 and 1 eV region are broader compared to 1 Pd case. The energies of the orbitals are better dispersed in the CB .The densities of the Ti 3d and Pd 3d is higher in the CB in 2 Pd case indicating a stronger hybridization. In the 3 Pd case as seen in Figure 5, there is strongest hybridization between Ti 3d and Pd 3d in CB and the intensity of Pd 3d significantly exceeds that of Ti 3d. The CB extends well within the 0.3 to 1.8 e V region without any distinct peaks as seen in 1 Pd case. The CB is now an extended region of hybridized orbitals in the 0.2 eV range and above. Hence narrowing of the band gap is definitely favoured by increased doping concentration of Pd .The Pd doping introduces new states in the TiO2 band gap and coupled with strong hybridization with Ti 3d orbitals results in band gap narrowing.

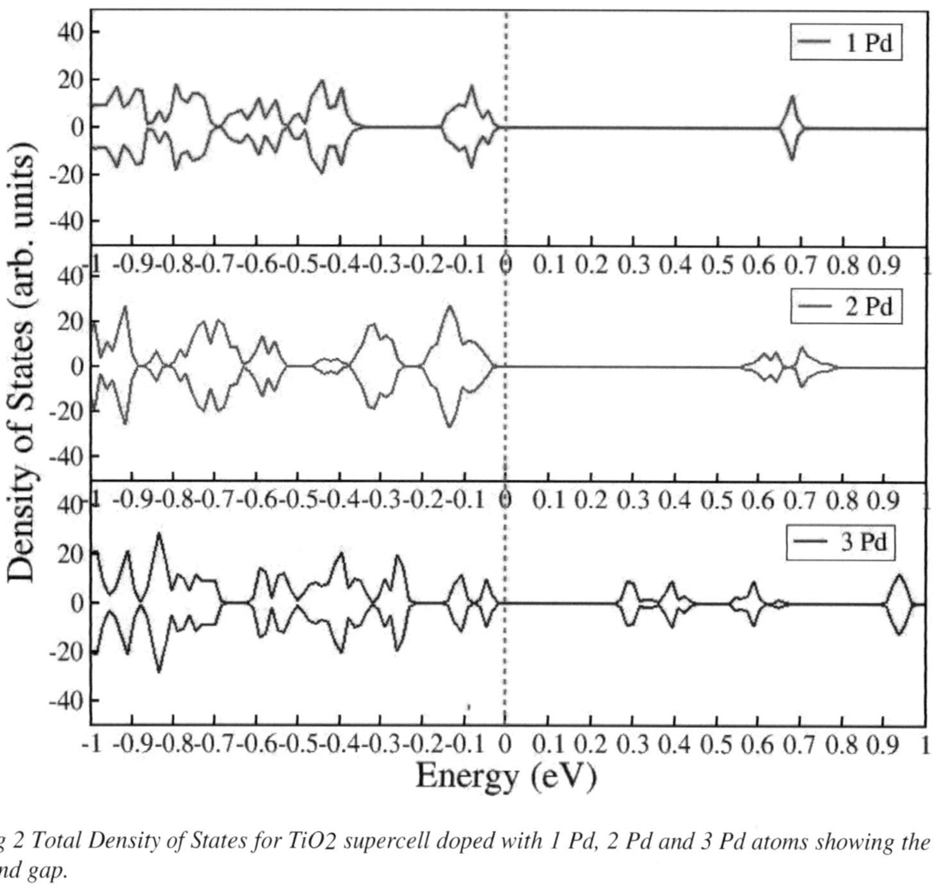

Fig 2 Total Density of States for TiO2 supercell doped with 1 Pd, 2 Pd and 3 Pd atoms showing the respective band gap.

IOP Conf. Series: Materials Science and Engineering 1219 (2022) 012041 doi:10.1088/1757-899X/1219/1/012041

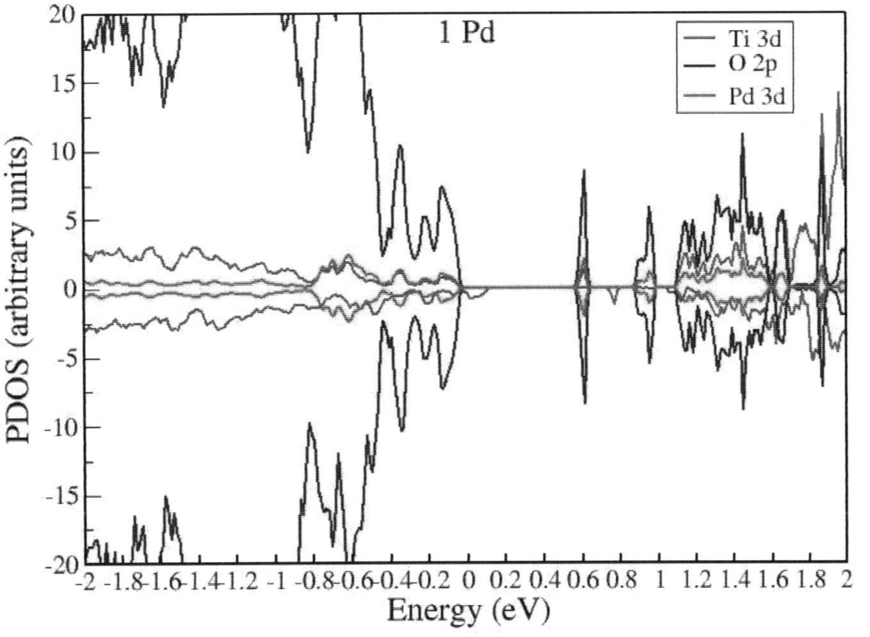

Figure 3 The spin polarized partial density of states for Ti7O16Pd1 supercell (4.17% Pd impurity).

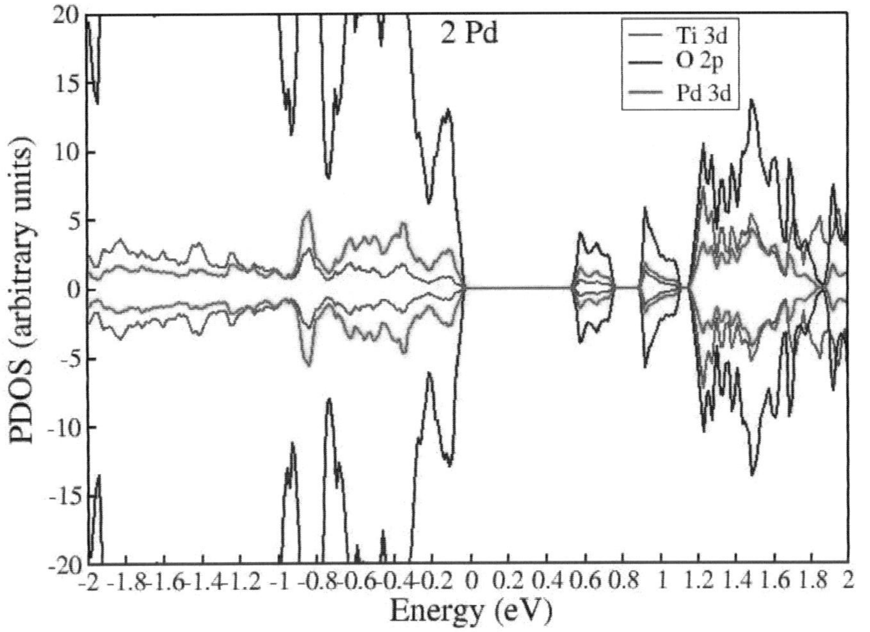

Figure 4 The spin polarized partial density of states for Ti6O16Pd2 supercell (8.33% Pd).

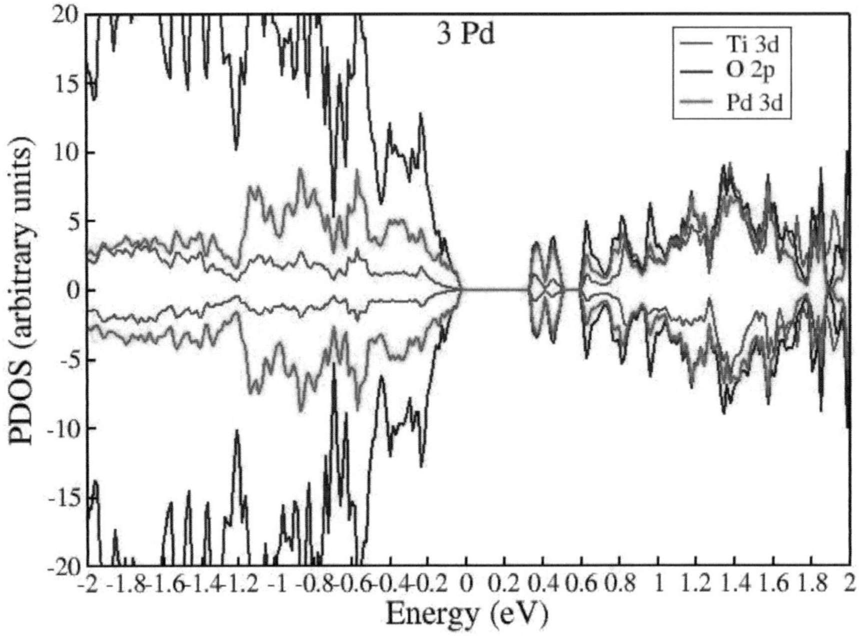

Figure 5 The spin polarized partial density of states for Ti5O16Pd3 supercell (12.5% Pd)

3.2 Charge Density Difference

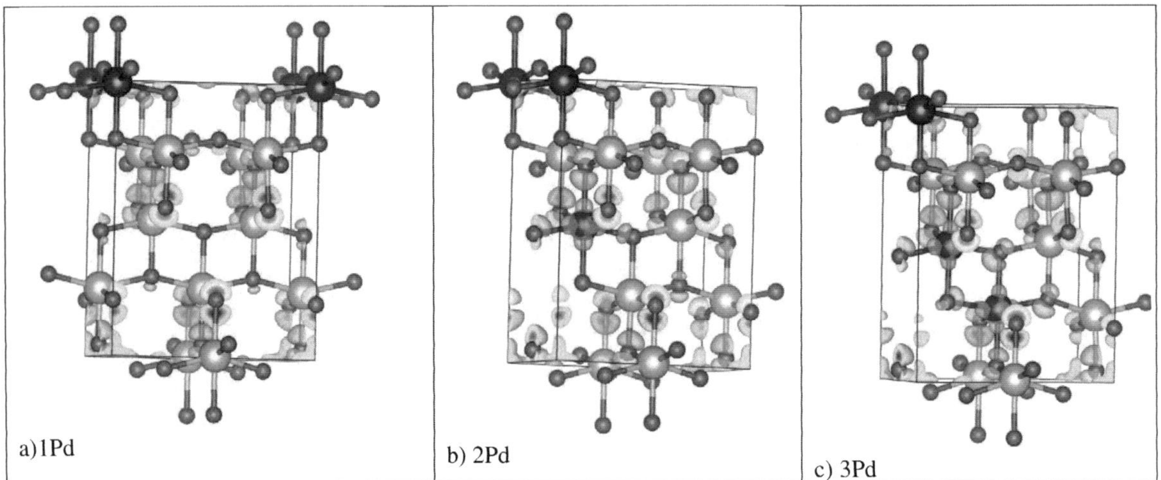

Fig 6 Isosurface charge density of Ti5O16Pd1-3 using isovalue of 0.005 e Å^{-3}. Cyan: charge depletion and Yellow: charge accumulation.

IOP Conf. Series: Materials Science and Engineering 1219 (2022) 012041 doi:10.1088/1757-899X/1219/1/012041

As shown in the charge density plot in Figure 6(a), we can find there is charge being transferred away from Pd atom. The neighbouring O atoms bonded to Pd also lose charge and we see a charge cloud forming in the Pd–O bond region. The O atoms in other positions show charge accumulation.

In Figure 6(b) the charge density plot reveals that both the Pd atoms show loss of significant amount of charge. Also the neighbouring O atoms of Pd lose charge each forming a charge cloud at Pd–O bond as shown. Fig 6 (c) shows all the three Pd atoms lose charge and the nearby oxygen atoms bonded to Pd atoms also lose charge thereby forming a covalent charge cloud in the Pd–O bond region. There is significant charge depletion around all the three dopant Pd atoms and their bonding O atoms. However, other O atoms in the structure show charge accumulation.

3.3 Optical Properties

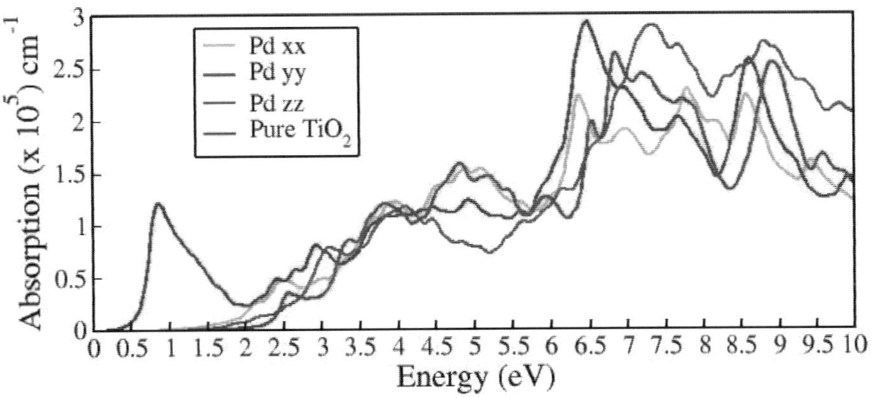

Fig 7 Absorption Spectra of TiO2 doped with 1Pd

In Figure 7, we have compared optical properties of 4.17% Pd doped TiO2 in its three directions with pure TiO2.We know the pure TiO2 form mostly absorbs in the UV range of the spectrum and its spectra here begins from 2 eV as seen in the figure . Upon doping with Pd, the absorption edge is being shifted towards the visible region and the doped structure show significant peaks in the visible spectrum. The maximum absorption in the visible region is around

$1.2 \times 10^5 cm^{-1}$ in the energy region 0.5-1.5 eV predicting that photocatalytic properties of Pd doped TiO2 will prevail in the visible region.

4 Conclusions

The electronic structure of Pd doped TiO2 reveal significant hybridization between Pd 3d and Ti 3d orbitals in the conduction band. As dopant amount increases, the orbitals are extended well across the conduction band region, thereby introducing many new hybrid orbitals into the conduction band, resulting in narrowing of the band gap. The charge density plots indicate charge depletion occurring for the dopant Pd atoms and their bonding O atoms. However charge accumulation is seen in O atoms elsewhere. Thus a covalent bond or charge cloud is prevalent in the Pd-O bond. This explains the narrowing of the band gap as dopant concentration increases from 4.17% to 8.33 % to 12.5%. Pd doped TiO2 has higher absorption coefficient compared to pure TiO2 in the visible region. We predict that optimum Pd doping concentration has successfully altered the electronic and optical properties of TiO2 to yield an efficient visible light photocatalyst.

IOP Conf. Series: Materials Science and Engineering 1219 (2022) 012041 doi:10.1088/1757-899X/1219/1/012041

References

[1] Fujishima A, Zhang X and Tryk D A 2008 TiO2photocatalysis and related surface phenomena *Surf. Sci. Rep.* **63** 515–82

[2] Wang Y, Zhang R, Li J, Li L and Lin S 2014 First-principles study on transition metal-doped anatase TiO2. *Nanoscale Res. Lett.* **9** 46

[3] Chen W, Yuan P, Zhang S, Sun Q, Liang E and Jia Y 2012 Electronic properties of anatase TiO 2 doped by lanthanides: A DFTU study *Phys. B Condens. Matter* **407** 1038–43

[4] Ismael M 2020 A review and recent advances in solar-to-hydrogen energy conversion based on photocatalytic water splitting over doped-TiO2 nanoparticles *Sol. Energy* **211** 522–46

[5] Dorothy A A, Subramaniam N G and Panigrahi P 2019 Tuning electronic and optical properties of TiO 2 with Pt/Ag doping to a prospective photocatalyst: a first principles DFT study *Mater. Res. Express* **6** 045913

[6] Bai X, Lv L, Zhang X and Hua Z 2016 Synthesis and photocatalytic properties of Palladium- loaded three dimensional flower-like anatase TiO2 with dominant {001} facets *J. Colloid Interface Sci.* **467** 1–9

[7] Li D, Jia J, Zheng T, Cheng X and Yu X 2016 Construction and characterization of visible light active Pd nano-crystallite decorated and C-N-S-co-doped TiO2nanosheet array photoelectrode for enhanced photocatalytic degradation of acetylsalicylic acid *Appl. Catal. BEnviron.* **188** 259–71

[8] Moon J, Park J A, Lee S J, Zyung T and Kim I D 2010 Pd-doped TiO2 nanofiber networks for gas sensor applications *Sensors Actuators, B Chem.* **149** 301–5

[9] Meng D, Yamazaki T and Kikuta T 2014 Preparation and gas sensing properties of undoped and Pd-doped TiO 2 nanowires *Sensors Actuators, B Chem.* **190** 838–43

[10] Kresse G and Furthmüller J 1996 Efficient iterative schemes for ab initio total-energy calculations using a plane-wave basis set *Phys. Rev. B - Condens. Matter Mater. Phys.* **54** 11169–86

[11] Perdew, J. P., Burke, K. & Ernzerhof M 1996 Generalized Gradient Approximation Made Simple *Phys. Rev. Lett.* **77** 3865–8

[12] Blöchl PE 1994 Projector augmented-wave method Phys. Rev. B 50 17953–79

[13] Muscat J, Swamy V and Harrison N M 2002 First-principles calculations of the phase stability of TiO2 *Phys. Rev. B - Condens. Matter Mater. Phys.* **65** 2241121–22411215

IOP Conf. Series: Materials Science and Engineering 1219 (2022) 012042 doi:10.1088/1757-899X/1219/1/012042

Polymer / Nanoparticle Hybrid Membrane for Fuel Cell Application

N. Ashwin Sidharth[1], K. Mani Rahulan[1], N. Angeline Little Flower [1*]

Department of Physics and Nanotechnology, SRM Institute of Science and Technology, Kattankulathur-603203, India

Corresponding author's email: angelinn@srmist.edu.in

Abstract The nanoparticle incorporated membranes were found to be optimistic versions of modified traditional polymeric membranes for fuel cell applications, with three main characteristics of enhanced diffusion, mechanical properties, and high resistance to thermal degradation. In Nafion membrane, Ion conductivity, which arises from the interaction between sulfone group and oxidized gas, the redox reactions involve thermal energy generation due to which water present inside the membrane evaporates thus significantly hinders the efficiency of membranes. In this present work, copper oxide and nickel tungstate nanoparticles were synthesized and nanoparticle immobilized polymer membranes were fabricated using the phase-inversion technique. The synthesized nanoparticles were characterized using Field Emission Scanning Electron Microscopy (FESEM) and X-Ray Diffraction (XRD) to analyze its surface morphology, crystallinity and elemental composition of materials were identified using Energy Dispersive X-Ray Analysis (EDAX). The membranes are studied using FESEM for its surface morphology and surface roughness of membrane by Atomic Force Microscopy (AFM) and wettability by Contact Angle measurement.

1. Introduction

The fuel cell expresses a similar energy conversion to that of a battery. The battery consists of chemical species that react to generate a potential difference. In common both the cells the reaction that takes place is an electrochemical reaction that involves electron transfer. Unlike batteries, where the fuel is enclosed into a unit and cannot be refilled. The fuel cell consists of an inlet and outlet valve through which the chemical molecules, i.e., the fuel is sent into the unit which consists of an electrode and electrolyte. Electrodes act as connecting terminals or leads, and solid support for the catalyst layer. The setup consists of a redox agent that reacts with the chemical species to produce heat, product, and most importantly electricity. Figure 1 depicts the basic schematic of the fuel cell. The most favoured fuel cell type is Proton Exchange Membranes (PEMs) fuel cell and reversible fuel cell in which hydrogen(H) and oxygen(O) react to produce water and electricity. The fuel cell setup consists of a series of anode-cathode and electrolyte media through which proton diffuses from the A-side and reacts with the oxygen on the B side. The negative charges i.e., electron flow via external circuit connected to the cell.

The charge transfer happens between anode to cathode. The positive (+) and negative (-) plates are the respective anode and cathode at which noble metal like platinum acts as the catalyst. The catalyst is supported on carbon paper which helps in preventing bulk usage of the catalytic material due to the cost reduction factors. The fuel cells work based on a chemical reaction known as a 'redox reaction'. 'Redox' stands for 'oxidation' and 'reduction' reaction in which, a catalyst reduces and oxidizes the chemical species which occurs at the respective catalytic site. This reaction causes anion cation generation by losing and gaining electrons between the chemical species and thereby a change in the oxidation state of the chemical species. The oxidized and reduced species react with each other causing a flow of current and by-product generation. The PEMs, which is additionally called 'polymer-electrolyte membrane (PEM)', is a selective ion-permeable membrane that only allows H+ ions to diffuse through. The overall fuel cell reaction is given below.

$$\text{Anode Reaction: } H_2 \rightarrow 2H_2^{+} + 4e^{-} \quad (1)$$

$$\text{Cathode Reaction: } O_2 + 4e^{-} \rightarrow 2O^{2-} \quad (2)$$

$$\text{Overall Cell Reaction: } O_2 + 4\,e^{-} + 4\,H^{+} \rightarrow 2H_2O \;\; E^0 = 1.23\text{ V} \quad (3)$$

The H+ ions which in the +1 oxidation state lacks electron and doesn't consist of neutrons and hence are called a proton [1]. These membranes are generally fabricated based on Arrhenius acid ionomers which conduct protons while restricting electron flow. The polymer membrane also acts as a reactant barrier in a fuel cell. An electrolyte is a substance that facilitates the flow of ions between a two-electrode setup. The electrolyte media can in the form of liquid or solid. The polymer can have the property of solid and liquid due to the presence of long chains. Electrolyte ionomers incorporated membranes reduce problems like bulkiness and spillage problems.

PEMs are either made from pure polymer membranes or composite material which are implanted in a polymer matrix. They are either made from pure polymer membranes or composite material which are implanted in a polymer matrix. The commercially available materials are the fluorocarbon-based polymer (PFSA) while Nafion is the commercial name of perfluorinated backbone like Teflon with an ionomer that conducts proton [2],[3].

In this present work, Polysulfone membranes immobilized with copper oxide and nickel tungstate nanoparticles were fabricated and studied. Polysulfone is a polymer with high dimensional stability with a glass transition temperature between 190°C -230°C. The strength and stiffness of the polymer are consistent and stable between −100 °C and 150 °C. Polysulfone polymer is highly hydrophobic in nature and this can be altered by inducing pores into the polymer. Sulfonated polysulfone membranes have recently attracted as a good alternative candidate to be used as proton exchange membrane [4],[5].

Nanoparticle incorporated polymer membrane had shown to be an optimistic version of traditional polysulfone membrane. Nickel tungaste and copper oxide nanoparticles is a p-type semiconducting material. Thereby incorporation of the nanoparticle in the membrane would increase the overall proton conductivity of the membrane.

PEMs Fuel cells use a solid polymer membrane (a thin membrane) which is the electrolyte. The proton conductivity is influenced by the saturation of water inside the polymer matrix but doesn't conduct electrons. For a good proton-conducting membrane, factors like conductivity(σ), permeability(P), and thermal stability are brought into consideration [6],[7].

Figure 1. Basic schematic of a fuel cell

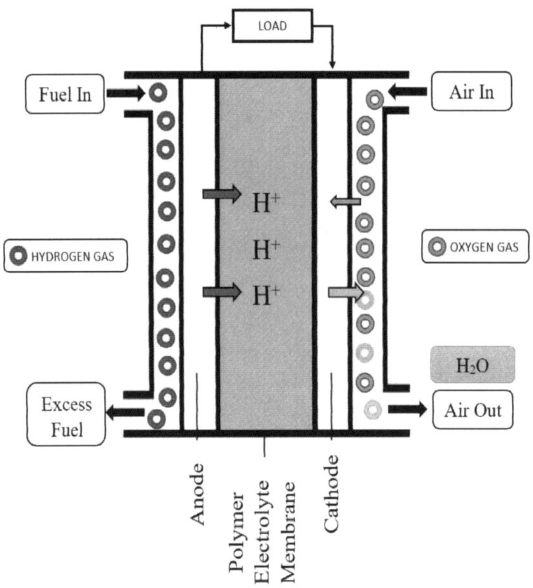

2. Experimental Procedure

Nickel tungstate

For the synthesis of nickel tungstate, nickel chloride of 0.1 M was taken in 20 ml of distilled water. and 0.1 M of Sodium Tungstate was taken in 20 ml distilled water separately. The sodium tungstate solution which was prepared was added slowly to the nickel chloride solution. The solution was stirred at 500 RPM for half an hour [8-12]. the green precipitate was washed at 8000 RPM with ethanol and water mixture in a 1:1 ratio three times and dried at 60°C for 24 hours and collected. The XRD analysis has shown the formation of nickel tungstate with an amorphous crystalline structure and Eds analysis has shown the presence of the desired elemental composition.

Copper oxide

To prepare copper oxide nanoparticles, copper (II)chloride of 0.1M was taken in 20ml of distilled water and 0.1M of sodium borohydride was taken in 20 ml distilled water separately. The sodium borohydride solution which was prepared was added slowly to the copper (II) chloride solution [13-16]. The solution was stirred at 500 rpm for half an hour. the black precipitate was washed at 8000 RPM with ethanol and water mixture in 1:1 ration three times and dried at 60°C for 24 hours and collected. The XRD analysis has shown the formation of copper oxide with a cubic structure.

Polysulfone membrane

The required amount of Polysulfone (PSF) is weighed according to its weight percentage and is dissolved in N-Methyl-2-pyrrolidone (NMP) at an adequate temperature, the solute and the solvent are mixed on a hot plate. After the mixture is fully dissolved, a previously weighed amount of Polyvinylpyrrolidone (PVP) is added and the mixture is stirred on a hot plate. The glass plate and the elcometer were cleaned using acetone. The membrane thickness is set by adjusting the pitch screw. The mixture is poured into the provided slot and is dragged across the glass plate. The glass plate is placed in a distilled water bath, which acts as a coagulant. After a period of 24 hours, the membrane is taken out and is pat to remove excess water using ultra soft tissue and keep at room temperature for drying [17-28]. The steps involved in membrane fabrication is shown in figure 2. In the same way polysulfone membranes immobilized with nanoparticles can also be fabricated by dispersing the nanoparticles in NMP to form nanoparticles immobilized polymer composite membranes.

Figure 2. Steps involved in membrane fabrication (a) Polysulfone in NMP, (b) PVP added to polysulfone, (c) Membrane dragging (d) After dragging (e) Phase Inversion

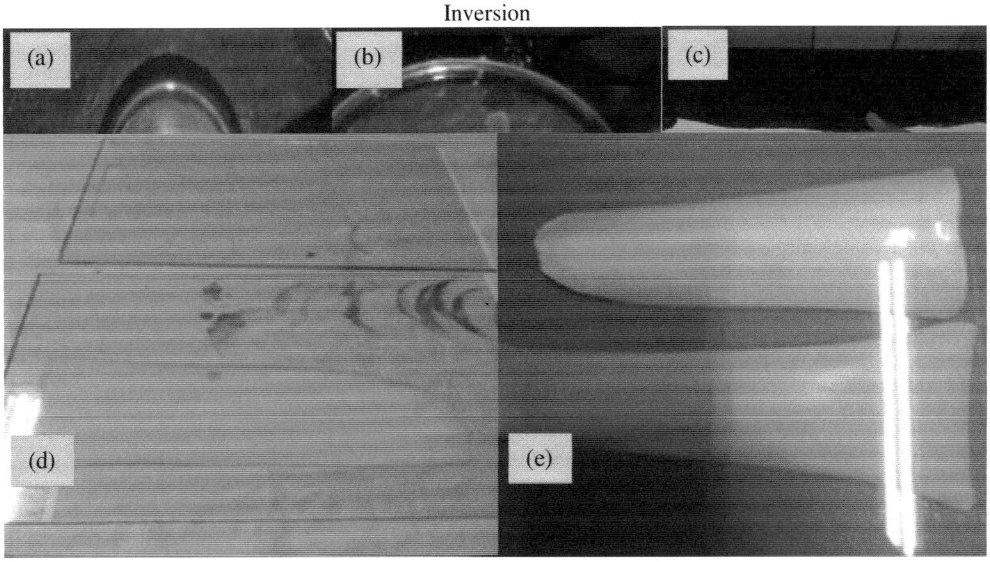

3. Results and discussion

X-ray diffraction study.

Figure 3: X-ray diffraction studies of synthesized materials (a) Copper oxide, (b)Nickel tungstate

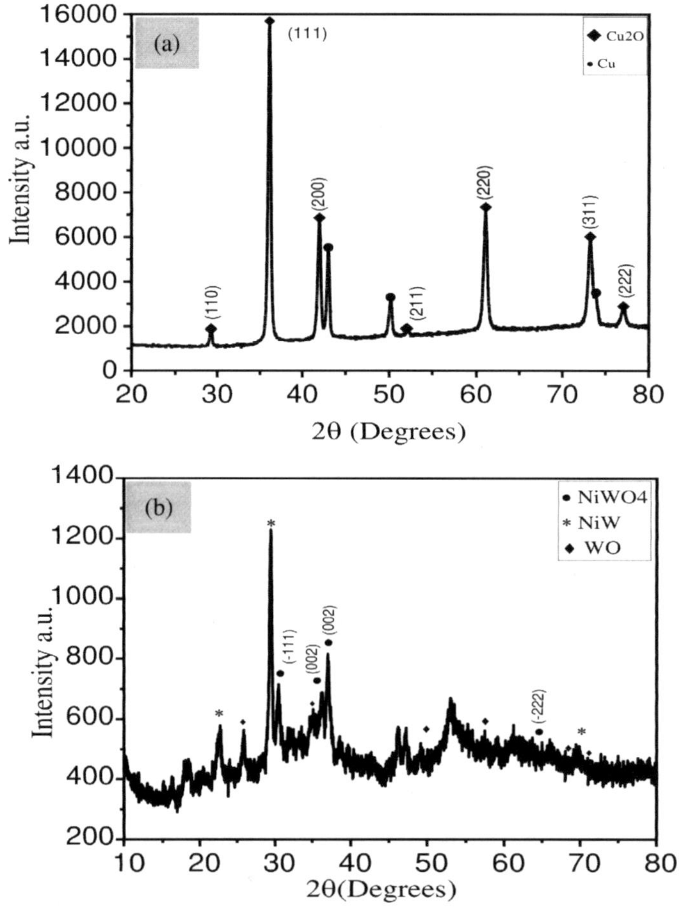

XRD pattern of Cu_2O nanoparticle is shown in Figure 3(a) confirms to be a single-phase with Cubic structure. The lattice parameters are a = 4.2580 Å, b = 4.2580 Å, c = 4.2580 Å. The peaks intensity and position are analogous to the reported values in (JCPDS file No.01-077-0199).

XRD pattern of Nickel tungstate nanoparticles was shown in Figure 3(b). confirms to be a single-phase with a Monoclinic structure. The lattice parameters area=4.6000 Å, b = 5.6650 Å, c = 4.9120 The peaks intensity and position are analogous to the reported values in (JCPDS file No.00-015-0755).

Figure 4. Scanning Electron Microscope images of (a) Nickel tungstate & (b)Polysulfone Membrane

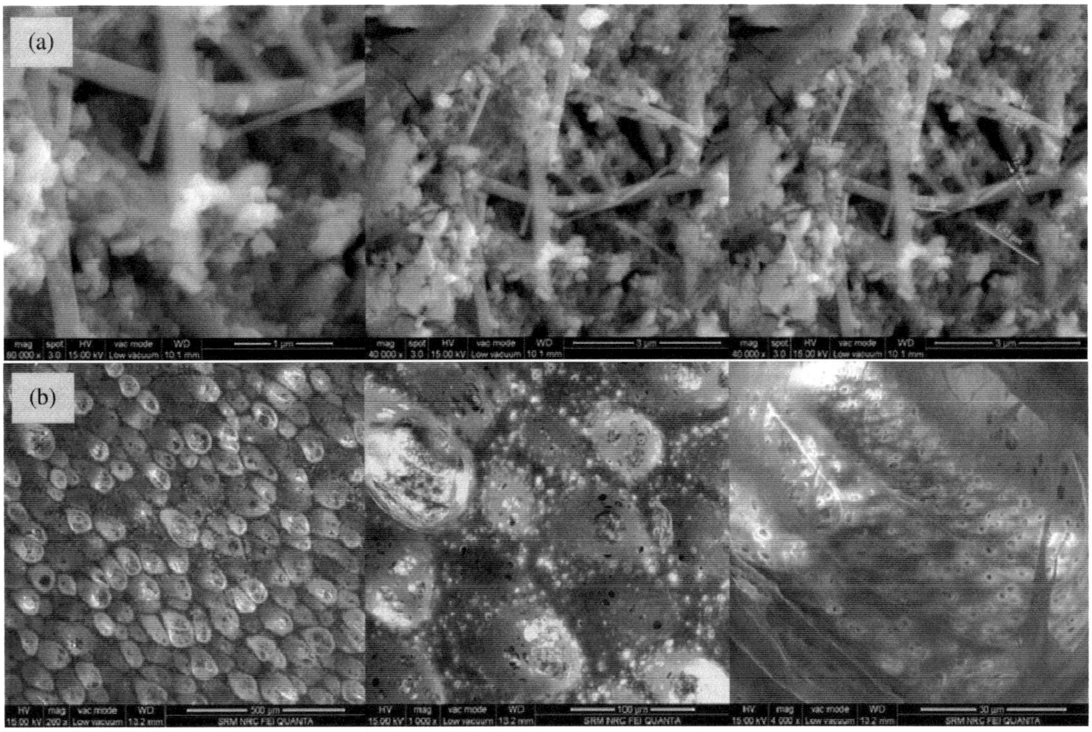

Polysulfone membrane with a porous structure is synthesized successfully by the phase inversion method. From the Scanning Electron Microscope (SEM) images, the pores confirmed to be circular with a mean size of $9.5 \mu m$.

Figure 5. EDS of (a) Nickel tungstate & (b)Polysulfone Membrane

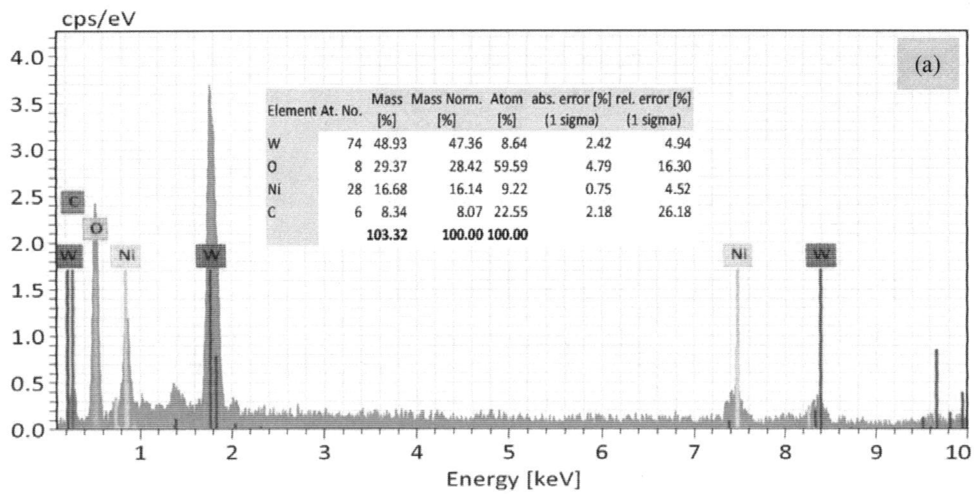

Element	At. No.	Mass [%]	Mass Norm. [%]	Atom [%]	abs. error [%] (1 sigma)	rel. error [%] (1 sigma)
W	74	48.93	47.36	8.64	2.42	4.94
O	8	29.37	28.42	59.59	4.79	16.30
Ni	28	16.68	16.14	9.22	0.75	4.52
C	6	8.34	8.07	22.55	2.18	26.18
		103.32	**100.00**	**100.00**		

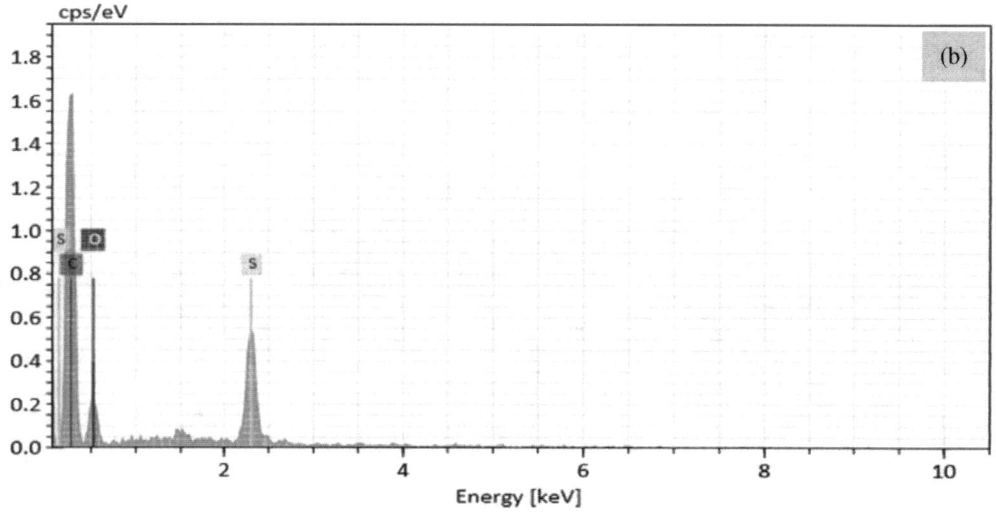

The elemental composition of the synthesized nickel tungstate and polysulfone was studied using EDX is shown in Figure 5(a) and 5(b) showing the absence of contaminants.

Figure 6. Contact angle image of Pure Polysulfone Membrane

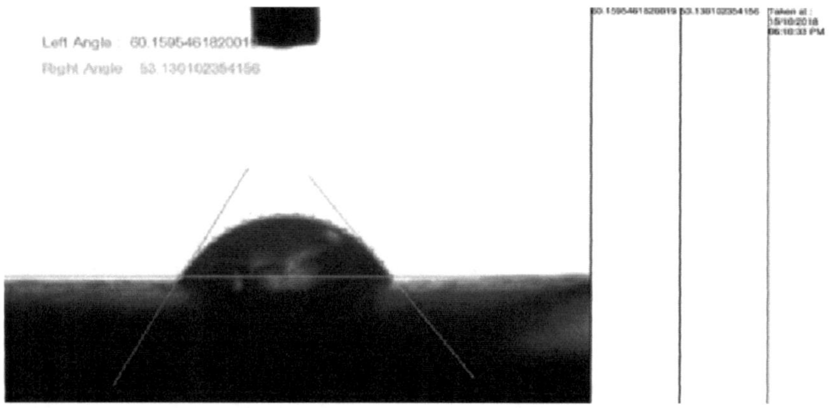

The contact angle of the synthesized polysulfone membrane was studied using the Sessile drop technique as shown in figure 6 where a water droplet is made to come in contact with the membrane surface. And the spreading of water is measured with respect to time and angle θ by which the surface energy of the membrane is calculated. Which in turn is used to classify the membrane as hydrophobic or hydrophilic

Table 1: Determination of surface energy and spreading coefficient

Matrix	Contact angle		Surface energy (N/M)	Wettability
	Advancing angle θa	Receding angle θr		
Polysulfone	60.15	53.13	0.73485888	High

The advancing angle θa and receding angle θr obtained using the Sessile drop technique is shown in table (1) The surface energy of the Polysulfone membrane is calculated to be 0.04704 N/m and spreading coefficient is measured to be 0.73485888 and equilibrium angle θc calculated to be 55.79 degrees. Since S>0 the membrane is classified as hydrophilic. The surface tension of water is 72.2mJ/m^2.

Calculation of surface energy of solid surface:

$$\gamma_S^{tot} = \frac{\gamma_L(1 + cos\theta_a)^2}{2 + cos\theta_r + cos\theta_a}$$

$$\gamma_S^{tot} = Surface\ free\ energy\ of\ solid\ substrate$$

$$\gamma_L = Surface\ tension\ of\ liquid\ used$$

$$\theta_a = advancing\ angle\ (contact\ angle\ measurement)$$

$$\theta_r = receding\ angle\ (contact\ angle\ measurement)$$

Surface Tension of water (72.2 **mJ/m^2**)

Figure 7. AFM Images of (a) polysulfone membrane with topography, (b)&(c) 2D image and 3D images of polysulfone

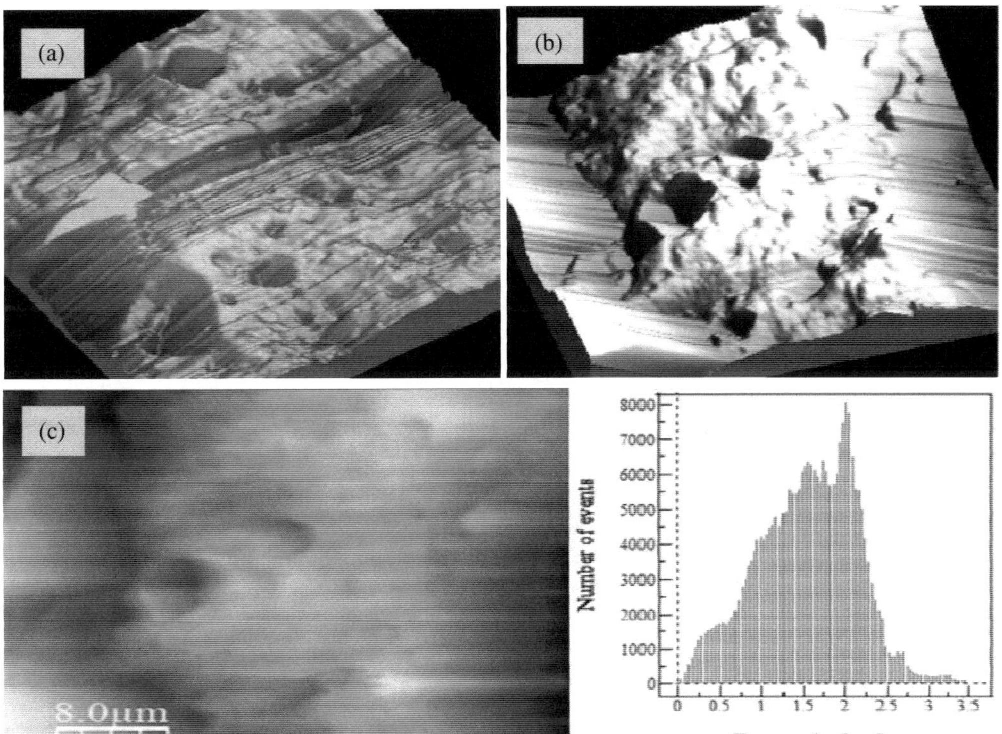

The surface roughness of the synthesized polysulfone membrane was studied using AFM. Figure 7 was obtained after an AFM tip is made to come in contact with the membrane surface and measures the surface area. the topography data gives the height of the peaks and troughs of the membrane which determines the

IOP Conf. Series: Materials Science and Engineering 1219 (2022) 012042 doi:10.1088/1757-899X/1219/1/012042

surface roughness. Roughness influences the hydrophilicity of the membrane. Figure 4.5(b) is the 2D and 3D images of the membrane surface. Average roughness was measured 0.4839 μm Root mean square (RMS) roughness was 0.5969μm.

4. Conclusions

In the current study, copper oxide and nickel tungstate nanoparticles were synthesized and characterized. Pure polysulfone membranes were also fabricated and studied using SEM and AFM which shows the morphology of the membranes with the highly porous nature and the average surface roughness . Surface roughness causes the suppression of the hydrophobic nature and has more wettability which was proved by contact angle measurement which leads to enormous application. Polysulfone has a high glass transition temperature due to which issues like carbonization below 100°C doesn't occur. Polysulfone membrane is highly resistant to acid attacks and has stable mechanical property even at temperatures higher than 60°C. Which makes it suitable for its application in a hydrogen fuel cell.

References

[1] Andrews, C. C. and Murphy, O. J. (2003). "Alternative electrochemical systems for ozonation of water.

[2] Kim, D. J., Jo, M. J., and Nam, S. Y. (2015). "A review of polymer–nanocomposite electrolyte membranes for fuel cell application." Journal of Industrial and Engineering Chemistry, 21,36–52.

[3] Lufrano, F., Squadrito, G., Patti, A., and Passalacqua, E. "Sulfonated polysulfone as promising membranes for polymer electrolyte fuel cells." 77(6),1250–1256.

[4] Hickner, Michael A., et al. "Alternative polymer systems for proton exchange membranes (PEMs)." Chemical reviews 104.10 (2004): 4587-4612.

[5] Prasad, M., Mohanty, S., and Nayak, S. K. "Polymer electrolyte membranes based on sulfonated polysulfone and functionalized layered silicate for direct methanol fuel cell applications.

[6] Yang, Z., Coutinho, D., Feng, F., Ferraris, J. P., and Balkus Jr, K. (2004). "Novel in-organic/organic hybrid electrolyte membranes." Prepr. Pap.-Am. Chem. Soc., Div.Fuel Chem, 49(2),599.

[7] Seyyedamirhossein Hosseini et al 2018 Semicond. Sci. Technol. 33 055008

[8] Pourmortazavi, Seied Mahdi, et al. "Procedure optimization for green synthesis of silver nanoparticles by aqueous extract of Eucalyptus oleosa." Spectrochimica Acta Part A: Molecular and Biomolecular Spectroscopy 136 (2015): 1249-1254.

[9] Adib, Kourosh, et al. "Facile chemical synthesis of cobalt tungstates nanoparticles as high performance supercapacitor." Journal of Materials Science: Materials in Electronics 27.5 (2016): 4541-4550.

[10]Habibi-Yangjeh, Aziz, and Maryam Shekofteh-Gohari. "Novel magnetic Fe3O4/ZnO/NiWO4 nanocomposites: enhanced visible-light photocatalytic performance through pn heterojunctions." Separation and Purification Technology 184 (2017): 334-346.

[11]Pirhashemi, Mahsa, and Aziz Habibi-Yangjeh. "ZnO/NiWO4/Ag2CrO4 nanocomposites with pnn heterojunctions: highly improved activity for degradations of water contaminants under visible light." Separation and Purification Technology 193 (2018): 69-80.

[12]Khanna, P. K., et al. "Synthesis and characterization of copper nanoparticles." Materials Letters 61.25 (2007): 4711-4714.

[13]Ruparelia, Jayesh P., et al. "Strain specificity in antimicrobial activity of silver and copper nanoparticles." Acta biomaterialia 4.3 (2008): 707-716.

[14]Dhas, N. Arul, C. Paul Raj, and Aharon Gedanken. "Synthesis, characterization, and properties of metallic copper nanoparticles." Chemistry of materials 10.5 (1998): 1446-1452.

[15]Mott, Derrick, et al. "Synthesis of size-controlled and shaped copper nanoparticles." Langmuir 23.10 (2007): 5740-5745.

[16]Kim, K. S., et al. "Surface modification of polysulfone ultrafiltration membrane by oxygen plasma treatment." Journal of membrane science 199.1-2 (2002): 135-145.

[17]Chakrabarty, B., A. K. Ghoshal, and M. K. Purkait. "Ultrafiltration of stable oil-in-water emulsion by polysulfone membrane." Journal of Membrane Science 325.1 (2008): 427-437.

IOP Conf. Series: Materials Science and Engineering 1219 (2022) 012042 doi:10.1088/1757-899X/1219/1/012042

[18]Fan, Huiqin, et al. "A facile strategy towards developing amphiphobic polysulfone membrane with double Re-entrant structure for membrane distillation." Journal of Membrane Science 602 (2020): 117933.

[19]Mohammadi, Maryam, and Shahram Mehdipour-Ataei. "Durable sulfonated partially f luorinated polysulfones as membrane for PEM fuel cell." Renewable Energy 158 (2020): 421-430.

[20]Tang, Hongying, et al. "Synthesis and properties of phosphonated polysulfones for durable high-temperature proton exchange membranes fuel cell." Journal of Membrane Science 605 (2020): 118107.

[21]Lufrano, E., et al. "Barrier properties of sulfonated polysulfone/layered double hydroxides nanocomposite membrane for direct methanol fuel cell operating at high methanol concentrations." International Journal of Hydrogen Energy 45.40 (2020): 20647-20658.

[22]Hu, Yidong, Liuming Yan, and Baohua Yue. "Sulfonation Mechanism of Polysulfone in Concentrated Sulfuric Acid for Proton Exchange Membrane Fuel cell Applications." ACS omega 5.22 (2020): 13219-13223.

[23]Simari, Cataldo, et al. "Highly-performing and low-cost nanostructured membranes based on Polysulfone and layered doubled hydroxide for high-temperature proton exchange membrane fuel cells." Journal of Power Sources 471 (2020): 228440.

[24]Zhang, Xinxin, et al. "Preparation of sulfonated polysulfone/sulfonated titanium dioxide hybrid membranes for DMFC applications." Journal of Applied Polymer Science 137.32 (2020): 48938.

[25]Yang, Z., Coutinho, D., Feng, F., Ferraris, J. P., and Balkus Jr, K. (2004). "Novel in- organic organic hybrid electrolyte membranes." Prepr. Pap.-Am. Chem. Soc., Div. Fuel Chem, 49(2), 599.

[26]C.N.R. Rao, S. Natarajan, R. Vaidhyanathan, Angew. Chem., Int. Ed. 43 (2004)1466. R. Matsuda, R. Kitaura, S. Kitagawa, Y. Kubota, R.V. Belosludov, T.C. Kobayashi, H. Sakamoto, T. Chiba, M. Takata, Y. Kawazoe, Y. Mita, Nature 436 (2005) 238.

[27]P. Wassweacheid, W. Keim, Angew. Chem., Int. Ed. 39 (2000) 3773.

[28]E.R. Copper, C.D. Andrew, P.S. Wheatley, P.B. Webb, P. Wormald, R.E. Morris, Nature 430(2004) 1012.T. Steiner, Semiconductor Nanostructures for Optoelectronic Applications, Artech House, London, 2004.

IOP Conf. Series: Materials Science and Engineering 1219 (2022) 012043 doi:10.1088/1757-899X/1219/1/012043

ONE POT POLYOL SYNTHESIS OF Fe_2O_3-Fe_3O_4 NANO COMPOSITES AND THEIR STRUCTURAL, OPTICAL, PROPERTY STUDIES

M. Jeyavenkatesh, J. Arunodaya, Trilochan Sahoo*

Department of Physics & Nanotechnology, SRM Institute of Science and Technology, SRM Nagar, Kattankulathur-603203, Kancheepuram, Chennai, TN, India

**Corresponding author email:* trilochs@srmist.edu.in

Abstract. The magnetic iron oxides are classified into three phases known as Magnetite (Fe_3O_4), Magnetite (-Fe_2O_3), Hematite (-Fe_2O_3).The ferric oxide synthesis with the excess of the ferrous oxide nanoparticles was carried out by the co-precipitation method. The precursors Ferric sulphate hydrate and Ferrous sulphate heptahydrate taken in the molar ratio of 1:2 in 100 ml of water and 30 ml of Hydrochloric acid added to initiate precipitation at 90 ^{O}C with vigorous stirring the ammonia solution was added. The prepared materials are characterized by XRD, and UV-Vis spectroscopy. The XRD showed formation of phase and crystallization of the nanomaterial is prepared. The UV-Vis spectroscopy used to determine the reflectance and absorptance as well as the optical bandgap of the nanomaterial, in the range of 800 nm to 200 nm. The analysis indicated formation of ferrous oxide impregnated ferric oxide nanoparticles with desired optical band gap.

Keywords: Ferrites, polyol synthesis, photocatalyst, nanocomposite

1. Introduction

Nanostructured materials have been studied for many technological, industrial and environmental applications. Nanoparticle exhibit many unique optical and chemical properties compared to their bulk counter parts [1]. Magnetic materials in many form and structure have been in application mainly as storage and high frequency sensor devices [2,3]. Magnetic nanoparticles have garnered a lot of attention for application in image contrast in MRI, magnetic labeling, and drug delivery [4-6]. Over the last few years magnetic materials have been studied for photo-catalysis and water splitting application. Magnetic nanoparticles as photo catalysis have the advantage over non-magnetic photo-catalysts that they can be recovered by magnetic separation and can be reused multiple times [7-10]. Ferrites such as ferric oxide and ferrous oxide are interesting materials for photo catalysis and hydrogen production application. Ferrites are soft magnetic materials could be easily separated by magnetic separation. Ferrites chemically stable, nontoxic and large quantities could be processed cost effectively. Cation distribution, surface area and morphology of ferrites play significant role in performance as catalyst [8,11]. Kaneko et al. prepared multicomponent zinc ferrites and studied their ability for hydrogen production in presence of sun light by water splitting. They demonstrated a 93% yield of the theoretical limit [12]. Mangrulkar and colleagues studied hydrogen production by nano ferrites in presence of Pt as co-catalyst. They observed that hydrogen production rate is strongly influenced by reactor configuration and temperature [13].

Ferrites could be prepared by a number of conventional synthesis processes such as, ball milling, spray pyrolysis, sol-gel, co-precipitation [14,15]. Manin et. al. demonstrated processing of well dispersed spherical cobalt ferrites. Zhang et al. reported preparation Ni doped ferrites by mechanical ball milling the precursors and followed by sintering at 750 ^{o}C, for magnetic studies [16]. Sharma et al. adopted sol-gel route for preparation of Zn doped nickel ferrite and studied their photo catalytic activity and observed an enhancement in performance of catalytic activity [18]. Among the various techniques co-precipitation is very powerful and interesting approach to prepare nanoscale and multicomponent or multiphase materials [19-21]. Joshi et al. prepared nanoparticle of Ni ferrite by co-precipitation method and performed structural analysis. They observed Ni^{2+} occupy octahedral and tetrahedral sites in the spinel structure. Phase formation and size control of the particle in co-precipitation processing depends on a number of factors. Further preparation of multicomponent system by co-precipitation

IOP Conf. Series: Materials Science and Engineering 1219 (2022) 012043 doi:10.1088/1757-899X/1219/1/012043

approach requires skillful control of the synthesis parameters. We present synthesis and analysis of ferric oxide with the excess of the ferrous oxide nanoparticles by the co-precipitation method

2. Materials and Method

Ferrous sulphate heptahydrate ($FeSO_4.7H_2O$) and Ferric sulphate hydrate ($Fe_2(SO_4)_3H_2O$) as the main precursors and additional to this water (H2O), Hydrochloric acid (HCl) and Ammonia solution (NH4OH) are used as starting precursor. Precursor solutions in 100ml of water was prepared by varying Fe^{2+} ion concentration and Fe^{3+} ion concentration. HCl (30 ml) was added to completely dissolve the solid precursors. The solution was stirred at 90°C with addition of drop wise addition ammonia solution which resulted in black precipitation. The precipitate cooled and was filtered and washed by ethanol and acetone. The precipitate was then dried at 90°C for three hours and characterized. The Fe^{2+} : Fe^{3+} concentration ratios of 1 and 4 was used for synthesis of the of nanoparticles.

3. Result and Discussion

X-ray Diffraction

The X-ray diffraction (XRD) pattern of Fe3O4 shown in Fig. 1 (a). The pattern show diffraction peaks from (220), (311), (400), (333), (440) planes of the Fe3O4. The pattern confirms that formation rhombohedral Fe3O4 nanoparticles. Crystallite size and strain in the nanoparticles calculated using the Scherrer formula. Figure 1(b) shows the X-ray diffraction pattern (XRD) of Fe_2O_3-Fe_3O_4 nanoparticles prepared by coprecipitation method. The XRD pattern exhibit diffraction peaks corresponding to (115), (118), (313), (316), (332), (408), (505), (525) plane of Fe2O3-Fe3O4 composite nanoparticle. Which indicate that the nanoparticles are crystallized in rhombohedral phase (JCPDS reference is 024 – 0072), The full width at half maximum of the (118) was measured to be 0.2754 and average crystallite size calculated to be 33.18 nm using Scherrer formula. The Table.1. show the crystallite size, strain of the two nanocrystals. When Fe^{2+} : Fe^{3+} =1, concentration maintained it resulted formation Fe_3O_4 and as the ration of concentration maintained to at 2, formation of Fe_2O_3/Fe_3O_4 composite nanoparticles formed.

Table.1. X-ray diffraction

	Fe3O4	Fe2O3/Fe3O4
Crystalline size D(nm)	8.536 nm	33.189 nm
Stress □	0.6644	0.1964
FWHM □	1.0296	0.2754
Structure	Rhombohedral	Rhombohedra

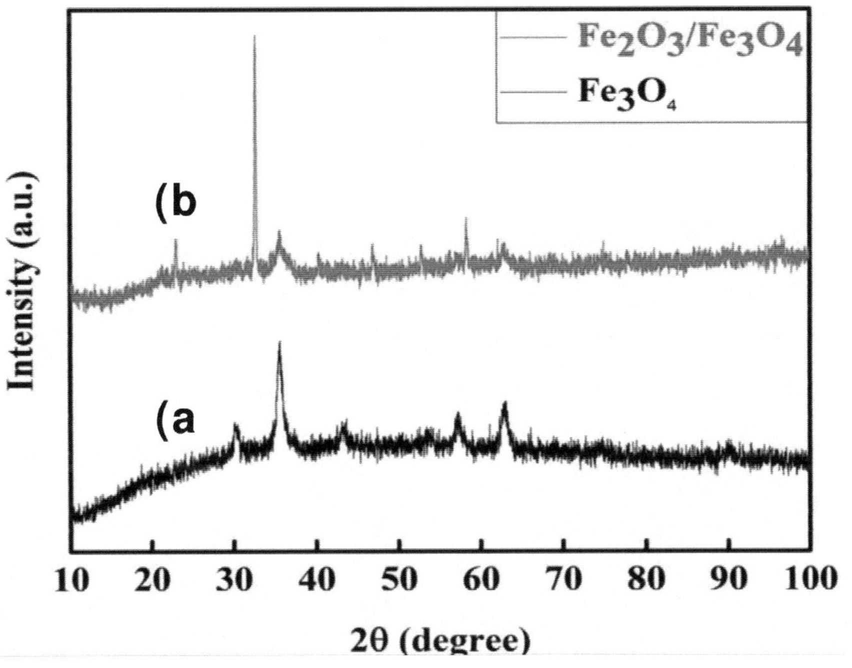

Figure 1. XRD pattern of (a) Fe3O4 and (b) Fe2O3/Fe3O4 composite nanoparticles prepared by coprecipitation technique.

UV-Vis Characterization

The UV-Vis absorption spectra of Fe3O4 and Fe2O3/Fe3O4 nanoparticle synthesized by co-precipitation technique. Fig.2(a) represent the UV-Vis spectrum of Fe3O4 nanoparticle and Fig.2(b) represent the UV-Vis spectrum of Fe2O3/Fe3O4 composite nanoparticle. From the spectra it can be clearly seen the absorption in the both type of nanoparticle has stronger absorption near UV band compared to visible band of the electromagnetic spectrum. Also, it can be observed that Fe3O4 has stronger absorbance than Fe2O3/Fe3O4 composite nanoparticles. Optical band of Fe3O4 and Fe2O3/Fe3O4 composite nanoparticles are calculated by plotting Tauc plot. Figure 3(a) and 3(b) show the Tauc plot of Fe3O4 and Fe2O3/Fe3O4 composite nanoparticles. The band gap of Fe3O4 and Fe2O3/Fe3O4 composite nanoparticles calculated to be 2.08 eV and 1.91 eV. The band gap found to reduce for Fe2O3/Fe3O4 composite nanoparticles marginally compared Fe3O4 nanoparticles.

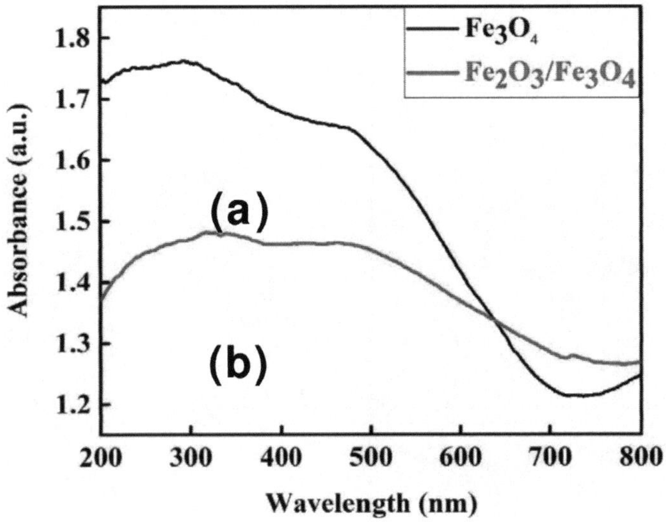

Figure 2. UV-Vis absorption spectrum of (a) Fe3O4 and (b) Fe2O3/Fe3O4 composite nanoparticles prepared by coprecipitation technique.

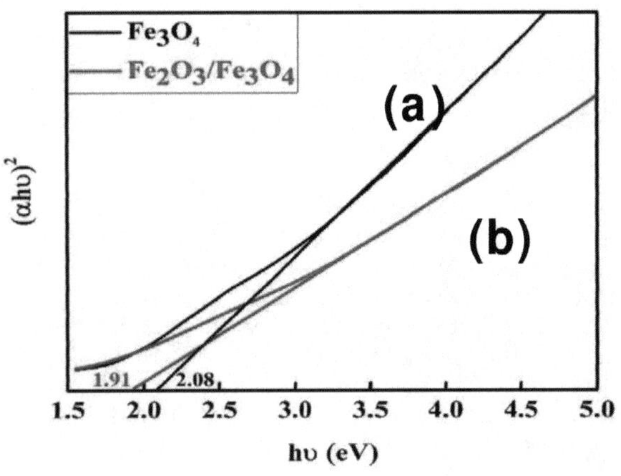

Figure 3. Tauc plot of (a) Fe3O4 and (b) Fe2O3/Fe3O4 composite nanoparticles prepared by coprecipitation technique.

Summary

Pure Fe3O4 and composite Fe2O3/Fe3O4 nanoparticles was prepared very simple coprecipitation technique. Just by simple variation ferric and ferrous ion concentration ratio simple Fe3O4 or Fe2O3/Fe3O4 nanoparticles are

IOP Conf. Series: Materials Science and Engineering 1219 (2022) 012043 doi:10.1088/1757-899X/1219/1/012043

synthesized. Structural analysis confirms formation of rhombohedral crystal structure. The band gap of the prepared sample was found be in the visible range which could be suitable for visible light photocatalysis.

References

1. P. I. Varghese, T. Pradeep, A textbook of Nanoscience and Nanotechnology, Mcgraw Hill Education (India) Pvt. 2017.
2. R. L. Comstock, Review- modern magnetic material in data storage, J. Mat. Sci.: Mat. Electronics, 13, (2002), 509–523.
3. G. V. Kurlyandskaya, S. V. Shcherbinin, S. O. Volchkov, S. M. Bhagat, E. Calle, R. Pérez, M. Vazquezbe, Soft magnetic materials for sensor applications in the high frequency range, J. Magn. Magn. Mater. 459, (2018),154158.
4. C. M. Fu, Y. F. Wang, Y. F. Guo, T. Y. Lin, and J. S. Chiu, " In Vivo Bio-Distribution of Intravenously Injected Tc-99 m Labeled Ferrite Nanoparticles Bounded with Biocompatible Medicals," IEEE Trans. Mag., 41, (2005), 4120– 2.
5. H. E. Horng, S. Y. Yang, Y. W. Huang, W. Q. Jiang, C. Y. Hong, and H. C. Yang, " Nanomagnetic Particles for SQUID-Based Magnetically Labeled Immunoassay," IEEE Trans. Appl. Supercond., 15, (2005), 668– 71 6. S. H. Gee, Y. K. Hong, D. W. Erickson, and M. H. Park, " Synthesis and Aging Effect of Spherical Magnetite (Fe3O4) Nanoparticles for Biosensor Applications," J. Appl. Phys., 93, (2003), 7560– 2.
7. H. Wang, L. Xu, C. Liu. Z. Jiang, Q. Feng, T. Wu, R. Wang, A novel magnetic photocatalyst Bi3O4Cl/SrFe12O19: Fabrication, characterization and its photocatalytic activity, 46 (2020), 460-467.
8. P. N. Anantharamaiah, S. Mondal, K. S. Manasa, S. Saha, M. Pai M, Enhancing the catalytic activity of recyclable nanocrystalline NiFe2O4 by replacing Ni by Cu, 46, (2020),1220-1226.
9. A. Goyal, S. Bansal, S. Singhal, Facile reduction of nitrophenols: comparative catalytic efficiency of MFe2O4 (M = Ni, Cu, Zn) nanoferrites, Int. J. Hydrogen Energy, 39 (2014), 4895-4908
10. M. Salavati-Niasari, F. Davar, T. Mahmoudi, A simple route to synthesize nanocrystalline nickel ferrite (NiFe2O4) in the presence of octanoic acid as a surfactant Polyhedron, 28 (2009), 1455-1458
11. C. Singh, A. Goyal, S. Singhal, Nickel-doped cobalt ferrite nanoparticles: efficient catalysts for the reduction of nitroaromatic compounds and photo-oxidative degradation of toxic dyes, Nanoscale, 6 (2014), 7959-7970.
12. H. Kaneko, N. Gokon, N. Hasegawa, Y. Tamaura, Solar thermochemical process for hydrogen production using ferrites, Energy, 30, (2005), 2171-2178
13. P. A. Mangrulkar, V. Polshettiwar, N. K. Labhsetwar, R. S. Varma, S. S. Rayalu, Nano-ferrites for water splitting: unprecedented high photocatalytic hydrogen production under visible light, Nanoscale, 4, (2012) 5202-5209
14. K. K. Kefeni, T. A. M. Msagati, B. B.Mamba, Ferrite nanoparticles: Synthesis, characterisation and applications in electronic device, Mat. Sci. Engg. B, 215 (2017) 37-55.
15. R. Srivastava, B. C. Yadav, Ferrite Materials: Introduction, Synthesis Techniques, and Applications as Sensors, Inter. J. Green Nanotech.4 (2012) 141-154.
16. R. V. Minin, V I Itin, V. A. Zhuravlev, V. A. Svetlichnyi, Synthesis of cubic ferrite CoFe 2O4 by spray pyrolysis, Journal of Physics Conference Series 1115(4) (2018) 042011.
17. Z. Zhang, G. Yao, X. Zhang, J, Ma, H. Lin, Synthesis and characterization of nickel ferrite nanoparticles via planetary ball milling assisted solid-state reaction, Ceram. Inter., 41,(2015)4523-4530
18. R. Sharma, S. Singhal, Structural, magnetic and electrical properties of zinc doped nickel ferrite and their application in photo catalytic degradation of methylene blue, Physica B: Cond. Matter, 414, (2013),83-90
19. I. M. Kolthoff, Theory of Coprecipitation. The Formation and Properties of Crystalline Precipitates, J. Phys. Chem. 36 (1932) 860–881
20. S. Joshi, M. Kumar, S. Chhoker, G. Srivastava, M. Jewariya, V.N.Singh, Structural, magnetic, dielectric and optical properties of nickel ferrite nanoparticles synthesized by co-precipitation method, J. Mol. Structure, 1076 (2014), 55-62.
21. Ch.Srinivas, B.V.Tirupanyamb, S.S.Meen, S.M.Yusuf, Ch.Seshu Babu, K.S.Ramakrishna, D.M.Potukuchi, D.L.Sastry, Structural and magnetic characterization of co-precipitated NixZn1−xFe2O4 ferrite nanoparticles, J. Magn. Magn. Mater. 407, (2016),135-141

Development of Al6061-B_4C Composite and Study the Effect of Heat Treatment on its Mechanical Properties

Rajvir Singh[a,*], Paawan Khanna[a], Ranvir Singh Panwar[b], Jagmohan Datt[c]

[a] *Research Scholar,* [b] *Assistant Professor,* [c] *Associate Professor,*
Department of Metallurgical and Materials Engineering, Punjab Engineering College (Deemed to be University), Sector-12 Chandigarh, 160012, India

E-mail ID: rajvirsingh.meta@pec.edu.in

Abstract. Aluminum composites are growing in demand as an advanced material in aerospace industries, automobile industries, marine applications and structural applications. In the current work, an attempt has been made to study the mechanical properties of the monolithic AA6061 alloy, it's composite by reinforcing B_4C (4wt%) particles and T6 heat treated composite on the same ground. Al6061-B_4C composite was developed by stir casting method and uniform distribution of particles were observed throughout the bulk which was confirmed by SEM-EDS. Subsequently T6 heat treatment was done on composite. Reinforcement in the matrix increase the hardness value of base alloy by 21.3%. The hardness value of heat-treated Al6061-B_4C composite was found to be increased by 20.5% than that of non-heat-treated composite which may be due to formation of coherent Mg_2Si precipitates. The presence of B_4C particles and gas porosity in the matrix of aluminum alloy tend to decrease the value of impact strength of composite and heattreated composite.

Keywords: *Aluminum composite, Boron Carbide, T6 Heat Treatment, Toughness, Impact Test*

1. Introduction

Aluminum alloys and its composites are widely used in aerospace, marine and automobile applications due to their high strength to weight ratio, wear resistant properties and high stiffness [1,2]. Aluminum Metal matrix Composites (AMMCs) can be developed by mainly two techniques enlisted in literature, one by stir casting method and another by powder metallurgy method. Stir casting process being economic and for mass production is mostly accepted process for developing AMMCs. The hardening phase in AMMCs can be in the form of particles, whiskers and fibers as reinforcement.

In literature, it was found that many researchers successfully enhanced the mechanical and tribological properties of monolithic aluminum alloys and AMMCs by using various processing like heat treatment, mechanical working, etc. Mazahery et al. found that the Al2024-B4C composite showed better hardness than monolithic Al2024 alloy[1]. Kalaiselvan et al. produced Al6061-B4C composite and concluded in their study that the microhardness was increased from 51.3 HV to 80.8HV with respect to addition of B4C particles and enhancement in tensile strength was observed from 185MPa to 215MPa by addition of reinforcement into matrix alloy[2]. Nagaral et al. successfully developed the Al2024-nano B4C composite and concluded in their study that the properties like hardness, yield strength and fracture strength can be enhanced by reinforcing nano B_4C particles into the Al2024 matrix[3].Nagaral et al. in their another study revealed that the elongation in the composite decreases with increasing boron carbide content thereby increasing the tensile strength because of high hardness of boron carbide[4].

In the present study an attempt has been done to investigate the mechanical properties for alloyin its three forms, as casted matrix alloy (MA), its composite (NHTC) and composite after heat treatment

IOP Conf. Series: Materials Science and Engineering 1219 (2022) 012044 doi:10.1088/1757-899X/1219/1/012044

(HTC). Matrix alloy has been processed by stir casting technique to make particle reinforced composite and further its T6 heat treatment has been done to know its response to this sequence of processes.This novel research work can be used to compare the properties of aforesaid three states of same aluminum alloy on a single platform.

2. EXPERIMENTAL DETAILS

The experimental procedure follows the flow chart a s shown in fig. 1.

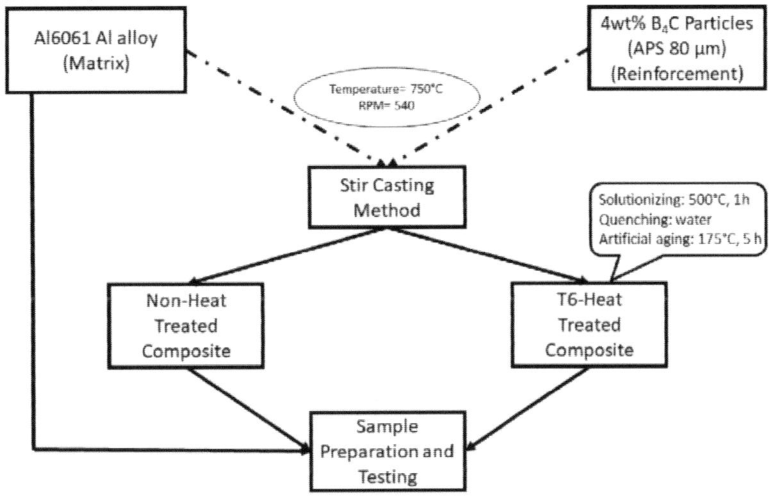

Fig. 1: Flow chart of work undertaken

Materials and Process

The matrix material, Al6061 is having a chemical composition enlisted in table 1.

Table 1: Chemical composition of Al6061

Alloying elements	Si	Mg	Fe	Cr	Mn	Cu	Ti	Other elements
Wt %	0.4-0.8	0.8-1.2	Max. 0.7	0.04-0.35	0.15	0.15-0.4	0.15	0.15

B_4C particles (APS 80 μm) is used as reinforcement in the composite. The XRD analysis as shown in fig. 2.confirmed that all the major peaks correspond to Boron carbide.

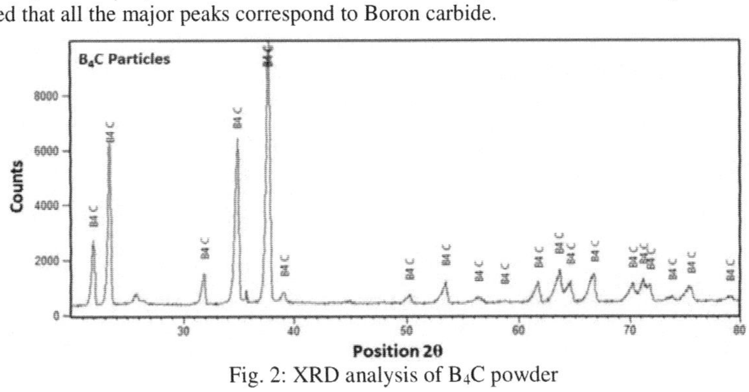

Fig. 2: XRD analysis of B_4C powder

Stir casting method was used for synthesis of composite in the current study. Firstly, A charge of 500g of aluminum alloy 6061 is melted in graphite crucible at a temperature of 750°C with help of the electric resistance heating furnace. The melt was agitated with the help of mechanical stirrer at 540 rpm to form a vortex. 2 wt% Mg metal was added to improve the wettability of reinforced particle with matrix.4 wt% B_4C particles preheated at 450°C for 1 hour were added into the molten vortex and stirred for 8 minutes for homogenous distribution of particles in the matrix. Then the stirrer was stopped and crucible was ejected from the furnace. The molten mixture was poured into moldpreheated at 400°Cand allowed to solidify in the mold as solid castings.

T6 heat treatment was used for Al6061-B_4C composite which include 3 steps. Firstly, solutionizing heat treatment was done by heating the specimen to 500°C in Programmable heating furnace and holding it at that temperature for 1 hour which is required to dissolve the intermetallic phases present in aluminum alloy in a single α-phase[5, 6]. The temperature should not be above the solidus temperature as it can lead to oxidation of the aluminum alloy. This oxidation can lead to a drastic decrease in ductility of material.The heating rate for heat treatment was set at 10°C/minute. Second step was to quickly water quench the samples and then removing from the water after cooling. This step is used to obtain a metastable supersaturated solution from the single-phase solid solution. Third step included artificial ageing at 175°C for 5 hours in programmable muffle furnace with a heating rate of 10°C/min.The process involves formation of embryo of dissolved intermetallic phase through thermal fluctuation and they subsequently precipitated out on the α-phase. As the time passes more and more precipitates form [7].

Mechanical properties were compared for casted base alloy (CBA), non-heat treated stir casted Al6061-B_4C composite (NHTC) and T6 heat treated stir casted Al6061-B_4C composite (HTC).

Characterization

Characterization of CBA, NHTC and HTC was done with the help of optical microscopy (Make: Radical, RXLr-4M), Scanning electron microscopy (SEM) and energy dispersive spectroscopy (EDS) (Make: JEOL, JSM-IT500).

Hardness

Micro-hardness of the CBA, NHTC and HTC samples was determined using Vicker's hardness tester(Make: Omni-tech, MVH-II). The surface of samples waswell-polishedby metallographic technique.The load taken was 500 gf for the dwell time of 10s.

Impact test

Izod test was carried out to measure toughness of theCBA, NHTC and HTC samples. The V-notched cylindrical samples were prepared according to ASTM E-23 standard. The samples were of 75 mm length and 11.4 mm diameter with a V-notch of 45° and depth of notch is 3.27 mm. The range of the blow hammer used was 0-160 Joules with the weight of 18.7 kg. Firstly, the hammer was blown to know the energy loss by air resistance and then it was blown with the specimen mounted in machine supported in cantilever position 90° to the blow hammer. Then difference of each energy is taken to know the energy absorbed by the specimen before fracture.

E_1= Energy absorbed by the air drag and friction, in kg-m

E_2= Energy absorbed by the air drag, friction and the specimen, in kg-m

Then, **Toughness, $E = E_1 - E_2$, in kg-m**

To convert it into Joules, multiply it with acceleration due to gravity, g (m/s^2)

And, **Notch Impact Strength = E/A**

Where, A= cross sectional area

3. results and discussion

Characterization

Optical micrographs of CBA, NHTC and HTCare shown in fig. 3. The dendritic growth of Si in CBA has been observed and developed composites showed nearly uniform distribution of B_4C particles in the matrix. This confirms the high efficiency of stir casting process and its respective chosen parameters.

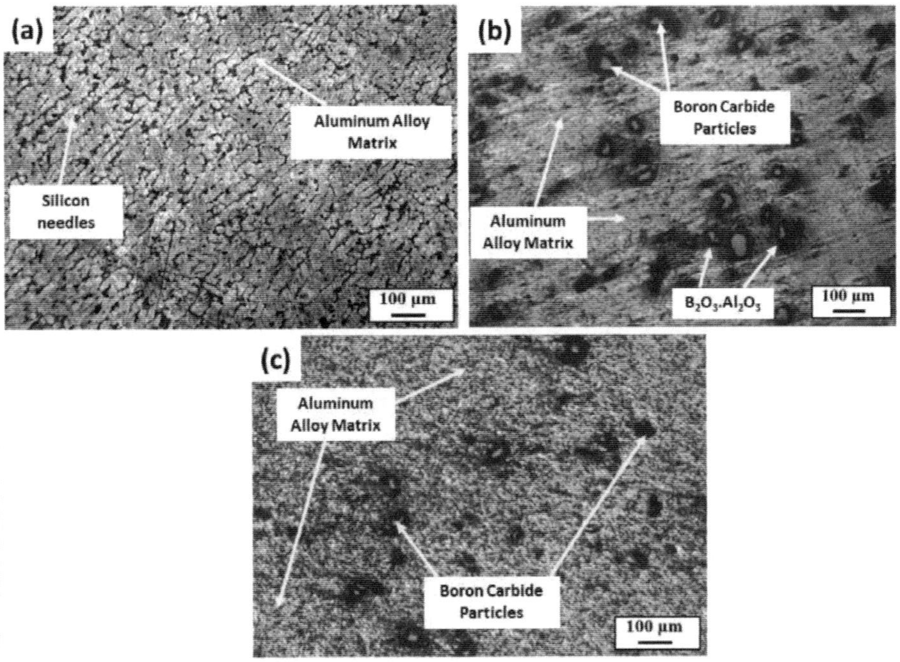

Fig. 3: Optical micrographs of (a) CBA,(b) NHTC and (c) HTC

In NHTC, the darker region around the particles is due to reactive nature of B_4C with air at elevated temperature, which in turn form B_2O_3 compound which helps to increase liquid metal interaction with the particle by increasing the interfacial energy between them. This was confirmed by EDS analysis as shown in fig. 4 (a).However this cannot be seen for another ceramic particle reinforcements (like SiC, TiB_2, Al_2O_3, etc) where oxide layer formation reduces the wettability, B_2O_3 compound when interacted with liquid aluminum the wettability enhances by the formation of $B_2O_3.Al_2O_3$ at the interface [8, 9].For HTC, there can be seen the dendritic structure corresponds to the coherent Mg_2Si precipitates also known as β'' which is confirmed by EDS analysis as indicated in fig. 4 (b).

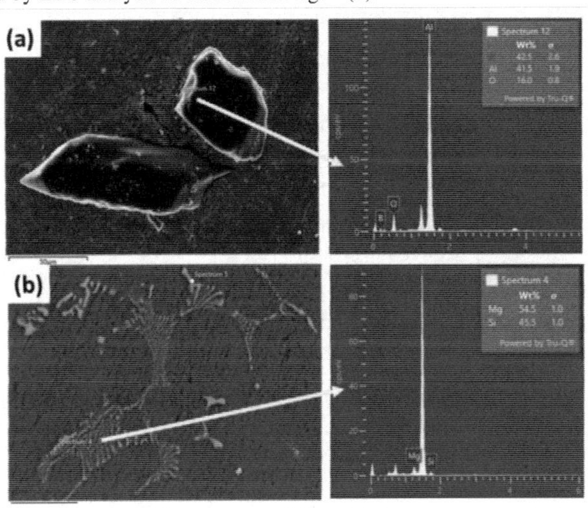

Fig. 4: EDS analysis confirming (a) the presence of $B_2O_3.Al_2O_3$at particle matrix interface in NHTC and (b) Mg_2Si precipitatesin HTC

IOP Conf. Series: Materials Science and Engineering 1219 (2022) 012044 doi:10.1088/1757-899X/1219/1/012044

SEM micrographs as shown in fig. 5 (a & b) reveal the homogenous dispersion of B_4C particles in aluminum matrix and good bonding between the particle and matrix. However, in some regions porosity and clustering of particles is observed. This observed surface porosity is not due to weak interface bonding but due to leaving of particles from matric while cutting or machining during sample preparation leaving pores in the casting [10].Fig 5 (c) taken in back-scattered mode indicated the dendrites of Mg_2Si precipitatein HTC.The B_4C particle shown is fully embedded and wetted by matrix. No porosity or any gap is obesrved between matrix and particle. It is clear that B_4C particles are forming good bonds with the matrix Al6061 alloy.

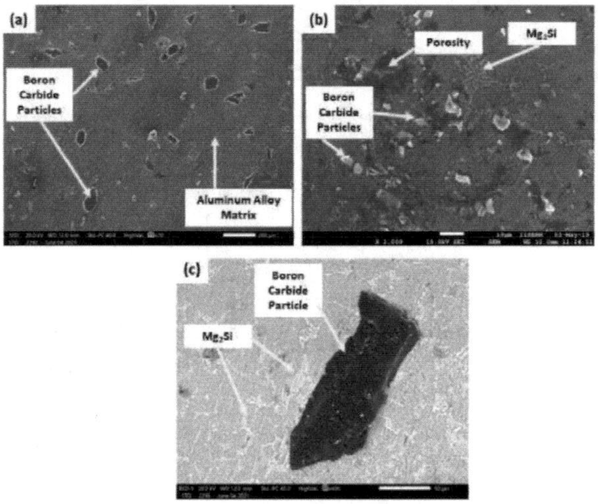

Fig. 5: SEM images showing (a) distribution of B_4C particles in NHTC; (b) B_4C particles and Mg_2Siprecipitatesin HTC and c) bonding at particle-matrix interface of HTC (back-scattered image)

Hardness test

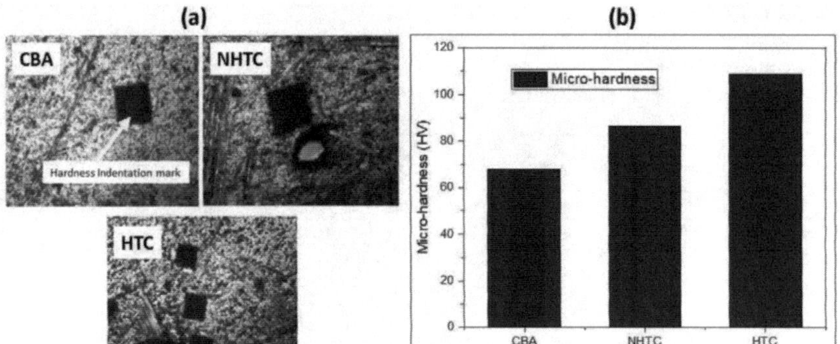

Fig.6:(a) Hardness indentation marks during Vicker's microhardness test and (b) Graph showing microhardness of CBA, NHTC and HTC.

As expected, hardness of the NHTC increased after the ageing process as shown in Fig.6.CBA showed lower value of hardness than that of NHTC and HTC. NHTC showed 21.3% more hardness than that of CBA and HTC showed 20.5% more hardness than that of NHTC. 37.4% increase in hardness was observed for HTC than that of CBA. The increase in hardness in HTC was attributed to the formation of coherent Mg_2Si precipitates also known as β''. It should be noted that B_4C reinforcement forms showed good wettability with aluminum matrix and precipitates of Mg_2Si formed in the matrix which is advantageous over the Al-SiC composite in which aluminum carbide is formed at matrixreinforcement interface during heat treatment which is highly undesirable. The Mg_2Si phases are responsible for increase in microhardness value [11].

Impact strength

The results showed in fig. 7 (a &b) are indicating that addition of B_4C particles in CBA and T6 heat treatment of the composite resulted in reduction in toughness and impact strength. It was observed that there is 33.3% decrease in toughness due to particle addition and 40.1% decrease due to T6 heat treatment. 60% decrease in toughness was observed for HTC than that of CBA.

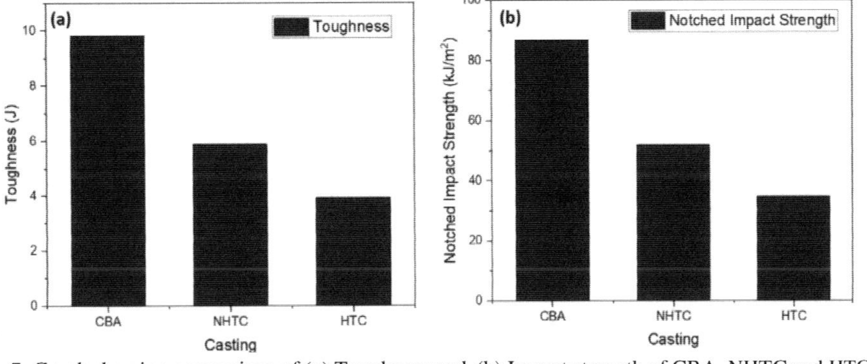

Fig. 7: Graph showing comparison of (a) Toughness and (b) Impact strength of CBA, NHTC and HTC

The lower toughness and impact strength of HTC can be attributed to the fact that precipitation of brittle Mg_2Si phase which hardens the composite also contribute to decrease in ductility of the composite after ageing. The decreased values in HTC can be attributed to increased stress concentrated areas by the formation of hardening precipitates in a composite which are already brittle in nature.Addition of B_4C particle although increases the hardness value but decreases the impact toughness value [12].

Acknowledgement

Authors would like to acknowledge the DRDO-ARMREB sponsored project for financial support for carrying out the current research work.

Conclusions

(i) Al6061-B_4C composite was successfully fabricated by the stir casting process. The optical and SEM-EDS analysis revealed the homogenous distribution of B_4C particles in the matrix. Subsequently T6 heat treatment was done on composite which resulted in the formation of brittle Mg_2Si phase.

(ii) Reinforcement in the matrix increase the hardness value of base alloy by 21.3%. The hardness value of heat-treated Al6061-B_4C composite was found to be increased by 20.5% than that of non-heat-treated composite which may be due to formation of Mg_2Si precipitates. Total of 37.4% increase in hardness was observed for T6 heat treated composite than that of base alloy.

(iii) The presence of B_4C particles and gas porosity in the matrix of aluminum alloy tend to decrease the value of impact strength of composite as it acts as stress concentrated sites in composite. The heat treatment further reduces the impact strength due to the formation of brittle Mg_2Si precipitates in the of aluminum matrix.

References

[1] A. Mazahery and M. O. Shabani, "Influence of the hard coated B_4C particulates on wear resistance of Al-Cu alloys," *Compos. Part B Eng.*, vol. 43, pp. 1302–1308, 2012, doi: 10.1016/j.compositesb.2012.01.011.

[2] K. Kalaiselvan, N. Murugan, and S. Parameswaran, "Production and characterization of AA6061-B_4C stir cast composite," *Mater. Des.*, vol. 32, pp. 4004–4009, 2011, doi: 10.1016/j.matdes.2011.03.018.

[3] M. Nagaral, B. K. Shivananda, V. Auradi, and S. A. Kori, "Development and mechanical-wear characterization of Al2024-nano B_4C composites for aerospace applications," *Strength, Fract. Complex.*, vol. 13, pp. 1–13, 2020, doi: 10.3233/SFC-190248.

[4] M. Nagaral, A. J. S, T. Jagadish, and M. Walikar, "Impact of Micro B_4C Particulates Addition on Mechanical Behavior of Al2024 Alloy Composites," *Int. J. Eng. Sci. Invent.*, vol. 7, pp. 78– 84, 2018, [Online]. Available: www.ijesi.org.

[5] F. Ozturk, A. Sisman, S. Toros, S. Kilic, and R. C. Picu, "Influence of aging treatment on mechanical properties of 6061 aluminum alloy," *Mater. Des.*, vol. 31, no. 2, pp. 972–975, 2010, doi: 10.1016/j.matdes.2009.08.017.

[6] C. Xu, H. He, W. Yu, and L. Li, "Influence of quenching temperature on peak aging time and hardness of Al-Mg-Si-Cu alloys strengthened by nano-sized precipitates," *Mater. Sci. Eng. A*, vol. 744, 2018, pp. 28–35, 2019, doi: 10.1016/j.msea.2018.11.149.

[7] S. Pogatscher, H. Antrekowitsch, H. Leitner, T. Ebner, and P. J. Uggowitzer, "Mechanisms controlling the artificial aging of Al-Mg-Si Alloys," *Acta Mater.*, vol. 59, pp. 3352–3363, 2011, doi: 10.1016/j.actamat.2011.02.010.

[8] S. Y. Oh, J. A. Cornie, and K. C. Russell, "Wetting of ceramic particulates with liquid aluminum alloys: Part I. Experimental techniques," *Metall. Trans. A*, vol. 20, pp. 527–532, 1989, doi: 10.1007/BF02653932.

[9] S. Y. Oh, J. A. Cornie, and K. C. Russell, "Wetting of ceramic particulates with liquid aluminum alloys: Part II. Study of wettability," *Metall. Trans. A*, vol. 20, pp. 533–541, 1989, doi: 10.1007/BF02653933.

[10] K. M. Shorowordi, T. Laoui, A. S. M. A. Haseeb, J. P. Celis, and L. Froyen, "Microstructure and interface characteristics of B_4C, SiC and Al_2O_3 reinforced Al matrix composites: A comparative study," *J. Mater. Process. Technol.*, vol. 142, pp. 738–743, 2003, doi: 10.1016/S0924-0136(03)00815-X.

[11] S. Sharma, T. Nanda and O. P. Pandey, "Investigation of T4 and T6 heat treatment on the wear properties of sillimanite reinforced LM30 aluminium alloy composites", *Wear* vol. 426–427, pp. 27–36, 2019, doi: 10.1016/j.wear.2018.12.065.

[12] S. Karabulut, H. Karakoc and R. Cıtak, "Influence of B_4C particle reinforcement on mechanical and machining properties of Al6061/B_4C composites", *Compos. Part BEng*. vol. 101, pp. 87– 98, 2016. doi: 10.1016/j.compositesb.2016.07.006.

Formation of Interfacial Reaction Products in Al-4Si-1Mg/SiC Composite

R.Raghu[1], M.J.Prathap[2], P.Ponpichai Pandiyan[2], R.Nandhakumar[2], A.Logeshwaran[2], T.Ragul[2]

[1]Assistant Professor, Sri Ramakrishna Engineering College, Coimbatore, India, 641022
[2]UG student, Sri Ramakrishna Engineering College, Coimbatore, India, 641022

Email: raghu.r@srec.ac.in, messiraghu.777@gmail.com

Abstract. In Al/SiC composite, Al and SiC reacts and forms Al_4C_3 at the interface according to the reaction $4Al + 3SiC \rightarrow Al_4C_3 + 3Si$. Hence in the present work, the Al-4Si-1Mg/SiC is fabricated through stir casting process. The fabricated composite is also ultrasonically treated prior to superheating to induce the interfacial reaction. The as cast composite and the ultrasonic treated composite are superheated at 900°C and computer aided cooling curve analysis is performed in the composite for different time to quantify the extent of interfacial reaction and to study the solidification behaviour. The Si released as a by-product can alter the composition and solidification behaviour. The salt flux reclamation technique is employed to extract the interfacial reaction products from the composite. The reclaimed powders are subjected to X-ray diffraction analysis and confirmed the presence of Al_4C_3 particles, which be used for grain refinement applications.

Keywords: Aluminium, Composite, Al_4C_3, Stir Casting, Ultrasonic.

1. Introduction

Composite is defined as the combination of two or more materials that has better properties than those of the individual materials used alone. Compared to metallic alloys, each material retains its separate chemical, physical, and mechanical properties in the composite. The key advantages of composite materials are high strength and stiffness combined with low density when compared with bulk material. They are used in various applications. The two constituents of the composites are matrix and reinforcement. The reinforcing phase delivers the strength and stiffness to the composite. Reinforcement is harder than matrix, usually particles. Metal Matrix Composites (MMCs) are widely used in several industries due to its numerous advantages. MMCs consist of a metallic matrix and dispersed ceramic or metallic phase. MMCs exhibit good mechanical properties however it depends upon the structure and chemical composition of the interface and the near-by interface region. The load transfer must take place from the matrix to the reinforcement through the matrix/reinforcement interface to achieve greater mechanical properties. The load transfer efficiency greatly depends upon the bond strength at the interface. Thus the composition of the interface plays greater role in determining the mechanical properties of the MMCs. Aluminium/Silicon Carbide (SiC) composites exhibit greater mechanical properties however the structure and chemical composition of the interfaces determines the level of properties. For greater mechanical properties, the load transfer should take place through the matrix/reinforcement interface. The formation of reaction products at the interface greatly affects mechanical properties of the composite. Various studies shown that Al and SiC reacts and forms Al_4C_3 at the interface according to the reaction

$$4Al + 3SiC \rightarrow Al_4C_3 + 3Si \quad (1.1)$$

The occurrence of such reaction has numerous detrimental effects on the properties of the composite. The resultant product Al_4C_3

1) Reduces the mechanical and physical properties since it is hygroscopic in nature.
2) Decreases the matrix/reinforcement interfacial strength.
3) Makes the composite to corrode under some environment such as in water, methanol and HCl due to its unstability.
4) Silicon is released due to the reaction which forms Al-Si at the interface and the grain boundary area which consequences in undesired properties of the composite.

To avoid the formation of Al_4C_3 in the composite, two methods commonly employed among various methods. Firstly, Si addition to the Al matrix and secondly, oxidizing SiC artificially to develop SiO_2 layer on the SiC. The extent of reaction also depends upon the processing method and their temperature and holding time. The formation of interfacial reaction product Al_4C_3 is disadvantageous in Al/SiC composite but it can be used as potential grain refiner in magnesium-based alloys for the enhancement of strength.

Magnesium alloys are gaining greater importance in automotive and aerospace sectors due to its low density, high specific strength and high specific stiffness. However, the range of application has been limited due to its poor processing ability and mechanical properties. Therefore, it is highly important to enhance its mechanical property for extensive usage in several applications. Grain refinement is an effective method to enhance the mechanical properties of Mg alloys. There is large number of grain refining methods existing for magnesium alloys but only very few commercial grain refiners available.

The various methods available are rapid cooling, super heating of melt, addition of zirconium and inoculation of carbon. Rapid cooling is not preferred for castings whereas superheating utilizes extra energy and increases the oxidation. Addition of zirconium to Mg-Al alloys result in poisoning of the alloy due to stable intermetallic formation. Carbon inoculation using different materials (C_2Cl_6, CCl_4, CaC_2) can be done for greater refining of magnesium alloys however the addition of such materials becomes difficult. Also, toxic gas is released while process which cause hazard to the environment. The greatly accepted concept of carbon inoculation for grain refinement of magnesium alloys is the addition of Al_4C_3 particles. The grain refiner Al_4C_3 can be obtained through the foresaid interfacial reaction in the Al/SiC composites. The Al_4C_3 can be separated from the Al/SiC composite through the salt flux addition technique and it can be used as potential grain refiner in magnesium alloys.

Jean Gabriel et al. [1] investigated on the reactions occurs at the particle matrix interface of compocast Al-5 % Mg alloys reinforced with as received and oxidized SiC. It is observed that extent of the reactions depends on the holding time of the composite beyond the liquidus temperature during remelting. The composite is remelted and kept at the temperature of 800°C for 20 min. There is formation of Al_4C_3 in as received SiC reinforced composite whereas in oxidized SiC particles reinforced composite, complex oxide of MgO and $MgAl_2O_4$ is formed. Laipeng Yan et al. [2] addressed the influence of Al_4C_3 formation on mechanical properties of carbon/aluminium (C/Al) composites reinforced Multi-Wall Carbon Nanotube (MWCNT) and grapheme nanosheet. Electrochemical dissolution method is developed to characterize the extent of interfacial reaction and it is concluded that upto 5 % interfacial reaction can be detected. Jongbok Park et al. [3] made a study on the effect of surface modification of the SiC tile on the interfacial strength and wettability of the Al 7075/SiC/SiC tile produced by liquid press process. The interfacial strength and wettability are measured by the push out method and sessile drop method respectively. It is observed that the oxidized SiC has greater wettability and interfacial strength with the Al7075/MMC due to the suppression of the Al_4C_3 formation at the interface.

Jae-chul lee et al. [4] addressed the interfacial phenomena in the Al/SiC composite through XRD, SEM and Transmission Electron Microscope (TEM). It is observed that Si and hexagonal platelet shaped Al_4C_3 are the two interfacial reaction products formed which increases with increase in temperature and holding time. It is also stated that formation of Al_4C_3 has been stopped when the equilibrium Si content reaches 7%. Chen et al. [5] studied the behaviour and structure of the matrix/reinforcement interface in $ZK60A/SiC_W/B_4C_P$ hybrid composite. From the results, it is stated that facetted cubo-octahedral shaped MgO particles found at the Mg/SiC interfaces and MgB_2 particles at the Mg/B_4C interface. Foo et al. [6] done electron microscope examination on the interface of Al 6061/SiC composite produced by powder metallurgy. Energy Dispersive X-Ray Spectroscopy (EDX) and Electron Energy Loss Spectroscopy (EELS) are carried out on the interface and concluded that intermetallic such as Mg_2Si and $FeSiAl_5$ has been formed during the sintering process and there is no formation of Al_4C_3 particles. Jae-Chul Lee et al. [7] studied the control of interfaces in the 2014 Al/oxidized SiC and Al + 2 Mg/oxidized SiC composite. It is described that SiO_2 layer at the SiC particles reacts with Al and Mg and forms $MgAl_2O_4$ and Si at the interface at the expense of SiO_2 layer. It is also stated that the formed $MgAl_2O_4$ and Si at the interface are found stable even at 720°C for 20h.

Li et al. [8] studied the mechanism of the interfacial reaction in the $Al6061/B_4C$ composite produced through powder metallurgy method. Detailed investigation employing optical microscopy, SEM, TEM and XRD is done. From the results, it is stated that numerous reactions occurred resulted in the formation of Al_3BC, $MgAl_2O_4$, MgB_7, $Mg_{0.78}Al_{0.75}B_{14}$, $AlB_{12}C_2$ and Al_4SiC_4. Durbadal Mandal and Srinath Viswanathan [9] investigated on the structure of the interface formed in 2124 Al/10 wt%. SiC composite during remelting at 800ºC and 900ºC for 10 minutes. XRD analysis shows that there is formation of Al_4C_3 and elemental Si at the interface due to the reaction of the liquid Al with SiC. Also formation of $MgAl_2O_4$ is observed at the interface due to segregation of Mg at the interface and its reaction with oxygen.

IOP Conf. Series: Materials Science and Engineering 1219 (2022) 012045 doi:10.1088/1757-899X/1219/1/012045

Salvo et al. [10] fabricated Al-Si-Mg matrix composites reinforced with as received and oxidized SiC particles through compocasting technique. The composite is remelted and the interfacial reactions are characterized by TEM. It is observed that $MgAl_2O_4$ has been formed due to reaction between the Al(Mg) with oxidized SiC during fabrication. Remelting of the composite at 800°C results in the formation of Al_4C_3 however the extent of the reaction is limited in case of oxidized SiC particles reinforced composite due to the protective $MgAl_2O_4$ layer formation during fabrication. Don Soo Shin et al. [11] investigated on the effect of processing methods on the formation of Al_4C_3 in Al2024/SiC composite. The reaction products have been extracted through the electrochemical dissolution technique and characterized through SEM, Auger electron spectrometry, TEM and XRD. The reaction products formed are hexagonal platelet shaped Al_4C_3 and Si, and their extent of the reaction products depends on the processing temperature and the holding time. It is also inferred that insignificant amount of interfacial reaction observed in spray-formed composite whereas considerable amount of interfacial reactions are observed in thixo-formed and compo-cast composites.

Tong Xiang Fan et al. (1998) studied the interfacial reaction characteristics in Al/SiC composite above liquidus during remelting by using liquid metal XRD and Differential Scanning Calorimeter (DSC). Through the XRD analysis, the presence of Al_4C_3 on the extracted SiC particles has been identified. Guang Yang et al. [12] investigated on the interfacial chemical reactions of the TiO_2 coated SiC reinforced Al composite through DSC. It is observed from the DSC results that extent of reaction has been reduced due to the TiO_2 coating during remelting at 850 °C however the reaction progresses after holding for 20 min.

Ravi et al. [13] attempted on separating the alloy and reinforcement from pure Al–SiCp composite through salt flux addition technique. It is stated that addition of 2.05 wt % of equimolar mixture of NaCl-KCl salt flux resulted in the metal and particle yield of 84 % and 50 % respectively. On addition of 5 wt % NaF to the equimolar mixture enhanced the yield to 91 and 73% respectively. It is also reported that the separated alloy can be used as low Si content alloy and the extracted SiC particles cannot be used due to its degradation on interfacial reaction during fabrication of the composite. Ravi et al. [14] made a study on reclaiming and characterizing the interfacial reaction products from the Al/SiC composites through salt flux addition technique. The composite is heat treated at 710°C, 740°C and 800°C for 2 hrs and observed that there is decrease in the liquidus temperature of the composite when the temperature is increased. This indicates that there is significant release of Si in the matrix due to the interfacial formation of Al_4C_3. It is concluded that Al_4C_3 formation can be identified even 0.21 wt pct of Si is released.

From the literature survey, it is observed that studies on extracting the interfacial reaction products from the Al/SiC composites and using it as grain refiner in the magnesium alloys has not been explored. Hence, the present research deals with the formation of Al_4C_3 in the Al/SiC composites and extracting it to use as potential grain refiner in the pure Mg, Mg-Zn and Mg-Al based alloys.

2. Materials and Methods

2.1. Materials

Al-4Si-1Mg alloy is prepared as the matrix. SiC particles (10 wt %) of an average size of 33 µm (Figure 1a) have been taken for reinforcing the prepared Al-4Si-1Mg alloy. The SEM analysis is performed on the as received particles to observed the morphology (Figure 1b)

Fig 1 a) As received SiC particles b) SEM micrograph of as received SiC particles

2.2. Fabrication Process of Composite

The Al-4Si-1Mg aluminium alloy is loaded in the graphite crucible and kept in the furnace (Figure 2) for melting. The SiC reinforcement particles are simultaneously preheated around 400 °C to remove the presence of moisture and other volatile impurities from the particles. On reaching the molten metal condition, the SiC particles are added to the molten metal and the stirring is carried out at 250 rpm simultaneously with the motor driven stirrer (Figure 3) to create vortex in the molten metal. After addition of reinforcement particles, the molten metal containing reinforcement particles in the graphite crucible is taken out from the furnace and poured at the temperature of 750°C into the metallic mould. Then the cast composite is ejected from the mould.

Fig 2 Furnace for melting of aluminium alloy.

Fig 3 Mechanical stirrer setup

2.3. Formation of Interfacial Reaction

Without ultrasonic treatment

The as cast Al-4Si-1Mg/SiC composite is remelted in the furnace and kept constant at 900 °C for different time to allow the interfacial reaction. The molten metal is taken from the furnace for different time of 0 min, 30 min, 1 hr, 2 hr, 3 hr and poured into the mould for the computer aided cooling curve analysis to quantify the amount of interfacial reaction. In computer aided cooling curve analysis, the poured molten metal is allowed to cool continuously to obtain the cooling curve. The quantification of the interfacial reaction is done by determining the liquidus temperature from the cooling curve. From the Al-Si phase diagram, for the obtained liquidus temperature at different time, the amount of Si released during the interfacial reaction can be determined.

With ultrasonic treatment

The as cast Al-4Si-1Mg/SiC composite is remelted and kept at 750°C for ultrasonically treatment. The ultrasonic probe (Figure 4) is coated to avoid dissolving into the molten metal. The probe is initially preheated to 750°C and ultrasonic treatment (20 kHz) is carried out for 5 min with the time interval of 10 min for every 1 min. The ultrasonic treatment is carried out in the molten metal to induce the extent of interfacial reaction. Then the molten metal is kept constant at 900 °C for different time to allow the interfacial reaction to take place. Then the computer aided cooling curve analysis is carried out at different time to obtain the cooling curve.

Fig 4 Ultrasonic stirrer setup

2.4. Reclamation of Interfacial Reaction Products

The reaction products formed in the composites should be reclaimed to identify the existence of interfacial reaction products through XRD. The salt flux (NaCl-KCl + 5 % NaF) is added to the melt and the vortex is created through the mechanically assisted stirring. On continuous stirring of the salt flux in the molten metal, the SiC particles along with the reaction products float over the melt. Skimming has to be done to remove the SiC particles along with the reaction products from the surface of the melt. The skimmed off particles are the reclaimed particles which are then sieved for the XRD analysis. The remaining melt in the crucible is poured into the mould as the reclaimed alloy.

2.5. Identification of the Interfacial Reaction Products

The reclaimed powders from the composites (ultrasonic and without ultrasonic) are subjected to XRD analysis to determine the presence of interfacial reaction products formed in the composite. The powders are also subjected to the SEM analysis to determine the morphology of the interfacial reaction products by observing the degradation of the SiC particles.

3. Results and Discussions

3.1. Al-4Si-1Mg/10SiC composite – without ultrasonic treatment

The Al-4Si-1Mg/10 SiC composite is successfully produced by stir casting route and superheating of the composite is performed at 900°C for various time interval to assess the interfacial reaction. The microstructure of the Al-4Si-1Mg/10 SiC composite (Figure 5) is observed through optical microscope. The microstructure reveals the distribution of the SiC particles in the composite. The surface is also observed with porosities and agglomerates due to the gas entrapment.

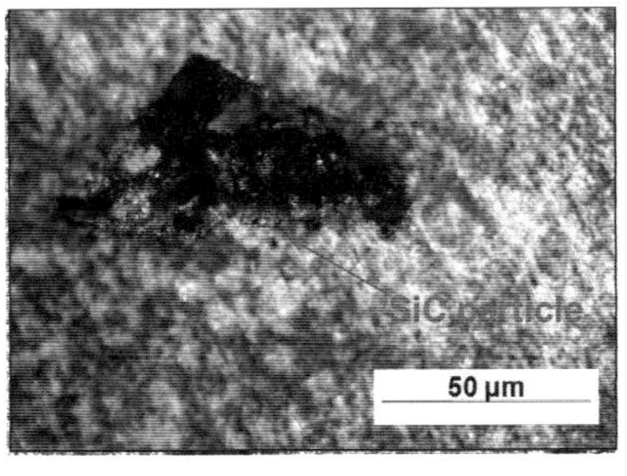

Fig 5 Optical micrograph of the Al-4Si-1Mg/10 SiC composite

The quantification of the interfacial reaction products is done through computer aided cooling curve analysis. The cooling curve of composite obtained for as cast condition and for different holding time (0 min, 30 min, 1 hr, 2 hr and 3 hr) through computer aided cooling curve analysis is shown in Figure 6.

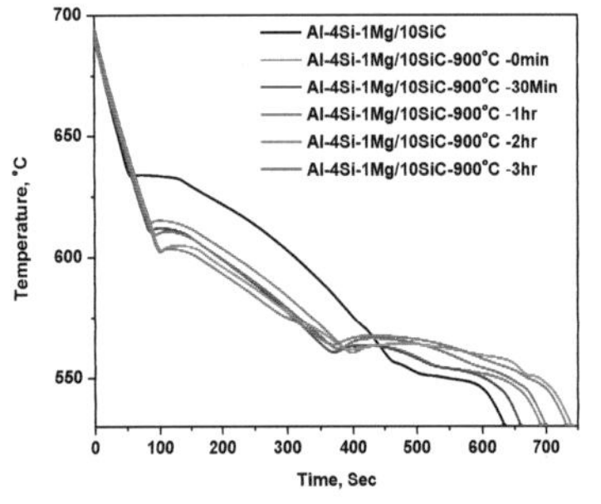

Fig 6 Computer aided cooling curve for Al-4Si-1Mg/10 SiC composite

From the figure 6, it is observed that liquidus temperature for the as cast composite is 633°C. Then the liquidus temperature of the composite is drastically decreased to 613°C when holding at 900°C for 0 min which confirms that there is formation of interfacial reaction products in the composite. Then the liquidus temperature of the composite is observed to be decreasing on increasing the holding time which is evident from the drop in liquidus temperature upto 602°C when holding for 3 hr. The drop in liquidus temperature is attributed to the formation of the Al_4C_3 particles and the release of the Si content in the matrix alloy.

Using the Al-Si phase diagram (Figure 7), the obtained liquidus temperature for the different time is converted into the Si content. The cooling curve analysis and the release of Si content at different time are shown in Figure 8(a-d).

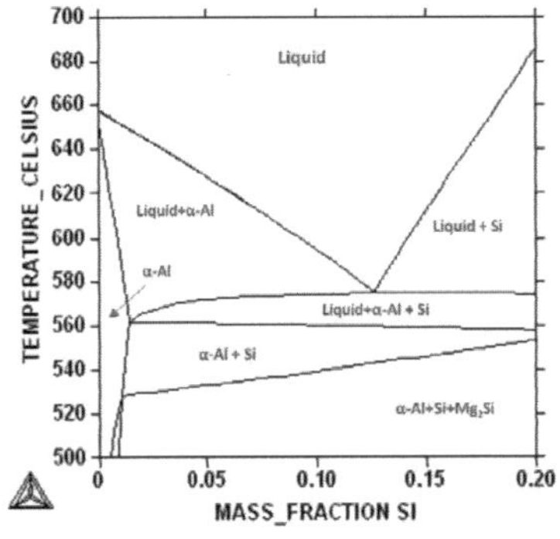

Fig 7 Al-1Mg-xSi phase diagram

From the cooling curve analysis (Figure 8d), the Si content in the as cast composite is found as 4.02 % and it increases to 5.9 % at 900 °C for holding time of 0 min. The release of Si is found gradually increasing and finally reaches 7.8 % Si at 900 °C for holding time of 3 hr. This is attributed to the significant formation of Al_4C_3 in the composite during superheating.

IOP Conf. Series: Materials Science and Engineering 1219 (2022) 012045 doi:10.1088/1757-899X/1219/1/012045

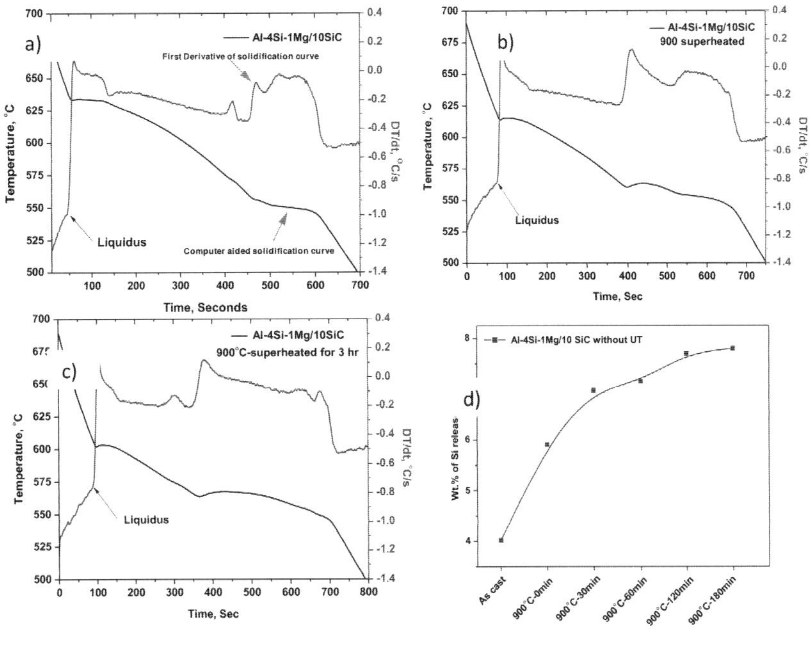

Fig 8 Computer aided cooling curve analysis a) As cast b) 900°C-0 min c) 900°C-3 hr and d) Amount of Si release at 900°C for different time.

3.2. Al-4Si-1Mg/10SiC COMPOSITE – WITH ULTRASONIC TREATMENT

The successfully fabricated Al-4Si-1Mg/10 SiC composite is treated with ultrasonic waves prior to the superheating. The ultrasonic stirring of the composite can increase the wettability by degasing and increasing the surface area by particle size reduction and thus may increase the rate of the reaction between the Al and SiC to form more interfacial reaction products. The microstructural analysis (Figure 9) of the ultrasonically treated Al-4Si-1Mg/10 SiC composite is done through optical microscope. The surface is observed with improved distribution of the SiC particles with size refinement and reduction of porosities and agglomerates which is attributed to the ultrasonic treatment of the composite. The SiC particles size gets reduced to 5.4 ± 2.6 μm after the ultrasonic treatment.

Fig 9 Optical micrograph of the ultrasonically treated Al-4Si-1Mg/10 SiC composite

Then the computer aided cooling curve analysis is performed on the ultrasonically treated composite at 700°C and superheated composite at 900°C for different holding time of 0 min, 30 min, 1 hr, 2 hr and 3 hr. The obtained cooling curve is displayed in Figure 10.

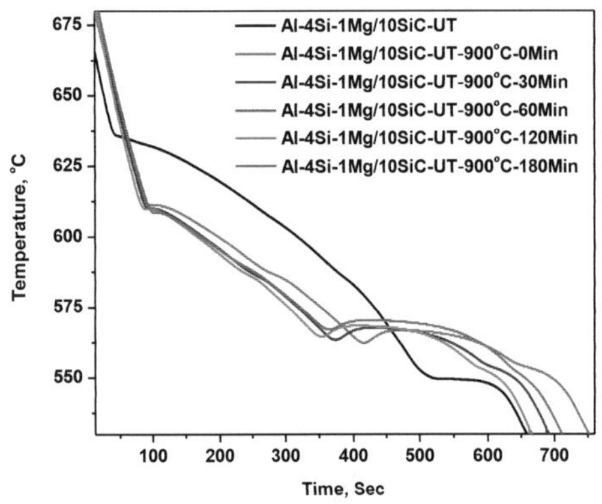

Fig 10 Computer aided cooling curve for ultrasonically treated Al-4Si-1Mg/10 SiC composite

From the cooling curve, it is observed that the liquidus temperature of the ultrasonically treated composite at 700°C is 633°C. Then the liquidus temperature is greatly reduced to 611°C when holding at 900°C (0 min) which is attributed to the formation of Al_4C_3 and release of Si content into the matrix. Hence it is understood that as the Si content increases in the alloy, liquidus temperature gets decreases. The liquidus temperature is found decreased to 608°C for holding upto 3 hr which confirms that there is formation of the interfacial reaction product. The quantification of the Al_4C_3 formation is done by converting the obtained different liquidus temperatures to Si content using the Al-Si phase diagram. The cooling curve analysis and the release of Si content for the ultrasonically treated Al-4Si-1Mg/10 SiC composite is shown in Figure 11(a-d). From the analysis, it is observed that ultrasonic treated composite has Si content of 4.2 % initially and it is greatly increased to 6.7 % at 900°C (0 min) which indicates that the rate of the reaction between the Al and SiC has been increased due to the treatment through ultrasonic waves. Then the Si content is found increased upto 7.15 % when holding for the time of 3 hr.

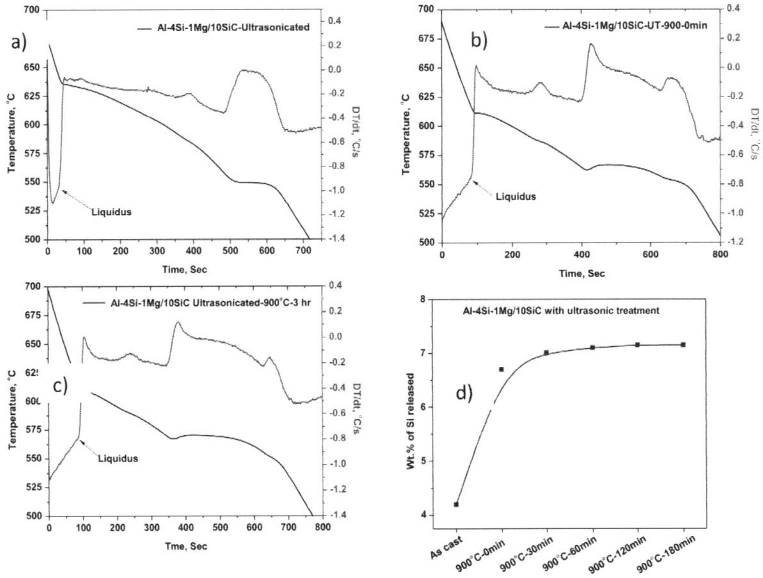

Fig 11 Computer aided cooling curve analysis of ultrasonically treated composite a) As cast b) 900°C-0 min c) 900°C-3 hr and d) Amount of Si release at 900°C for different time.

3.3. COMPARISON OF INTERFACIAL REACTION IN THE COMPOSITES

The interfacial reaction products are reclaimed from the Al-4Si-1Mg/10 SiC composite and ultrasonically treated Al-4Si-1Mg/10 SiC composite through the salt flux (NaCl-KCl + 5 % NaF) reclamation process. XRD analysis (Figure 12) is performed on the reclaimed powders to confirm the presence of the interfacial reaction product (Al_4C_3 particles) in it. From the Figure 12, it is observed that the peaks of Al_4C_3 are present which indicates that the salt flux (NaCl-KCl + 5 % NaF) has greater efficiency in reclaiming the interfacial reaction products from the composite.

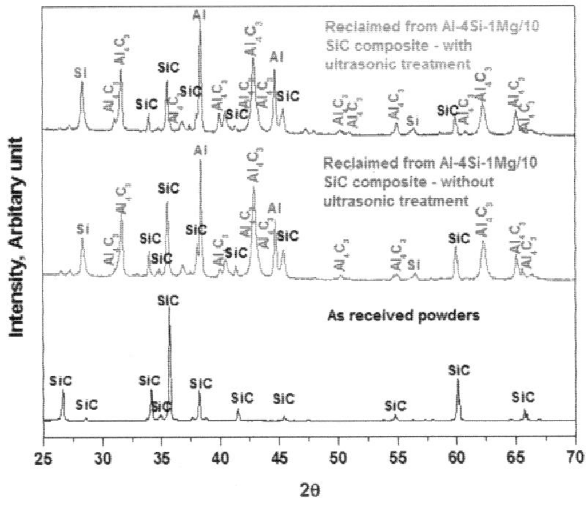

Fig 12 XRD of the reclaimed powders from the Al-4Si-1Mg/10 SiC composite and ultrasonically treated Al-4Si-1Mg/10 SiC composite

The amount of Si release in the as cast Al-4Si-1Mg/10 SiC composite and ultrasonically treated Al-4Si-1Mg/10 SiC composite are compared in the Figure 13. The amount of Si release for the as-cast composite is gradually increased from 4.02 % to 7.8 % on holding for 3 hr. In case of the ultrasonically treated composite, the amount of Si release increases drastically from 4.2 % in as cast condition to 6.7 % at 900°C (0 min).

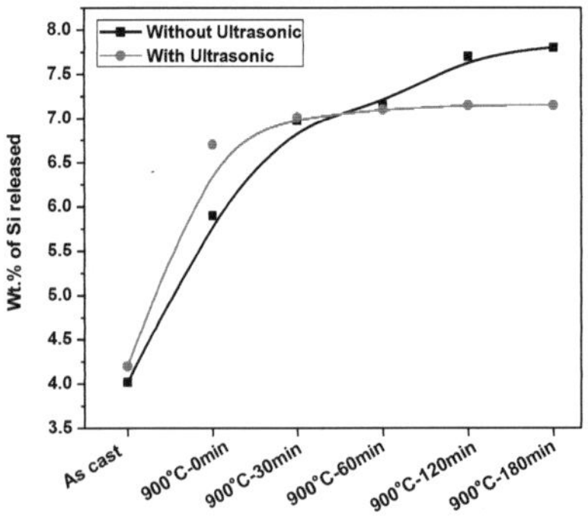

Fig 13 Comparison of Si release in Al-4Si-1Mg/10 SiC composite and ultrasonically treated Al-4Si-1Mg/10 SiC composite.

This indicates that the treatment through ultrasonic waves increased the rate of the reaction between the Al and SiC within small amount of time. Then the Si release is found stabilized from 30 min to 3 hr which may be due to inadequate amount of SiC particles in the composite. Further optimisations are required.

4. Conclusion

The Al-4Si-1Mg/10 wt% SiC composite is successfully fabricated through stir casting process and it is superheated (with and without ultrasonic treatment) to form the Al_4C_3 interfacial reaction product. Interfacial reaction product has been quantified with computer aided cooling curve analysis and determined that liquidus temperature of the composites gets decrease with the release of the Si content in the matrix. Salt flux reclamation technique is approached for successful reclamation of the Al_4C_3 particles. The X-ray diffraction analysis on the reclaimed particles confirmed the presence of Al_4C_3 particles. The reclaimed particles are also subjected to Scanning Electron Microscopy analysis and the morphology has been observed.

References

1. Jean-Gabriel Legoux, Gilles Lesperance, and Harve Ribes, and Suery M, 1989, "Characterization of the interface reactions in Al-SiC composites and of their effect on the composition of the Al-Mg matrix," Metallurgical Society of Canadian Institute of Mining and Metallurgy, pp. 324-336.
2. Yan Laipeng, Tan Zhanqiu, Ji Gang, Li Zhiqiang, Fan Genlian, Schryvers D, Shan Aidang, and Zhang Di, 2016, "A quantitative method to characterize the Al_4C_3-formed interfacial reaction: the case study of MWCNT/Al composites," Materials Characterization, Vol. 112, pp. 213-218.
3. Jongbok Park, Junho Lee, Ilguk Jo, Seungchan Cho, Sang Kwan Lee, Sang Bok Lee, Ho Jin Ryu, and Soon Hyung Hong, 2016, "Surface modification effects of SiC tile on the wettability and interfacial bond strength of SiC tile/Al7075-SiC_p hybrid composites," Surface & Coatings Technology, Vol. 307, pp. 399–406.
4. Jae-Chul LeE, Ji-Young Byun, Sung-Bae Park, and Ho-In Lee, 1998, "Prediction of Si contents to suppress the formation of Al_4C_3 in the SiC_p/Al composite," Acta Materilia, Vol. 46, No. 5, pp. 1771-178
5. Chen Y X., Li D X., and Zhang G D., 2000, "Electron microscopy studies of interfacial reaction in a (SiC_w-B_4C_p)/ZK60A hybrid composite," Scripta Materilia, Vol. 43, pp. 337-341.
6. Foo K S., Banks W M., Craven A J., and Hendry A., 1994, "Interface characterization of an SiC particulate/6061 aluminium alloy composite," Composites, Vol. 25, No. 7, pp. 677-683.

7. Jae-Chul Lee, Jae-Pyoung Ahn, Jae-Hyeok Shim, and Zhongliang Shi, and Ho-In Lee,1999, "Control of the interface in SiC/Al composites," Scripta Materialia, Vol. 41, No. 8, pp. 895-900.
8. Li Y Z., Wang, Wang W G., Xiao B L., and Ma Z Y., 2015, "Interfacial reaction mechanism between matrix and reinforcement in B_4C/6061Al composites," Materials Chemistry and Physics, Vol.154, pp. 107-117.
9. Mandal Durbadal and Viswanathan Srinath, 2013, "Effect of re-melting on particle distribution and interface formation in SiC reinforced 2124 Al matrix composite," Materials Characterization, Vol. 86, pp. 21-27.
10. Salvo L., Esperance G L., Suery M., and Legoux J.G., 1994, "Interfacial reactions and age hardening in Al-Mg-Si metal matrix composites reinforced with SiC particles," Materials Science and Engineering A, Vol. 177, pp.173-183.
11. Don-Soo Shin, Jae Chul Lee, Eui-Pak Yoon, and Ho-In Lee, 1997, "Effect of the processing methods on the formation of Al_4C_3 in SiC/2024 Al composites," Materials Research Bulletin, Vol. 32, No. 9, pp. 1155-1163.
12. Guang Yang, Tongxiang Fan, and Di Zhang, 2004, "Interfacial reaction of Al matrix composites reinforced with TiO_2-coated SiC particles during remelting," Materials Letters, Vol. 58, pp. 1546-1552.
13. Ravi K R., Pillai R M., Pai B C., and Chakraborty M., 2007, "Separation of matrix alloy and reinforcement from aluminum metal matrix composites scrap by salt flux addition," Bulletin of Material Science, Vol. 30, No. 4, pp. 393-398.
14. Ravi K R., Pillai R M., Pai B C., and Chakraborty M., 2007, "A novel approach for extracting and characterizing interfacial reaction products in Al-SiC_p composites," Metallurgical and Materials Transactions A, Vol. 38A, pp.1666-1670.

IOP Conf. Series: Materials Science and Engineering 1219 (2022) 012046 doi:10.1088/1757-899X/1219/1/012046

STRUCTURAL ANALYSIS OF AA2024 INTERACTION REINFORCED WITH CARBON NANOTUBES AND SILICON NANOCOMPOSITES STUDIED BY FOURIER TRANSFORM INFRARED SPECTROSCOPY

M. Muniyappan and N. Iyandurai*

Department of Physics, Thiruvalluvar Govt. Arts College, Rasipuram, Namakkal - 637 401, TamilNadu, India.

*E-mail:: ayandurai15@gmail.com

Abstract: AA2024 is used in applications where a high strength to weight ratio is required. In the present work, Infrared spectroscopy had been used to expand our knowledge about AA2024-carbon nanotubes-silicon interactions at various proportions. The analysis of Infrared data supported the subsistence of structural specificities in the interaction. From the observed results, the effect of CNTs and silicon on AA2024 is clearly explained. For example, the infrared band observed at 619 cm^{-1} is assigned to the Al-O-Si stretching of AA2024. This band is shifted to 612 cm^{-1} when 4% CNTs is added and 617 cm^{-1} when 4% silicon is added with AA2024, which shows strong interaction is present. The FTIR ban observed at 771 cm^{-1} is assigned to silicon-carbon stretching mode, which is shifted to 775 cm^{-1}. This suggests that sturdy interaction is present.

Keywords: AA2024, Carbon Nanotubes, Silicon, Hybrid Metal Matrix Nanocomposites

1. Introduction

Aluminum alloy 2024 (AA2024) are widely used in aeroplanes, aerospace, automobiles, and other industries due to their excellent strength-to-weight ratio. The biggest drawback of AA2024 alloy is that they exhibit poor tribological properties. Therefore, it is preferred to build up a replacement metal matrix composite materials with improved tribological properties such as wear rate and coefficient of friction [1-5].

Carbon nanotubes (CNTs) are a type of material that has opened up new possibilities for composite materials. Carbon nanotubes are tremendously infrequent, used as reinforcements in the MMCs manufacturing industry because of its excellent electrical and mechanical properties [6, 7]. CNTs are having high elastic modulus, yield strength in addition to exceptional thermal and chemical stability [8].

Silicon is a crystalline solid that is hard and brittle. Since it has both metal and non-metal properties, it is called tetravalent metalloid. Silicon increases the resistance of materials such as aluminium, magnesium, and copper [9].

The prepared nanocomposites were characterized with the help of element present and chemical bonding analysis for Fourier transform infrared spectroscopy (FTIR) was evaluated.

2. Materials and Methods

Materials

Aluminium Alloy 2024 were purchased from Plus metals, Mumbai, India. Carbon Nanotubes (CNTs) were purchased from IENT Inc, Erode, Tamil Nadu, India. Silicon (Si) was purchased from Coimbatore Metals, Coimbatore, Tamil Nadu, India.

Methods

Silicon nanoparticles from silicon powder was synthesized through high energy ball milling technique [10] and hybrid metal matrix nanocomposite materials were fabricated by two-stage stir casting technique [11]. In this research, AA2024 - 2% Carbon nanotubes, AA2024 - 4% Carbon nanotubes, AA2024 - 2% Silicon, AA2024 - 4% Silicon, AA2024 - 2% Carbon nanotubes - 2% Silicon, AA2024 - 2% Carbon nanotubes - 4% Silicon, AA2024 - 4% Carbon nanotubes - 2% Silicon,and AA2024 - 4% Carbon nanotubes - 4% Silicon, was prepared. The carbon nanotubes, silicon nanoparticles and their combination are preheated to 300°C for three hours to take out dampness. The aluminium alloy 2024 (AA2024) was melted in a resistance furnace. Then the temperature was increased to 750°C, and the melt was stirred. The stirring was maintained between 15 min at an impeller speed of 325 rpm. The temperature of the melt was maintained at 700°C during the addition of CNTs, Silicon and their mixtures. The molten melt with reinforced particulates was transferred into the sand mould. The molten melt has been then allowed to solidify in the mould.

High energy ball milling

Fritsch, Germany make, planetary mono mill classic line pulverisette-6 model was used for the synthesis of silicon nanoparticles. The dimensions of the instrument are 500 × 370 × 530 mm (hight × width × depth) and weight is 67 kg. Commercially available silicon powder was milled at 350 rpm in a tempered steel grinding bowl of size 250 mL using tempered steel grinding balls (6 balls, diameter 30 mm) in an ambient atmosphere for 1 hour. The milled materials were used directly.

Fourier Transform Infrared Spectroscopy

FTIR spectra were recorded in Perkin Elmer "Spectrum Two" FTIR spectrometer. This instrument uses a typical, high-performance, room-temperature $LiTaO_3$ (lithium tantalate) MIR detector, standard optical system with KBr windows for data collection over a spectral range of 8300 - 350 cm^{-1} at a best resolution of 0.5 cm^{-1}, optional ZnSe windows, to be used in exceptionally humid environments for data collection over a spectral range of 6000 - 550 cm^{-1}, and a standard AVC with optional APV/AVI configuration. During this work, spectra of molecular fingerprint of the sample were recorded within the spectral range from 4000 cm^{-1} to 400 cm^{-1}.

3. Results and Discussion

Fourier Transform Infrared Spectroscopy (FTIR) Analysis

Fig. 1 (a-k) and Fig. 2 shows an FTIR spectrum of AA2024, Carbon Nanotubes, Silicon, AA 2024 - Carbon Nanotubes , AA2024 - Silicon, AA2024 - Carbon Nanotubes - Silicon, AA2024 - Silicon - Carbon Nanotubes (various proportions) Nanocomposites. Overall spectral assignments are tabulated as shown in Table 1. FTIR spectrum is recorded in the range of 4000 - 400 cm^{-1}.

The infrared band observed at 1406 cm^{-1} is associated with boehmite Al=O vibration [13] and the band observed at 947 cm^{-1} is due to Al-O vibration [15]. FTIR band observed at 619 cm^{-1} is due to R-O-Si (R=Al, Mg, Fe) stretching mode [17] and also the band observed at 514 cm^{-1} is due to Si-O-Al symmetric vibration [17]. The band observed at 471cm^{-1} is due to Si-O asymmetrical bending vibration and asymmetric Si-O-C deformation vibration [19, 20] and the FTIR band observed at 428 cm^{-1} is due to Si-O-Mg bending mode vibration [21].

IOP Conf. Series: Materials Science and Engineering 1219 (2022) 012046 doi:10.1088/1757-899X/1219/1/012046

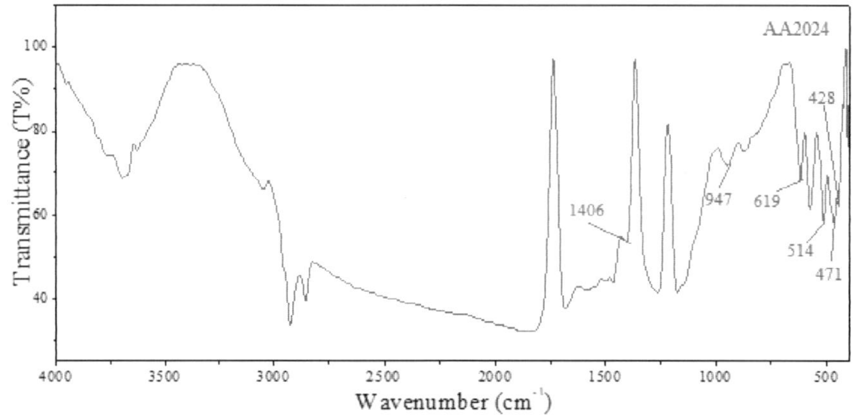

(a) FTIR Spectra of AA 2024

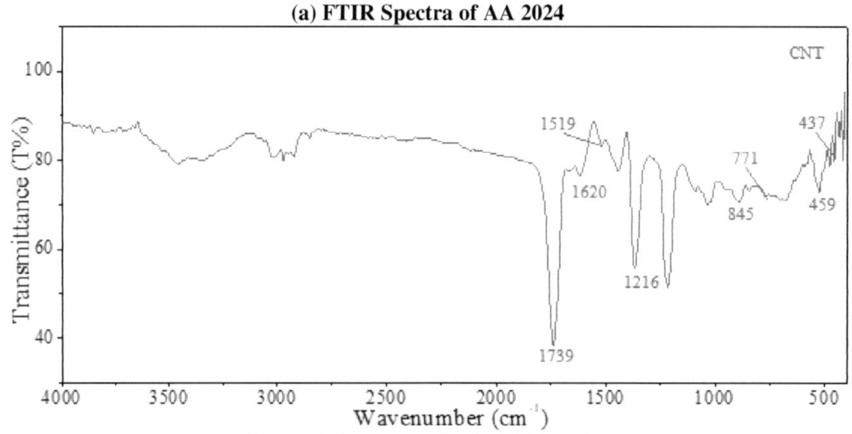

(b) FTIR Spectra of Carbon Nanotubes

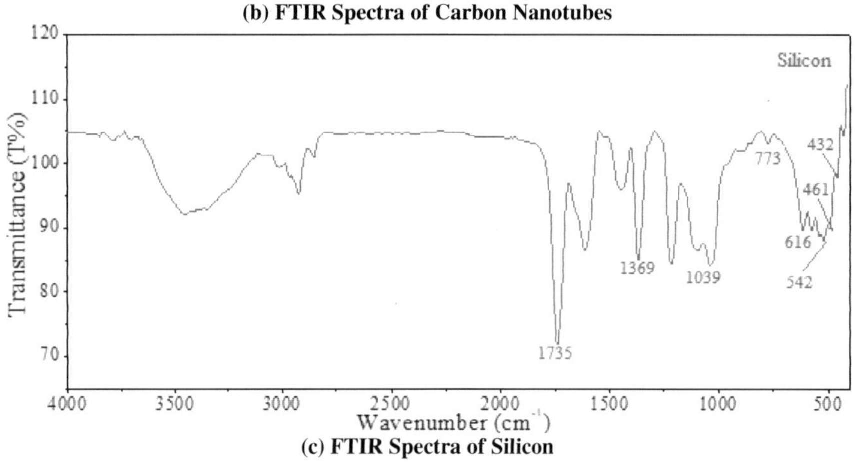

(c) FTIR Spectra of Silicon

Table 1 FTIR spectral assignments of AA 2024 HMMNCs

S. No	Wavenumber (cm^{-1})											Assignment
	AA 2024	CNT	Si	AA2024+ 2% CNT	AA2024 + 4% CNT	AA2024 + 2%Si	AA2024 + 4%Si	AA2024 +2% CNT +2%Si	AA2024+2% CNT +4%Si	AA2024+4% CNT +2%Si	AA2024 +4% CNT +4%Si	
1	-	-	-	1579	1591	-	1580	1585	1591	1594	1587	C=C Stretching[12]
2	1406	-	-	1405	1406	-	-	1403	1405	1409	1404	Boehmite Al=O vibration[13]
3	-	-	-	1172	1174	-	1172	1170	1174	1176	1175	C-O-C Symmetrical stretching vibration in carbohydrate[14]
4	947	-	-	949	945	961	942	949	944	949	929	Al - O vibration[15]
5	-	771	773	-	-	775	682	681	680	685	-	Si-C stretching mode[16]
6	619	-	616	612	616	616	617	618	618	621	618	R-O-Si With R=Al, Mg, Fe [17]
7	-	-	542	-	537	-	541	-	541	545	-	Si-O-Al, Fe-O stretching[18]
8	514	-	-	517	513	512	515	516	514	516	515	Si-O-Al Vibration[17]
9	471	459	461	475	474	-	474	470	473	477	476	Si-O asymmetrical bending vibration, asymmetric Si-O-C deformation vibration[19,20]
10	428	437	432	436	432	427	429	-	428	431	429	Si-O-Mg bending mode vibration[21]

An FTIR bands observed at 1665cm^{-1} and 1620 cm^{-1} are due to C=C stretching vibrations respectively[22,12]. FTIR band observed at 1519 cm^{-1} is due to backbone C=C skeletal stretching of CNTs [23]. The band observed at 1216 cm^{-1} is due to C-O stretching vibration [23]. The band observed at 1739 cm^{-1} is due to C=O stretching bonds [23] and the infrared band observed at 845 cm^{-1} is due to C-O bending vibration [15].

The band observed at 1039 cm^{-1} is due to Si-O stretching vibration [21] and the band observed at 461 cm^{-1} is due to Si-O asymmetrical bending vibration [23]. FTIR band observed at 542 cm^{-1} is due to Si-O-Si deformation [24] and the band observed at 432 cm^{-1} is due to Si-O deformation [24].

IOP Conf. Series: Materials Science and Engineering 1219 (2022) 012046 doi:10.1088/1757-899X/1219/1/012046

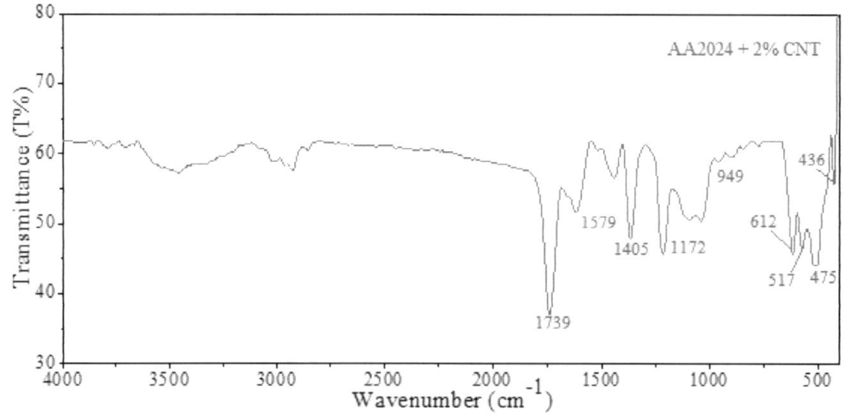

(d) FTIR Spectra of AA 2024 + 2% CNT

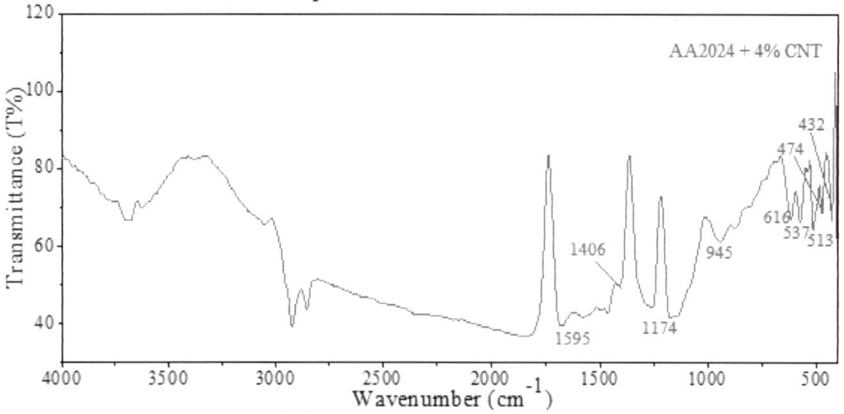

(e) FTIR Spectra of AA 2024 + 4% CNT

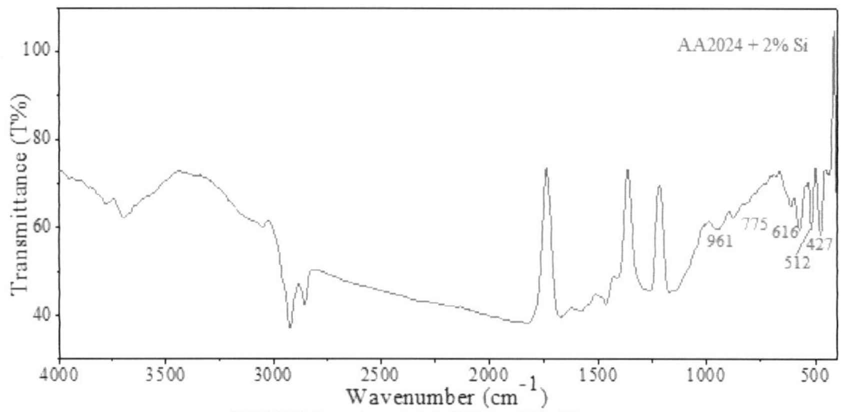

(f) FTIR Spectra of AA 2024 + 2% Si

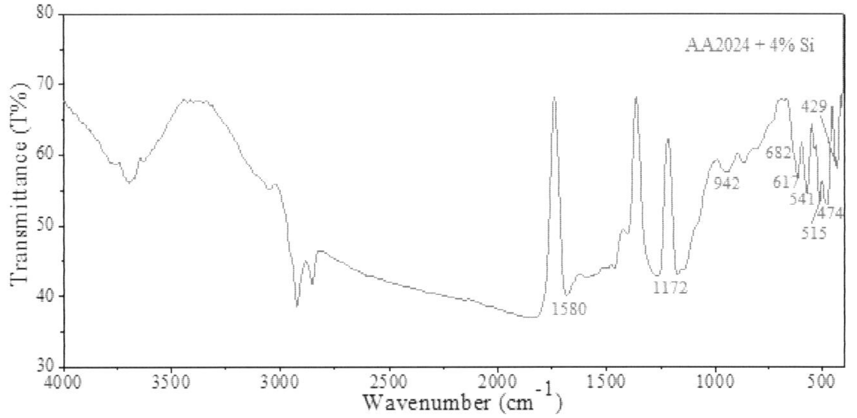

(g) FTIR Spectra of AA 2024 + 4% Si

FTIR spectra have been taken to study the interactions among AA2024, carbon nanotubes and silicon nanoparticles. The infrared band observed at 1579 - 1594 cm^{-1} is due to C=C stretching. The band observed at 1403 - 1409 cm^{-1} is associated with boehmite Al=O vibration. For our study, this band was observed at 1406 cm^{-1}, which is shifted to 1405 cm^{-1} in the presence of 2% CNTs, 1403 cm^{-1} in the presence of 2% CNTs with 2% silicon, 1405 cm^{-1} in the presence of 2% CNTs with 4% silicon, 1409cm^{-1} in the presence of 4% CNTs with 2% silicon, 1404 cm^{-1} in the presence of 4%CNTs with 4% silicon. FTIR band observed at 1172 - 1176 cm^{-1} is due to C-O-C asymmetrical stretching vibration in carbohydrate. The band observed at 929 - 949 cm^{-1} is due to Al-O vibration. For our study, this band was observed at 947 cm^{-1}. This band is shifted to 949 cm^{-1} in the presence of 2% CNTs and also 4% CNTs with 2% silicon, 929 cm^{-1} in the presence of 4% CNTs with 4% silicon.

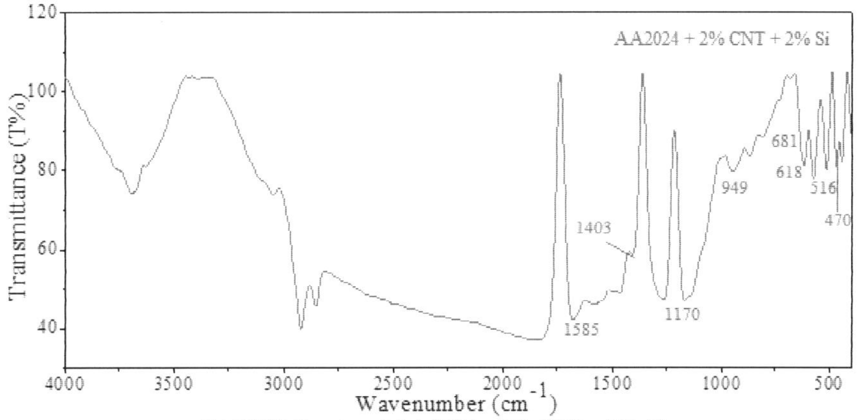

(h) FTIR Spectra of AA 2024 + 2% CNT + 2% Si

IOP Conf. Series: Materials Science and Engineering 1219 (2022) 012046 doi:10.1088/1757-899X/1219/1/012046

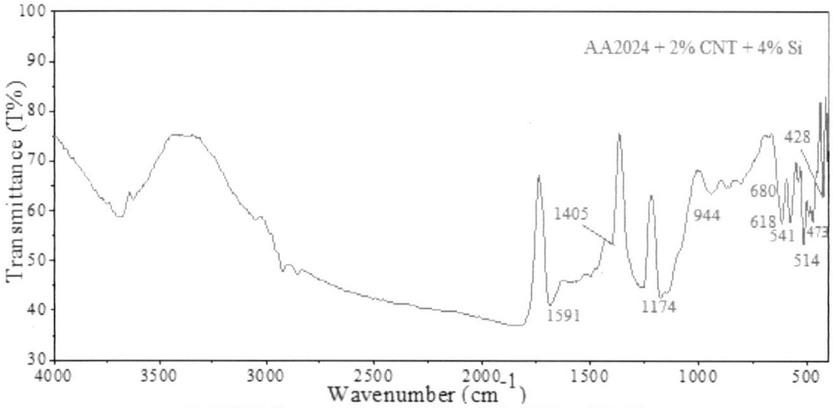

(i) FTIR Spectra of AA 2024 + 2% CNT + 4% Si

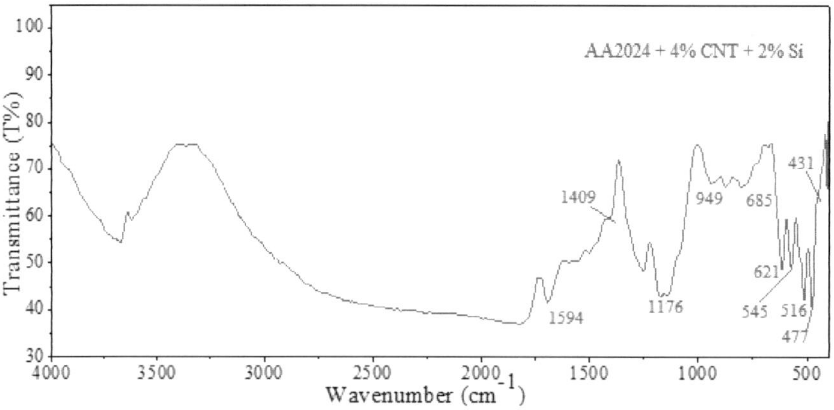

(j) FTIR Spectra of AA 2024 + 4% CNT + 2% Si

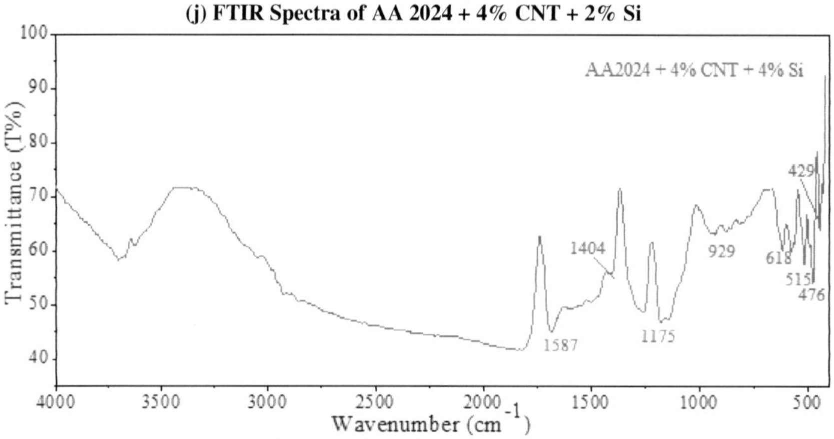

(k) FTIR Spectra of AA 2024 + 4% CNT + 4% Si

Fig. 1 FTIR Spectra of AA 2024 Hybrid Metal Matric Nanocomposite

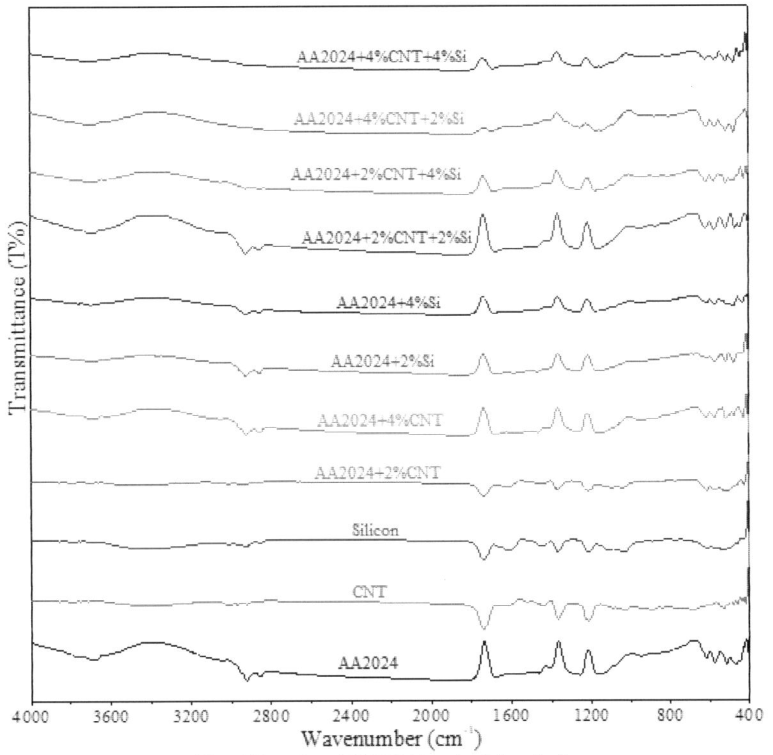

Fig. 2 Results of FTIR Spectral Analysis

Infrared band observed at 680 - 775 cm^{-1} is due to Si-C stretching mode. FTIR band observed at 619 cm^{-1} is due to R-O-Si(R=Al, Mg, Fe) stretching mode. This band is shifted downwards by several cm^{-1} due to the interactions between AA 2024 and reinforcements. But this band is shifted upward by 2 cm^{-1}, due to the antisymmetrical stretching vibrations. The band observed at 537 - 541 cm^{-1} is due to Si-O-Al, Fe-O stretching vibration. The band observed at 512 - 516 cm^{-1} is due to Si-O-Al vibration. We observed such a band at 514 cm-1, which shows no major shifts.

Infrared band observed at 471 cm^{-1} is due to Si-O asymmetrical bending vibration and asymmetric Si-O-C deformation vibration. This shifts in the order, 471 → 475 → 474 → 470 → 473 → 477 → 476 cm-1 in the presence of 2% CNTs, 4% CNT and 4% silicon, 2% CNTs with 2% silicon, 2% CNTs with 4% silicon, 4% CNTs with 2% silicon and 4% CNTs with 4% silicon respectively. The band observed at 427 - 437 cm^{-1} is due to Si-O-Mg bending mode vibration. This band is shifted to 431 cm^{-1} in the presence of 4% CNTs with 2% silicon.

Conclusion:

AA2024 - 2% Carbon nanotubes, AA2024 - 4% Carbon nanotubes, AA2024 - 2% Silicon, AA2024 - 4% Silicon, AA2024 - 2% Carbon nanotubes - 2% Silicon, AA2024 - 2% Carbon nanotubes - 4% Silicon, AA2024 - 4% Carbon nanotubes - 2% Silicon,and AA2024 - 4% Carbon nanotubes - 4% Silicon metal matrix nanocomposites were effectively fabricated by two-stage stir casting method.

The subsequent results have been made:

- The FTIR analysis confirms the presence of species present in the prepared hybrid metal matrix nanocomposites.
- It is noted that when 2% CNTs, 4% CNTs, 2% silicon, 4% silicon, 2% CNTs with 2% silicon, 2% CNTs with 4% silicon and 4% CNTs with 4% silicon is added with

AA 2024, frequency is shifted to the lower frequency side by several cm^{-1}. This designates that the strong interaction is present.

- It is significant to note that the wavenumbers of 4% CNTs with 2% silicon go to the higher shift.

References:

[1] Ravindran P, Manisekar K, Rathika P and Narayanasamy P 2013 Mater. Des. 45 561-570.
[2] Govind Nandipati, Ravindra kommineni, Nageswara Rao Damera and Ramanaiah Nallu 2013 IOSR J. Mech. Civ. Eng. 7 01-07.
[3] Murthy N V, Prasad Reddy A, Selvaraj N and Rao C S P 2016 Mater. Sci. Eng. 149 012106.
[4] Sourabh Bhaskar, Mukesh Kumar and Amar Patnaik, Silicon, 12 (2020) 843-865.
[5] CH S Vidyasagar and D B Karunakar 2020 Trans. Nonferrous Met. Soc. China 30 288-302.
[6] Mingfang Qian, Xuexi Zhang, Jianchao Li and 2019 Xiang Gao, Mater. Chem. Phys. 226, 344-349.
[7] Benyamin Abbasipour, Behzad Niroumand, Sayed Mahmoud Monir Vaghefi and Mohammad Abedi 2019 Trans. Nonferrous Met. Soc. China 29 1993-2004.
[8] Sohail M. A. K. Mohammed and Daolun L. Chen 2020 Adv. Eng. Mater. 22 1901176.
[9] Kurt E. Petersen 1982 Proceedings of the IEEE 70 420-457.
[10] Benedicte Eikeland Nilssen and Rolf Arne Kleiv 2020 Silicon 12 2413-2423.
[11] Arun Kumar Sharma, Rakesh Bhandari, Amit Aherwar and Camelia Pinca-Bretotean 2020 Materials Today: Proceedings 27 1608-1612.
[12] Seyyed Salman Seyyed Afghahi, Reza Peymanfar and Shahrzad Javanshir 2017 J. Magn. Magn. Mater. 423 152-157.
[13] Mrad M, Ben Amor Y, Dhouibi L and Montemor M F 2017 Surf. Interface Anal 1-11.
[14] Tengfei Li, Gaojin Lyu, Yu Liu, Rui Lou and Lucian A. Lucia 2017 Int. J. Mol. Sci. 18 2266.
[15] Le Bozec N, Persson D, Nazarov A and Thierry D 2002 J. Electrochem. Soc. 149 B403-B408.
[16] Xiaodong Zhang, Xiaoxiao Huang and Guangwu Wen 2010 Nanotechnology 21 385601.
[17] Zorica P. Tomic, Svetlana B. Antic Mladenovic and Biljana M. Babic, J. Agric. Sci. 56 2011.
[18] Bhaskar J. Saikia and Gopalakrishnarao P 2010 J. Mod. Phys. 1 206-210.
[19] Bera A, Mandal A and Kumar T 2015 Petro. Sci. Tech. 33 542-549.
[20] Cubillana-Aguilera L M, Palacios-Santander J M and Naranjo-Rodr´ıguez I 2006 J. Sol-Gel Sci. Technol. 40 55-64.
[21] Giuseppe Falini, Elisabetta Foresti, Massimo Gazzano and Alessandro F. Gualtieri, 2004 Chem. Eur. J. 10 3043-3049.
[22] Laskowskal A, Zaborskill M, Boiteux G, Gain O and Marzec A 2014 EXPRESS Polym. Lett. 8 374-386.
[23] Muhammad Azhar Hayat Nawaz, Sajid Rauf, Gaelle Catanante and Mian Hasnain Nawaz 2016 Sensors 16 1651.
[24] Hongxia Zhang, Zhiwei Niu and Zhi Liu 2015 J. Radioanal. Nucl. Chem. 303 87-97.

IOP Conf. Series: Materials Science and Engineering 1219 (2022) 012047 doi:10.1088/1757-899X/1219/1/012047

An optimum design of one dimensional photonic crystal for solar cell applications

Karthika Sankar[1], Rahna Manoharan[2], Safna Saif[1], Priya Rose T[1,3]*

[1] International School of Photonics, Cochin University of Science and Technology, Kochi, Kerala, India- 682022
[2] Department of Physics, Kannur University, Kannur, Kerala, India-670567
[3] Inter University Center for Nanomaterials and Devices, Cochin University of Science and Technology, Kochi, Kerala, India-682022

*E-mail: priyarose@cusat.ac.in

Abstract. Efficient anti reflection (AR) coatings and back reflectors are crucial components for the better performance of a solar cell. Hence perfect designing of the same is very important. We propose the design of both these structures based on a multilayer dielectric system, a one dimensional photonic crystal, with maximum efficiency on the basis of optical interference transfer matrix theory. Optical parameters were optimized and transmission and reflection spectra were obtained using transfer matrix formulation via a personal computer using MATLAB program. Theoretical analysis has shown that a triple layer AR coating and a DBR based back reflector effectively enhance the efficiency of a solar cell.

Keywords: Solar cell, one dimensional photonic crystal, anti reflection coating, back reflector, transfer matrix method

1 Introduction

At present the experimental realization of low cost, high efficiency solar cells (SCs) is being one of the active research areas. The real challenge is to cope with the increasing demand of global energy without adverse impact on nature. Among all the known renewable energy sources, solar energy- being easily accessible, clean, infinitely available, inexhaustible has been greatly appreciated as an efficient substitute for the current energy crisis[1-4]. Thin film solar cells are very promising candidates for cost effective and highly efficient photovoltaics. The widely accepted thin film solar cell technologies using organic materials or thin silicon layers require light trapping facilities to achieve high efficiencies. Light trapping refers to the techniques, which help to capture the major portion the light in the desired part of a solar cell without any leakage[5-8]. Introduction of surface-textured substrates and the use of special layers called back reflectors are the two important techniques for trapping light within the structure. Reflection at the textured back contact has a disadvantage, since the conventional metal reflectors like aluminium (Al) and silver (Ag) suffer from undesired plasmon absorption, thus limiting the long wavelength response of the solar cell[7,9]. The performance of the SC can be significantly enhanced by the proper light management within the device. Photonic crystals (PCs) are one of the best candidates for this purpose since the wavelength selective manipulation of reflection and transmission can be realized using PCs[10-13].

Photonic crystals are periodic optical structures designed to affect the motion of photons just like the periodicity of a semiconductor crystal affects the motion of electrons. Periodicity can vary from single-dimensional (1D) to three-dimensional (3D) with a period of the order of the wavelength of light, and according to this, PC can be categorized into one dimensional (1D), two dimensional (2D) and three dimensional (3D) photonic crystals. The fingerprint feature of a PC is the existence of a photonic band gap (PBG), where frequencies within the band gap are not allowed to propagate through the structure resulting complete reflection[14-20]. This signature property can be utilized to enhance the efficiency of solar cells.

Theoretical as well as experimental investigations based on PCs to enhance the performance of a solar cell have already attracted researcher's interest. Several groups have reported designing and fabrication of multilayer anti reflection (AR) coatings and PC based back reflectors for solar cell applications. Design of a triple layer AR coating composed of three different dielectric media with excellent transmission characteristics has been reported by B Geetha Priyadarshini and A K Sharma[21]. Jaker Hossain et al have reported a theoretical model for the optimization of AR coatings for PEDOT:PSS/c-Si hetero junction solar cell[22]. A PC based light trapping approach is analyzed by Peter Bermel et al[12] to enhance the solar cell efficiency. Using finite difference time domain method (FDTD) it has been shown that a solar cell with distributed Bragg reflector (DBR) structure improves the performance of the device[23]. Biswas et al[24] have reported an enhanced absorption of a nanocrystalline silicon solar cell with a PC back reflector. Recently 1DPC with a broad bandwidth in the wavelength range 580-1200 nm has been fabricated and incorporated into an amorphous silicon thin film solar cell as the back reflector and the optical performance of the device has been studied[25]. Herein, employing transfer matrix method, we demonstrate that 1DPC based back reflectors and AR coatings can significantly enhance the performance of a solar cell.
One dimensional photonic crystals otherwise called Bragg stacks or distributed Bragg reflectors are having the simplest structure where the refractive indices periodically vary only along one direction. When light propagates through such a structure, constructive and destructive interferences occur, leading to the wavelength-selective reflectance or transmittance behavior. 1DPC can be designed for maximum reflectance for a broad region of wavelengths ($\Delta\lambda$) centered at the Bragg's wavelength (λ_B), which is referred as the photonic band gap since transmittance is almost negligible in this wavelength range.
The Bragg wavelength is given by,

$$\lambda_B = 4n_l d_l = 4n_h d_h \tag{1}$$

and the corresponding reflectance of the structure is given by,

$$R(\lambda_B) = \left[\frac{n_0(n_l)^{2N} - n_s(n_h)^{2N}}{n_0(n_l)^{2N} + n_s(n_h)^{2N}}\right]^2 \tag{2}$$

Where n_l and n_h are the low and high refractive indices of alternating layers with d_l and d_h are the respective thicknesses, n_0 and n_s are the refractive indices of incident medium and substrate and N is the number of stacked pairs[10].

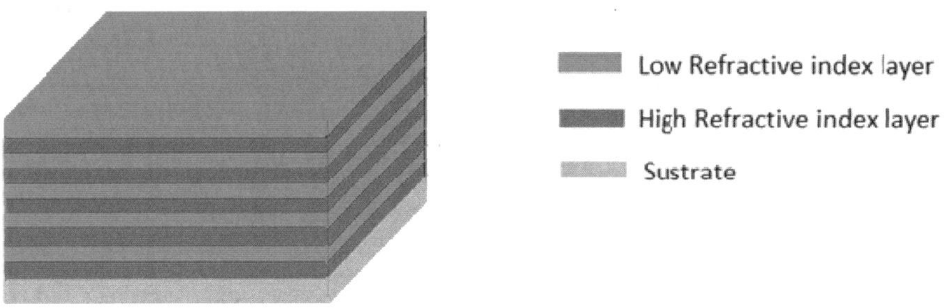

Figure 1 Schematic representation of one dimensional photonic crystals.

IOP Conf. Series: Materials Science and Engineering 1219 (2022) 012047 doi:10.1088/1757-899X/1219/1/012047

2 Theoretical Model

Transfer Matrix Method (TMM)[26-29] is used for the simulation analysis using MATLAB software to obtain the reflection and transmission spectra of the designed structures. Comparing to other computational techniques such as Finite Difference Time Domain (FDTD)[30] method and Finite Element Method (FEM)[31], TMM is relatively simple technique that does not require complicated computational analysis. The TMM is usually used as an efficient analytical tool for calculation of wave propagation in optical media. The characteristic matrix of each dielectric layer can be written as,

$$M_j = \begin{pmatrix} \cos\delta_j & \left(-\frac{i}{\eta_j}\right)\sin\delta_j \\ -i\eta_j \sin\delta_j & \cos\delta_j \end{pmatrix} \quad (3)$$

Where,

$$\delta_j = \frac{2\pi}{\lambda}\, n_j\, d_j\, cos\theta_j \quad (4)$$

Here λ is the wavelength, n_j, d_j and θ_j are the refractive index, thickness and refraction angle of jth layer. The parameter η_j for the jth layer is given by,

$$\eta_j = \sqrt{\frac{\varepsilon_0}{\mu_0}}\, n_j\, cos\theta_j \text{ , for TE Mode} \quad (5)$$

and,

$$\eta_j = \sqrt{\frac{\varepsilon_0}{\mu_0}}\, \frac{n_j}{cos\theta_j} \text{ , for TM Mode.} \quad (6)$$

where ε_0 and μ_0 are the free space permittivity and permeability respectively. Using the above basic ideas, we can construct the transfer matrix of the entire photonic crystal structure as given by,

$$M = M_1 . M_2 . M_3 ......M_N = \begin{pmatrix} m_{11} & m_{12} \\ m_{21} & m_{22} \end{pmatrix} \quad (7)$$

The reflection and transmission coefficients can be determined as,

$$r = \frac{\eta_0 (m_{11}+m_{12}\eta_m)-(m_{21}+m_{22}\eta_m)}{\eta_0(m_{11}+m_{12}\eta_m)+(m_{21}+m_{22}\eta_m)} \quad (8)$$

$$t = \frac{2\eta_0}{\eta_0(m_{11}+m_{12}\eta_m)+(m_{21}+m_{22}\eta_m)} \quad (9)$$

where η_0 and η_m are the parameters for the air and output layers, respectively. We can directly determine the reflectivity R and transmission T from these expressions[32],

$$R = | r |^2 \quad (10)$$

$$T = \frac{\eta_m}{\eta_0} | t |^2 \quad (11)$$

Both the AR coatings and the back reflectors for solar cells have been designed with appropriate transmittance and reflectance characteristics. For an ideal AR coating, the magnitude of the front surface reflections must be equal to the sum of all other partial reflections in the thin film layers. This can be made possible by choosing the refractive index of the single layer(n_l) coating equals to the square root of the product of refractive indices of the substrate (n_s) and the medium, generally air (n_0)[33]

$$n_l = \sqrt{n_s n_0} \quad (12)$$

But such a single layer AR coating provides minimum reflectance only for a single wavelength, $\lambda/4n_l$. Hence for solar cell applications conducive method is a multilayer AR coating having minimum reflectance for the entire visible range. The amplitude of the reflected beam at the air-coating interface and coating-substrate interface has to be 180° out of phase, giving rise to destructive interference, for realising an optimal anti reflection effect in the spectral range of interest. Magnesium fluoride (MgF_2), zinc sulphide (ZnS) and aluminium oxide (Al_2O_3) multilayers were considered for designing AR coating with maximum transmittance in the range 400 – 700 nm. Refractive index values for MgF_2, ZnS and Al_2O_3 were obtained from materials library database. Reference wavelength was fixed at 510 nm and normal incidence of light on the coating has been considered. Microscopic glass and air were considered as the substrate and the surrounding medium respectively[21].

IOP Conf. Series: Materials Science and Engineering 1219 (2022) 012047 doi:10.1088/1757-899X/1219/1/012047

3 Results and Discussions

3.1 Design of Back Reflector

An efficient back reflector enhances the performance of a solar cell and hence the perfect designing of the same is of great importance. Solar cells based on organic semi conductors (OSC) have less than 100 nm thick active layer due to poor charge carrier mobilities. In semi transparent organic solar cells (ST OSC), the thin semi transparent electrodes are not able to deliver adequate specular reflection, as a result, the active layer can absorb only a part of the incident radiation. The remaining part pass directly through the device resulting zero contribution to the power conversion efficiency (PCE). To address this problem, suitably designed 1DPCs can be incorporated into the structure. Using 1DPC as a distributed Bragg reflector with PBG matches with the wavelength range that does not get absorbed by the active layer, will be reflected back onto the active layer and thus significantly enhances the performance. The active layer in an OSC is having a maximum absorption in 460-580 nm range, hence for better performance of the solar cell, 1DPC with PBG in this wavelength range is required. Material combinations of MgF_2/ZnS and Al_2O_3/ZnS were considered to design DBR with maximum reflectance centered around 500 nm. Using the reference wavelength 500 nm, calculated the thicknesses of each layers and using the developed MATLAB code, reflection spectrum has been obtained. The reflectance spectrum of 2,4,6,8 pairs of Al_2O_3/ZnS and MgF_2/ZnS systems are shown in Fig. 2. The optical thickness of each layer is calculated so that it equals the quarter of the reference wavelength. As evident from the spectrum, the reflectance increases with the number of constituting layers in the multilayer structure. The reason is that light gets partially reflected at each interface, and since the spacing is periodic the multiple reflections of the incident light interfere destructively to eliminate the forward propagating wave[14].

3.2 Design of AR Coating

Fig. 3 gives the transmittance spectrum of designed AR coatings. Fig. 3(a) corresponds to the transmittance curve for a single layer of MgF_2. Calculated quarter wave optical thickness for MgF_2 layer for minimum reflectance at around 510 nm is about 92 nm. The maximum transmittance occurred at 510 nm and it is about 98.5%. Maximum transmittance resulted only for a very short range of wavelengths centered at the designed wavelength. To overcome this issue, a double layer quarter wave stack composed of ZnS and MgF_2 is considered. The corresponding transmittance spectrum is given in Fig. 3(b). Even though the transmittance has increased, the wavelength range with maximum transmittance is still observed to be very short. The optical thickness of ZnS layer was then increased, equals to the half of the reference wavelength, keeping quarter wavelength thickness of MgF_2. It is evident from the Fig. 3(c) that the range of wavelengths with maximum transmittance has increased considerably compared with the first and second cases. Fig. 3(d) corresponds to a triple layer coating where the constituent layers are MgF_2, ZnS and Al_2O_3. Here a half wave optical thickness ZnS layer was sandwiched between quarter wave stacks of MgF_2 and Al_2O_3. 100% transmittance was achieved using this configuration. It is observed that sandwiching a high refractive index layer between two low index layers ($n_{MgF_2} < n_{ZnS} > n_{Al_2O_3}$) leads to minimum reflection and maximum transmittance. The details regarding the design of multilayer AR coatings is given in Table 1. In every structure materials are listed in the same order as they appear from the top of the multilayer system. Q and H correspond to the optical thicknesses equal to the quarter and half of the reference wavelength respectively.

Table 1 Design of AR coatings.

Design	Materails	Refractive Index	Thickness (nm)	Pattern
1. Single Layer	MgF_2	1.38	92.39	Q
2. Double Layer	MgF_2	1.38	92.39	Q
	ZnS	2.32	54.95	Q
3. Double Layer	MgF_2	1.38	92.39	Q
	ZnS	2.32	109.91	H
4. Triple Layer	MgF_2	1.38	92.39	Q
	ZnS	2.32	109.91	H
	Al_2O_3	1.62	78.70	Q

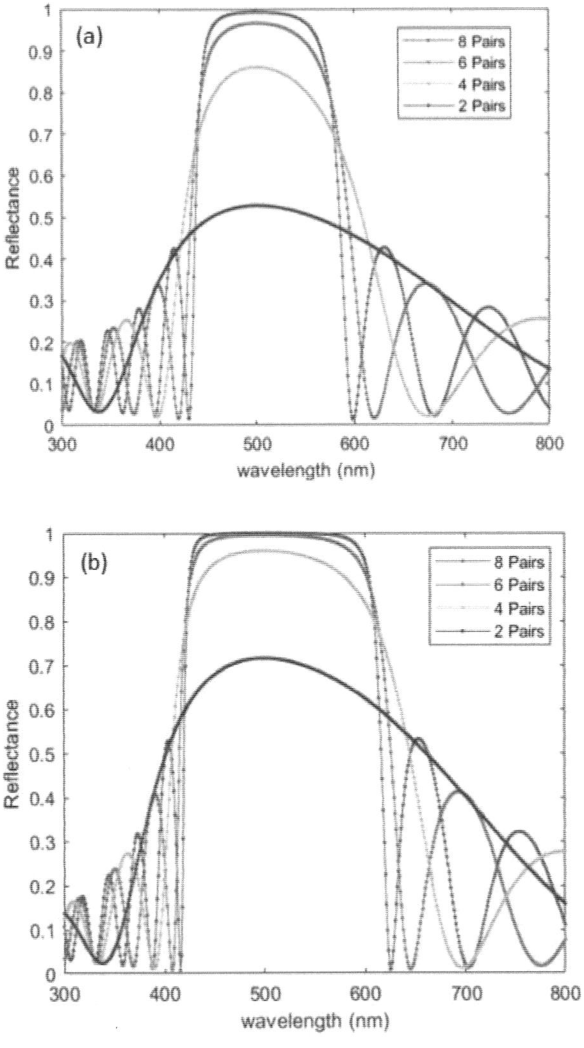

Figure 2 Reflectance spectrum of (a) Al_2O_3/ZnS and (b) MgF_2/ZnS 1DPCs

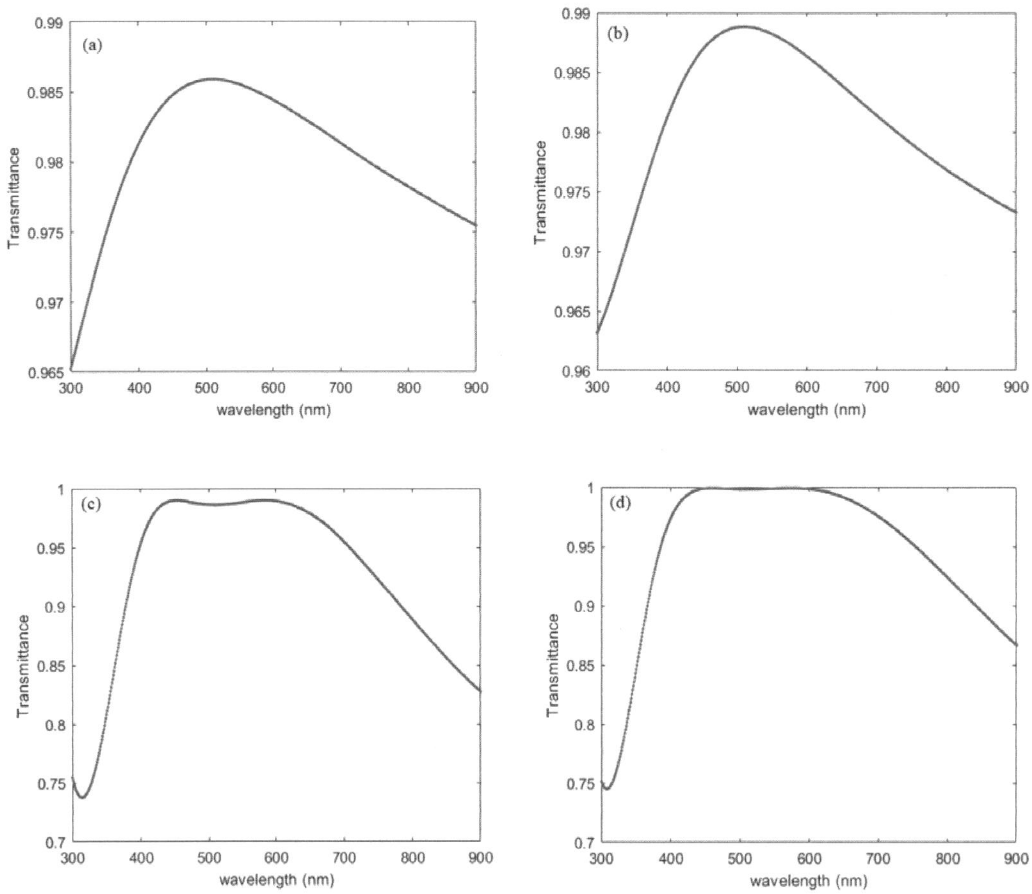

Figure 3 Transmittance spectrum of (a) single layer (Q) (b) double layer (QQ) (c) double layer (QH) and (d) triple layer (QHQ) AR coatings

4 Conclusion

Photonic crystal based multilayer AR coatings and back reflectors are recently getting great attention for enhancing the power conversion efficiency of a solar cell. In this study, a theoretical model for the optimization of multilayer antireflection coating as well as back reflectors for solar cell has been presented based on optical interference transfer matrix theory. Multilayer AR coating with 100% transmission capacity in the range 400-700 nm is designed using MATLAB program. Sandwiching a high refractive index layer in between two low index layers and employing Q-H-Q thickness pattern was found to be an excellent design for the perfect anti reflection applications. An ideal back reflector is also designed based on one dimensional photonic crystal which can act as a DBR. As expected, the reflectance of DBR enhanced with a greater number of constituting layers.

Acknowledgements
KS and SS would like to thank CSIR, Council of Scientific and Industrial Research, India for the financial support.

Author credits:
KS and RM have contributed equally to the work.

References

[1] Choubey P C, Oudhia A and Dewangan R 2012 *Recent Res. Sci. Technol.* **4** 8 99-101
[2] Sharma S, Jain K K and Sharma A 2015 *Mater. Sci. Appl.* **6** 12 1145-55
[3] Asim N, Sopian K, Ahmadi S, Saeedfar K, Alghoul M A, Saadatian O and Saleem H Zaidi 2012 *Renewable and Sustainable Energy Reviews* **16** 5834-47
[4] Parida B, Iniyan S and Goic R 2011 *Renewable and sustainable energy reviews* **15** 3 1625-36
[5] Green M A 1982 Solar Cells: Operating Priciples, Technology and System Applications (Prentice-Hall, Englewood Cliffs, NJ)
[6] Niggemann M, Riede M, Gombert A and Leo K 2008 *Physica status solidi a* **205** 12 2862-74
[7] Springer J, Poruba A, Müllerova L, Vanecek M, Kluth O and Rech B 2004 *J. Appl. Phys.* **95** 3 1427-29
[8] Moulin E, Paetzold U W, Bittkau K, Owen J, Kirchhoff J, Bauer A and Carius R 2013 *Progr. Photovol.* **21** 5 1236–47
[9] Krc J, Zeman M, Campa A and Smole F 2006 *MRS Online Proceedings Library Archive* **910** A25-A
[10] Isabella O, Lipovšek B, Krč J and Zeman M 2009 Mater. Res. Soc. Symp. Proc. Vol 1153 (San Francisco, California)
[11] Zeman Miro, and Krc J 2008 *J. Mater. Res.***23** 4 889-98
[12] Bermel P, Luo C, Zeng L, Kimerling L C and Joannopoulos J D 2007 *Optics express* **15** 25 16986-17000
[13] Kim D I, Lee J W, Jeong R H, Yu J H, Yang J W, Nam S H and Boo J H 2019 *Appl. Surf. Sci.* **487** 409-15
[14] Meade R D, Winn J N and Joannopoulos J D 1995 Photonic crystals: Molding the flow of light (Princeton University Press)
[15] John S 1987 *Phys. Rev. Lett.* **58** 23 2486-89
[16] Yablonovitch E 1987 *Phys. Rev. Lett.* **58** 20 2059-62
[17] Lopez C 2003 *Adv. Mater.* **15** 20 1679-1704
[18] Weissman J M, Sunkara H B, Tee A S and Asher S A 1996 *Science* **274** 959–65
[19] Holtz J H and Asher S A 1997 *Nature* **389** 6653 829-32
[20] Chong Yao, Jiaoyu Ren, Cihui Liu, Tao Yin, Yanxi Zhu, and Liqin Ge 2014 *ACS Appl. Mater. & Interfaces* **6** 19 16727-733
[21] Priyadarshini B G and Sharma A K 2016 *Bull. Mater. Sci.* **39** 3 683-89
[22] Hossain J, Mondal B K, Mostaque S K, Al Ahmed S R and Shirai H 2019 *Mater. Res. Express* **7** 1 015502
[23] Dubey R S, Saravanan S, and Kalainathan S 2014 *AIP Advances* **4** 12 127121
[24] Biswas R, Bhattacharya J, Lewis B, Chakravarty N and Dalal V *Sol. Energy Mat. Sol. Cells* **94** 12 2337-42
[25] Soman A and Antony A 2018 *Solar Energy* **162** 525-32
[26] Yeh P 1988 Optical waves in layered media (John Wiley & Sons.)
[27] Qi L, Yang Z and Fu T 2012 *Phys. Plasmas* **19** 1 012509
[28] Orfanidis S J 2016 Electromagnetic Waves and Antennas (Rutgers University)
[29] Miao L, Shi J, Wang J, Zhao D, Chen Z and Wang Q 2016 *Optical Engineering* **55** 5 057101

[30] Fang Y. T, Zhou J 2007 *J. Russ. Laser Res.* **28** 4 372-376

[31] Kim W. J, O'Brien J. D. 2004 *JOSA B* **21** 2 289-295

[32] Mohebbi M 2015 *J. Sens. Sens. Syst.* **4** 1 209-15

[33] Hadley L N and Dennison D M 1947 *J. Opt. Soc. Am* **37** 6 451-65

SCAPS simulated $FASnI_3$ and $MASnI_3$ based PSC solar cells: A comparison of device performance

M.Vishnuwaran[1], K. Ramachandran[1*] Priyanka Roy[2] and Ayush khare[2]

[*1]Department of Physics, SRM Institute of Science and Technology, Chennai-26, India

[2]Department of Physics, National Institute of Technology, Raipur 492010 India

E-mail id: kaviramach76@gmail.com

Abstract. In this paper, we have designed simulated, compare, and analyzed the high efficiency of $FASnI_3$ and $MASnI_3$ based PSC solar cell. In recent times, organic and inorganic Perovskite-based solar cells are turning to be hopeful devices due to their unique proprieties in the PV field. In our present work, we were concentrating several studies on Perovskite solar cell-based $FASnI_3$ and $MASnI_3$ Perovskite layer. Most of the Perovskite solar cell is made by Pb material for getting the high efficiency. However, the factor of toxicity and stability of the Pb-constructed devices is the main challenge to progress in commercial manufacture. In this study, numerical modeling of PSC used the alternative materials which are Sn-based Perovskite solar cells. In this work, we investigated tin-based perovskite solar cells with different absorber layers and also we investigated using high thickness in ETL and HTL what are change could be happen these are the things also we discussed in this paper. If we increase the thickness of the ETL and HTL the efficiency also increased we got some good results on it. Here we report simulation of lead-free Tin based PSC processed on $FASnI_3$ and $MASnI_3$ two different Perovskite layers. In our present work, we using 1D-SCAPS for numerical analysis.

Keywords: Lead-free, Perovskite solar cell, SCAPS-1D Simulation

1. Introduction

$CH_3NH_3PbI_3$ based PSC has received major attention in the PV industry due to its significant improvement in power conversion efficiency (PCE) [1, 2]. At present, the power conversion efficiency of PSC reached 25.2% which was initially just 3.8% in 2009[3, 4]. This major feat brought Perovskite solar cell as a tough competitor of Silicon made a solar cell in the photovoltaic marketplace.

The foremost concern from an environmental viewpoint is the presence of vast noxious material like Pb concomitant with the whole lifespan of PSC [5, 6]. For that reason, considerable required to introduce non-toxic materials in perovskite solar cell. Fortuitously, various types of less toxic cations such as Tin, Bismuth, Antimony, and Germanium have been substantiated as a replacement of Pb in PSCs. The recent implement of lead free material with an approximate direct bandgap in PSC has given a new way towards nontoxic Perovskite solar cells. Tin based Perovskite is got more considerations in the PV field because of its similar properties like Pb-based perovskite solar cell with good power conversion efficiency. There will be some disadvantages in the tin based solar cell like stability issues and other things at ambient condition [7, 8, 9]. Till now, 9% is the best power conversion efficiency of tin-based PSC [10]. In our present simulation, we show an effective different absorber layer with Tin-based Perovskite solar cell by using FASnI3 and MASnI3 with the same ETL and HTL layer and achieved efficiency as 23.74% has been achieved. In this present work, we analyze the impact of different temperatures at different thicknesses of the absorber layer on the device's performance.

IOP Conf. Series: Materials Science and Engineering 1219 (2022) 012048 doi:10.1088/1757-899X/1219/1/012048

2. Device Simulation Parameters

2.1. Device structure

PC-1-D, AFORS-HET, AMPS_1D, ASA, and SCAPS-1D are the simulation software to simulate solar cell applications. In this work, we have done AM1.5G illumination in solar cell capacitance simulator-1D. SCAPS -1D was designed by the University of Gent, Belgium. Simulation calculations, i.e., Poisson, hole continuity, and electron continuity equation are three fundamental equations mainly derived and carried out inside SCAPS-1D. [11, 12, 13]. Previously this simulation software is used to study cadmium telluride and copper indium gallium selenide-based solar cells only but now this technique was used to simulate perovskite-based solar cells and their results are good to a consensus with experimental results. Simulation work can be performed in both dark and light conditions. This SCAPS software can be designed with different layers [14, 15, and 16].

2.2. Simulation method

The simulated perovskite solar cell device configuration of FTO/P3HT/FASnI3/PCBM/Au and FTO/P3HT/MASnI3/PCBM/Au. The absorber layer of FASnI3 and MASnI3 are sandwiched between the Electron transport layer and hole transport layer here PCBM acts as HTL and P3HT act as ETL respectively in figure 1 (a, b). All physical parameters are collected from various reported papers, listed in Table 1. Some of the parameters are expected to be identical for the perovskite solar cell layer and fixed as 10 cm/s.

Using the initial simulation study and parameters have been shown. Device output with Voc of 3.20V, Jsc of 29.39 mA/cm2, FF of 24.52%, and PCE of 23.11 for FASnI3 and open-circuit voltage of 1.081V, the current density of 34.64 mA/cm2, fill factor 63.36 %, and PCE of 23.74 for MASnI3. This work was carried out concurrently with different temperatures at different thicknesses. At simulation time we increase the thickness of the ab-layer simultaneously we increase the temperature.

3. Results and Discussion

3.1 Effect in the Thickness of Absorbing Layer

The absorbing layer plays a vital role in the Perovskite solar cells to enhance the efficiency of the device. In this present work, the cell optimized with different thickness of the ab- layer with different temperature to enhance the PSC in PCE. The thickness of the ab-material FASnI3 and MASnI3 has been varied from 400 to 1200nm at different temperatures from 320 to 400k. When the thickness of the absorber layer will increase simultaneously increase the temperature. The parameters are taken from the reported paper are shown in Table 1, 2 continuous for PSC structure FTO/PCBM/absorbing layer/P3HT/Au. Figure 2(a,b,c,d) shows that the thickness of the absorber layer. Increasing the thickness of the ab-layer more photons are absorbed of higher wavelength. From the result, a Hugh number of electron-hole pairs were generated mainly due to the enhanced current density, open-circuit voltage, and PCE [19].

Figure 2(a,b,c,d) show increase in the thickness of ab-layer from 400-1200nm for both tin-based perovskite solar cell. Though, the efficiency of the simulated perovskite ab-layer of FASnI3 and MASnI3 were obtained at 23.11% and 23.74%. from the simulation 600nm be the thickness of the absorber layer with the temperature at 340k respectively for both simulated tin-based cells.

3.2. Consequence of Variation In Temperature

Perovskite solar cells function with the temperature-based device with highly operating parameters. The ambient temperature of the simulating device has been fixed at 320k to 400k. The device performance of perovskite solar cell is highly affected at an atmospheric temperature [20, 21]., a simulation study is carried out between 320k to 400k for both FASnI3 and MASnI3 absorbing layers to find temperature variation. According to the diode circuit model, the open-circuit voltage can be expressed as:

$$\text{Voc} = \frac{nKT}{q} \log\left(\frac{Jsc}{J0} + 1\right) \quad (1)$$

Here n, k, T, J_0, and q be the ideality factor, Boltzmann constant, temperature, reverse saturation current, and charge respectively.

Due to the rise the temperature, the value of J_0 increases which leads to reduce the Voc [22]. From Figure. 3(a), it could be detected that the temperature is varied from 320k-400k, a large decrement has been observed from 7.89v to 0.85 of FASnI3 and a slight decrease is noticed in Voc from 1.623v - 0.764v for MASnI3. But, in figure 3(b) Jsc is increased marginally from 27.26 to 31.07 for FASnI3 and a slight increase is observed for MASnI3 at 32.50 to 36.10 due to increasing in temperature, the bandgap of perovskite solar cell reduced to generating a huge number of electron-hole pairs consequential a better Jsc figure 3(c) shows the FF is increased from 16.08% - 79.95% for FASnI3: FF is initially increased from 44.99% to 74.30% for the various temperature range from 320K to 400k. Some other parameters affected due to increasing temperatures resultant in a decrease in device efficiency. Figure 3(d) shown that cell efficiency is moderate for the different temperatures at the different thicknesses. Temperature 340k at 600nm the efficiency of the cell is 23.11 for FASnI3 and 23.74 for MASnI3 respectively.

3.3 Analysis of QE at various thickness

In this simulation studies, the thickness of ab layers of FASnI3 and MASnI3 are various from 400nm to 1200nm respectively wavelength range 300nm – 900 nm to knows the impact of thickness on QE. FASnI3 Figure.4(a) shows the quantum efficiency of the device at 600 nm thickness which was a start to increase in 't' and reached a range of quantum efficiency up to approximately 99.95 when the wavelength range from 400nm - 1200nm. An increase in the thickness of the ab-layer MASnI3 exhibits a constant maximum QE of 99.9%. The improved value of QE of absorbing materials determines the huge number of electron collection and hole pairs. Quantum efficiency for MASnI3 is 69.55% at a thickness of 100 nm and reached 96.51 at 600 nm as showed in Figure 4. The performance of Perovskite solar cell with ab-materials FASnI3 and MASnI3 are got nearly at 600 nm could recommend that the perovskite light ab-materials are active more in the observable region that make it a possible contender of PV application for the future. [24,25].

3.4. Analysis of final device parameters values

In this work, we kept the optimized parameter are constant to get better cell performance. Figure 2(b) shows the current density of improved devices with Sn-based perovskite sensitizers as FASnI3 and MASnI3. The reported theoretical studies for ab-layers were listed in the table. From that simulation reports the outcomes of these simulation studies are getting good results. After a proper optimization of various parameters the cell exhibited superior with open-circuit voltage, current density, fill factor, and PCE of 3.20V, 29.39mA/cm2, 24.52%, and 23.11%respectively for FASnI3 and 1.081V, 34.64mA/cm2, 63.36 %, and 23.74% respectively MASnI3 with these observed results, it could be indicated that the Sn-based perovskite sensitizer material can be a challenger for commercial use in PV.

4. Conclusions

Through SCAPS-1D we have done our present work with cell construction of front contact/PCBM/perovskite layer/P3HT/back contact. The consequence of variation in thickness and temperature of ab-layers and also in ETL and HTL are inspected to increase the cell performance. All these parameters are detected simultaneous ways the cell showed power conversion efficiency of 23.11% for FASnI3 and 23.74% for MASnI3 based absorbing layer in 600nm at 360k for both PSC. The informed simulation results can provide a unique way to improve the efficiency of Tin-based PSC.

IOP Conf. Series: Materials Science and Engineering 1219 (2022) 012048 doi:10.1088/1757-899X/1219/1/012048

Reference

[1] P.Roy, N.Kumar Sinha, and A.khare, An investigation on the impact of temperature variation over the performance of tin-based perovskite solar cell: A numerical simulation approach, Materials today: proceesings,https://doi.org/10.1016/j.matpc.2020.09.281

[2] L. Serrano-Lujan, N. Espinosa, T.T. Larsen-Olsen, J. Abad, A. Urbina, F.C. Krebs, Tin- and lead-based perovskite solar cells under scrutiny: an environmental perspective, Adv. Energy Mater. 5 (20) (2015), 1501119.

[3] National Renewable Energy Laboratory Best Research-Cell Efficiency Chart. https://www.nrel.gov/pv/cell-efficiency.html.

[4] Kojima, A.; Teshima, K.; Shirai, Y.; Miyasaka, T. Organometal Halide Perovskites as Visible-Light Sensitizers for Photovoltaic Cells. J. Am. Chem. Soc. 2009, 131 (17), 6050–6051.

[5] L. Serrano-Lujan, N. Espinosa, T.T. Larsen-Olsen, J. Abad, A. Urbina, F.C. Krebs, Tin- and lead-based perovskite solar cells under scrutiny: an environmental perspective, Adv. Energy Mater. 5 (20) (2015), 1501119.

[6] Y. Ogomi, A. Morita, S. Tsukamoto, T. Saitho, N. Fujikawa, Q. Shen, et al., CH3NH3SnxPb(1-x)I3 perovskite solar cells covering up to 1060 nm, J. Phys. Chem. Lett. 5 (6) (2014) 1004–1011.

[7] F. Hao, C.C. Stoumpos, R.P.H. Chang, M.G. Kanatzidis, Anomalous bandgap behavior in mixed Sn and Pb perovskites enables broadening of absorption spectrum in solar cells, J. Am. Chem. Soc. 136 (22) (2014) 8094–8099.

[8] N.K. Noel, S.D. Stranks, A. Abate, C. Wehrenfennig, S. Guarnera, A.- A. Haghighirad, et al., Lead-free organic-inorganic tin halide perovskites for photovoltaic applications, Energy Environ. Sci. 7 (9) (2014) 3061–3068.

[9] Shuyan Shao, Jingjin Dong, Duim Herman, H. Gert, ten Brink, Graeme R. Blake, Giuseppe Portale, Maria Antonietta Loi, Enhancing the crystallinity and perfecting the orientation of formamidinium tin iodide for highly efficient Sn-based perovskite solar cells, Nano Energy 60 (2019) 810–816.

[10] W. Liao, D. Zhao, Y. Yu, C.R. Grice, C. Wang, A.J. Cimaroli, et al., Lead-free inverted planar formamidinium tin triiodide perovskite solar cells achieving power conversion efficiencies up to 6.22, Adv. Mater. 28 (42) (2016) 9333–9340.

[11] D. Hui-Jing, W. Wei-Chao, Z. Jian-Zhuo, Device simulation of lead-free CH3NH3SnI3 perovskite solar cells with high efficiency, Chin. Phys. B 25 (10) (2016), 108802.

[12] A. Niemegeers, M. Burgelman (Eds.), Numerical Modelling of AC-Characteristics of CdTe and CIS Solar Cells, Conference Record of the Twenty-Fifth IEEE Photovoltaic Specialists Conference, 1996, pp. 13–17.

[13] M.D. Stamate, On the dielectric properties of dc magnetron TiO2 thin films, Appl. Surf. Sci. 218 (1) (2003) 318–323.

[14] Evaluation of new materials for electron and hole transport layers in perovskite-based solar cells through SCAPS-1D simulations, in S. Bansal, P. Aryal (Eds.), IEEE 43rd Photovoltaic Specialists Conference (PVSC) June 2016, pp. 5–10.

[15] D. Poplavskyy, J. Nelson, Nondispersive hole transport in amorphous films of methoxy-spirofluorene-arylamine organic compound, Appl. Phys. 93 (1) (2003) 341–346.

[16] D. Liu, T.L. Kelly, Perovskite solar cells with a planar heterojunction structure prepared using room-temperature solution processing techniques, Nat. Photon. 8 (2013) 133.

[17] F. Hao, C.C. Stoumpos, D.H. Cao, R.P.H. Chang, M.G. Kanatzidis, Lead-free solid-state organic-inorganic halide perovskite solar cells. Nat. Photon. 8(6), 489–494 (2014). https://doi.org/10.1038/nphoton.2014.82 71. M.H. Kumar, S. Dharani, W.L. Leong, P.P. B

[18] T.M. Koh, T. Krishnamoorthy, N. Yantara, C. Shi, W.L. Leong, P.P. Boix, et al., Formamidinium tin-based perovskite with low Eg for photovoltaic applications, J. Mater. Chem. 3 (29) (2015) 14996–15000.

[19] Z. Zhao, F. Gu, Y. Li, W. Sun, S. Ye, H. Rao, et al., Mixed-organic-cation tin iodide for lead-free perovskite solar cells with an efficiency of 8.12, Adv. Sci. 4 (11) (2017), 1700204.

[20] C. Kim, T.D. Huan, S. Krishnan, R. Ramprasad, A hybrid organic-inorganic perovskite dataset, Sci. Data 4 (2017), 170057. [26] L.M. Herz, Charge-carrier mobilities in metal halide perovskites: fundamental mechanisms and limits, ACS Energy Lett. 2 (7) (2017) 1539–1548.

[21] Mandadapu U, Victor Vedanayakam S, Thyagarajan K (2017) Numerical simulation of CH3NH3PbI3-xClx perovskite solar cell using SCAPS-1D. Int J Eng Sci Invention 2:40–45

[22]. Zhao P, Liu Z, Lin Z, Chen D, Jie S, Zhang C, Zhang J, Chang J, Hao Y (2018) Device simulation of inverted CH3NH3PbI3-xClx perovskite solar cells based on PCBM electron transport layer and NiO hole transport layer. Sol Energy 169:11–18

[23] Minemoto T, Murata M (2014) Device modeling of perovskite solar cells based on structural similarity with thin film inorganic semiconductor solar cells. J Appl Phys 116:054505. https://doi. org/10.1063/1.4891982

[24] Wei X et al (2017) Numerical simulation and experimental validation of inverted planar perovskite solar cells based on NiOx hole transport layer. Superlattices Microstruct 112:383–393 17

[25]. M. I. Hossain, F. H. Alharbi, and N. Tabet, "Copper oxide as inorganic hole transport material for lead halide perovskite based solar cells," Sol. Energy 120, 370–380 (2015).

An Investigation on Structural Properties of Radiation Damaged $ZrSiO_4$ Using X-Ray Diffractometry and DFT

Deepankara Vrushabhadas Shastri.[1,2], John Donald Raj J[2], Kantha D Arunachalam[1*]

[1]Centre for Environmental Nuclear Research, SRM Institute of Science and Technology, Kattankulathur, Tamilnadu, India, 60320

[2]Department of Physics & Nanotechnology, SRM Institute of Science and Technology, Kattankulathur, Tamilnadu, India, 603203

**Corresponding author email: kanthad.arunachalam@gmail.com*

Abstract. Naturally occurring Zirconium minerals contain interstitial impurities of radioactive elements. Over a period of time due to alpha decay, the structure of minerals degrades through an internal bombardment process called metamictization. These metamict minerals proposed to be used as reference signatures for Naturally Occurring Radioactive Materials (NORM) Monitoring because of their unique parameters in structural properties. We have investigated structural damage of partially metamict Zirconium Silicate samples ($ZrSiO_4$), collected from Lambapur - Peddagattu region, Andhra Pradesh, India known for larger Uranium deposits. X-Ray Diffractometry data was used to study structural parameters experimentally. Structural properties of $ZrSiO_4$ belonging to space group I41/amd have been modeled and studied theoretically using Quantum ATK within local density approximation and generalized gradient approximation in-order to compare the structural properties obtained from the experimental data. We have investigated that the lattice parameters increased with a = 0.033 Å and c = 0.051 Å in the sample collected from proposed Uranium sites. We propose to use the lattice parameter variability of $ZrSiO_4$ as a tracer tool in nuclear forensics.

Keywords: Uranium mines, Radiation damage, NORM, Alpha decay

1. Introduction

Zirconium, an abundant element most of the time found as Zirconium Silicate $ZrSiO_4$ belongs to space group of I41/amd. $ZrSiO_4$ is an orthosilicate and the structure this category comprises of a separate from Silicate (SiO_4) tetrahedra and Zirconium (ZrO_8) polyhedra. The SiO_4 and ZrO_8 usually form a chain type structure which is parallel to the C axis by sharing the edges. Typically, Zircon hosts radioactive interstitial impurity like uranium and thorium in its crystal lattice, Amount of impurity vary anywhere between 5 ppm to 4000 ppm for Uranium and 2 ppm to 2000 ppm for Thorium depending on age and location of the deposits [1]. Radioactive elements like ^{238}U, ^{235}U, and ^{232}Th always carries its daughter nuclides, due to combination of parent and daughter nuclides zircon show significantly high radioactivity compared to average minerals. Alpha decay is a significant event which causes structural damage to the zircon.

It is to worth note that U-rich zircon, contains up to 12.9 weight percentage of U (0.1 U per formula unit) occurring in the melt of silicate that has been formed in the process at the Chernobyl nuclear plant during the year 1986. During α-decay two concurrent processes will take place one is the release of high energy 4-5 MeV α-particle and also ~100 Kev recoil atom.[2–4]. Because of the recoil interaction zircon transforms into partially metamict over geological times[4–7].

Fundamentally because of the damage in the crystal structure, metamict zircon can be described as a combination of the two phases of crystalline and amorphous based on the amount of radiation damage by alpha decay. These metamict zircons can be characterized by Scanning electron microscopy (SEM) [8], X-ray diffraction (XRD) [9], Raman spectroscopy [8,10] and Laser-Induced Confocal Photoluminescence Spectroscopy [11]. But structure of the domains damaged by alpha decay is not well known because of the aperiodicity in the system.

Zircon with its silicate structure shows a variation in unit cell parameters with an increase in pressure, but unit cell parameters increase with increasing temperature [12,6,8] even though in metamict zircons the overall shape of SiO_4 tetrahedra remains the same [13]. Between SiO_4 and ZrO_8 polyhedra, there are open channels and small voids which are the significant sites for impurity of radioactive elements. Such sites are responsible for accommodating U, Th, etc.

This study examines the structural variations in lattice parameters, bond length, volume in zircon collected from proposed uranium sites and compares it to experimentally stable zircon [14,15]. We have attempted to study and compare various structural parameters using Density functional theory (DFT). We propose structural variations of zircon can be used as tracer tools in nuclear forensics for the applications of radioactive material monitoring or to study any nuclear fallout.

2. Experimental Methods

2.1 Sampling methods

Samples have been collected from the locations of Lambapur Peddagattu region, Andhra Pradesh where Uranium mines had been proposed . Systematic and Grid Sampling approach was followed to identify sample location. Samples with regularly spaced intervals over space were considered. An initial location is selected randomly and thereafter the rest of the sampling locations are considered at regular intervals in the respective area.

2.2 Computational Calculation

In this work, the investigation of structural properties is carried out with the Density Functional Theory (DFT). We have used QuantumATK Version 2019.03 - SP1 to model Zircon structure based on the Linear Combination of Atomic Orbitals (LCAO) as well as the Plane Wave (PW) method. Local density approximations (LDA) of Perdew and Zunger and generalized gradient approximations (GGA) of Perdew-Burke-Ernzerhof (PBE) which are the functional exchange correlation techniques have been used. A single unit cell of the Zircon structure is used as the initial structure for the geometry optimization. The plane-wave (linear combination of atomic orbitals) basis with a mesh cut-off of 75 Hartree with a $5 \times 5 \times 5$ Monkhorst-Pack K-point mesh is used for the sampling of Brillouin Zone. The geometry optimization has been carried out till the tolerance force on every ion was less than 0.05 $eVÅ^{-1}$ and the total energy convergence is within the value of 1×10^{-4} eV.

2.3 X-Ray Diffractometry

The X-Ray Diffraction spectra had been obtained at a room temperature by Cu-Kα radiation in a Bruker D8 Advance Model- Davinci Design diffractometer with the help of detector SSD160, LYNXEYE. The data were obtained in the range of 10°to 70° range (step size 0.02° and 6s counting time is considered for each step).

The background was subtracted using the Chebyshev function. Peaks were modeled by using a pseudo-Voigt function with one Gaussian (GW) and two Lorentzian coefficients.

3. Results

3.1 Theoretical Investigations

The structure of zircon will be seen as it consists of (SiO_4) anions and Zr_4 cations, as depicted in Fig.1. Ground state structural parameters of $ZrSiO_4$ have been determined and compared to the ideal structure of $ZrSiO_4$ in Table 1. Where Plane wave GGA-PBE Overestimated values with error of 4.2 %, Planewave LDA-PZ underestimated values by 0.68%. In another basis set LCAO, GGA-PBE overestimated values with maximum error of 4.46% and LDA-PZ with least error of 0.1%.

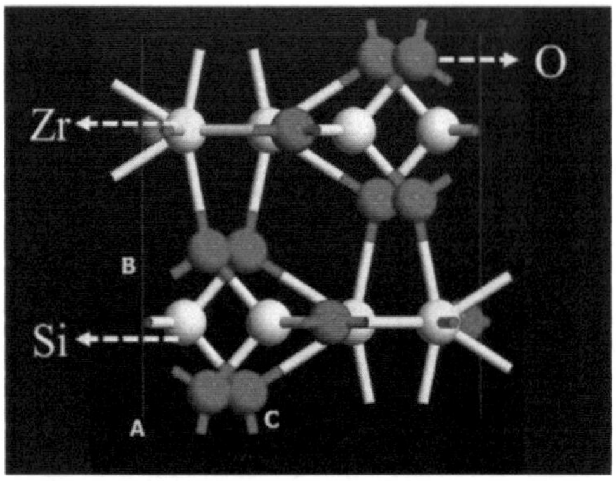

Figure 1: Experimental $ZrSiO_4$ Crystal Structure built using X-Ray diffractometry lattice parameter data increased bond distance resulting in lesser energy

The *a* and *c* are the unit-cell parameters that increase linearly with the increasing unit-cell volume, shown in Table 1. Such linear relations were also observed in other minerals [16]

For Implementing DFT many basis sets and exchange correlation functions have been used to investigate closer experimental value. In that by using basis set Linear combination of atomic orbitals (LCAO) and exchange correlation Perdew and Alex Zunger method (LDA- PZ) closest to our experimental studies with least error of +0.1%. Compared to ideal structure from experiment our studied sample increased volume shown in Table 1. Si–O bond length (A°) is increased by 0.029A°, and Zr–O bond length increased by 0.13A°. Increase in bond length is passive way of justifying loss of long-range translational symmetry which shows ongoing process of metamictization.

Table 1: Calculated structural parameters of $ZrSiO_4$ compared to experimental values Error

Method	**Basis Set**	**Exchange Correlation functionals**	**Lattice constants**		**Volume (Å^3)**	**Si–O bond Length (A°)**	**Zr–O bond length (A°)**
			***a* (Å)**	***c* (Å)**			
DFT	PW	GGA-PBE	6.896	6.111	145.346	1.633	2.263
		LDA- PZ	6.562	5.918	127.441	1.613	2.113
	LCAO	GGA-PBE	6.902	6.174	147.096	1.649	2.258
		LDA- PZ	6.614	5.957	130.332	1.625	2.134
This Work	-	-	6.64	6.03	132.93	1.651	2.260
Ideal Structure-Experiment	-	-	6.607	5.982	130.564	1.622	2.130

There is directly proportional correlation between lattice parameters and volume of the cell has been observed when lattice parameter increased by $a = 0.033$ Å, $c = 0.051$ Å both volume and bond length is increased, further weakening the crystal structure. We propose degree of variation in lattice parameter with combination of other structural parameters such as volume swealing, Bond length variation can be used signature of the proposed uranium sites of that location. Further investigations of other uranium sites have to confirm the study.

3.2 XRD Analysis

The variations of the unit-cell volume are shown in table 1 and computationally built crystal structure is shown in Fig 1. XRD analysis is done to identify different minerals and to extract lattice parameters data to compute structural parameters. Structural parameters in table 1 shows Zircon from uranium sites is partially metamict resulting from radiation damage due to alpha decay.

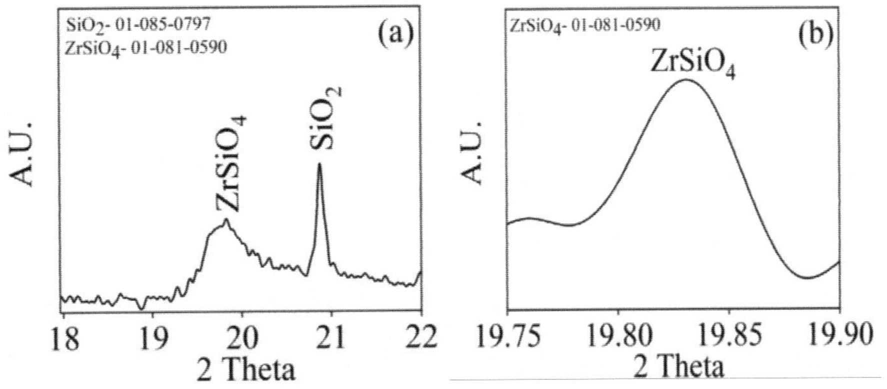

Figure 2: XRD Spectrum of Zirconium Silicate

In the XRD spectrum Fig 2A $ZrSiO_4$ is identified in the hkl plane {101} and 2θ value of 19.868. Broadening in peak is the effect of partial metamictization, and decrease in crystallinity due to radiation damage. In Fig 2B clearly indicates broadening of the peak further justifying partial metamictization.

4. Discussions

This study shows the different trends among the structural variations in zircon samples obtained from an area of proposed uranium mines, Study further confirms partial metamictization by showing increased volume of the cell and experimentally broadening of XRD peak of Zircon. One of the Nuclear Forensics signatures we can extract from the study is the increased bond length, Volume and lattice parameters of zircon can indicate high background radiation or any kind of actinide deposition in that geographical location. Further radioactivity data will be correlated to investigate any other relation between alpha decay and variations in structural parameters.

Acknowledgment

Department of Atomic Energy, Board of Research in Nuclear Science, Bhabha Atomic Research Center, Mumbai is acknowledged. We thank the Department of Science and Technology (DST) FIST, Government of India (via Project No. SR/FST/PSI-155/2010)
for providing the computational infrastructure. Authors are also thankful to SRM-HPCC, of SRM Institute of Science and Technology.

References

[1] S. Ellsworth, A. Navrotsky, R.C. Ewing, Physics and Chemistry of Minerals 21 (1994) 140–149.
[2] E. Jäger, J.C. Hunziker, Lectures in Isotope Geology, Springer Science & Business Media, 2012.
[3] E.B. Anderson, B.E. Burakov, E.M. Pazukhin, Radiochim. Acta 60:2-3 (1993).
[4] (2012) 563–600.
[5] M. Zhang, E.K.H. Salje, I. Farnan, A. Graeme-Barber, P. Daniel, R.C. Ewing, A.M. Clark, H. Leroux, Journal of Physics: Condensed Matter 12 (2000) 1915–1925.
[6] S. Ríos, E.K.H. Salje, M. Zhang, R.C. Ewing, Journal of Physics: Condensed Matter 12 (2000) 2401–2412.
[7] S. Ríos, T. Malcherek, E.K.H. Salje, C. Domeneghetti, Acta Crystallographica Section B Structural Science 56 (2000) 947–952.
[8] A. Kaiser, M. Lobert, R. Telle, Journal of the European Ceramic Society 28 (2008) 2199–2211.
[9] M. Zaman, S. Antao, Minerals 10 (2020) 947.
[10] Chem. Geol. 538 (2020) 119494.
[11] C. Lenz, E. Belousova, G.R. Lumpkin, Minerals 10 (2020) 83.

[12] S. Mitra, High Pressure Geochemistry & Mineral Physics: Basics for Planetology and Geo-Material Science, Elsevier, 2004.

[13] J.M. Hanchar, P.W.O. Hoskin, Zircon, Walter de Gruyter GmbH & Co KG, 2018.

[14] J.M. Hanchar, A Geochemical Investigation of Zircon, 1996.

[15] F.J. Berry, D. Eadon, J. Holloway, L.E. Smart, J. Mater. Chem. 6 (1996) 221–225. [16] S.M. Antao, Phys. Chem. Miner. 40 (2013) 705–716.

Improvement in property of waste brick admixtured cement cured with different water conditions

R. Gopalakrishnan, A.Geetha*

*Department of Physics and Nanotechnology, SRM Institute of Science and Technology, Kattankulathur- 603203, Tamilnadu, India

E mail: geethaa@srmist.edu.in

Abstract: The purpose of this report is to study the use of waste brick as an additive for Portland cement, in which 5, 10, 15 and 20% (by weight) of Portland cement is substituted by waste brick. The specimens are subsequently hardened for a period of 7, 28, 90 and 180 days under curing with the condition of distilled water, groundwater and seawater. The waste brick additives prompt the initial hydration of Portland cement and promote the hydration reaction. In the next stage of hardening (after 28, 90 and 180 days of hydration), waste brick supplement (up to 20 wt.%) enhances the compressive strength of the specimens, illustrating this effect as the development of calcium silicate hydrates. It is to be mentioned that waste brick activates the hydric properties of Portland cement. In the initial stages, additional seawater promotes cement hydration and, in the later stage, retards the same.

Keywords: Cement: waste brick: waters: strength: porosity: hydric parameters

1. Introduction

At present, the planet suffers from extreme heat conservation and radiation control. The construction industry contributes to the exploitation of global heat and soil emissions to quite an extent. Practical rules for breaking down actual construction restrictions, including energy-efficient building units and the elimination of the volume of numerous dusts. Therefore, in the construction industry, the supplementary cementitious materials are used to reduce the heat energy and increase the durability [1]. In addition to this, some studies [2, 3] have shown that the brick powder has pozzolanic properties. As an implementation of crystalline structures of calcium silicates in amorphous compounds in which it represents clay as large heats ranging from 600 °C to 1000 °C, it gives its pozzolanic activity in the construction process of bricks. Other tests regarding mortar and concrete brick powder [4-7] have shown that up to 20% of the brick powder as cement replacement could be acceptable and it would not alter the environmental and mechanical properties of the brick. The particle size of the brick powder has been found to be a significant component of this excellent performance improvement [8], and it again requires the behavior of optimum mechanical strength that is obtained when the particle size is less than 0.06 mm [9]. Schackow et al. [10] concluded that the calcium hydroxide reaction among the compounds of amorphous brick powder produces more C-S-H due to the pozzolanic reaction.

The main objective of this research is to examine the long-term (up to 180 days) hydration and improve the use of waste brick powder as cementitious materials. In this present study, cement has been replaced with up to 20% waste brick powder. The research also studies the application of mortars using ordinary Portland cement without any additions. In addition to that, the durability-relevant parameters, dynamic porosity and water absorption and its coefficients were used in order to ensure that the brick powder did not lose those properties during its improvement. In order to determine the mechanical properties of the mortars, their compressive strengths are also explored.

2. Experimental

2.1. Materials

Ordinary Portland cement (OPC) was used, the Blaine finesse of which is 445 m2/kg and the density is 3050 kg/m3. The waste brick powder was collected from a nearby manufacturing company. The density of waste brick powder (WB) is 2498 kg/m3 and the fineness of Blaine is 414 m2/kg, which is equivalent to that of cement. The chemical composition of the cement and waste brick (WB) powder and the chemical analysis of distilled water(DW), groundwater (GW) and seawater (SW) were treated from the standard values.

2.2. Methods

2.2.1. Mixing procedure

The OPC was replaced by 0%, 5%, 10%, 15% and 20% of the WB. For all the samples, the water-to-cement ratio is 0.4 and cured under the condition of distilled water, groundwater and seawater.

2.2.2. Determination of setting time

The OPC and WB admixtured samples were used in this measurement using the Vicat apparatus in accordance with the appropriate procedure [11], the results of which are presented in Fig 1.

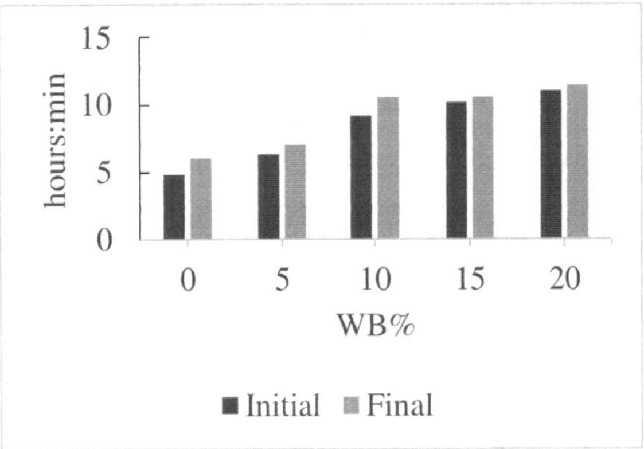

Fig.1a

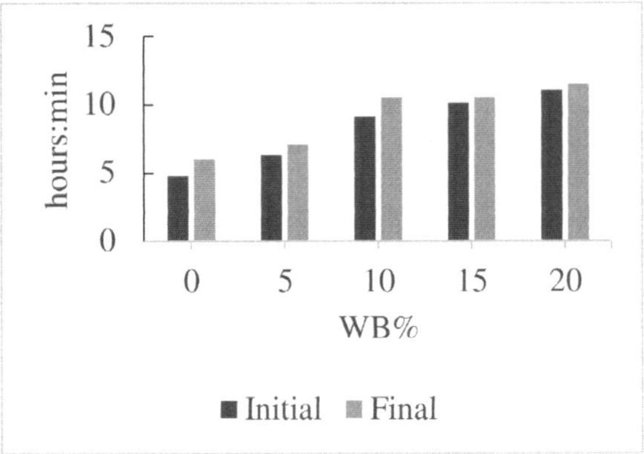

Fig.1b

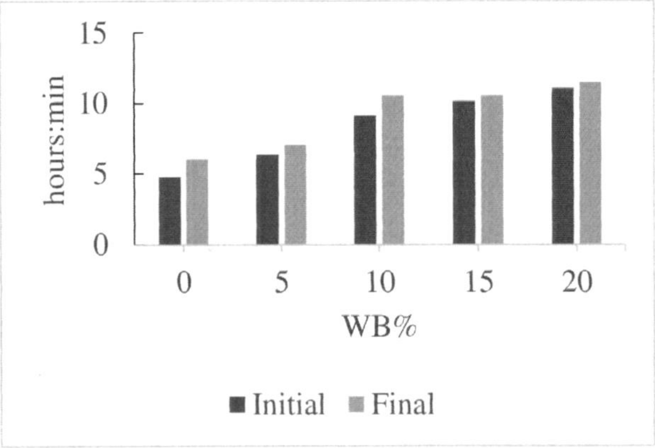

Fig.1c

Fig.1.Setting time of WB admixtured cement paste treated with (a) distilled Water (b) ground water (c) seawater

2.2.3. Determination of compressive strength

The compressive strength of OPC and admixtured samples was measured at 7-day, 28-day, 90-day and 180-day mortar cubes as per the standard procedure [11]. The specimen was mixed with DW, GW and SW. The final result of compressive strength value is found to be mean of the three samples, which is presented in Fig 2.

IOP Conf. Series: Materials Science and Engineering 1219 (2022) 012050 doi:10.1088/1757-899X/1219/1/012050

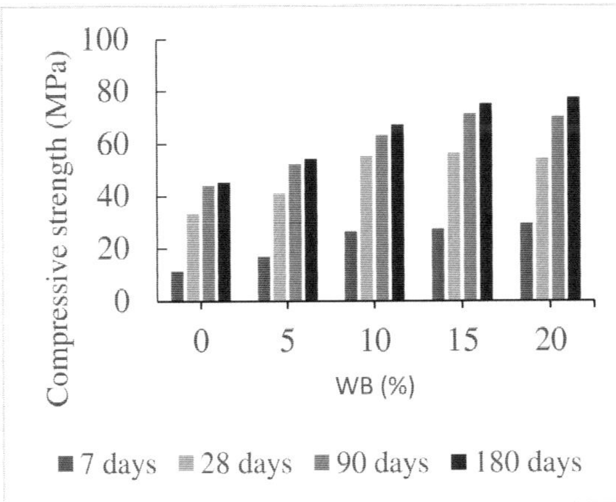

Fig.2a

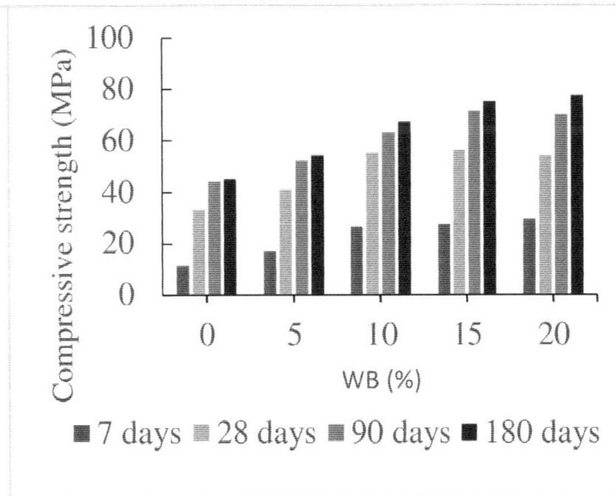

Fig.2b

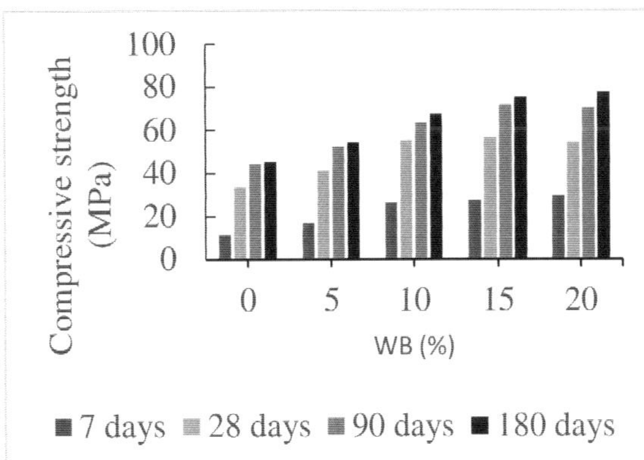

Fig.2c

Fig.2.Compressive strength of WB admixtured cement mortar treated with (a) distilled water (b)ground water (c)seawater

2.2.4. Determination of hydric parameters

Using the oven-drying process, the porosity of the mortar specimen was measured by the equation $((M_s - M_o)/(M_s - M_h)) \times 100$. The following hydric properties carried out applying the formula: water absorption $(100 \times (M_t - M_o)/M_o)$, the water absorption coefficient $((M_t - M_o)/t)$, apparent density $(M_o/(M_s - M_h))$ and true density $(M_o/(M_o - M_h))$ – where M_t is the mass of the wet sample, M_s is the mass of the saturated test sample, M_o is the mass of the dried test sample, and M_h is the hydrostatic weight of the saturated test sample [12]. The results of these formulae are given in Table 1.

Table 1. Hydric parameters of WB admixtured cement treated with different water

Water	WB%	P			WA			WAC			ρ_t			
		7day	28 days	90days	7day	28 days	90days	7day	28 days	90days	7day	28 days	90days	
DW	0	26.44	20.14	17.13	13.22	12.11	11.52	13.29	2.23	0.51	1.51	1.42	1.32	1.09
	5	23.55	21.32	15.14	8.23	7.14	6.32	10.74	1.52	0.29	1.44	1.35	1.27	1.05
	10	24.46	17.29	13.14	8.02	6.14	5.23	9.89	1.49	0.21	1.12	1.21	1.14	1.02
	15	23.24	16.32	12.14	7.23	6.01	5.33	8.54	1.23	0.19	1.19	1.16	1.01	1.01
	20	21.36	15.34	14.23	7.15	5.66	5.21	8.12	1.19	0.17	1.18	1.13	0.98	0.97
GW	0	25.43	19.11	16.12	13.01	11.58	10.55	11.23	1.4	0.35	1.44	1.28	1.32	1.05
	5	23.21	19.03	16	7.99	7	6.23	11	1.22	0.23	1.33	1.25	1.23	0.99
	10	22.14	17.05	13.11	7.85	6.02	5.05	1002	0.87	0.14	1.24	1.19	1.05	0.95
	15	21.03	15.23	12.01	7.1	5.55	5	9.45	0.81	0.11	1.21	1.15	1.01	0.84
	20	20.21	14.12	11.05	7.01	5.2	4.52	9.05	0.75	0.09	1.12	1.09	0.99	0.75
SW	0	21.48	18.23	19.23	11.23	11.14	12.15	11	1.35	1.56	1.34	1.27	1.51	1.11
	5	21.21	17.1	18.12	7.52	6.58	7.02	10.52	1.22	1.46	1.28	1.21	1.39	1.04
	10	20.05	15.01	17.14	7.12	5.54	6.05	10.04	1.19	1.25	1.21	1.17	1.24	1.01
	15	19.11	14.11	14.36	7	5.21	5.85	9.23	1.13	1.21	1.17	1.11	1.1	1
	20	18.12	13.25	13.25	6.56	5.12	5.24	9.14	1.12	1.02	1.14	1.02	1.06	0.96

Hydric parameters [(P = Porosity (%), WA = Water absorption (%),

WAC = Water absorption coefficient ρ_t = True density (g cm^{-3})

3. Results and discussion

3.1. Setting time measurements

The results of initial and final setting time of the OPC samples and WB admixtured samples are shown in Fig. 1. With the addition of recycled WB, the initial setting and final setting time for both of the samples are extended. For the 1 wt.% WB mixture, the setting time is greater than that of the ordinary Portland cement samples. The initial setting time for OPC is 5 hours 10 minutes, while 6 hours 25 minutes for the final setting after mixing. As the sum of WB increase by up to 20%, the initial set decreases to 11 hours 10 minutes, while the final set decreases to 11 hours 54 minutes. The OPC samples have shown comparatively a lower setting time than the WB mixtures. A high content of alkali occurs in the flash environment but does not adopt a flash setting in this case because the dissolution of the alkali from WB is not quick enough, hence only contributing to a minimum impact on the phase of crystallization. It relates the essence of the environment to the dispersal and hydration of cement particles. The altered cement mass in pastes occur in the state of less hydrate formation because of the cement along with recycled WB contents. At early ages, the report showed that the cementitious effects created by the pozzolanic reaction show very limited changes.

For the SW-treated samples, the setting time is low because the concentrations of chloride, sulphate and sodium ions are greater in the seawater. These ions induce the hydration process and produce more hydration products. From the findings, it has been understood that the interaction between cement, WB and water could be seen, thus changing the setting times. Groundwater has lower ion content than seawater but higher than distilled water. Therefore the response time between the admixtured cement paste treated with distilled water and seawater is intermediate.

3.2. Compressive strength measurements

Figure 2 demonstrate the experimental effects of compressive strength at various ages and different replacement quantities of cement. It has been observed that with the increasing substitution rate of cement by waste brick powder after 7 days of compressive strength improved. It showed the same pattern as that of 90 days of mortar healing after 28 days of healing. Such growth was marginal at later ages (180 days). Compared to 5% and 15%, 20% of cement substituted by brick powder produces the greatest conduct. These findings are established via hydration and the Portlandite rate adopted. Compressive strength generated from a predetermined relationship with the output of C-S-H and porosity. The processing of C-S-H and the production of portlandite are carried out by the hydration of cement, which dictates that the portlandite shift contributes to the rate of C-S-H. This leads to the contraction of the porosity and thus to the expansion of the strength by increasing the amount of higher hydrated products.

Navratilova and Rovnaníková [13] noted that the rates of high surface area and wide amorphous phase of brick powders increase the pozzolanic reaction and improve the compressive strength of the seawater-treated WB admixtured cement. The early-age compressive strength is higher compared to DW and SW because seawater speeds up cement hydration again in a supplement to WB because of the presence of chloride and sulphate ions. Using reaction production and crystallisation salt, these ions close up the pores, which increases the reaction of hydration and decreases the time of thickening, so that the successive changes in strength can take place.

3.3. Hydric test

In SW more such ions are present, which decrease the level of porosity. At 7 days, it increases the compressive strength for all WB replacements. The porosity increases at 28 days onwards due to the retardation process. In DW- and GW-cured samples, WA decreases (Table 1) as the rate of WB replacement and hydration time increase. When compared to DW and GW, the WA and WAC values in seawater samples (Table 1) decrease by up to 28 days, causing the hydration reaction to boost the vision of WB and SW and filling of the pores by the inventions of reaction products. It indicates that the water absorption is low for WB specimens. In the samples, the concentration was seen at 7 days and there is a significant difference in the WAC and true density for various WB ratios. The variance due to the porous structure is considerably reduced at 28 days. Between 7 and 28 days of seawater curing specimen, the values of WA and WAC increase. By implementing seawater and WB, the microstructure gets loose, hence increasing porosity, WA and WAC.

4. Conclusion

1. There are three basic sections that specify the addition to which WB forms: when it replaces cement in a mortar, mortar strength. These are the influence of the filler effect. OPC hydration acceleration and WB pozzolanic reaction with C-H. The WB and altered both setting time and compressive strength.

2. According to experimental results, the hydration reaction of SW-treated sample is faster up to 7 days because of chloride and sulphate ions, but after 28 days retarded the hydration reaction as these ions are disappeared.

3. Irrespective of water, with a rise in the percentage of WB, a gradual decrease in water absorption and water absorption coefficient was observed.

IOP Conf. Series: Materials Science and Engineering 1219 (2022) 012050 doi:10.1088/1757-899X/1219/1/012050

4. The compressive strength, porosity and hydric studies are responsible for accelerating the effects of WB treated with SW.

5. Compressive strength, water absorption, porosity, apparent density of all the mixtures containing waste brick powder provided the conditions of OPC. The addition of waste brick increased the ductility of the OPC mixtures.

References

[1]. Brown N O W, Malmqvist T, Bai W and Molinari M 2013 *Build. Envir.* **61(1)** 140–148

[2]. Pereira-De-Oliveira L A, Castro-Gomes J P, Santos P M S 2012 *Const. Build. Mat* **3** 1197–203

[3]. Navrátilová E, Rovnaníková P 2016 *Const. Build. Mat.* **120** 530–539

[4]. Aliabdo A A, Abd-Elmoaty A E M, Hassan H H 2014 *Alexandria Eng. J.* **53** 151-168

[5]. Kulovaná T, Vejmelková E, Keppert M, Rovnaníková P, Keršner Z, Cˇerny´ R 2016 *Structural Concrete* **17** 105–115

[6]. Lavat A E, Trezza M A, Poggi M 2009 *Waste Management* **29** 1666–1674

[7]. Ge Z, Gao Z, Sun R, Zheng L 2012 *Const. Build. Mat.* **31** 289–293

[8]. Kwon Y H, Kang S H, Hong S G, Moon J 2017 *Applied Science* **7(5)** 522-533

[9]. Pereira-De-Oliveira A, Castro-Gomes J P, Santos P M S 2012 *Const. Build. Mat.* **31** 197–203

[10]. Schackow A, Stringari D, Senff L, Correia S L, Segadães A M 2015 *Cem. Concr. Comp.* **62** 82–89.

[11]. Shetty M S. Concrete Technology. New Delhi: S. Chand and Co. Ltd., 2004

[12]. Navrátilová E, Rovnaníková P, 2016 *Const. Build. Mat.* **120** 530-539

[13] Govindarajan D, Gopalakrishnan R, 2009 *Frontiers of Mat. Sci. in China* **3(4)** 395-402

A Review on Properties of Concrete with Magnetizing Cement

Kalyani Gurram[1], Pannirselvam N[2*]

[1] Research Scholar, Department of Civil Engineering
[2] Associate Professor, Department of Civil Engineering
Department of Civil Engineering, SRM Institute of Science and Technology, Potheri, Kattankulathur, Chengalpattu District, Chennai - 603 203, Tamil Nadu, India

E-mail: [1]gurramk@srmist.edu.in [2*]pannirsn@srmist.edu.in

Abstract: The incorporation of strong magnetic powders (NdFeB) into cement mixtures has produced improved mechanical properties of cement pastes. The magnetic cement (cementitious magnet) used a simple and inexpensive cement process to form complex forms of the NdFeB magnet. Particulate NdFeB has been distributed uniformly in the cement matrix. The magnetic cement composites increased in density and compressive strength with the NdFeB concentration. Magnetization (stable up to one year) also increased with NdFeB powder but there was no improvement in coercivity. The introduction of magnetic particles into cement matrices has also been another important method in recent years to enhance the mechanical and other properties of cement-based materials. Nanoparticles (IONs) made from iron oxide (Fe_2O_3, Fe_2O_4) are among the most popular cement fillers. Cement composites mixed with IONs showed enhanced compressive strength, as well as other enhanced functions. This paper presents literature analysis of a study of magnetizing cement on concrete properties.

Keywords: Concrete Properties, Magnetic cement, NdFeB Powder

1. Introduction

Recent times cement has been modified for many a purpose by adding suitable admixtures/additive. One such recent focus is magnetization of cement. Enhancing magnetic properties can enhance the absorption of cement by electromagnetic waves. In addition, few of the application of enhancing the electromagnetic properties of cement are asfollows:

1. Nano-Fe_3O_4 Magnetic fluid demonstrates an immense future use in the design of electromagnetic wave disturbance shielding structures (He et al.,2018) [40].
2. The persistent magnetic character of cements has been shown for the lifting / moving of metal elements or different magnetic cement items. The ability for revolutionary technologies happens in areas such as electromagnetic protection, magnetic moving, advanced manufacturing, innovative structure layout, decoration and toys. (Chuewangkam et al., 2019) [2]
3. Dynamic wireless charging
4. Metaresonator antennas (Sum et al., 2017) [3]
5. Stable under chloride and sulphate environment (Bragança et al.,2016) [6]
6. For the enhancement of piezoresistive behavior (Vipulanandan and Mohammed,2015) [7]
7. Ease of mixing pumping and placing (Nair et al.,2014) [5]
8. To analysis the quality of cement bonding (Nair et al.,2017) [4]
9. In oil and gas well application (Wu et al.,2015) [15]

Further this literature study is divided into followingheads

- Magnetized water in cement andconcrete
- Materials for magnetization ofcement
- Fresh properties of magnetized cement(MC)
- Hardened properties of magnetized cement

IOP Conf. Series: Materials Science and Engineering 1219 (2022) 012051 doi:10.1088/1757-899X/1219/1/012051

- Gap inliterature
-

Summary

Literature shows that additions of magnetized water have the following effect on cement paste/concrete

1. Increase inworkability
2. Increase in compressive strength, especially early strength
3. Reduction in the dosage of super plasticizer for the required strength ordurability
4. Reduction in shrinkagecrack

However all these effect depends on the strength of magneticfield.

2. Magnetized water in cement and concrete

Water performs a key position within the achievement of the desired cement properties. The cement hydration process is controlled by water. Through the application of magnetic fluids, the fluids ions can quickly reach the cement pieces, enabling a more total hydration cycle to take place and improving the hydraulic strength of the concrete. Such results can rely heavily on the strength of the magnetic field.

Su et al., 2000 [16] Study of the influence of magnetic water on the compressive strength and the workability of mortar and concrete. Tests indicate that the compressive power of mortar samples mixed with magnetic water by 0.8–1.35 T, 9–19 percent more than specimens mixed with tap water. Likewise, the compressive power of magnetic water-prepared concrete 10–23 per cent more relative to standard water. For fact, the highest improvement can be achieved when the magnetic water is of 0.8 or 1.2 T. Gholhaki et al., 2018 [18] The quantity of water amount decreased with the use of magnetic power. The research was conducted on Self-Compacting Concrete (SCC). The findings reveal that magnetic water and pozzolanic substances in SCC will enhance self-compatibility requirements in terms of float ability and viscosity. In contrast, the magnetic water-containing mixture of SCC and 20 percent of silica smoke can be known as a most effective blending configuration at the age of 28 days because the compressive strength and tensile electricity stepped forward to 49 per cent and 41 per cent, respectively, and the water absorption value decreased to 55 per cent. Further, magnetic water can reduce the amount of High Variety Water Reducer (HRWR) required for SCC with the aid of up to 45per cent.

Wei et al., 2017 [21]Analysis of the effect of magnetic water on the diminishing crack of concrete. Results indicate that the strain intensity factor and the overall cracking region are decreased while magnetic water was used. This suggests that the early-age cracking strength of concrete combined with magnetic water is strengthened than that of tap water. The effect was found to differ with magnetic power. The highest improvement in early-age cracking resistance of concrete is obtained when the magnetic power reaches 260mT and the magnetic field duration reaches 280 mm.

Afshin et al., 2010 [23] Magnetic water used in high-strength concrete was found to increase the workability and consistency of the concrete. Magnetic water has been shown to induce a reduction in the cement material needed for the stated compressive strength volume. Concrete with magnetic water reported a higher downturn of up to 45 percent, a higher compressive power of up to 18 percent and a decrease of up to 28 percent in cement.

Esfahani et al., 2018 [20] Throughout their analysis, magnetic water and specific doses of super plasticizer were used in SCC and found a 10 per cent decrease in the amount of water needed and a 12 per cent improvement in compressive intensity and a 30 per cent decrease in the dosage of the super plasticizer when utilizing magnetic water. Su et al., 2003 [22] Investigated the workability and compressive strength of mortar and concrete, which were blended with Magnetic Field Treated Water

(MFTW) and contained fly ash. Outcomes show that the compressive strength of mortar samples combined with MFTW is higher than the ones prepared with tap water.The best compressive strength increase of concrete is achieved when the magnetic strength of treated water is of 0.8 and 1.2 T.

Khorshidi et al., 2014 [10] Conducted laboratory experiments to test fluidity, compressive power, cement paste settling period with magnetic water. The force and position of the magnetic field, the velocity and duration of the water moving through the magnetic system, and the quantity and form of the colloidal particles have a direct impact on the properties of the magnetic material and the usage of such material to paste the cement improves its fluidity and compressive strength by up to 10%. Abdel-Raouf et al., 2009 [11] This has been demonstrated that the magnetization degree of the water influences the workability and power of the concrete. This research used water at magnetization rates of 0, 0.1, 0.2, 0.3 and 0.4 Tesla. The results show that magnetized water improves concrete workability, although the improvement varies depending on the mixing parameters. Few signs of changes in concrete quality have also been found. Nevertheless, decreases in concrete mechanical properties have been reported in many mixtures. Note should be taken when choosing a magnetic field to ensure minimum adverse impacts.

Summary

Incorporation of magnetic field to the cement is achieved by using various types of nano material. Very few literatures exist at the have a look at of rheological conduct of cement paste. The fabric is found to show off shear thinning conduct.

3. Magnetization ofcement

Magnetizations of cement for various applications as mentioned in above section have been carried out in many ways. In this section, literature related to the cement magnetization is presented.

Materials used for magnetization ofcement

Magnetic susceptibility of the cement can be improved by the addition of nano particles and the following are the various materials that were added to cement for this purpose are listed in table 1. Table 1: Material for magnetization of cement

Material	Reference
Nano-Fe_3O_4	He et al., 2018 [40]
Hard magnetic powders (NdFeB)	Chuewangkam et al., 2019 [2]
Copper slag (for improving the conductivity)	Wang et al., 2016 [8]
Carbon encapsulated Cobalt	Blyszko, 2008 [9]
Carbon encapsulated nickel andironnanoparticles	Guskos, 2010 [14]

Influence of magnetization on freshcement

Few workability and rheological studies had been carried out to study the influence of magnetization on fresh cement properties.

Nair et al., 2016 [17] The rheological response of cement paste was studied by incorporating magnetic particles into the cement paste and by varying the magnitude of the magnetic field strength. Magnetic fields have been altered in a low-high-low and high-low-high configuration to determine the impact of cyclical magnetic field fluctuations on rheological behavior. A cement-based magnetic fluid has been shown to demonstrate field responsiveness in its rheological behavior while the magnetic field is varied. At an early age where the magnetic field was varying, the rheological behavior was shown to be less reliant on the time period in which the magnetic field was applied and more related to the amplitude of the magnetic field. In later years, in the presence of a cyclic magnetic field, the rheological reaction would be determined by the magnetorheological reaction and aging processes (e.g., hydration and thixotropy) that have an effect on the stiffening of cement paste Nair et al., 2014 [39] in another study examined on demand for fluid application in the mixing, pumping and positioning of concrete.

Jiao et al., 2019 [12] Used a Small Amplitude Oscillatory Shear (SAOS) technique to study the structural build-up of cementite paste with Nano-Fe_3O_4 under time-varying magnetic fields. Many types of magnetic fields, such as continuous, abrupt shift and longitudinal transition, have been added to cement paste. The findings revealed that the structural build-up of the cementite paste relied on the magnetizing period and the strength of the magnetic field. Applying continuous magnetic fields, the liquid-like behavior changed within the first few minutes and then the solid-like properties changed. Both the unexpected rise in magnetic fields and the rapid decline of magnetic fields culminated in a significant drop in data modules. The linearly increasing magnetic field resulted in a small rise in storage modulus with a higher fluid-like behavior. Once the magnetic field was linearly decreased from 0.5 T to approx. 0.25 T, the systemic build-up has been improved.

Vipulanandan et al., 2020 et al., [13] Investigated methods that could potentially be easily followed within the area to modify the rheological residences of cement slurry in situ in the course of the installation of oil and fuel wells. The effects of magnetic discipline strengths and temperatures at the rheological properties of an oil well cement (Class H) changed with nano particles of iron oxide (Nano-Fe_3O_4) were studied. The sensitivity of the usage of electric resistance inside the monitoring of changes in cement slurry has additionally been investigated. The water-to-cement ratio used on this evaluation turned into 0.38 as it is miles a common ratio used in oil nicely cementation.The contents of Nano-Fe_3O_4(common particle length of 30 nm) in cement slurry ranged by up to 1% by means of weight of cement. The temperature ranged from 25°C to 85°C. Magnetic field electricity ranged to 0.6 T. The preliminary electric resistance of the cement slurry modified with Nano-Fe_3O_4reduced through 6 to 52 percent on the basis of Nano-Fe_3O_4material, temperature and magnetic discipline strength. The shear thinning activity of cement slurry with and without Nano-$Fe_3O_4$was quantified using the Vipulanandan rheological model and contrasted to the Herschel-Bulkley model. The rheological properties of cement slurry adjusted with Nano-Fe_3O_4under various magnetic field strengths is compared with the electrical resistance of cement slurry using nonlinear resistance and Vipulanandan structure correlation models.

4. Influence of magnetization on the hardened properties ofcement

Wang et al., 2016 [8] an effort was made to improve the magnetic and electrical conductivity of cement using copper slag. It has been noted that both the actual and the theoretical portions of the permittivity rise with an improvement in the copper slag content and the minimum value was found to be 30%. Mortars with a 50 percent copper slag content show higher compressive strength and better microwave absorption efficiency. Soto-bernal et al., 2015 [35] performed an experiment to expose fresh cement paste to static magnetic fields at three specific magnetic induction strengths: 19.07, 22.22 and 25.37 Gauss.

Microstructural characterization shows that there are variations in the quantity and morphology of C–S–H gel; the quantity of C–S–H is bigger and its morphology will become denser and much less porous with higher magneto static induction strengths; it also indicates proof of modifications within the mineralogical composition of hydrated cement paste. The temperature rise does not have detrimental effects on cement paste compressive power, because the magneto static force stimulates the hydration cycle by a molecular reforming mechanism, which helps cement paste strengthen micro structurally, with decreased porosity and higher mechanical efficiency.

He et al., 2018 [40] Nano-Fe_3O_4used to magnetize cement. The co-precipitation process is used to magnetize cement. Co-precipitation is an easy and most traditional process for extracting IONPs (Fe_3O_4 and Fe_2O_3) from ferric / ferrous salt solution with a 1:2 molar ratio by applying specific solutions at room temperature or at high temperature. The optimum amount was determined to be 5 percent Nano-Fe_3O_4in the form of gas. It also improved cement's early compressive ability.

Chuewangkam et al., 2019 [2] Recently magnetized cement utilizing strong magnetic powders. The density and compressive strength of the magnetic cement composites were calculated with various NdFeB concentrations and it was found that the NdFeB concentration improved the density and the compressive intensity (Figure 1). Nevertheless, the inclusion of NdFeB did not alter the oppressive land. Magnetic materials are also found to remain intact for up to 300 days.

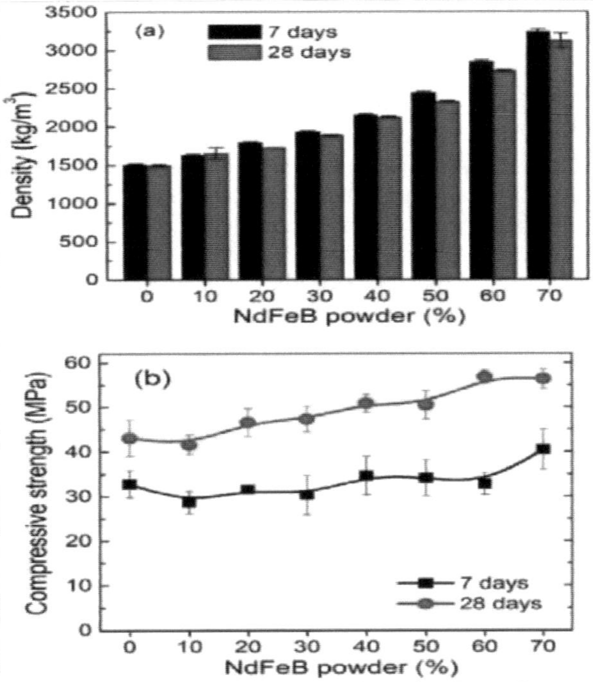

Figure1. Properties of magnetic cement for different concentration of NdFeB (Chuewangkam etal., 2019) [2]

Guskos et al., 2008 [33] Four types of cement mortars with out and with a small quantity of magnetic nano debris of fcc-nickel and 5-007-iron encapsulated in carbon have been prepared for their study. Samples have been analyzed using Ferro Magnetic Resonance (FMR) and compressive energy techniques. FMR spectra of the subsequent samples had been recorded: cement mortar powder, cement mortar powder / nano particles and cement mortar powder / nano particles / waterNo variations were found in the FMR spectra of the two former forms. Samples with magnetic nano particles displayed a large and extreme almost symmetrical FMR graph, while a sample with additional water molecules displayed a lower range of strength. For the former sample, a compressive strength test of the nickel encapsulated nano particles with or without carbon showed an increase of more than 15 percent, while the opposite effect for concrete was observed with α-iron nano particles. The existence and association of magnetic nano particles embedded in non-magnetic materials may dramatically change any of their physical properties.

Guskos et al., 2010 [14] Observed that there is a strong effect on the device magnet resonance spectra particularly at higher temperatures due to low concentration of cobalt nano particles in a microsilica / cement matrix.

Amin et al., 2013 [31] The effect on the hydration properties of both Ordinary Portland Cement (OPC) and High-Slag Cement (HSC), on the application of magnetite nano particles was observed. Compressive strength, chemical mixture of water content, and study of X-ray diffraction, and differential calorimeter were studied in the admixed magnet cement pastes. The tests of the compressive resistance showed that OPC and HSC reinforced pastes with varying concentrations of magnetic fluid had higher compressive values than the OPC and HSC concrete pastes. Calcium silicate hydrate, portlandite and calcium sulfoaluminatehydrate are the major hydration compounds detected by XRD diffractograms. In addition to one major fundamental distinction, namely the production of calcium iron silicate as a hydraulic substance with a fair hydrational character, the introduction of magnetic fluid nano particles to both OPC and HSC did not affect the principal hydration properties of the clean OPC or HSC Cement.

Li et al., 2004 [34] Experimental analysis for Nano-$Fe_3O_4$and nano-SiO_2cement mortar mechanical properties was carried out. Reports of the tests revealed that on the 7th day and the 28th day of concrete mortar combined with nano subjects the compressive and bending power was stronger than a regular cement mortar. Therefore, nano-particles can be added to improve the mechanical characteristics of concrete. The SEM evaluation of microstructures among cement mortar and cement mortar discovered that the Nano-$Fe_3O_4$and nano-SiO_2 occupy the pores and decrease $CaOH_2$ between the hydrates.

Kwon et al., 2010 [38] Dielectric constant measured and conductivity in cement mortar with five distinct saturation situations. Durability measures including porosity, diffusion of chloride, air permeability, twists, and water diffusion are carried out with the same mixture proportions. The tests under 60 percent of saturation, which indicates consistent effects, are chosen and are summed with one value from continuously measured EM properties within a 5–20 GHz Frequency spectrum for various saturation values. Together with reliability measures are the estimates of measurements based on performance below 60% of the saturation.

Pantin and Moir 1961 [37] The magnet concrete used as an artificial tracer indicates that such currents will carry sand grains at least 11/2 mm in diameter

Paya et al., 1996 [36] Sample magnetized with fly ash, and the fluid and mechanical properties of cement paste were tested. The non-magnetic fractions created mortar that was equivalent or higher than mortar that comprised the corresponding magnetic fractions. On the other hand, compressive strength development studies have demonstrated that the non-magnetic fractions are more pozzolanic than the magnet fractions, and that, for non-mechanically treated samples, the optimum replacement of cement percentages with fly ash fractions was 45%, whereas for non-magnetic and magnetic soil fractions, optimal values were 30%.

Gopalakrishnan et al., 2012 [32] Fly ash has also been documenting the effect on the properties of Portland cement hydrated by magnetic susceptibility experiments of groundwater and marine water. The magnetic sensitivity was calculated with Faraday Curie balance procedure at various hydration ages and is associated with changes in the set-up period and the calculation of compressive power. Regardless of the atmosphere, the observed findings suggest that, with increasing fly ash substitute cement levels, the magnetic sensitivity rises.

Govindarajan et al., 2009 [28] Magnetic susceptibility analysis identified the function of metakaolin in the properties of Portland cement hydrated with soil and sea water. Faraday Curie was determined by the sensitivity of various hydration times and was related to the increase in the setting time of modified cement compressive power. The results show that the magnetic susceptibility increases with an increasing level of metakaolin substitution in cement regardless of water.

Tsukada et al., 2014 [27] The procedure for measuring moisture content and hydration state as a magnetic mixture medium has been developed for the magnetic calculation. Such components have been tested by means of a specially built high-temperature superconductor supranational quantum interference system for the magnetization-magnetic field curves and mortar with different humidity materials. The tests from the MRI demonstrate that the free water concentration in the mortar is related to a decrease in magnetic response.

Wang et al., 2005 [24] Research was conducted on magnetic resistance and turbidity of the hydrothermally synthesized calcium silicate (C–S–H) gel under different environments. Experimental results show that the C–S–H gel with diatomite and lime is optimally prone to hydrothermal synthesis (m) and turbidity (±). The hydrothermal C–S–H gel synthesized for the best time is more sensitive and less turbid. Susceptibility and turbidity are apparent and can be used as a quality assurance measure for C–S–H gel synthesis.

Summary

- Magnetization modifies the hydration process ofcement.
- Considerable increase in the strength was observed.
- No study relating the concentration of nano particle and influence of magnetic field strength.
- More studies required to understand the internal structural changes that occur due to magnetization and its effect in mechanicalcharacteristics.

Conclusions:

The studies that have been carried out to study the mechanical properties of cement paste/ concrete with magnetic water mainly focused on only few aspects of the mechanical properties of paste/cement. More studies are needed on the complete characterization of the material especially on the influence of magnetic water on the internal structural change in the material, its conductive properties. Also, different level of magnetic strength that influences these properties needs focus.Very few literatures exist on the study of rheological behavior of cement paste. More focus is needed to study the influence of percentage of nano fluids/particles and magnitude of magnetic field on the influence of flow behavior. Also, no study exists in correlating the flow behavior with internal structuralchanges.More studies required to understand the internal structural changes that occur due to magnetization and its effect in mechanical characteristics of magnetic cement. No study relating the concentration of nano particle and influence of magnetic fieldstrength.It is acknowledged from this literature evaluate that a cementitious course can be used to supply permanent NdFeB magnets. The approach is rather easy and inexpensive. Cementitious magnets can be produced in diverse methods, permitting complex and net formed goods to be manufactured. Certain benefits include low fabrication costs and no special tooling requirements. The magnetic debris of the goods has been determined to be allotted homogeneously within the cement matrix. The magnetic cement paste's rheological flow was more difficult for NdFeB powders at rates above 50 weight percent. Because of the better density and strength of NdFeB the density and strength of the magnetic cement composites elevated with the material of NdFeB. Both cement composites had a better degree of magnetization than natural Ni or cubic ferrites, even though their coercion turned into extra than difficult ferrites. The magnetic properties are very stable for as much as three hundred days, which suggests the toughness of these composites with magnetic cement. The (BH) peak of the products became nevertheless smaller than everlasting commercial magnets; however it could sincerely be stepped forward. Such cements were proven to have a permanent magnetic character for lifting / transferring metal elements or different magnetic cement elements.

References

[1] L. Jung-Kun *et al.*, "Use of magnetic nanoparticles for smart drilling fluids," *Am. Assoc. Drill. Eng.*, pp. 0–4, 2009.

[2] N. Chuewangkam, S. Pinitsoontorn, and P. Chindaprasirt, "Properties of NdFeB magnetic cement," *Cem. Concr. Compos.*, vol. 103, no. May, pp. 204–212, 2019, doi: 10.1016/j.cemconcomp.2019.05.010.

[3] Y. L. Sum, V. Rheinheimer, B. H. Soong, and P. J. M. Monteiro, "Effects of Cement Paste Enhanced with Iron-Based Magnetic Particles on an Embedded Small Resonator Antenna," *Sci. Rep.*, vol. 7, no. 1, pp. 1–7, 2017, doi: 10.1038/s41598-017-15289-6.

[4] R. D. Srinair, T. W. Patzek, and E. Van Oort, "Detecting poor cement bonding and zonal isolation problems using magnetic cement slurries," *Proc. - SPE Annu. Tech. Conf. Exhib.*, 2017, doi: 10.2118/187047-ms.

[5] S. D. Nair and R. D. Ferron, "Set-on-demand concrete," *Cem. Concr. Res.*, vol. 57, pp. 13–27, 2014, doi: 10.1016/j.cemconres.2013.12.001.

[6] M. O. G. P. Bragança, K. F. Portella, M. M. Bonato, E. Alberti, and C. E. Cláudia, "Performance of Portland cement concretes with 1% nano-Fe3O4 addition: Electrochemical stability under chloride and sulfate environments," *Constr. Build. Mater.*, vol. 117, no. 2016, pp. 152–162, 2016, doi: 10.1016/j.conbuildmat.2016.05.033.

[7] C. Vipulanandan and A. Mohammed, "Smart cement modified with iron oxide nanoparticles to enhance the piezoresistive behavior and compressive strength for oil well applications," *Smart Mater. Struct.*, vol. 24, no. 12, p. 125020, 2015, doi: 10.1088/0964-1726/24/12/125020.

[8] Z. Wang, T. Zhang, and L. Zhou, "Investigation on electromagnetic and microwave absorption properties of copper slag-filled cement mortar," *Cem. Concr. Compos.*, vol. 74, pp. 174–181, 2016, doi: 10.1016/j.cemconcomp.2016.10.003.

[9] J. Blyszko *et al.*, "Study of mechanical properties of concrete with low concentration of magnetic nanoparticles," *J. Non. Cryst. Solids*, vol. 354, no. 35–39, pp. 4515–4518, 2008, doi: 10.1016/j.jnoncrysol.2008.06.101.

[10] N. Khorshidi, M. Ansari, and M. Bayat, "An investigation of water magnetization and its influence on some concrete specificities like fluidity and compressive strength," *Comput. Concr.*, vol. 13, no. 5, pp. 649–657, 2014, doi: 10.12989/cac.2014.13.5.649.

[11] M. Abdel-Raouf and M. N. Abou-Zeid, "Properties of concrete incorporating magnetized water," *Transp. Res. Rec.*, no. 2113, pp. 62–71, 2009, doi: 10.3141/2113-08.

[12] D. Jiao, K. El Cheikh, C. Shi, K. Lesage, and G. De Schutter, "Structural build-up of cementitious paste with nano-Fe3O4 under time-varying magnetic fields," *Cem. Concr. Res.*, vol. 124, no. August, p. 105857, 2019, doi: 10.1016/j.cemconres.2019.105857.

[13] A. S. Mohammed, "Vipulanandan model for the rheological properties with ultimate shear stress of oil well cement modified with nanoclay," *Egypt. J. Pet.*, vol. 27, no. 3, pp. 335–347, 2018, doi: 10.1016/j.ejpe.2017.05.007.

[14] N. Guskos, G. Zolnierkiewicz, J. Typek, J. Blyszko, W. Kiernozycki, and U. Narkiewicz, "Ferromagnetic resonance and compressive strength study of cement mortars containing carbon encapsulated nickel and iron nanoparticles," *Rev. Adv. Mater. Sci.*, vol. 23, no. 1, pp. 113–117, 2010.

[15] "Magnetorheological Fluids for Oil and Gas Well Application APPROVED BY SUPERVISING COMMITTEE :," 2015.

[16] N. Su, Y. H. Wu, and C. Y. Mar, "Effect of magnetic water on the engineering properties of concrete containing granulated blast-furnace slag," *Cem. Concr. Res.*, vol. 30, no. 4, pp. 599– 605, 2000, doi: 10.1016/S0008-8846(00)00215-5.

[17] S. D. Nair and R. D. Ferron, "Real time control of fresh cement paste stiffening: Smart cementbased materials via a magnetorheological approach," *Rheol. Acta*, vol. 55, no. 7, pp. 571–579, 2016, doi: 10.1007/s00397-016-0923-x.

[18] M. Gholhaki, A. kheyroddin, M. Hajforoush, and M. Kazemi, "An investigation on the fresh and hardened properties of self-compacting concrete incorporating magnetic water with various pozzolanic materials," *Constr. Build. Mater.*, vol. 158, pp. 173–180, 2018, doi: 10.1016/j.conbuildmat.2017.09.135.

[19] B. J. Goluchowska, "Some factors affecting an increase in magnetic susceptibility of cement dusts," *J. Appl. Geophys.*, vol. 48, no. 2, pp. 103–112, 2001, doi: 10.1016/S09269851(01)00083-0.

[20] A. R. Esfahani, M. Reisi, and B. Mohr, "Magnetized Water Effect on Compressive Strength and Dosage of Superplasticizers and Water in Self-Compacting Concrete," *J. Mater. Civ. Eng.*, vol. 30, no. 3, p. 04018008, 2018, doi: 10.1061/(asce)mt.1943-5533.0002174.

[21] H. Wei, Y. Wang, and J. Luo, "Influence of magnetic water on early-age shrinkage cracking of concrete," *Constr. Build. Mater.*, vol. 147, pp. 91–100, 2017, doi: 10.1016/j.conbuildmat.2017.04.140.

[22] N. Su and C. F. Wu, "Effect of magnetic field treated water on mortar and concrete containing fly ash," *Cem. Concr. Compos.*, vol. 25, no. 7, pp. 681–688, 2003, doi: 10.1016/S09589465(02)00098-7.

[23] H. Afshin, M. Gholizadeh, and N. Khorshidi, "Improving Mechanical properties of high strength concrete by magnetic water technology," *Sci. Iran.*, vol. 17, no. 1 A, pp. 74–79, 2010.

[24] Z. Wang, L. Sun, Y. Hou, and E. Tang, "In situ monitoring of C-S-H gel processing by magnetic susceptibility and turbidity measurements," *J. Am. Ceram. Soc.*, vol. 88, no. 7, pp. 1799–1802, 2005, doi: 10.1111/j.1551-2916.2005.00423.x.

[25] "EMR / ESR / EPR Spectroscopy for Characterization of Nanomaterials Basic Concepts of Iron and Steel Making," p. 2436, 2020.

[26] C. Stations and I. Charging, "Magnetic cement enables dynamic wireless charging," pp. 1–2, 2020.

[27] K. Tsukada *et al.*, "Magnetic susceptibility and magnetic resonance measurements of the moisture content and hydration condition of a magnetic mixture material," *J. Appl. Phys.*, vol. 115, no. 17, pp. 6–9, 2014, doi: 10.1063/1.4852075.

[28] D. Govindarajan and R. Gopalakrishnan, "Magnetic susceptibility measurements on metakaolin admixtured cement hydrated with ground water and sea water," *Int. J. Miner. Metall. Mater.*, vol. 16, no. 3, pp. 349–354, 2009, doi: 10.1016/S1674-4799(09)60063-7.

[29] N. Guskos *et al.*, "Ageing effect in carbon-coated cobalt nanoparticles embedded in a cement matrix studied by FMR," *Rev. Adv. Mater. Sci.*, vol. 14, no. 2, pp. 130–134, 2007.

[30] N. Makul, P. Rattanadecho, and D. K. Agrawal, "Applications of microwave energy in cement and concrete - A review," *Renew. Sustain. Energy Rev.*, vol. 37, pp. 715–733, 2014, doi: 10.1016/j.rser.2014.05.054.

[31] M. S. Amin, S. M. A. El-Gamal, and F. S. Hashem, "Effect of addition of nano-magnetite on the hydration characteristics of hardened Portland cement and high slag cement pastes," *J. Therm. Anal. Calorim.*, vol. 112, no. 3, pp. 1253–1259, 2013, doi: 10.1007/s10973-012-2663-1.

[32] R. Gopalakrishnan, S. Barathan, and D. Govindarajan, "Magnetic Susceptibility Measurements on Fly Ash Admixtured Cement Hydrated with Groundwater and Seawater," *Am. J. Mater. Sci.*, vol. 2, no. 1, pp. 32–36, 2012, doi: 10.5923/j.materials.20120201.06.

[33] N. Guskos *et al.*, "Magnetic properties of the micro-silica/cement matrix with carbon-coated cobalt nanoparticles and free radical DPPH," *J. Non. Cryst. Solids*, vol. 354, no. 35–39, pp. 4510–4514, 2008, doi: 10.1016/j.jnoncrysol.2008.06.067.

[34] H. Li, H. G. Xiao, J. Yuan, and J. Ou, "Microstructure of cement mortar with nano-particles," *Compos. Part B Eng.*, vol. 35, no. 2, pp. 185–189, 2004, doi: 10.1016/S1359-8368(03)00052-0.

[35] J. J. Soto-Bernal, R. Gonzalez-Mota, I. Rosales-Candelas, and J. A. Ortiz-Lozano, "Effects of Static Magnetic Fields on the Physical, Mechanical, and Microstructural Properties of Cement Pastes," *Adv. Mater. Sci. Eng.*, vol. 2015, 2015, doi: 10.1155/2015/934195.

[36] J. Paya, J. Monzo, M. V. Borrachero, and E. Peris-Mora, "Comparisons among magnetic and non-magnetic fly ash fractions: Strength development of cement-fly ash mortars," *Waste Manag.*, vol. 16, no. 1–3, pp. 119–124, 1996, doi: 10.1016/S0956-053X(96)00034-7.

[37] H. M. Pantin, "Magnetic concrete as an artificial tracer mineral," *New Zeal. J. Geol. Geophys.*, vol. 4, no. 4, pp. 424–433, 1961, doi: 10.1080/00288306.1961.10420131.

[38] S. J. Kwon, M. Q. Feng, and S. S. Park, "Characterization of electromagnetic properties for durability performance and saturation in hardened cement mortar," *NDT E Int.*, vol. 43, no. 2, pp. 86–95, 2010, doi: 10.1016/j.ndteint.2009.09.002.

[39] S. D. Nair and R. D. Ferron, "Set-on-demand concrete," *Cem. Concr. Res.*, 2014, doi: 10.1016/j.cemconres.2013.12.001.

[40] Y. He, L. Lu, K. Sun, F. Wang, and S. Hu, "Electromagnetic wave absorbing cement-based composite using Nano-Fe3O4 magnetic fluid as absorber," *Cem. Concr. Compos.*, vol. 92, no. May 2017, pp. 1–6, 2018, doi: 10.1016/j.cemconcomp.2018.05.004.

Influence of sintering temperature on structural, morphological and optical properties of nanocrystalline bismuth ferrite thin films via sol-gel processing

A.Deepak Sharma[1], H. Basantakumar Sharma[2]

[1]Department of Physics, Manipur University, Canchipur, Imphal, India, 795003
[2]Department of Physics, Manipur University, Canchipur, Imphal, India, 795003

Email: adeepaksh@manipuruniv.ac.in

Abstract

Nanocrystalline thin films of pure bismuth ferrite were synthesized by sol-gel deposition technique on the corning glass substrate. The influence of sintering temperature on microstructure, morphology, and optical parameters has been systematically studied. Optimization of sintering temperature for single-phase nanocrystalline bismuth ferrite fabrication was carried out and found to be at 550 °C. X-ray diffraction confirms the presence of distorted rhombohedral perovskite structure of bismuth ferrite oriented along (hkl) plane of (101). Surface morphology and Root mean square (RMS) surface roughness of the thin films prepared were determined from the micrographs of atomic force microscopy (AFM). Several optical constants viz. bandgap energy, extinction coefficient, Urbach energy, has been determined using Ultraviolet-Visible spectroscopy of the thin films within the range 200 -800 nm. With the increase in sintering temperature, there is a decrease in the optical bandgap of the material which would be essential for the wide range of applications in thin-film solar cells along with optoelectronic devices.

Keywords: Sol-gel; grain size; bandgap;

1. Introduction

Multiferroic materials possessing at least two ferroic orders (ferroelectric, anti-ferroelectric and ferroelastic) have gained a huge amount of attention in recent years owing to the coupling between electric and magnetic ferroic orders. This characteristic property makes it viable to influence electric polarization by magnetic field or to control magnetism by electric field for potential applications in low cost highly sensitive magnetometers, memory devices, spintronics devices, sensors, magnetically controlled transducers, etc. [1,2]. Bismuth ferrite (BFO) is a room-temperature multiferroic material having T_c ~ 1103 K & T_N ~ 643K having potential applications in various ways. Another characteristic feature of bismuth ferrite attracting more attention is its good optical response with a narrow energy bandgap of 2.3-2.7 eV (approx.), which is low as compared to other ferroelectric materials.[3]. $BiFeO_3$ has been systematically studied theoretically and experimentally by various other researchers focusing on the fabrication of thin films, ceramics, nanoparticles, and nanorods in the last two decades [4-7].

In the past decades, smooth multilayered films of BFO have been deposited onto various substrates by a variety of techniques available that includes magnetron sputtering using radio-frequency (RF), pulsed laser deposition (PLD), chemical vapor deposition (CVD) using metallo-organic precursors, dip coating, and sol-gel spin coating technique [8]. Sol-gel chemical deposition method has certain advantages over other prevailing techniques which include a large-area of deposition, cost-effectiveness, a fast fabrication process, better composition control, and hence compatible with industrial production. Successful fabrication of crack-free BFO thin films onto corning glass substrate depends on several controllable parameters like sintering temperature, RPM setting of the spin coater, duration of sintering, precursors' mole concentration used, etc. Optimization of these parameters is essential for the production of defect-free thin films for both research and commercial significance.

In general, the optical performance of semiconductors is co-related to the electronic band structure of the semiconductor. Thus, systematic analysis of optical parameters could lead to new applications in optical devices [9]. Many doped bismuth ferrite materials are extensively studied by various other workers focusing on improving optical parameters like refractive index, optical bandgap, transmittance and absorption spectra, etc. [10-12]. However, depositing single-phase pristine bismuth ferrite thin films by spin coating onto a corning glass substrate is still challenging due to the development of secondary phases and the difficulty to optimize suitable sintering temperature. Narrow bandgap having high absorption coefficient are essential parameters for thin-film solar cell applications as it would absorb maximum solar radiation. In this work, the influence of sintering temperature on the microstructural, morphological, and optical parameters of BFO thin films has been studied systematically for finding useful applications in optoelectronic devices. BFO thin films deposited by programmable spin coater onto glass substrate were sintered at three different temperatures, say 450^0C, 500^0C, and 550^0C, and investigated to improve optical parameters viz. bandgap, extinction coefficients, Urbach energy, and their corresponding morphological and microstructural properties.

2.Materials and methods

Bismuth ferrite thin films (BFO) have been synthesized using a sol-gel deposition technique by bismuth nitrate pentahydrate ($Bi(NO_3)_3 . 5H_2O$); ≥ 98% purity) and iron nitrate nanohydrate ($Fe(NO_3)_3 . 9H_2O$; ≥ 98% purity) as chemical precursors. A stock solution of pure bismuth ferrite was prepared in 2-methoxy ethanol ($CH_3C_2H_4OH$) by dissolving the stoichiometric amounts of the metal precursors. To the stock solution 2ml of acetic acid anhydrate was added followed by continuous stirring on a hot plate at 60°C for 30min. The resulting stock solution is filtered using a syringe filter (0.45 µm). Before depositing the film, the corning glass substrates were dipped in HNO_3 (30% conc.) for 12h and then rinsed with CH_3OH (methanol). Finally, the substrates were sonicated in acetone and de-ionised (DI) water for 5 min each to make the surface of the substrates clean and ready for deposition. The sol was spin-coated on the surface of pre cleansed substrates using spin coater whose revolution per minute (rpm) of was programmed to increase stepwise cautiously and finally set at 3000 rpm for 30s. The as-prepared thin films were heated at 350°C for 5mins on a hot plate and subsequently cooled before the next layer deposition. The 7-layered films have been finally annealed at three different temperatures of 450°C, 500°C and 550°C for 2h of constant heating in a furnace to form the uniform nanocrystalline BFO thin films.

3. Results and Discussion

3.1 Structural analysis: The structural properties of nanocrystalline $BiFeO_3$ thin films deposited on the corning glass substrates were determined from the X-Ray diffraction (XRD) data measured by X'PERT PRO PANalytical. The crystallite sizes of the thin films annealed at three different temperatures were estimated using Debye's Scherrer formula $D = \frac{0.9\lambda}{\beta cos\theta}$ where λ is the wavelength of radiation (Cu_α X- rays) used, and θ is Bragg's angle [11].

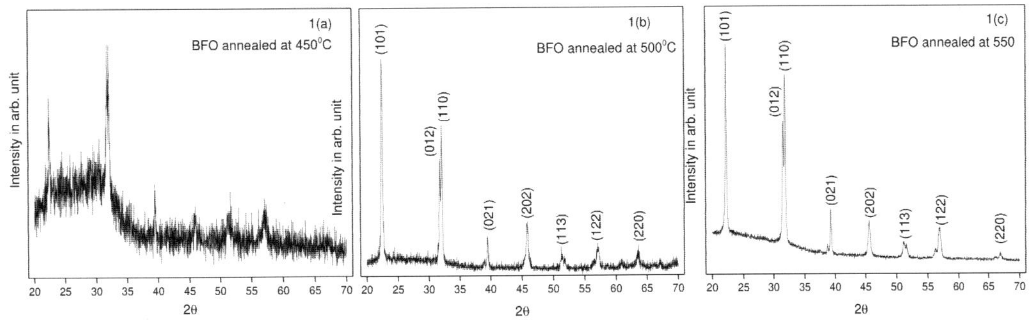

Figure 1(a),(b), (c): XRD pattern of bismuth ferrite thin films sintered at 450^{0}C, 500^{0}C and 550^{0}C as indicated

Figure 1(a)-(c) shows the XRD pattern of thin films annealed at 450^{0}C, 500^{0}C, and 550^{0}C. It can be seen that the crystallinity of the bismuth ferrite thin films increases with the rise in sintering temperature of the BFO sample. The development of pristine bismuth ferrite thin films could be seen at 550^{0}C with no other impurity phases. XRD pattern is found to be matched with rhombohedral perovskite structures ICCD card No. 00-020-0169 with preferred orientation along (101) plane. The lattice parameters estimated from the XRD diffraction peaks and their crystallite sizes were reported in Table I.

Table I: Lattice parameters

Samples	Lattice constants		Unit cell volume in Å^3	Crystallite size in nm
	'a' in Å	'c' in Å		
BFO 450	5.576	6.917	186.3	15.2 nm
BFO 500	5.576	6.931	186.6	37.8 nm
BFO 550	5.586	6.970	188.4	41.2 nm

The lattice parameters *a* and *c* are found to be increased with the increase in sintering temperature of the thin films, which is supported by the increase in crystallite size of the BFO nanostructure.

3.2 Morphological Studies:

To investigate the surface morphology of the thin films, atomic force microscopy (AFM)-Park XE 70 has been used. **Fig. 2(a)-(c)** represent the AFM surface micrographs of BFO 7-layer thin films annealed at different temperatures, 450 °C, 500 °C, 550 °C (assigned as BFO 450, BFO 500, BFO 550 hereafter). As evident from **Fig. 2(a)-(c)**, the grains in the BFO 550 are larger as compared to other BFO films annealed at lower temperatures viz. BFO 450 and BFO 500 owing to the agglomeration of crystallites with a rise in annealing temperature. It was observed from the **Fig 2**. that the grain size and RMS surface roughness value increase with the change in annealing temperature of the films from 450°C to 550°C and the values are reported in **Table 2.**

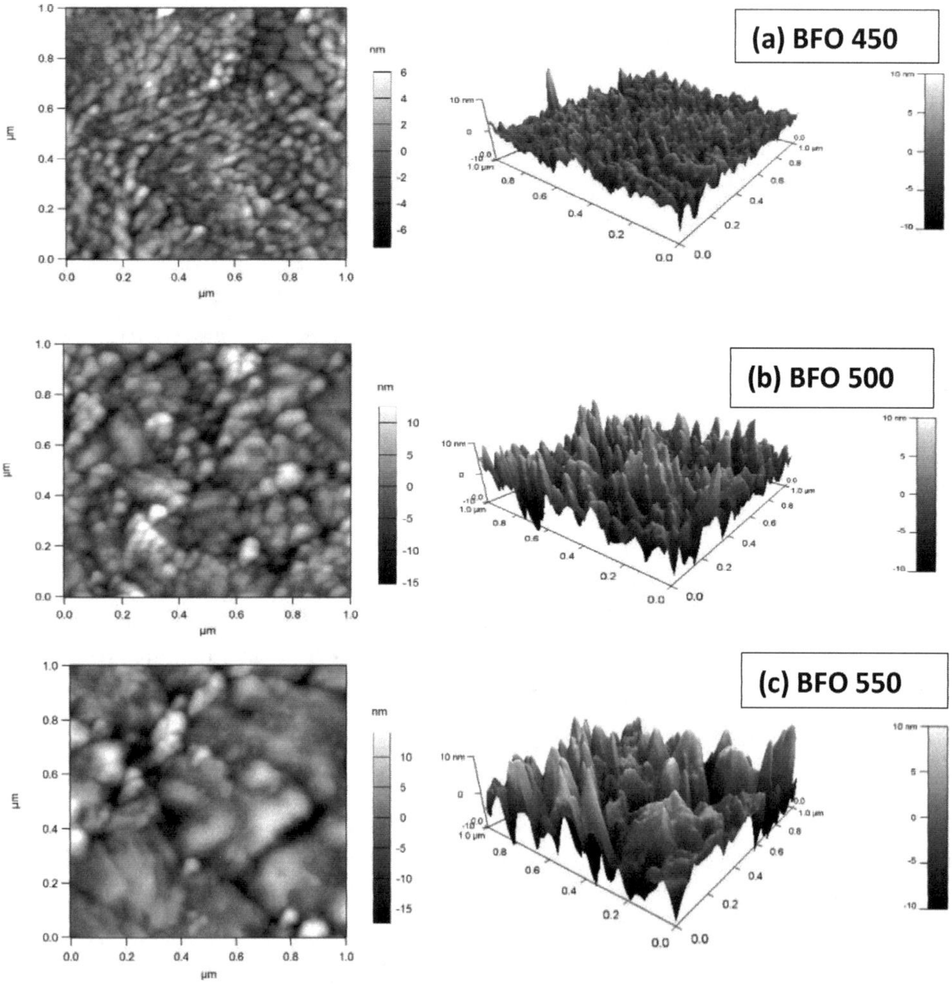

Figure 2(a),(b),(c): AFM micrograph of BFO 450, BFO 500 and BFO 550 respectively

Table 2. Crystallite sizes, Grain Sizes and Surface roughness values for BFO 450, BFO 500 and BFO 550 thin films

Samples -$BiFeO_3$	Crystallite size by (nm)	Grain Size (nm)	RMS Surface Roughness (nm)
BFO 450	15.2	35.7	1.847
BFO 500	37.8	73.5	4.436
BFO 550	45.7	118.1	5.239

Higher differences in the values of crystallite size as compared to the grain size reveal that the polycrystalline thin films so form consist of many crystallites in each grain.

3.3 Optical properties:

Bismuth ferrite is a potential candidate for thin-film solar cells owing to its relatively narrow bandgap. The thin films prepared were subjected to an Ultra-Violet spectrophotometer having a wavelength range from 200 nm to 800nm. The optical bandgap of the bismuth ferrite thin films was calculated using classical Tauc's law considering the direct inter-band electron transition between the conduction band and valence band [3]

$$(\alpha h\nu)^2 = A(h\nu - E_g) \quad (1)$$

where α is the absorption coefficient, A is a constant, $h\nu$ is the photon energy of the incident light, and E_g is the optical bandgap. E_g was estimated by plotting $(\alpha h\nu)^2$ along x-axis and $h\nu$ along the y-axis and extrapolating the linear region to $(\alpha h\nu)^2 = 0$ as depicted in figure 3(a)-(c).

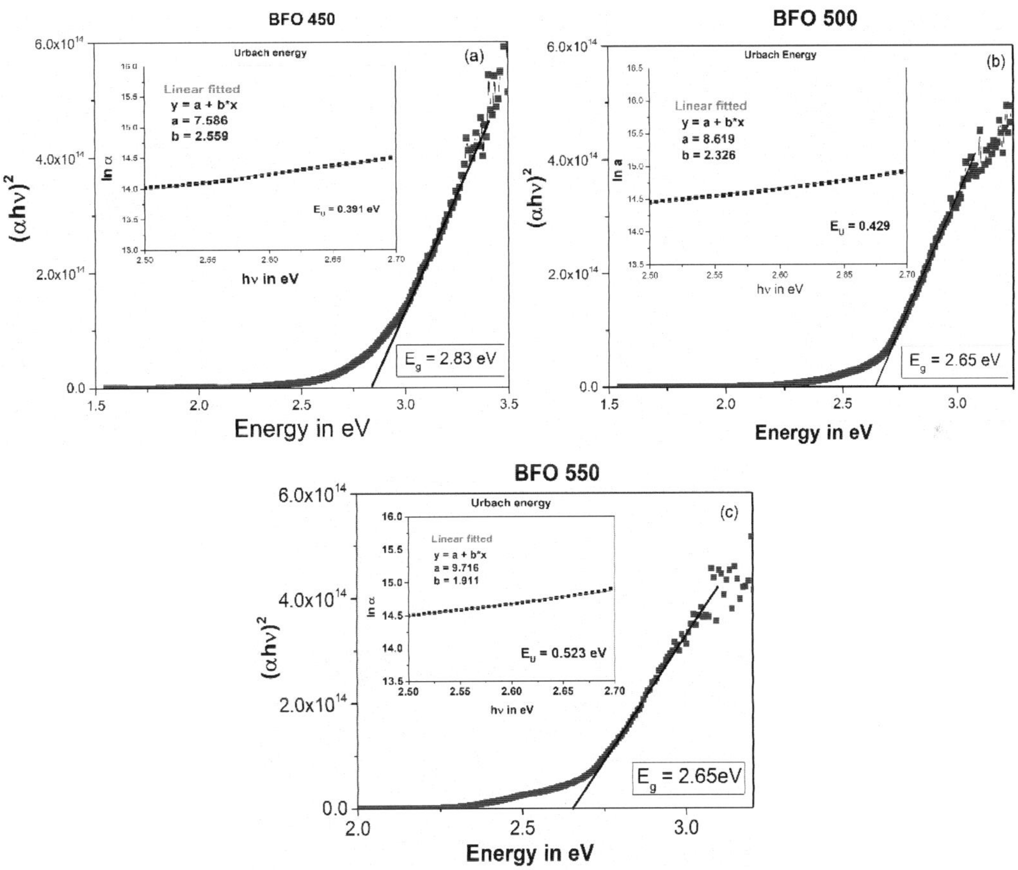

Figure 3(a)-(c): The $(\alpha h\nu)^2$ vs $h\nu$ plots of BFO 450, BFO 500 and BFO 550 thin films

As evident, the bandgap is found to be decreased as we increase the sintering temperature from 450^{0}C to 500^{0}C. The insets of graphs 3(a)-(c) denote the linear fitted graph near the absorption edge from which the Urbach energy is calculated by taking the reciprocal of the slope of the graph. The Urbach region near the adsorption edge denotes the structural and thermal distortion and is found to be increased from 0.391eV to 0.523 eV with the increase in sintering temperature [13]. The redshift of the optical bandgap might be attributed to the variation of crystallite size, lattice distortion, and oxygen vacancies incorporated due to the increase in annealing temperature, which might alter the band structure of BFO [14]

Fig 4(a) shows the thin film transmission spectra of BFO 450, BFO 500 and BFO 550 in between 300-800 nm wavelength of light. The interference pattern is observed in all the BFO thin films due to multiple reflections from multilayered structures. Transmission percentage decreases with the increase in sintering temperature from 450 to 550^{0}C, as evident from fig 4(a). The higher value of k (calculated using $k = \frac{\alpha\lambda}{4\pi}$) in fig 4(b) denotes the non-transparent nature of the thin films at lower wavelengths which may be attributed to high absorbance and increased scattering [15].

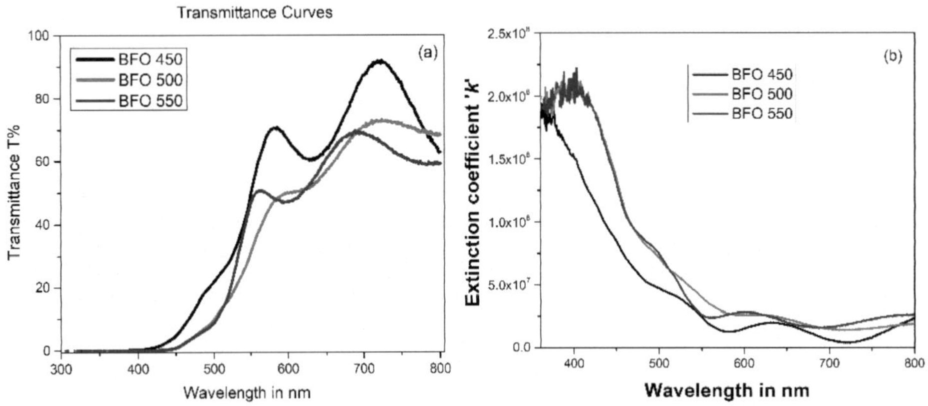

Figure 4(a): Transmittance spectra and (b) extinction coefficient variation of BFO 450, BFO 500 and BFO 550 thin films

4. Conclusions

Nanocrystalline bismuth ferrite multilayered thin films were successfully deposited onto corning glass substrates via sol-gel processing technique and sintered at 450^{0}C, 500^{0}C and 550^{0}C. Optimization of sintering temperature was observed at 550^{0}C with improved optical and structural properties. With the increase in sintering temperature, the lattice parameters and cell volume tend to increase with the growth of crystallite sizes of the samples. This distortion was confirmed by the increase in grain size estimated from AFM micrographs due to agglomeration with the temperature rise. Tauc's plot of the BFO 450, BFO 500, and BFO 550 determine that the optical bandgaps(E_g) of the thin films fabricated and the values of E_g decrease with higher sintering temperature. However, Urbach energy estimated from the Urbach tail region of the Tauc's plot was found to be increased from 0.391 eV to 0.523 eV, thereby denoting the possibility of thermal distortion and narrowing the optical bandgap. Low transmission percentage, narrow optical bandgap with a high extinction coefficient of BFO 550 would find useful applications in solar cells and other optoelectronic devices.

Acknowledgments
One of the authors, A. Deepak Sharma is very thankful to DST-INSPIRE for providing a full-time Ph.D. doctorial fellowship. The authors extend sincere thanks to IIT Kanpur and Manipur university for the experimental measurements.

References:

1. Y. Sun, W. Cai, R. Gao, X. Cao, F. Wang, T. Lei, X. Deng, G. Chen, H. He, C. Fu 2017 J. Mater. Sci. Mater. Electron. 28 12039–12047.
2. S. Gupta, M. Tomar, V. Gupta, A.R. James, M. Pal, R. Guo, A. Bhalla 2014 J. Appl. Phys. 115.
3. W. Cai, C. Fu, R. Gao, W. Jiang, X. Deng, G. Chen 2014 J. Alloys Compd. 617 240–246.
4. Q. Zhang, D. Sando, V. Nagarajan, J. 2016 Mater. Chem. C. 4 4092.
5. A.D. Sharma, H.B. Sharma 2019 Integr. Ferroelectr. 203 81–90.
6. W.A. Wani, S. Kundu, K. Ramaswamy, H. Venkataraman 2020 J. Alloys Compd. 156334.
7. M.M. Seyfouri, D. Wang 2020 Crit. Rev. Solid State Mater. Sci. 0 1–26.
8. Z. Chaodan, Z. Duanming, Y. Jun, Y. Bin, W. Yunyi, W. Longhai, W. Yunbo, Z. Wenli 2007 Integr. Ferroelectr. 94 23–30.
9. H. Gao, J. Tian, H. Zheng, F. Tan, W. Zhang 2014 J. Mater. Sci. Mater. Electron. 26 700–704.
10. X. Deng, Z. Zeng, R. Gao, Z. Wang, G. Chen, W. Cai, C. Fu 2020 J. Alloys Compd. 831 154857.
11. A.R. Chavan, S.D. Birajdar, R.R. Chilwar, K.M. Jadhav 2018 J. Alloys Compd. 735 2287–2297.
12. C. Anthonyraj, M. Muneeswaran, S. Gokul Raj, N. V. Giridharan, V. Sivakumar, G. Senguttuvan 2015 J. Mater. Sci. Mater. Electron. 49–58.
13. A. Tumuluri, K. Lakshun Naidu, K.C. James Raju 2014 Int. J. ChemTech Res. 6 3353–3356.
14. R. Bhatt, I. Bhaumik, S. Ganesamoorthy, A.K. Karnal, M.K. Swami, H.S. Patel, P.K. Gupta 2012 Phys. Status Solidi Appl. Mater. Sci. 209 176–180.
15. A. Essahlaoui, H. Essaoudi, A. Hallaoui, M. Bouhadda, A. Labzour, A. Housni 2018 J. Mater. Environ. Sci. 9 228–234

IOP Conf. Series: Materials Science and Engineering 1219 (2022) 012053 doi:10.1088/1757-899X/1219/1/012053

Synthesis and functional properties of ZnO/CuO nanocomposite for gas sensing applications

P. Bharathi[a,b], S. Karthigeyan[b], M. Krishna Mohan[a,b*]

[a] Functional Materials and Energy Devices Laboratory, Department of Physics and Nanotechnology, SRM Institute of Science and Technology, Kattankulathur-603 203, India
[b] Department of Physics and Nanotechnology, SRM Institute of Science and Technology, Kattankulathur-603 203, India

E-mail: krishnam1@srmist.edu.in

Abstract: ZnO/ CuO nanocomposite is one of the promising materials for gas sensing application. The p-n heterojunction formation plays vital role by providing more gas interaction sites with the analyte gas molecules. Formation of hetero junction improves the charge transport and reduces the operating temperature. Herein we report the simple synthesis and characterization of ZnO nanoparticles and ZnO/CuO nanocomposite using facile hydrothermal synthesis method. The preparedsamples confirm the hexagonal crystal structure. The morphological micrograph confirms the uniform growth of CuO nanoparticles on the ZnO matrix.

Key words: ZnO nanoparticle, ZnO/CuO nanocomposite, gas sensing.

1. Introduction

The majority of hazardous environmental gases and conservatory gases are released from power plants and chemical industries, vehicles. This causes the acid rain and affects the living species.Thus, it is essential to have a precise and fast method to detectthe hazardous gases.A gas sensor is a working device to warn us against dangerous gasses in our surroundings.Their reliable and effective functions prevent unexpected injury or damage to our property or our lives.Thus, human safety can be said to be closely linked to sensitivity and selectivity, which are features of gas sensors [1-3].

In the case of chemical sensors, the semiconducting metal oxides were generally used, due to the high strength, therapeutic and electronic properties and can be used for high working temperatures. In order to increase susceptibility and selectivity of the sensor, it can be easily optimized using doping of foreign elements etc. Among various types of metal oxide semiconductors SnO_2, CuO, TiO_2, NiO, WO_3 and ZnO show high selectivity and sensitivity at the high operating temperature [4,5]. Among them, the ZnOhave gained much attentiondue to its, available innature, thermal stability, exceptional electrical property, andpromising gas selectivity with sensitivity.To further increase the sensor response, the formation of heterostructure with *p* and *n*- type provides a possible solution. For example, ZnO/CuO hybrid nanostructures prepared by Fuadi *et al.*in which ZnO was covered with small clusters of CuO and provides enhanced sensing response[6].Snowflake like ZnO structures have been synthesized by the Gao *et al.* to sense the NO_xgas molecules using ionic liquid.Kumar *et al.* prepared nanocomposites with simple hydrolysis ZnO/reduced Graphene Oxide for the detection of NO_2.Since the graphene oxide is reduced, 2D structure has a large surface area and the bond edge is useful for the connection with nanoparticles on its surface [7].

In this present work, ZnO and ZnO/CuO were synthesized using one simple hydrothermal method and co-precipitation method. The morphology of ZnO and ZnO/CuO composite was directedusing hydrothermal method. The surface morphological and structuralanalyses of the prepared samples were analyzedusingField Emission Scanning Electron Microscope (FE-SEM) and X-Ray Diffraction. Raman Spectroscopy used to analyses the vibrational modes, lattice disorders.

IOP Conf. Series: Materials Science and Engineering 1219 (2022) 012053 doi:10.1088/1757-899X/1219/1/012053

2. Experimental Section

Materials Used

Zinc nitrate hexahydrate($Zn(NO_3)_2 \cdot 6H_2O$), copper acetate dihydrate ($Cu(CO_2CH_3)_2 \cdot 2H_2O$), sodium hydroxide (NaOH), and ethanol (C_2H_5OH), has been purchased from Sigma-Aldrich with an analytical grade and further used without purification.

Synthesis Procedure

To prepare ZnO, 1:1 ratio of sodium hydroxide (0.2 M) and Zinc nitrate hexahydrate(0.2 M) was dissolved into 50 mL of DI-water.The obtained solution was transfer into Teflon-lined autoclave and maintained at 100 °C for 5h. The obtained product was washed and centrifuged with DI-water and ethanol for several times. The final product was labelled as Z1. Same procedure was repeated as ethanol solvent and denoted as Z2. To synthesize ZnO/CuO nanocomposite co-precipitation method was used. Where 0.2 M of Zinc source dissolved in 25 mL of ethanol and named as solution A. 0.2 M of Copper Acetate dihydrate was dissolved in 25 mL of ethanol and mentioned as solution B. And then solution A was added into the solution B by dropwise and allowed to stir for 1h. The solution aged for overnight and centrifuged using DI-water and ethanol. Final product named as ZC.

Characterization techniques:

X-Ray Diffraction analysis was performed using the PANalytical X-Ray Diffractometer with Cu Kć (λ = 1.5406 Å) radiation with the scanning rate of 2°per min. Raman bands were notedusingmicro-Raman spectrometer (HORIBA France, LABRAM HR Evolution) with an Nd: YAG laser (532 nm) with the range of 200 cm^{-1} to 500 cm^{-1}. The morphology of the as-prepared samples was analysed by a Field Emission Scanning Electron Microscope (F E I Quanta FEG 200) with an accelerating voltage of 15 kV.

3. Results and Discussion

The X-Ray Diffraction pattern of pristine ZnO(Z1 and Z2) was shown in Fig. 1. The peaks at 31.74°, 34.42°, 36.24°, 47.48°, 56.52°,62.80°, 66.42°, 67.84°, and 69.06° are corresponds to the (100),(002), (101), (102), (110), (103), (200), (112), and (201) planes respectively as shown in Fig. 1(a). The indexed peaks were showed hexagonal wurtzite structure and itwas accordance with the JCPDS file no.89-7102, with lattice parameter a = *3.27 Å,c= 5.24 Å*.The crystallite size of (0 0 2)plane was calculated by Scherrer's equation (1)

$$D = \frac{k\lambda}{\beta \cos\theta} \qquad (1)$$

IOP Conf. Series: Materials Science and Engineering 1219 (2022) 012053 doi:10.1088/1757-899X/1219/1/012053

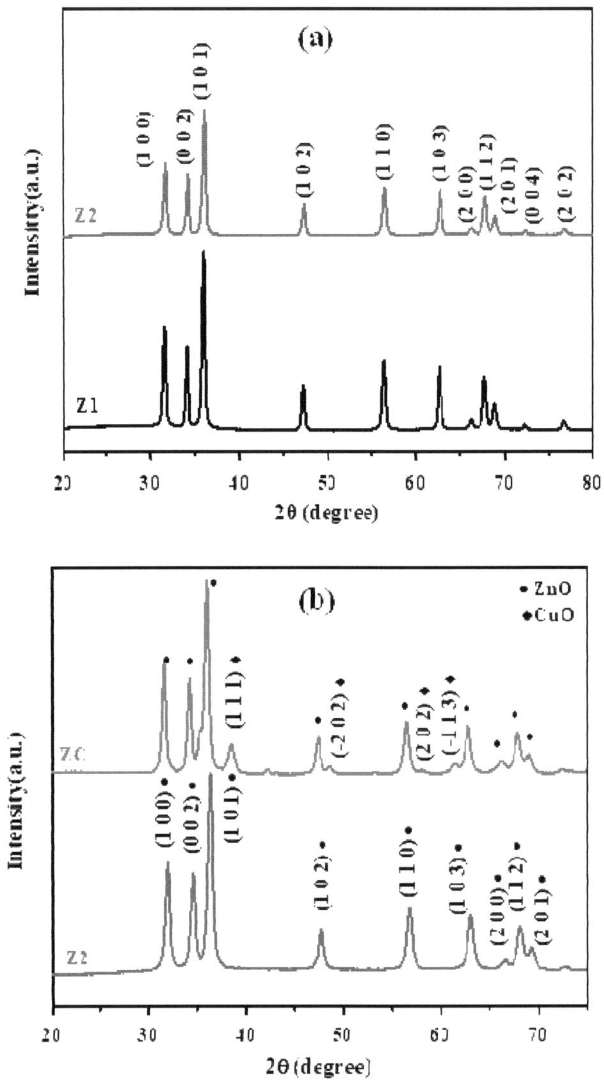

Figure 1. XRD patterns of (a) pristine ZnO, (b) pristine ZnO, ZnO/CuO nanocomposite.

where K is Scherrer constant (0.89), λ is X-ray wavelength, β is the FWHM of the corresponding peaks, θ is the incidentangle of the peak. The crystallite size of pristine ZnO (Z1) is 23 nm and for Z2 is 25 nm [8, 9, 10].

The XRD patterns of ZnO/CuO composite are shown in Fig. 1(b). The spectra showed peaks corresponding to ZnO with the additional peaks of CuO. This result confirms the monoclinic structure of CuO with Lattice parameter $a = 4.684$ Å, $b = 3.425$ Å, $c = 5.129$ Å, and well matched with the JCPDS card No. 05 – 0661(CuO).The calculated crystallite size D = 17 nm. There are no additional peaks have been observed. After compositing with CuO, the crystallite was decreased, which plays vital role in the sensing characteristics. The less crystallite size has been achieved by ZnO/CuO nanocomposite [11].

Figure 2 showed the morphology of the prepared samples with different magnifications. Z1 showed sheet like morphology of ZnO. This is due to hydroxyl groups presence in the samples, which

IOP Conf. Series: Materials Science and Engineering 1219 (2022) 012053 doi:10.1088/1757-899X/1219/1/012053

leads to the growth in same direction. Z2 sample showed sphere-like morphology. In this ethanol used as solvent which acts as stabilizer. This inhibits the growth of the nucleation, which leads to the formation of sphere-like morphology. This helps to increase the sensor performance to due large surface to volume ratio and provides more interaction sites to the gas molecules. ZC sample showed agglomerated particle like morphology. Where ZnO nanospheres are counted as spheres and CuO particles tends to grow on the spheres. This morphology leads to drastically change in the sensing performance by forming the heterostructure[12,13]. The formation of heterostructure provides more pathways for the gas to interact with the matrix, and the band bending of ZnO and CuO causes changes in the sensing response. Figure 3 showed the EDAX spectra of prepared samples, where it confirms the elemental distribution over the sample.

Figure 2. FE-SEM Micrograph of pristine ZnO, ZnO/CuO nanocomposite

Theformation of the nanostructures was denoted by the following equations,

$$Zn(NO_3)_2.6H_2O \rightarrow Zn^{2+} + 2NO^{3-} + 6H^+ + 6OH^- (2)$$
$$2Zn^{2+} + 4OH^- \rightarrow 2Zn(OH)_2 (3)$$
$$Zn(OH)_2 + 2OH^- \rightarrow Zn(OH)_4^{2-} (4)$$
$$Zn(OH)_4^{2-} \rightarrow ZnO + H_2O + 2OH^- \quad (5)$$

When zinc nitrate hexahydrate and sodium hydroxide was dissolved, the formation of Zn^{2+} ions and OH^- ions takes place as stated in eqn. (2). Then, the disassociated Zn^{2+} ions are creating a bond with OH^-, hence it leads to the formation of ZnO nuclei (3-5). Sodium hydroxide ions responsible for the concentration of OH^-ions. This increases the interaction between Zn^{2+} ions and OH^- ions to produce ZnO nuclei. This premises to the createthe agglomerated nanoplates-like morphology [14, 15].

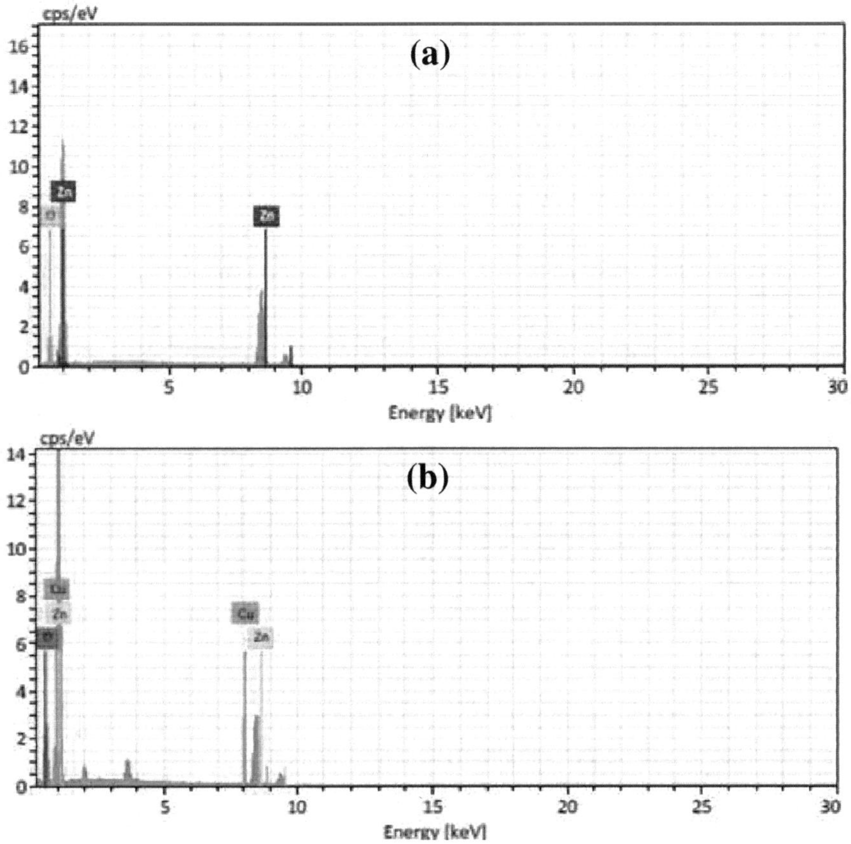

Figure 3. EDAX of(a) pristine ZnO, (b ZnO/CuO nanocomposite

The vibrational modes of Z1, Z2 and ZC wereanalyzed by micro-Raman spectroscopy as shown in Fig. 4. TheRaman spectra of Z1, Z2, exhibits the peaks at 333 and 438 cm^{-1} [16]. The peak at 438 cm^{-1}related tonon-optical phonon vibration of E_2H mode of O and Zn lattice.

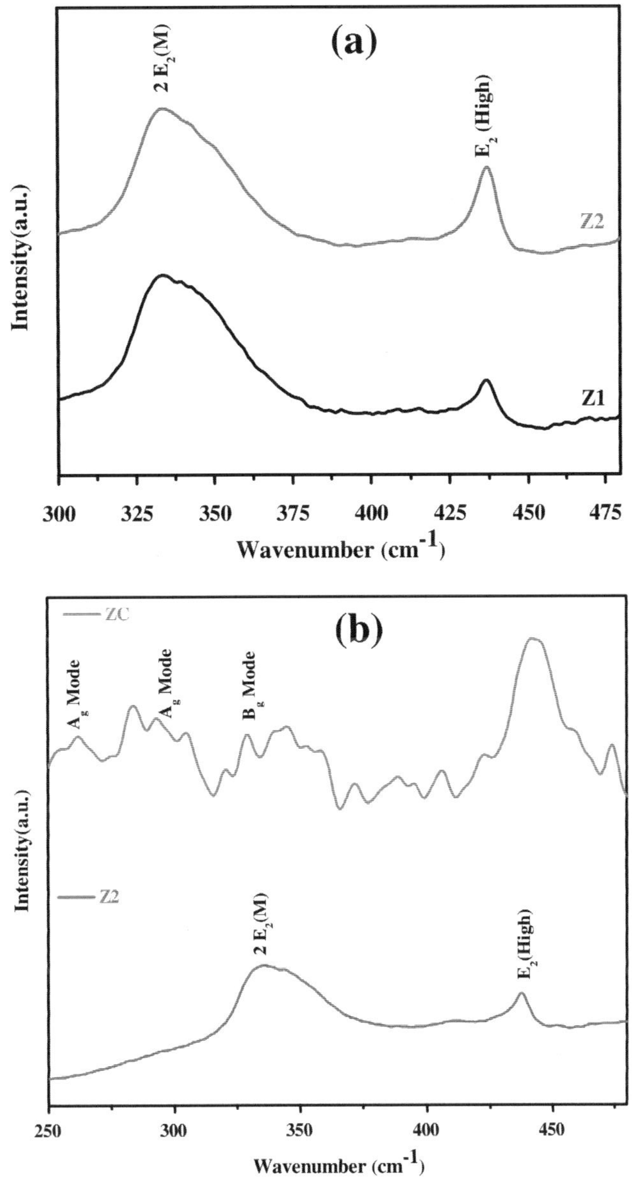

Figure 4. Raman Spectra of (a) pristine ZnO, (b) pristine ZnO, ZnO/CuO nanocomposite

The peak at $333cm^{-1}$was attributed to multi-phonons scattering of E_{2H} - E_{2L} and A_1 phonons, which indicates the high crystallinity nature of the sample. In the ZC sample along with ZnO scattering modes three peaks were presence in the samples. The peak located at 262, 293 cm^{-1} was indicated by Ag mode. At 328 cm^{-1} was attributed by Bg mode [17].

Conclusion:

ZnO and ZnO/ CuO nanocomposites were preparedusing hydrothermal and precipitation method. XRD analysis confirms that the prepared samples are in good crystallinenatureand confirms the hexagonal phase. EDAX spectrum clearly shows Zn, O and Cu presencein the sample. Morphological analysis

IOP Conf. Series: Materials Science and Engineering 1219 (2022) 012053 doi:10.1088/1757-899X/1219/1/012053

confirmed the nanoparticle over the sphere-like morphology. The small grains provide more reactive sites for the interaction of gas species.

Reference:

1. R. Kumar, O. Al-Dossary, G. Kumar, A.Umar, Zinc oxide nanostructures for NO_2 Gas–Sensor Applications: A Review, Nano-Micro Lett. 7(2) (2015) 97–120.
2. A. Mirzaei, S.S. Kim, H.W. Kim, Resistance-based H_2S gas sensors using metal oxide nanostructures: A review of recent advances, J. Hazard. Mater. 357 (2018) 314-331.
3. L. Zhu, W. Zeng, H. Ye, Y. Li, Volatile organic compound sensing based on coral rock-like ZnO, Mater. Res. Bull. 100 (2018) 259-264.
4. Qi Qi, Tong Zhang, Shujuan Wang, Xuejun Zheng, Humidity sensing properties of KCl-doped ZnO nanofibers with super-rapid response and recovery, Sensors and Actuat. B-Chem. 137 (2009) 649–655.
5. H. Long, W. Zeng, H. Wang, M. Qian, Y. Liang, Z. Wang, Self-assembled biomolecular 1D nanostructures for aqueous sodium-ion battery, Adv. Sci. 5 (2018) 1700634.
6. Fuadi, M. K., D. Yang, C. Park, and I. Park. "Highly sensitive NO 2 gas sensor based on zinc oxide/copper oxide hybrid-nanostructures." In *2013 Transducers & Eurosensors XXVII: The 17th International Conference on Solid-State Sensors, Actuators and Microsystems (TRANSDUCERS & EUROSENSORS XXVII)*, pp. 1641-1644. IEEE, 2013.
7. Kumar, Nagesh, Arvind Kumar Srivastava, Hari Shankar Patel, Bipin Kumar Gupta, and Ghanshyam Das Varma. "Facile synthesis of ZnO–reduced graphene oxide nanocomposites for NO2 gas sensing applications." *European Journal of Inorganic Chemistry* 2015, no. 11 (2015): 1912-1923.
8. Lerner, Immanuel, Michael Trus, Roy Cohen, Ofer Yizhar, Itzhak Nussinovitch, and Daphne Atlas. "Ion interaction at the pore of Lc-type Ca2+ channel is sufficient to mediate depolarization-induced exocytosis." Journal of neurochemistry 97, no. 1 (2006): 116-127.
9. Faisal, M., Adel A. Ismail, Ahmed A. Ibrahim, Houcine Bouzid, and Saleh A. Al-Sayari. "Highly efficient photocatalyst based on Ce doped ZnO nanorods: Controllable synthesis and enhanced photocatalytic activity." Chemical engineering journal 229 (2013): 225-233.
10. G. Vijayaprasath , R. Murugan , Y. Hayakawa , G. Ravi, Optical and magnetic studies on Gd doped ZnO nanoparticles synthesized by co-precipitation method, J. Lumin 178 (2016) 375–383.
11. Poloju, Madhukar, Nagabandi Jayababu, and MV Ramana Reddy. "Improved gas sensing performance of Al doped ZnO/CuO nanocompositebased ammonia gas sensor." Materials Science and Engineering: B 227 (2018): 61-67.
12. Wahab, Rizwan, S. G. Ansari, Young Soon Kim, Minwu Song, and Hyung-Shik Shin. The role of pH variation on the growth of zinc oxide nanostructures, Appl. Surf. Sci. 255 (2009): 4891-4896.
13. Wahab, Rizwan, Farheen Khan, Yogendra Kumar Mishra, Javed Musarrat, and Abdulaziz A. Al-Khedhairy. Antibacterial studies and statistical design set data of quasi zinc oxide nanostructures. Rsc Adv. 6, no. 38 (2016): 32328-32339.
14. X. Chen, X. Jing, J. Wang, J. Liu, D. Song, L. Liu, Fabrication of spindle-like ZnO architectures for highly sensitive gas sensors, Superlattice. Microst. 63 (2013) 204–214.
15. Wahab, Rizwan, Young-Soon Kim, and Hyung-Shik Shin. Fabrication, characterization and growth mechanism of heterostructured zinc oxide nanostructures via solution method. Curr. Appl. Phys. 11, no. 3 (2011): 334-340.
16. K. A. Alim, V. A. Fonoberov, M. Shamsa, and A. A. Balandin, Micro-Raman investigation of optical phonons in ZnO nanocrystals, J. Appl. Phys. 97 (2005) 124313.
17. Ahmed, Faheem, Nishat Arshi, M. S. Anwar, Rehan Danish, and Bon Heun Koo. Morphological evolution of ZnO nanostructures and their aspect ratio-induced enhancement in photocatalytic properties. RSC Adv. 55 (2014): 29249 29263.

AUTHOR INDEX